Modern Oxidation Methods

Edited by
Jan-Erling Bäckvall

Further Reading from Wiley-VCH

R. Mahrwald (Ed.)

Modern Aldol Reactions, 2 Vols.

2004
ISBN 3-527-30714-1

A. de Meijere, F. Diederich (Eds.)

Metal-Catalyzed Cross-Coupling Reactions, 2 Vols., 2nd Ed.

2004
ISBN 3-527-30518-1

B. Cornils, W. A. Herrmann (Eds.)

Aqueous-Phase Organometallic Catalysis, 2nd Ed.

Concepts and Applications

2004
ISBN 3-527-30712-5

M. Beller, C. Bolm (Eds.)

Transition Metals for Organic Synthesis, 2 Vols., 2nd Ed.

Building Blocks and Fine Chemicals

2004
ISBN 3-527-30613-7

Modern Oxidation Methods

Edited by
Jan-Erling Bäckvall

WILEY-VCH Verlag GmbH & Co. KGaA

Editor:

Professor Dr. Jan-Erling Bäckvall
Department of Organic Chemistry
Arrhenius Laboratory
Stockholm University
SE 106 91 Stockholm
Sweden

■ This book was carefully produced. Nevertheless, authors, editor and publisher do not warrant the information contained therein to be free of errors. Readers are advised to keep in mind that statements, data, illustrations, procedural details or other items may inadvertently be inaccurate.

Library of Congress Card No.: applied for

British Library Cataloguing-in-Publication Data: A catalogue record for this book is available from the British Library.

Bibliographic information published by Die Deutsche Bibliothek
Die Deutsche Bibliothek lists this publication in the Deutsche Nationalbibliografie; detailed bibliographic data is available in the Internet at <http://dnb.ddb.de>.

© 2004 Wiley-VCH Verlag GmbH & Co. KGaA, Weinheim

All rights reserved (including those of translation into other languages). No part of this book may be reproduced in any form – nor transmitted or translated into a machine language without written permission from the publishers. Registered names, trademarks, etc. used in this book, even when not specifically marked as such, are not to be considered unprotected by law.

Printed in the Federal Republic of Germany
Printed on acid-free paper

Composition   ProSatz Unger, Weinheim
Printing   Strauss GmbH, Mörlenbach
Bookbinding   Litges & Dopf Buchbinderei GmbH, Heppenheim

ISBN   3-527-30642-0

Preface

Oxidation reactions play an important role in organic chemistry and there is an increasing demand for selective and mild oxidation methods in modern organic synthesis. During the last two decades there has been a spectacular development in the field and a large number of novel and useful oxidation reactions have been discovered. Significant progress has been achieved within the area of catalytic oxidations, which has led to a range of selective and mild processes. These reactions may be based on organocatalysis, metal catalysis or biocatalysis. In this regard enantioselective catalytic oxidation reactions are of particular interest.

Due to the rich development of oxidation reactions in recent years there was a need for a book covering the area. The purpose of this book on "Modern Oxidation Methods" is to fill this need and provide the chemistry community with an overview of some recent developments in the field. In particular some general and synthetically useful oxidation methods that are frequently used by organic chemists are covered. These methods include catalytic as well as non-catalytic oxidation reactions in the science frontier of the field. Today there is an emphasis on the use of environmentally friendly oxidants ("green" oxidants) that lead to a minimum amount of waste. Examples of such oxidants are molecular oxygen and hydrogen peroxide. Many of the oxidation methods discussed and reviewed in this book are based on the use of "green" oxidants.

In this multi-authored book selected authors in the field of oxidation provide the reader with an up to date of a number of important fields of modern oxidation methodology. Chapter 1 summarizes recent advances on the use of "green oxidants" such as H_2O_2 and O_2 in the osmium-catalyzed dihydroxylation of olefins. Immobilization of osmium is also discussed and with these recent achievements industrial applications seem to be near. Another important transformation of olefins is epoxidation. In Chapter 2 transition metal-catalyzed epoxidations are reviewed and in Chapter 3 recent advances in organocatalytic ketone-catalyzed epoxidations are covered. Catalytic oxidations of alcohols with the use of environmentally benign oxidants have developed tremendously during the last decade and in Chapter 4 this area is reviewed. Aerobic oxidations catalyzed by *N*-hydroxyphtalimides (NHPI) are reviewed in Chapter 5. In particular oxidation of hydrocarbons via C–H activation are treated but also oxidations of alkenes and alcohols are covered.

Modern Oxidation Methods. Edited by Jan-Erling Bäckvall
Copyright © 2004 WILEY-VCH Verlag GmbH & Co. KGaA, Weinheim
ISBN: 3-527-30642-0

In Chapter 6 ruthenium-catalyzed oxidation of various substrates are reviewed including alkenes, alcohols, amines, amides, β-lactams, phenols, and hydrocarbons. Many of these oxidations involve oxidations by "green oxidants" such as molecular oxygen and alkyl hydroperoxides. Chapter 7 deals with heteroatom oxidation and selective oxidations of sulfides (thioethers) to sulfoxides and tertiary amines to amine oxides are discussed. The chapter covers stoichiometric and catalytic reactions including biocatalytic reactions. Oxidations catalyzed by polyoxymetalates have increased in use during the last decade and this area is covered in Chapter 8. Oxidations with various monooxygen donors, peroxides (including hydrogen peroxide) and molecular oxygen are reviewed. Also, recent attempts to heterogenize homogeneous polyoxymetalate catalysts are discussed. Chapter 9 comprises an extensive review on oxidation of ketones with some focus on recent advances in Baeyer-Villiger oxidations. Catalytic as well as stoichiometic reactions are covered. Finally, in Chapter 10 manganese-catalyzed hydrogen peroxide oxidations are reviewed. The chapter includes epoxidation, dihydroxylation of olefins, oxidation of alcohols and sulfoxidation.

I hope that this book will be of value to chemists involved in oxidation reactions in both academic and industrial research and that it will stimulate further development in this important field.

Stockholm, July 2004   *Jan-E. Bäckvall*

Contents

Preface V
List of Contributors XIII

1 Recent Developments in the Osmium-catalyzed Dihydroxylation of Olefins 1
Uta Sundermeier, Christian Döbler, and Matthias Beller
1.1 Introduction 1
1.2 Environmentally Friendly Terminal Oxidants 2
1.2.1 Hydrogen Peroxide 2
1.2.2 Hypochlorite 5
1.2.3 Oxygen or Air 7
1.3 Supported Osmium Catalyst 12
1.4 Ionic Liquids 16
References 17

2 Transition Metal-catalyzed Epoxidation of Alkenes 21
Hans Adolfsson
2.1 Introduction 21
2.2 Choice of Oxidant for Selective Epoxidation 22
2.3 Epoxidations of Alkenes Catalyzed by Early Transition Metals 23
2.4 Molybdenum and Tungsten-catalyzed Epoxidations 23
2.4.1 Homogeneous Catalysts – Hydrogen Peroxide as the Terminal Oxidant 24
2.4.2 Heterogeneous Catalysts 27
2.5 Manganese-catalyzed Epoxidations 28
2.6 Rhenium-catalyzed Epoxidations 32
2.6.1 MTO as an Epoxidation Catalyst – Original Findings 35
2.6.2 The Influence of Heterocyclic Additives 35
2.6.3 The Role of the Additive 38
2.6.4 Other Oxidants 39
2.6.5 Solvents/Media 41
2.6.6 Solid Support 42
2.6.7 Asymmetric Epoxidations Using MTO 43

Modern Oxidation Methods. Edited by Jan-Erling Bäckvall
Copyright © 2004 WILEY-VCH Verlag GmbH & Co. KGaA, Weinheim
ISBN: 3-527-30642-0

2.7	Iron-catalyzed Epoxidations 44
2.8	Concluding Remarks 46
	References 47

3 Organocatalytic Oxidation. Ketone-catalyzed Asymmetric Epoxidation of Olefins 51

YIAN SHI

- 3.1 Introduction 51
- 3.2 Early Ketones 52
- 3.3 C_2 Symmetric Binaphthyl-based and Related Ketones 53
- 3.4 Ammonium Ketones 58
- 3.5 Bicyclo[3.2.1]octan-3-ones 59
- 3.6 Carbohydrate Based and Related Ketones 60
- 3.7 Carbocyclic Ketones 72
- 3.8 Ketones with an Attached Chiral Moiety 75
- 3.9 Conclusion 76

Acknowledgments 78

References 78

4 Modern Oxidation of Alcohols Using Environmentally Benign Oxidants 83

I. W. C. E. ARENDS and R. A. SHELDON

- 4.1 Introduction 83
- 4.2 Oxoammonium-based Oxidation of Alcohols – TEMPO as Catalyst 83
- 4.3 Metal-mediated Oxidation of Alcohols – Mechanism 87
- 4.4 Ruthenium-catalyzed Oxidations with O_2 88
- 4.5 Palladium-catalyzed Oxidations with O_2 100
- 4.6 Copper-catalyzed Oxidations with O_2 105
- 4.7 Other Metals as Catalysts for Oxidation with O_2 109
- 4.8 Catalytic Oxidation of Alcohols with Hydrogen Peroxide 111
- 4.9 Concluding Remarks 113

References 114

5 Aerobic Oxidations and Related Reactions Catalyzed by N-Hydroxyphthalimide 119

YASUTAKA ISHII and SATOSHI SAKAGUCHI

- 5.1 Introduction 119
- .2 NHPI-catalyzed Aerobic Oxidation 120
- 5.2.1 Alkane Oxidations with Dioxygen 120
- 5.2.2 Oxidation of Alkylarenes 125
- 5.2.2.1 Oxidation of Alkylbenzenes 125
- 5.2.2.2 Synthesis of Terephthalic Acid 127
- 5.2.2.3 Oxidation of Methylpyridines and Methylquinolines 129
- 5.2.2.4 Oxidation of Hydroaromatic and Benzylic Compounds 131
- 5.2.3 Preparation of Acetylenic Ketones by Direct Oxidation of Alkynes 132
- 5.2.4 Oxidation of Alcohols 133

5.2.5	Epoxidation of Alkenes Using Dioxygen as Terminal Oxidant 136
5.2.6	Baeyer-Villiger Oxidation of KA-Oil 137
5.2.7	Preparation of ε-Caprolactam Precoursor from KA-Oil 138
5.3	Functionalization of Alkanes Catalyzed by NHPI 139
5.3.1	Carboxylation of Alkanes with CO and O_2 139
5.3.2	First Catalytic Nitration of Alkanes Using NO_2 140
5.3.3	Sulfoxidation of Alkanes Catalyzed by Vanadium 142
5.3.4	Reaction of NO with Organic Compounds 144
5.3.5	Ritter-type Reaction with Cerium Ammonium Nitrate (CAN) 145
5.4	Carbon–Carbon Bond Forming Reaction *via* Generation of Carbon Radicals Assisted by NHPI 147
5.4.1	Oxyalkylation of Alkenes with Alkanes and Dioxygen 147
5.4.2	Synthesis of α-Hydroxy-γ-lactones by Addition of α-Hydroxy Carbon Radicals to Unsaturated Esters 148
5.4.3	Hydroxyacylation of Alkenes Using 1,3-Dioxolanes and Dioxygen 149
5.4.4	Hydroacylation of Alkenes Using NHPI as a Polarity-reversal Catalyst 150
5.5	Conclusions 152
	References 153

6	**Ruthenium-catalyzed Oxidation of Alkenes, Alcohols, Amines, Amides, β-Lactams, Phenols, and Hydrocarbons 165**
	SHUN-ICHI MURAHASHI and NARUYOSHI KOMIYA
6.1	Introduction 165
6.2	RuO_4-promoted Oxidation 165
6.3	Oxidation with Low-valent Ruthenium Catalysts and Oxidants 169
6.3.1	Oxidation of Alkenes 169
6.3.2	Oxidation of Alcohols 172
6.3.3	Oxidation of Amines 175
6.3.4	Oxidation of Amides and β-Lactams 179
6.3.5	Oxidation of Phenols 181
6.3.6	Oxidation of Hydrocarbons 183
	References 186

7	**Selective Oxidation of Amines and Sulfides 193**
	JAN-E. BÄCKVALL
7.1	Introduction 193
7.2	Oxidation of Sulfides to Sulfoxides 193
7.2.1	Stoichiometric Reactions 194
7.2.1.1	Peracids 194
7.2.1.2	Dioxiranes 194
7.2.1.3	Oxone and Derivatives 195
7.2.1.4	H_2O_2 in "Fluorous Phase" 195
7.2.2	Chemocatalytic Reactions 196
7.2.2.1	H_2O_2 as Terminal Oxidant 196
7.2.2.2	Molecular Oxygen as Terminal Oxidant 205

7.2.2.3 Alkyl Hydroperoxides as Terminal Oxidant 207
7.2.2.4 Other Oxidants in Catalytic Reactions 209
7.2.3 Biocatalytic Reactions 209
7.2.3.1 Haloperoxidases 209
7.2.3.2 Ketone Monooxygenases 210
7.3 Oxidation of Tertiary Amines to N-Oxides 211
7.3.1 Stoichiometric Reactions 212
7.3.2 Chemocatalytic Oxidations 213
7.3.3 Biocatalytic Oxidation 216
7.3.4 Applications of Amine N-oxidation in Coupled Catalytic Processes 216
7.4 Concluding Remarks 218
References 218

8 Liquid Phase Oxidation Reactions Catalyzed by Polyoxometalates 223
Ronny Neumann
8.1 Introduction 223
8.2 Polyoxometalates (POMs) 224
8.3 Oxidation with Mono-oxygen Donors 226
8.4 Oxidation with Peroxygen Compounds 231
8.5 Oxidation with Molecular Oxygen 238
8.6 Heterogenization of Homogeneous Reaction Systems 245
8.7 Conclusion 247
References 248

9 Oxidation of Carbonyl Compounds 253
Jacques Le Paih, Jean-Cédric Frison and Carsten Bolm
9.1 Introduction 253
9.2 Oxidations of Aldehydes 253
9.2.1 Conversions of Aldehydes to Carboxylic Acid Derivatives by Direct Oxidations 253
9.2.1.1 Metal-free Oxidants 254
9.2.1.2 Metal-based Oxidants 255
9.2.1.3 Halogen-based Oxidants 257
9.2.1.4 Sulfur- and Selenium-based Oxidants 258
9.2.1.5 Nitrogen-based Oxidants 259
9.2.1.6 Miscellaneous 259
9.2.2 Conversions of Aldehydes into Carboxylic Acid Derivatives by Aldehyde Specific Reactions 259
9.2.2.1 Dismutations and Dehydrogenations 259
9.2.2.2 Oxidative Aldehyde Rearrangements 261
9.2.3 Conversions of Aldehyde Derivatives into Carboxylic Acid Derivatives 263
9.2.3.1 Acetals 263
9.2.3.2 Nitrogen Derivatives 263
9.2.3.3 Miscellaneous Substrates 264
9.2.4 Oxidative Decarboxylations of Aldehydes 265

9.3	Oxidations of Ketones	265
9.3.1	Ketone Cleavage Reactions	265
9.3.1.1	Simple Acyclic Ketones	265
9.3.1.2	Simple Cyclic Ketones	266
9.3.1.3	Functionalized Ketones	267
9.3.2	Oxidative Rearrangements of Ketones	267
9.3.2.1	Baeyer-Villiger Reactions	267
9.3.2.2	Ketone Amidations	272
9.3.2.3	Miscellaneous Rearrangements	275
9.3.3	Willgerodt Reactions	276
9.4	Conclusions	277
	References	277

10 Manganese-based Oxidation with Hydrogen Peroxide 295
Jelle Brinksma, Johannes W. de Boer, Ronald Hage, and Ben L. Feringa 295

10.1	Introduction	295
10.2	Biomimetic Manganese Oxidation Catalysis	296
10.3	Bleaching Catalysis	298
10.4	Catalytic Epoxidation	298
10.4.1	Manganese Porphyrin Catalysts	299
10.4.2	Manganese–salen Catalysts	302
10.4.3	Mn-1,4,7-triazacyclononane Catalysts	305
10.4.4	Miscellaneous Catalysts	311
10.5	*cis*-Dihydroxylation	314
10.6	Alcohol Oxidation to Aldehydes	317
10.7	Sulfide to Sulfoxide Oxidation	318
10.8	Conclusions	321
	References	321

Subject Index 327

List of Contributors

Hans Adolfsson
Department of Organic Chemistry
Arrhenius Laboratory
Stockholm University
106 91 Stockholm
Sweden

Isabel W. C. E. Arends
Delft University of Technology
Biocatalysis and Organic Chemistry
Julianalaan 136
2628 BL Delft
The Netherlands

Jan-E. Bäckvall
Department of Organic Chemistry
Arrhenius Laboratory
Stockholm University
106 91 Stockholm
Sweden

Matthias Beller
Institut für Organische Katalyse-
forschung an der Universität Rostock e. V.
(IfOK)
Buchbinderstrasse 5–6
18055 Rostock
Germany

Johannes W. de Boer
Laboratory of Organic Chemistry
Stratingh Institute
University of Groningen
Nijenborgh 4
9747 AG Groningen
The Netherlands

Carsten Bolm
Institute of Organic Chemistry
RWTH Aachen
Professor-Pirlet-Str. 1
52056 Aachen
Germany

Jelle Brinksma
Laboratory of Organic Chemistry
Stratingh Institute
University of Groningen
Nijenborgh 4
9747 AG Groningen
The Netherlands

Christian Döbler
Institut für Organische Katalyse-
forschung an der Universität Rostock e. V.
(IfOK)
Buchbinderstrasse 5–6
18055 Rostock
Germany

List of Contributors

Ben L. Feringa
Laboratory of Organic Chemistry
Stratingh Institute
University of Groningen
Nijenborgh 4
9747 AG Groningen
The Netherlands

Jean-Cédric Frison
Institute of Organic Chemistry
RWTH Aachen
Professor-Pirlet-Str. 1
52056 Aachen
Germany

Ronald Hage
Unilever R&D
Po Box 114
3130 AC Vlaardingen
The Netherlands

Yasutaka Ishii
Department of Applied Chemistry
Faculty of Engineering
Kansai University
Suita
Osaka 564-8680
Japan

Naruyoshi Komiya
Department of Chemistry
Graduate School of Engineering Science
Osaka University
1-3, Machikaneyama
Toyonaka
Osaka 560-8531
Japan

Jacques Le Paih
Institute of Organic Chemistry
RWTH Aachen
Professor-Pirlet-Str. 1
52056 Aachen
Germany

Shun-Ichi Murahashi
Department of Applied Chemistry
Okayama University of Science
1-1 Ridai-cho
Okayama
Okayama 700-0005
Japan

Ronny Neumann
Department of Organic Chemistry
Weizmann Institute of Science
Rehovot
76100 Israel

Satoshi Sakaguchi
Department of Applied Chemistry
Faculty of Engineering
Kansai University
Suita
Osaka 564-8680
Japan

Roger A. Sheldon
Delft University of Technology
Biocatalysis and Organic Chemistry
Julianalaan 136
2628 BL Delft
The Netherlands

Yian Shi
Department of Chemistry
Colorado State University
Fort Collins
Colorado 80523
USA

Uta Sundermeier
Institut für Organische Katalyse-
forschung an der Universität Rostock e. V.
(IfOK)
Buchbinderstrasse 5–6
18055 Rostock
Germany

1
Recent Developments in the Osmium-catalyzed Dihydroxylation of Olefins

UTA SUNDERMEIER, CHRISTIAN DÖBLER, and MATTHIAS BELLER

1.1
Introduction

The oxidative functionalization of olefins is of major importance for both organic synthesis and the industrial production of bulk and fine chemicals [1]. Among the different oxidation products of olefins, 1,2-diols are used in a wide variety of applications. Ethylene- and propylene-glycol are produced on a multi-million ton scale per annum, due to their importance as polyester monomers and anti-freeze agents [2]. A number of 1,2-diols such as 2,3-dimethyl-2,3-butanediol, 1,2-octanediol, 1,2-hexanediol, 1,2-pentanediol, 1,2- and 2,3-butanediol are of interest in the fine chemical industry. In addition, chiral 1,2-diols are employed as intermediates for pharmaceuticals and agrochemicals. At present 1,2-diols are manufactured industrially by a two step sequence consisting of epoxidation of an olefin with a hydroperoxide or a peracid followed by hydrolysis of the resulting epoxide [3]. Compared with this process the dihydroxylation of C=C double bonds constitutes a more atom-efficient and shorter route to 1,2-diols. In general the dihydroxylation of olefins is catalyzed by osmium, ruthenium or manganese oxo species. The osmium-catalyzed variant is the most reliable and efficient method for the synthesis of *cis*-1,2-diols [4]. Using osmium in catalytic amounts together with a secondary oxidant in stoichiometric amounts various olefins, including mono-, di-, and trisubstituted unfunctionalized, as well as many functionalized olefins, can be converted into the corresponding diols. OsO_4 as an electrophilic reagent reacts only slowly with electron-deficient olefins, and therefore higher amounts of catalyst and ligand are necessary in these cases. Recent studies have revealed that these substrates react much more efficiently when the pH of the reaction medium is maintained on the acidic side [5]. Here, citric acid appears to be superior for maintaining the pH in the desired range. On the other hand, in another study it was found that providing a constant pH value of 12.0 leads to improved reaction rates for internal olefins [6].

Since its discovery by Sharpless and coworkers, catalytic asymmetric dihydroxylation (AD) has significantly enhanced the utility of osmium-catalyzed dihydroxylation (Scheme 1.1) [7]. Numerous applications in organic synthesis have appeared in recent years [8].

Modern Oxidation Methods. Edited by Jan-Erling Bäckvall
Copyright © 2004 WILEY-VCH Verlag GmbH & Co. KGaA, Weinheim
ISBN: 3-527-30642-0

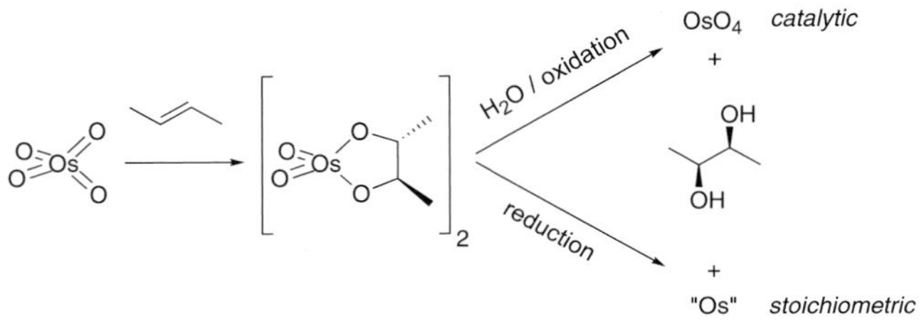

Scheme 1.1 Osmylation of olefins

While the problem of enantioselectivity has largely been solved through extensive synthesis and screening of cinchona alkaloid ligands by the Sharpless group, some features of this general method remain problematic for larger scale applications. Firstly, the use of the expensive osmium catalyst must be minimized and an efficient recycling of the metal should be developed. Secondly, the applied reoxidants for Os^{VI} species are expensive and lead to overstoichiometric amounts of waste.

In the past several reoxidation processes for osmium(VI) glycolates or other osmium(VI) species have been developed. Historically, chlorates [9] and hydrogen peroxide [10] were first applied as stoichiometric oxidants, however in both cases the dihydroxylation often proceeds with low chemoselectivity. Other reoxidants for osmium(VI) are *tert*-butyl hydroperoxide in the presence of Et$_4$NOH [11] and a range of N-oxides, such as N-methylmorpholine N-oxide (NMO) [12] (the Upjohn process) and trimethylamine N-oxide. K$_3$[Fe(CN)$_6$] gave a substantial improvement in the enantioselectivities in asymmetric dihydroxylations when it was introduced as a reoxidant for osmium(VI) species in 1990 [13]. However, even as early on as 1975 it was already being described as an oxidant for Os-catalyzed oxidation reactions [14]. Today the "AD-mix", containing the catalyst precursor K$_2$[OsO$_2$(OH)$_4$], the co-oxidant K$_3$[Fe(CN)$_6$], the base K$_2$CO$_3$, and the chiral ligand, is commercially available and the dihydroxylation reaction is easy to carry out. However, the production of overstoichiometric amounts of waste remains as a significant disadvantage of the reaction protocol.

This chapter will summarize the recent developments in the area of osmium-catalyzed dihydroxylations, which bring this transformation closer to a "green reaction". Hence, special emphasis is given to the use of new reoxidants and recycling of the osmium catalyst.

1.2
Environmentally Friendly Terminal Oxidants

1.2.1
Hydrogen Peroxide

Ever since the Upjohn procedure was published in 1976 the N-methylmorpholine N-oxide-based procedure has become one of the standard methods for osmium-catalyzed dihydroxylations. However, in the asymmetric dihydroxylation NMO has not

1.2 Environmentally Friendly Terminal Oxidants

been fully appreciated since it was difficult to obtain high ee with this oxidant. Some years ago it was demonstrated that NMO could be employed as the oxidant in the AD reaction to give high ee in aqueous tert-BuOH with slow addition of the olefin [15].

In spite of the fact that hydrogen peroxide was one of the first stoichiometric oxidants to be introduced for the osmium-catalyzed dihydroxylation it was not actually used until recently. When using hydrogen peroxide as the reoxidant for transition metal catalysts, very often there is the big disadvantage that a large excess of H_2O_2 is required, implying that the unproductive peroxide decomposition is the major process.

Recently Bäckvall and coworkers were able to improve the H_2O_2 reoxidation process significantly by using N-methylmorpholine together with flavin as co-catalysts in the presence of hydrogen peroxide [16]. Thus a renaissance of both NMO and H_2O_2 was induced. The mechanism of the triple catalytic H_2O_2 oxidation is shown in Scheme 1.2.

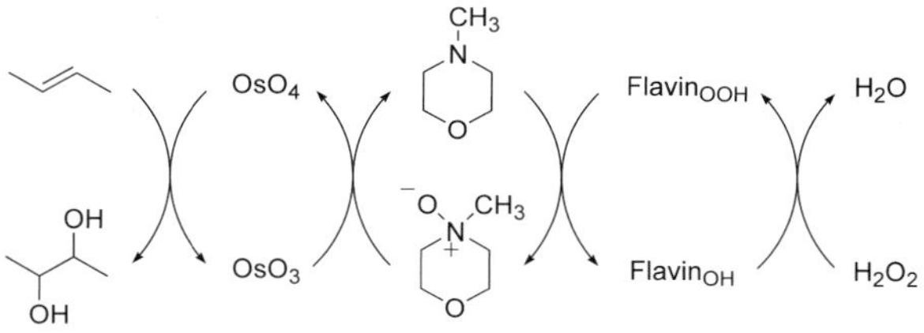

Scheme 1.2 Osmium-catalyzed dihydroxylation of olefins using H_2O_2 as the terminal oxidant

The flavin hydroperoxide generated from flavin and H_2O_2 recycles the N-methylmorpholine (NMM) to N-methylmorpholine N-oxide (NMO), which in turn reoxidizes the Os^{VI} to Os^{VIII}. While the use of hydrogen peroxide as the oxidant without the electron-transfer mediators (NMM, flavin) is inefficient and nonselective, various olefins were oxidized to diols in good to excellent yields employing this mild triple catalytic system (Scheme 1.3).

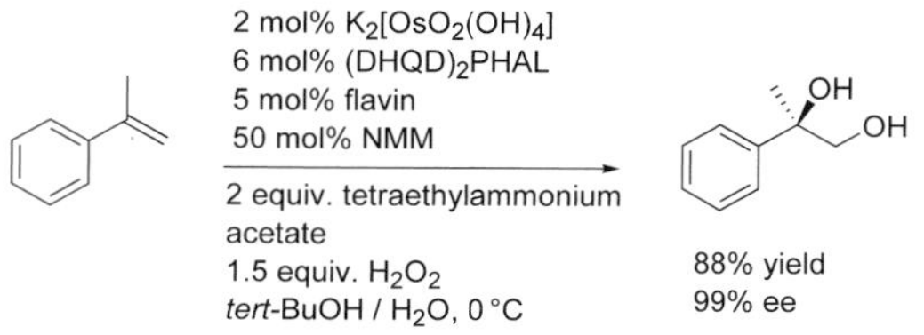

Scheme 1.3 Osmium-catalyzed dihydroxylation of α-methylstyrene using H_2O_2

By using a chiral Sharpless ligand high enantioselectivities were obtained. Here, an increase in the addition time for olefin and H_2O_2 can have a positive effect on the enantioselectivity.

Bäckvall and coworkers have shown that other tertiary amines can assume the role of the N-methylmorpholine. They reported on the first example of an enantioselective catalytic redox process where the chiral ligand has two different modes of operation: (1) to provide stereocontrol in the addition of the substrate, and (2) to be responsible for the reoxidation of the metal through an oxidized form [17]. The results obtained with hydroquinidine 1,4-phthalazinediyl diether (DHQD)$_2$PHAL both as an electron-transfer mediator and chiral ligand in the osmium-catalyzed dihydroxylation are comparable to those obtained employing NMM together with (DHQD)$_2$PHAL. The proposed catalytic cycle for the reaction is depicted in Scheme 1.4.

The flavin is an efficient electron-transfer mediator, but rather unstable. Several transition metal complexes, for instance vanadyl acetylacetonate, can also activate hydrogen peroxide and are capable of replacing the flavin in the dihydroxylation reaction [18].

More recently Bäckvall and coworkers developed a novel and robust system for osmium-catalyzed asymmetric dihydroxylation of olefins by H$_2$O$_2$ with methyltrioxorhenium (MTO) as the electron transfer mediator [19]. Interestingly, here MTO catalyzes oxidation of the chiral ligand to its mono-N-oxide, which in turn reoxidizes Os^{VI} to Os^{VIII}. This system gives vicinal diols in good yields and high enantiomeric excess up to 99%.

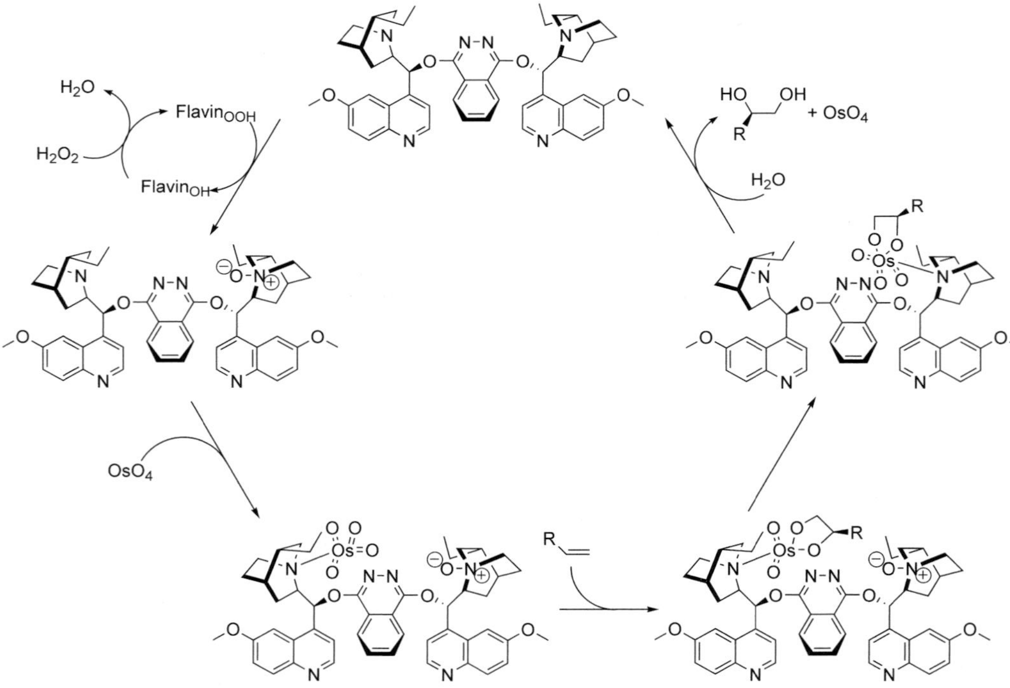

Scheme 1.4 Catalytic cycle for the enantioselective dihydroxylation of olefins using (DHQD)$_2$PHAL for oxygen transfer and as a source of chirality

1.2.2
Hypochlorite

Apart from oxygen and hydrogen peroxide, bleach is the simplest and cheapest oxidant that can be used in industry without problems. In the past this oxidant has only been applied in the presence of osmium complexes in two patents in the early 1970s for the oxidation of fatty acids [20]. In 2003 the first general dihydroxylation procedure of various olefins in the presence of sodium hypochlorite as the reoxidant was described by us [21]. Using α-methylstyrene as a model compound, 100% conversion and 98% yield of the desired 1,2-diol were obtained (Scheme 1.5).

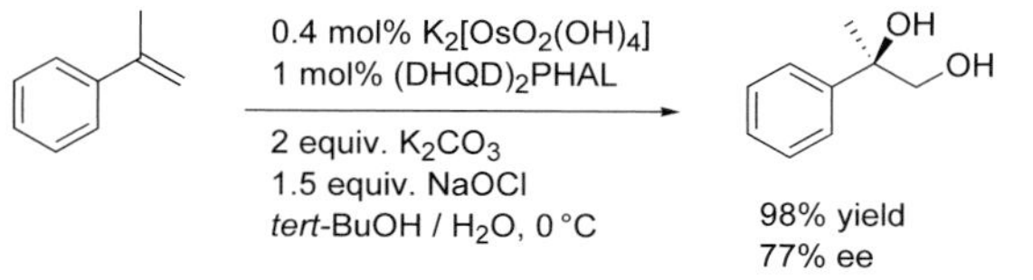

Scheme 1.5 Osmium-catalyzed dihydroxylation of α-methylstyrene using sodium hypochlorite

Interestingly, the yield of 2-phenyl-1,2-propanediol after 1 h was significantly higher using hypochlorite compared with literature protocols using NMO (90%) [22] or $K_3[Fe(CN)_6]$ (90%) at this temperature. The turnover frequency was 242 h^{-1}, which is a reasonable level [23]. Under the conditions shown in Scheme 1.5 an enantioselectivity of only 77% ee is obtained, while 94% ee is reported using $K_3[Fe(CN)_6]$ as the reoxidant. The lower enantioselectivity can be explained by some involvement of the so-called second catalytic cycle with the intermediate Os^{VI} glycolate being oxidized to an Os^{VIII} species prior to hydrolysis (Scheme 1.6) [24].

Nevertheless, the enantioselectivity was improved by applying a higher ligand concentration. In the presence of 5 mol% $(DHQD)_2PHAL$ a good enantioselectivity of 91% ee is observed for α-methylstyrene. Using tert-butylmethylether as the organic co-solvent instead of tert-butanol, 99% yield and 89% ee with only 1 mol% $(DHQD)_2PHAL$ are reported for the same substrate. This increase in enantioselectivity can be explained by an increase in the concentration of the chiral ligand in the organic phase. Increasing the polarity of the water phase by using a 10% aqueous NaCl solution showed a similar positive effect. Table 1.1 shows the results of the asymmetric dihydroxylation of various olefins with NaOCl as the terminal oxidant.

Despite the slow hydrolysis of the corresponding sterically hindered Os^{VI} glycolate, trans-5-decene reacted fast without any problems. This result is especially interesting since it is necessary to add stoichiometric amounts of hydrolysis aids to the dihydroxylation of most internal olefins in the presence of other oxidants.

With this protocol a very fast, easy to perform, and cheap procedure for the asymmetric dihydroxylation is presented.

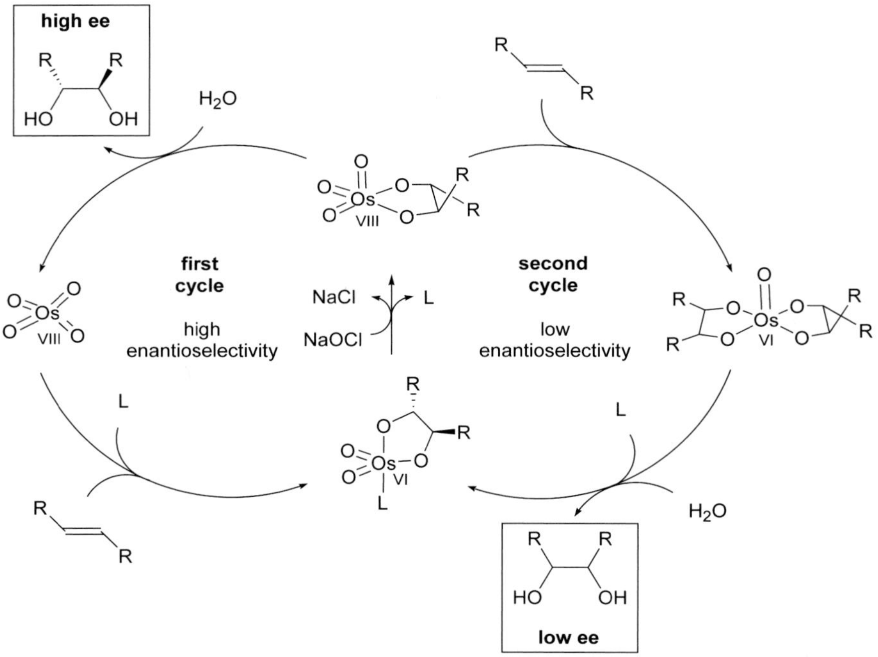

Scheme 1.6 The two catalytic cycles in the asymmetric dihydroxylation

Tab. 1.1 Asymmetric dihydroxylation of different olefins using NaOCl as terminal oxidant[a]

Entry	Olefin	Time (h)	Yield (%)	Selectivity (%)	ee (%)	ee (%) Ref.
1	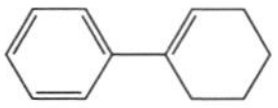 (phenylcyclohexene)	1	88	88	95	99
2	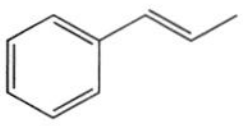 (β-methylstyrene)	2	93	99	95	97
3	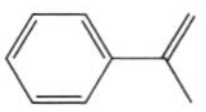 (α-methylstyrene)	1	99	99	91	95
4	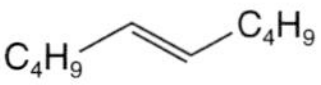 C$_4$H$_9$—=—C$_4$H$_9$	1	92	94	93	97
5	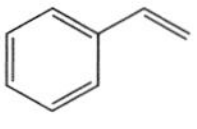 (styrene)	1	84	84	91	97
6	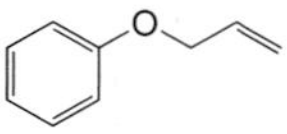 (allyl phenyl ether)	2	88	94	73	88

[a] Reaction conditions: 2 mmol olefin, 0.4 mol% K$_2$[OsO$_2$(OH)$_4$], 5 mol% (DHQD)$_2$PHAL, 10 mL H$_2$O, 10 mL ^{t}BuOH, 1.5 equiv. NaOCl, 2 equiv. K$_2$CO$_3$, 0 °C.

Tab. 1.1 (continued)

Entry	Olefin	Time (h)	Yield (%)	Selectivity (%)	ee (%)	ee (%) Ref.
7	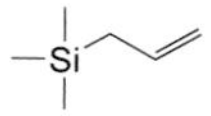 —Si⟨⟩	2	87	93	80[b]	
8	 C_6H_{13}	2	97	97	73	
9	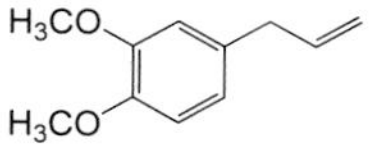 H_3CO-aryl-allyl, H_3CO	2	94	96	34[b]	
10	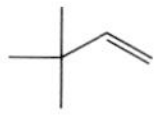 t-Bu-vinyl	2	97	>97	80[b]	92

[b] 5 mol% $(DHQD)_2PYR$ instead of $(DHQD)_2PHAL$.

1.2.3
Oxygen or Air

In the past it has been demonstrated by several groups that in the presence of OsO_4 and oxygen mainly non-selective oxidation reactions take place [25]. However, in 1999 Krief et al. published a reaction system consisting of oxygen, catalytic amounts of OsO_4 and selenides for the asymmetric dihydroxylation of α-methylstyrene under irradiation with visible light in the presence of a sensitizer (Scheme 1.7) [26]. Here, the selenides are oxidized to their oxides by singlet oxygen and the selene oxides are able to re-oxidize osmium(VI) to osmium(VIII). The reaction works with similar yields and ee values to those of the Sharpless-AD. Potassium carbonate is also used, but only one tenth of the amount present in the AD-mix. Air can be used instead of pure oxygen.

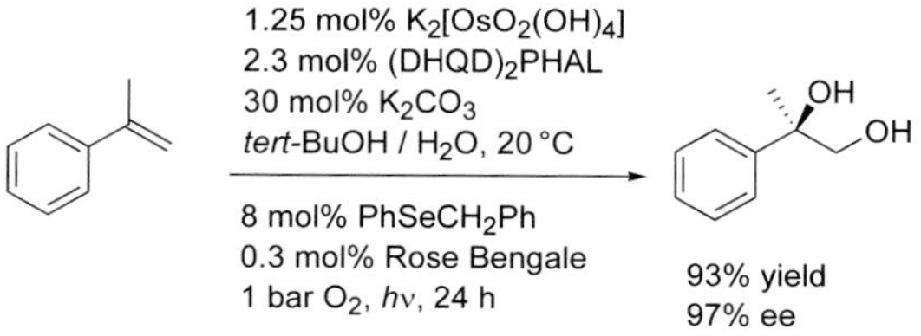

Reaction conditions:
1.25 mol% $K_2[OsO_2(OH)_4]$
2.3 mol% $(DHQD)_2PHAL$
30 mol% K_2CO_3
tert-BuOH / H_2O, 20 °C

8 mol% $PhSeCH_2Ph$
0.3 mol% Rose Bengale
1 bar O_2, hv, 24 h

93% yield
97% ee

Scheme 1.7 Osmium-catalyzed dihydroxylation using $^1O_2$ and benzyl phenyl selenide

The reaction was extended to a wide range of aromatic and aliphatic olefins [27]. It was shown that both yield and enantioselectivity are influenced by the pH of the reaction medium. The procedure was also applied to practical syntheses of natural product derivatives [28]. This version of the AD reaction not only uses a more ecological co-oxidant, it also requires much less matter: 87 mg of matter (catalyst, ligand, base, reoxidant) are required to oxidize 1 mmol of the same olefin instead of 1400 mg when the AD-mix is used.

Also in 1999 there was the first publication on the use of molecular oxygen without any additive to reoxidize osmium(VI) to osmium(VIII). We reported that the osmium-catalyzed dihydroxylation of aliphatic and aromatic olefins proceeds efficiently in the presence of dioxygen under ambient conditions [29]. As shown in Table 1.2 the new dihydroxylation procedure constitutes a significant advancement compared with other reoxidation procedures. Here, the dihydroxylation of α-methylstyrene is compared using different stoichiometric oxidants. The yield of the 1,2-diol remains good to very good (87–96%), independent of the oxidant used. The best enantioselectivities (94–96% ee) are obtained with hydroquinidine 1,4-phthalazinediyl diether [(DHQD)$_2$PHAL] as the ligand at 0–12 °C (Table 1.2, entries 1 and 3).

The dihydroxylation process with oxygen is clearly the most ecologically favorable procedure (Table 1.2, entry 5), when the production of waste from a stoichiometric reoxidant is considered. With the use of K$_3$[Fe(CN)$_6$] as oxidant approximately 8.1 kg of iron salts per kg of product are formed. However, in the case of the Krief (Table 1.2, entry 3) and Bäckvall procedures (Table 1.2, entry 4) as well as in the presence of NaOCl (Table 1.2, entry 6) some byproducts also arise due to the use of cocatalysts and co-oxidants. It should be noted that only salts and byproducts formed

Tab. 1.2 Comparison of the dihydroxylation of α-methylstyrene in the presence of different oxidants

Entry	Oxidant	Yield (%)	Reaction conditions	ee (%)	TON	Waste (oxidant) (kg/kg diol)	Ref.
1	K$_3$[Fe(CN)$_6$]	90	0 °C K$_2$[OsO$_2$(OH)$_4$] ^{t}BuOH/H$_2$O	94[a]	450	8.1[c]	[7b]
2	NMO	90	0 °C OsO$_4$ acetone/H$_2$O	33[b]	225	0.88[d]	[22]
3	PhSeCH$_2$Ph/O$_2$ PhSeCH$_2$Ph/air	89 87	12 °C K$_2$[OsO$_2$(OH)$_4$] ^{t}BuOH/H$_2$O	96[a] 93[a]	222 48	0.16[e] 0.16[e]	[26a] [26a]
4	NMM/flavin/H$_2$O$_2$	93	RT OsO$_4$ acetone/H$_2$O	–	46	0.33[f]	[16a]
5	O$_2$	96	50 °C K$_2$[OsO$_2$(OH)$_4$] ^{t}BuOH/aq. buffer	80[a]	192	–	[29]
6	NaOCl	99	0 °C K$_2$[OsO$_2$(OH)$_4$] ^{t}BuOH/H$_2$O	91[a]	247	0.58[g]	[21]

[a] Ligand: Hydroquinidine 1,4-phthalazinediyl diether. [b] Hydroquinidine p-chlorobenzoate.
[c] K$_4$[Fe(CN)$_6$]. [d] N-Methylmorpholine (NMM). [e] PhSe(O)CH$_2$Ph. [f] NMO/flavin-OOH. [g] NaCl.

from the oxidant have been included in the calculation. Other waste products have not been considered. Nevertheless the calculations presented in Table 1.2 give a rough estimation of the environmental impact of the reaction.

Since the use of pure molecular oxygen on a larger scale might lead to safety problems it is even more advantageous to use air as the oxidizing agent. Hence, all current bulk oxidation processes, e.g., the oxidation of BTX (benzene, toluene, xylene) aromatics or alkanes to give carboxylic acids, and the conversion of ethylene into ethylene oxide, use air and not pure oxygen as the oxidant [30]. In Table 1.3 the results of the dihydroxylation of α-methylstyrene as a model compound using air as the stoichiometric oxidant are shown in contrast to that with pure oxygen (Scheme 1.8; Table 1.3) [31].

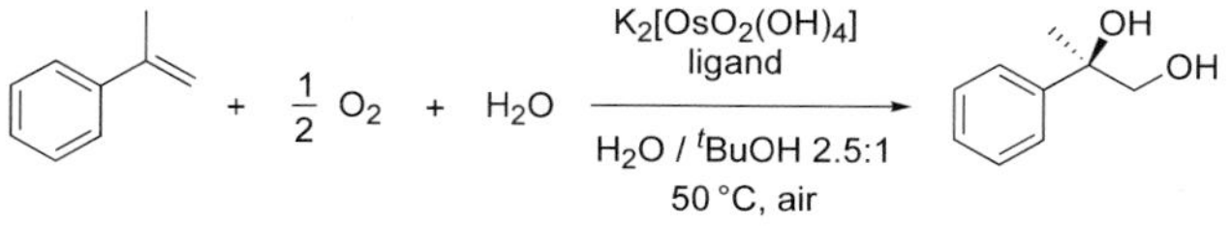

Scheme 1.8 Osmium-catalyzed dihydroxylation of α-methylstyrene

The dihydroxylation of α-methylstyrene in the presence of 1 bar of pure oxygen proceeds smoothly (Table 1.3, entries 1–2), with the best results being obtained at pH 10.4. In the presence of 0.5 mol% $K_2[OsO_2(OH)_4]$/1.5 mol% DABCO or 1.5 mol% $(DHQD)_2PHAL$ at pH 10.4 and 50 °C total conversion was achieved after 16 h or 20 h depending on the ligand. While the total yield and selectivity of the reaction are excellent (97% and 96%, respectively), the total turnover frequency of the catalyst is comparatively low (TOF = 10–12 h^{-1}). In the presence of the chiral cinchona ligand

Tab. 1.3 Dihydroxylation of α-methylstyrene with air[a]

Entry	Pressure (bar)[c]	Cat. (mol%)	Ligand	L/Os	[L] (mmol L^{-1})	Time (h)	Yield (%)	Selectivity (%)	ee (%)
1	1 (pure O_2)	0.5	DABCO[d]	3:1	3.0	16	97	97	–
2	1 (pure O_2)	0.5	$(DHQD)_2PHAL$[e]	3:1	3.0	20	96	96	80
3	1	0.5	DABCO	3.1	3.0	24	24	85	–
4	1	0.5	DABCO	3.1	3.0	68	58	83	–
5	5	0.1	DABCO	3:1	0.6	24	41	93	–
6	9	0.1	DABCO	3:1	0.6	24	76	92	–
7	20	0.5	$(DHQD)_2PHAL$	3:1	3.0	17	96	96	82
8	20	0.1	$(DHQD)_2PHAL$	3:1	0.6	24	95	95	62
9	20	0.1	$(DHQD)_2PHAL$	15:1	3.0	24	95	95	83
10[b]	20	0.1	$(DHQD)_2PHAL$	3:1	1.5	24	94	94	67
11[b]	20	0.1	$(DHQD)_2PHAL$	6:1	3.0	24	94	94	78
12[b]	20	0.1	$(DHQD)_2PHAL$	15:1	7.5	24	60	95	82

[a] Reaction conditions: $K_2[OsO_2(OH)_4]$, 50 °C, 2 mmol olefin, 25 mL buffer solution (pH 10.4), 10 mL ^{t}BuOH. [b] 10 mmol olefin, 50 mL buffer solution (pH 10.4), 20 mL ^{t}BuOH. [c] The autoclave was purged with air and then pressurized to the given value. [d] 1,4-Diazabicyclo[2.2.2.]octane. [e] Hydroquinidine 1,4-phthalazinediyl diether.

(DHQD)$_2$PHAL an *ee* of 80% is observed. Sharpless et al. reported an enantioselectivity of 94% for the dihydroxylation of α-methylstyrene with (DHQD)$_2$PHAL as the ligand using K$_3$[Fe(CN)$_6$] as the reoxidant at 0 °C [32]. Studies of the ceiling *ee* at 50 °C (88% *ee*) show that the main difference in the enantioselectivity stems from the higher reaction temperature. Using air instead of pure oxygen gas gave only 24% of the corresponding diol after 24 h (TOF = 1 h^{-1}; Table 1.3, entry 3). Although the reaction is slow, it is important to note that the catalyst stays active, as shown by the fact that 58% of the product is obtained after 68 h (Table 1.3, entry 4). Interestingly the chemoselectivity of the dihydroxylation does not significantly decrease after a prolonged reaction time. At 5–20 bar air pressure the turnover frequency of the catalyst is improved (Table 1.3, entries 5–11).

Full conversion of a α-methylstyrene is achieved at an air pressure of 20 bar in the presence of 0.1 mol% of osmium, which corresponds to a turnover frequency of 40 h^{-1} (Table 1.3, entries 8–11). Thus, by increasing the air pressure to 20 bar, it was possible to reduce the amount of osmium catalyst by a factor of 5. A decrease of the osmium catalyst *and* the ligand leads to a decrease in the enantioselectivity of from 82% to 62% *ee*. This is easily explained by the fact that the ligand concentration determines the stereoselectivity of the dihydroxylation reaction (Table 1.3, entries 7 and 9).

While the reaction at higher substrate concentration (10 mmol instead of 2 mmol) proceeds only sluggishly at 1 bar even with pure oxygen, full conversion is achieved after 24 h at 20 bar of air (Table 1.3, entries 10 and 11, and Table 1.4, entries 17 and 18). In all experiments performed under air pressure the chemoselectivity of the dihydroxylation remained excellent (92–96%).

Table 1.4 shows the results of the osmium-catalyzed dihydroxylation of various olefins with air.

As depicted in Table 1.4 all olefins gave the corresponding diols in moderate to good yields (48–89%). Applying standard reaction conditions, the best yields of diols were obtained with 1-octene (97%), 1-phenyl-1-cyclohexene (88%), *trans*-5-decene (85%), allyl phenyl ether (77%) and styrene (76%). The enantioselectivities varied from 53 to 98% *ee* depending on the substrate. It is important to note that the chemoselectivity of the reaction decreases under standard conditions in the following substrate order: α-methylstyrene = 1-octene > 1-phenyl-1-cyclohexene > *trans*-5-decene > *n*-C$_6$F$_{13}$CH=CH$_2$ > allyl phenyl ether > styrene >> *trans*-stilbene. A correlation between the chemoselectivity of the reaction and the sensitivity of the produced diol towards further oxidation is evident, with the main side reaction being the oxidative cleavage of the C=C double bond. Aromatic diols with benzylic hydrogen atoms are especially sensitive to this oxidation reaction. Thus, the dihydroxylation of *trans*-stilbene gave no hydrobenzoin in the biphasic mixture water/*tert*-butanol at pH 10.4, 50 °C and 20 bar air pressure (Table 1.4, entry 9). Instead of dihydroxylation a highly selective cleavage of stilbene to give benzaldehyde (84–87% yield) was observed. Interestingly, changing the solvent to isobutyl methyl ketone (Table 1.4, entry 12) makes it possible to obtain hydrobenzoin in high yield (89%) and enantioselectivity (98%) at pH 10.4.

The mechanism of the dihydroxylation reaction with oxygen or air is presumed to be similar to the catalytic cycle presented by Sharpless et al. for the osmium-cata-

1.2 Environmentally Friendly Terminal Oxidants | 11

Tab. 1.4 Dihydroxylation of various olefins with air[a]

Entry	Olefin	Cat. (mol%)	Ligand	L/Os	[L] (mmol L^{-1})	Time (h)	Yield (%)[b]	Selectivity (%)[b]	ee (%)
1		0.5	(DHQD)$_2$PHAL	3:1	3.0	24	42	42	87
2		0.5	(DHQD)$_2$PHAL	3:1	3.0	16	66	66	86
3		0.5	(DHQD)$_2$PHAL	3:1	3.0	14	76	76	87
4		0.5	(DHQD)$_2$PHAL	3:1	3.0	24	88	88	89
5		0.5	(DHQD)$_2$PHAL	3:1	3.0	24	63	63	67
6		0.5	(DHQD)$_2$PHAL	3:1	3.0	18	68	68	68
7		0.5	(DHQD)$_2$PHAL	3:1	3.0	14	67	67	66
8		0.5	(DHQD)$_2$PHAL	3:1	3.0	9	77	77	68
9		0.5	–	–	–	24	0 (84)	0 (84)	–
10[c]		1.0	DABCO	3:1	1.5	24	4 (77)	5 (87)	–
11[c, d]		1.0	(DHQD)$_2$PHAL	3:1	1.5	24	40 (35)	48 (42)	86
12[c, e]		1.0	(DHQD)$_2$PHAL	3:1	1.5	24	89 (7)	89 (7)	98
13[d]	C$_4$H$_9$–CH=CH–C$_4$H$_9$	1.0	(DHQD)$_2$PHAL	3:1	6.0	24	85	85	82
14		0.5	(DHQD)$_2$PHAL	3:1	3.0	18	96	96	63
15		0.1	(DHQD)$_2$PHAL	3:1	0.6	24	95	95	44
16	C$_6$H$_{13}$	0.1	(DHQD)$_2$PHAL	15:1	3.0	24	97	97	62
17[f]		0.1	(DHQD)$_2$PHAL	3:1	1.5	24	94	94	47
18[f]		0.1	(DHQD)$_2$PHAL	6:1	3.0	24	95	95	62
19	C$_6$F$_{13}$	2.0	(DHQD)$_2$PYR[g]	3:1	12.0	24	55	–	68

[a] Reaction conditions: K$_2$[OsO$_2$(OH)$_4$], 50 °C, 2 mmol olefin, 20 bar air, pH = 10.4, 25 mL buffer solution, 10 mL ^{t}BuOH; entries 9–12: 15 mL buffer solution, 20 mL ^{t}BuOH, entries 17–18: 50 mL buffer solution, 20 mL ^{t}BuOH. [b] Values in parentheses are for benzaldehyde. [c] 1 mmol olefin. [d] pH = 12. [e] Isobutyl methyl ketone instead of ^{t}BuOH. [f] 10 mmol olefin. [g] Hydroquinidine 2,5-diphenyl-4,6-pyrimidinediyl diether.

lyzed dihydroxylation with K$_3$[Fe(CN)$_6$] as the reoxidant (Scheme 1.9). The addition of the olefin to a ligated Os^{VIII} species proceeds mainly in the organic phase. Depending on the hydrolytic stability of the resulting Os^{VI} glycolate complex, the rate determining step of the reaction is either hydrolysis of the Os^{VI} glycolate or the reoxidation of Os^{VI} hydroxy species. There must be a minor involvement of a second catalytic cycle, as suggested for the dihydroxylation with NMO. Such a second cycle would lead to significantly lower enantioselectivities, as the attack of a second olefin molecule on the Os^{VIII} glycolate would occur in the absence of the chiral ligand. The observed enantioselectivities for the dihydroxylation with air are only slightly lower than the data previously published by the Sharpless group, despite the higher reaction temperature (50 °C vs. 0 °C). Therefore the direct oxidation of the Os^{VI} glycolate to an Os^{VIII} glycolate does not represent a major reaction pathway.

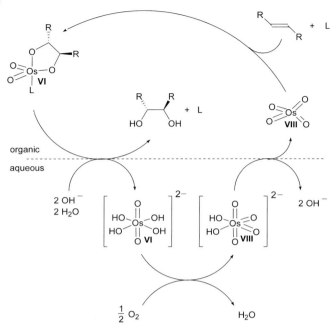

Scheme 1.9 Proposed catalytic cycle for the dihydroxylation of olefins with OsO$_4$ and oxygen as the terminal oxidant

1.3
Supported Osmium Catalyst

Hazardous toxicity and high costs are the chief drawbacks to reactions using osmium tetroxide. Besides the development of procedures where catalytic amounts of osmium tetroxide are joined with a stoichiometrically used secondary oxidant continuously regenerating the tetroxide, these disadvantages can be overcome by the use of stable and nonvolatile adducts of osmium tetroxide with heterogeneous supports [33]. They offer the advantages of easy and safe handling, simple separation from the reaction medium, and the possibility to reuse the expensive transition metal. Unfortunately, problems with the stability of the polymer support and leaching of the metal generally occur.

In this context Cainelli and coworkers had already reported, in 1989, the preparation of polymer-supported catalysts: here, OsO$_4$ was immobilized on several amine type polymers [34]. Such catalysts have structures of the type OsO$_4$ · L with the N-group of the polymer (= L) being coordinated to the Lewis acidic osmium center. Based upon this concept, a catalytic enantioselective dihydroxylation was established by using polymers containing cinchona alkaloid derivatives [35]. However, since the amine ligands coordinate to osmium under equilibrium conditions, recovery of the osmium using polymer supported ligands was difficult. Os-diolate hydrolysis seems to require detachment from the polymeric ligand, and hence causes leaching.

Herrmann and coworkers reported on the preparation of immobilized OsO_4 on poly(4-vinyl pyridine) and its use in the dihydroxylation of alkenes by means of hydrogen peroxide [36]. However, the problems of gradual polymer decomposition and osmium leaching were not solved.

A new strategy was published by Kobayashi and coworkers in 1998: they used microencapsulated osmium tetroxide. Here the metal is immobilized onto a polymer on the basis of physical envelopment by the polymer and on electron interactions between the π-electrons of the benzene rings of the polystyrene based polymer and a vacant orbital of the Lewis acid [37]. Using cyclohexene as a model compound it was shown that this microencapsulated osmium tetroxide (MC OsO_4) can be used as a catalyst in the dihydroxylation, with NMO as the stoichiometric oxidant (Scheme 1.10).

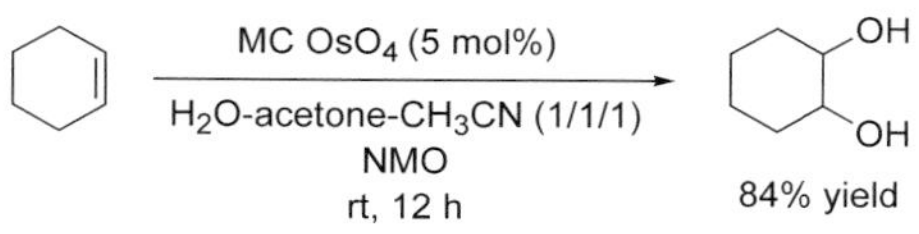

Scheme 1.10 Dihydroxylation of cyclohexene using microencapsulated osmium tetroxide (MC OsO_4)

In contrast to other typical OsO_4-catalyzed dihydroxylations, where H_2O-^{t}BuOH is used as the solvent system, the best yields were obtained in H_2O/acetone/CH_3CN. While the reaction was successfully carried out using NMO, moderate yields were obtained using trimethylamine N-oxide, and much lower yields were observed using hydrogen peroxide or potassium ferricyanide. The catalyst was recovered quantitatively by simple filtration and reused several times. The activity of the recovered catalyst did not decrease even after the fifth use.

A study of the rate of conversion of the starting material showed that the reaction proceeds faster using OsO_4 than using the microencapsulated catalyst. This is ascribed to the slower reoxidation of the microencapsulated osmium ester with NMO, compared with simple OsO_4.

Subsequently acryronitrile/butadiene/polystyrene polymer was used as a support based on the same microencapsulation technique and several olefins, including cyclic and acyclic, terminal, mono-, di-, tri-, and tetrasubstituted, gave the corresponding diols in high yields [38]. When $(DHQD)_2PHAL$ as a chiral source was added to the reaction mixture enantioselectivities up to 95% ee were obtained. However, this reaction requires slow addition of the olefin. After running a 100 mmol experiment, more than 95% of the ABS-MC OsO_4 and the chiral ligand were recovered.

Recently Kobayashi and coworkers reported on a new type of microencapsulated osmium tetroxide using phenoxyethoxymethyl-polystyrene as the support [39]. With this catalyst, asymmetric dihydroxylation of olefins has been successfully performed using $(DHQD)_2PHAL$ as a chiral ligand and $K_3[Fe(CN)_6]$ as a cooxidant in H_2O/acetone (Scheme 1.11).

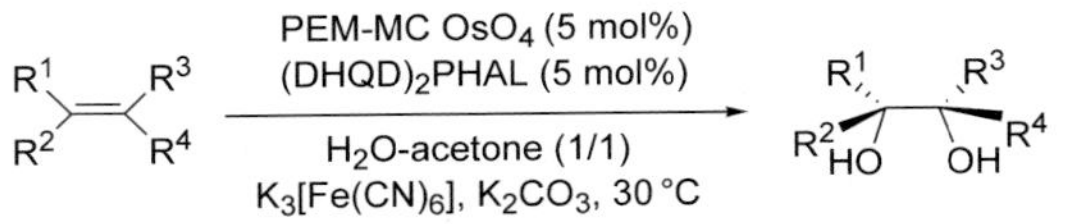

Scheme 1.11 Asymmetric dihydroxylation of olefins using PEM-MC OsO$_4$

In this instance the dihydroxylation does not require slow addition of the olefin, and the catalyst can be recovered quantitatively by simple filtration and reused without loss of activity.

Jacobs and coworkers published a completely different type of heterogeneous osmium catalyst. Their approach is based on two details from the mechanism of the cis-dihydroxylation: (1) tetrasubstituted olefins are smoothly osmylated to an osmate(VI) ester, but these esters are not hydrolyzed under mild conditions, and (2) an Os^{VI} monodiolate complex can be reoxidized to cis-dioxo Os^{VIII} without release of the diol; subsequent addition of a second olefin results in an Os bisdiolate complex. These two properties make it possible to immobilize a catalytically active osmium compound by the addition of OsO$_4$ to a tetrasubstituted olefin that is covalently linked to a silica support. The tetrasubstituted diolate ester which is formed at one side of the Os atom is stable, and keeps the catalyst fixed on the support material. The catalytic reaction can take place at the free coordination sites of Os (Scheme 1.12) [40].

The dihydroxylation of monosubstituted and disubstituted aliphatic olefins and cyclic olefins was successfully performed using this heterogeneous catalyst and

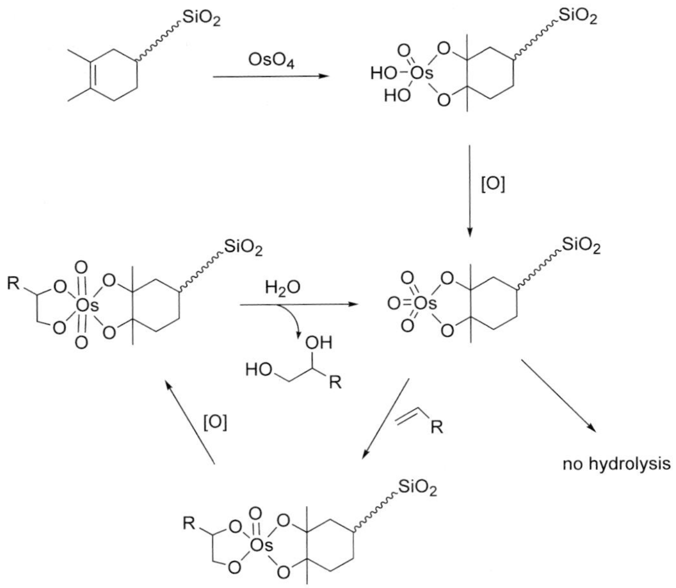

Scheme 1.12 Immobilization of Os in a tertiary diolate complex, and proposed catalytic cycle for cis-dihydroxylation

NMO as the cooxidant. With respect to the olefin, 0.25 mol% Os was needed and the excellent chemoselectivity of the homogeneous reaction with NMO is preserved. However, somewhat increased reaction times are required. The development of an asymmetric variant of this process by addition of the typical chiral alkaloid ligands of the asymmetric dihydroxylation should be difficult since the reactions performed with these heterogeneous catalysts are taking place in the so-called second cycle. With alkaloid ligands high *ee* values are only achieved in dihydroxylations occurring in the first cycle. However, recent findings by the groups of Sharpless and Adolfsson show that even second-cycle dihydroxylations may give substantial *ee* results [41]. Although this process must be optimized, further development of the concept of an enantioselective second-cycle process offers a perspective for a future heterogeneous asymmetric catalyst.

Choudary and his group reported, in 2001, on the design of an ion-exchange technique for the development of recoverable and reusable osmium catalysts immobilized on layered double hydroxides (LDH), modified silica, and organic resin for asymmetric dihydroxylation [42]. An activity profile of the dihydroxylation of *trans*-stilbene with various exchanger/OsO_4 catalysts revealed that LDH/OsO_4 displays the highest activity and that the heterogenized catalysts in general have higher reactivity than $K_2[OsO_2(OH)_4]$. When *trans*-stilbene was added to a mixture of LDH/OsO_4, $(DHQD)_2PHAL$ as the chiral ligand (1 mol% each), and NMO in $H_2O/^tBuOH$, the desired diol is obtained in 96% yield with 99% *ee*. Similarly, excellent *ee* results are obtained with resin/OsO_4 and SiO_2/OsO_4 in the same reaction. All of the prepared catalysts are recovered quantitatively by simple filtration and reused for five cycles with consistent activity. With this procedure, various olefins ranging from mono- to trisubstituted and from activated to non-activated are transformed into their diols. In most cases, the desired diols are formed in higher yields, albeit with almost similar *ee* values as reported in homogeneous systems. Slow addition of the olefin to the reaction mixture is warranted to achieve higher *ee*. This LDH/OsO_4 system presented by Choudary and coworkers is superior in terms of activity, enantioselectivity and scope of the reaction in comparison with that of Kobayashi.

Although the LDH/OsO_4 shows excellent activity with NMO, it is deactivated when $K_3[Fe(CN)_6]$ or molecular oxygen is used as the co-oxidant [43]. This deactivation is attributed to the displacement of OsO_4^{2-} by the competing anions, which include ferricyanide, ferrocyanide, and phosphate ions (from the aqueous buffer solution). To solve this problem resin/OsO_4 and SiO_2/OsO_4 were designed and prepared by the ion-exchange process on the quaternary ammonium-anchored resin and silica, respectively, as these ion-exchangers are expected to prefer bivalent anions rather than trivalent anions. These new heterogeneous catalysts show consistent performance in the dihydroxylation of α-methylstyrene for a number of recycles using NMO, $K_3[Fe(CN)_6]$ or O_2 as reoxidant. The resin/OsO_4 catalyst, however, displays higher activity than the SiO_2/OsO_4 catalyst. In the presence of Sharpless ligands various olefins were oxidized with high enantioselectivity using these heterogeneous systems. Very good *ee* results were obtained with each of the three co-oxidants. Equimolar ratios of ligand to osmium are sufficient to achieve excellent *ee* results. This is in contrast to the homogeneous reaction in which a 2–3 molar excess of the expensive chiral ligand to osmium is usually employed. These studies indicate that the binding ability of these heterogeneous osmium catalysts with the chiral ligand is greater than the homogeneous analogue.

Incidentally, this forms the first report of a heterogeneous osmium-catalyst mediated AD reaction of olefins using molecular oxygen as the co-oxidant. Under identical conditions, the turnover numbers of the heterogeneous catalyst are almost similar to the homogeneous system.

Furthermore, Choudary and coworkers presented a procedure for the application of a heterogeneous catalytic system for the AD reaction in combination with hydrogen peroxide as co-oxidant [44]. Here a triple catalytic system composed of NMM and two heterogeneous catalysts was designed. A titanium silicalite acts as the electron transfer mediator to perform oxidation of NMM that is used in catalytic amounts with hydrogen peroxide to provide *in situ* NMO continuously for AD of olefins, which is catalyzed by another heterogeneous catalyst, silica gel-supported cinchona alkaloid [SGS-(DHQD)$_2$PHAL]-OsO$_4$. Good yields were observed for various olefins. Again very good *ee* results have been achieved with an equimolar ratio of ligand to osmium, but slow addition of olefin and H$_2$O$_2$ is necessary. Unfortunately, recovery and reuse of the SGS-(DHQD)$_2$PHAL-OsO$_4$/TS-1 revealed that about 30% of the osmium had leached during the reaction. This amount has to be replenished in each additional run.

1.4
Ionic Liquids

Recently ionic liquids have become popular as new solvents in organic synthesis [45, 46]. They can dissolve a wide range of organometallic compounds and are miscible with organic compounds. They are highly polar but non-coordinating. In general ionic liquids exhibit excellent chemical and thermal stability along with ease of reuse. It is possible to vary their miscibility with water and organic solvents simply by changing the counter anion. Advantageously they have essentially negligible vapor pressure.

In 2002 olefin dihydroxylation by recoverable and reusable OsO$_4$ in ionic liquids was published for the first time [47]. Yanada and coworkers described the immobilization of OsO$_4$ in 1-ethyl-3-methylimidazolium tetrafluoroborate [47a]. They chose 1,1-diphenylethylene as a model compound and found that the use of 5 mol% OsO$_4$ in [emim]BF$_4$, 1.2 equiv. of NMO · H$_2$O, and room temperature were the best reaction conditions for good yield. After 18 h 100% of the corresponding diol was obtained. OsO$_4$-catalyzed reactions with other co-oxidants such as hydrogen peroxide, sodium percarbonate, and *tert*-butyl hydroperoxide gave poor results. With anhydrous NMO only 6% diol was found. After the reaction the 1,2-diol can be extracted with ethyl acetate and the ionic liquid containing the catalyst can be reused for further catalytic oxidation reaction. It was shown that even in the fifth run the obtained yield did not change. This new method using immobilized OsO$_4$ in an ionic liquid was applied to several substrates, including mono-, di-, and trisubstituted aliphatic olefins, as well as to aromatic olefins. In all cases, the desired diols were obtained in high yields.

The group working with Yao developed a slightly different procedure. They used [bmim]PF_6 (bmim = 1-n-butyl-3-methylimidazol)/water/^{t}BuOH (1:1:2) as the solvent system and NMO (1.2 equiv.) as the reoxidant for the osmium catalyst [47b]. Here 2 mol% osmium are needed for efficient dihydroxylation of various olefins. After the reaction, all volatiles were removed under reduced pressure and the product was extracted from the ionic liquid layer using ether. The ionic liquid layer containing the catalyst can be used several times with only a slight drop in catalyst activity. In order to prevent osmium leaching, 1.2 equiv. of DMAP relative to OsO_4 have to be added to the reaction mixture. This amine forms stable complexes with OsO_4, and this strong binding to a polar amine enhances its partitioning in the more polar ionic liquid layer. Recently, Song and coworkers reported on the Os-catalyzed dihydroxylation using NMO in mixtures of ionic liquids (1-butyl-3-methylimidazolium hexafluorophosphate or hexafluoroantimonate) with acetone/H_2O [48]. They used 1,4-bis(9-O-quininyl)phthalazine [(QN)$_2$PHAL] as the chiral ligand. (QN)$_2$PHAL will be converted into a new ligand bearing highly polar residues (four hydroxy groups in the 10,11-positions of the quinine parts) during AD reactions of olefins. The use of (QN)$_2$PHAL instead of (DHQD)$_2$PHAL afforded the same yields and ee results and, moreover, resulted in drastic improvement in recyclability of both catalytic components. In another recent report Branco and coworkers described the $K_2OsO_2(OH)_4$/$K_3Fe(CN)_6$/(DHQD)$_2$PHAL or (DHQD)$_2$PYR system for the asymmetric dihydroxylation using two different ionic liquids [49]. Both of the systems used, [bmim][PF_6]/water and [bmim][PF_6]/water/$tert$-butanol (bmim = 1-n-butyl-3-methylimidazol), are effective for a considerable number of runs (e.g., run 1, 88%, ee 90%; run 9, 83%, ee 89%). Only after 11 or 12 cycles was a significant drop in the chemical yield and optical purity observed.

In summary, it has been demonstrated that the application of an ionic liquid provides a simple approach to the immobilization of an osmium catalyst for olefin dihydroxylation. It is important to note that the volatility and toxicity of OsO_4 are greatly suppressed when ionic liquids are used.

References

[1] M. Beller, C. Bolm, *Transition Metals for Organic Synthesis*, Wiley-VCH, Weinheim, **1998**.

[2] Worldwide production capacities for ethylene glycol in 1995: 9.7 million tons per annum; worldwide production of 1,2-propylene glycol in 1994: 1.1 million tons per annum; K. Weissermel, H. J. Arpe, *Industrielle Organische Chemie*, 5th edn., Wiley-VCH, Weinheim, **1998**, p. 167 and p. 302.

[3] (a) H. H. Szmant, *Organic Building Blocks of the Chemical Industry*, Wiley, New York, **1989**, p. 347; (b) G. Pohl, H. Gaube in *Ullmann's Encyclopedia of Industrial Chemistry*, Vol. A1, VCH, Weinheim, **1985**, p. 305.

[4] Reviews: (a) M. Schröder *Chem. Rev.* **1980**, *80*, 187; (b) H. C. Kolb, M. S. Van Nieuwenhze, K. B. Sharpless, *Chem. Rev.* **1994**, *94*, 2483; (c) M. Beller, K. B. Sharpless in B. Cornils, W. A. Herrmann (Eds.), *Applied Homogeneous Catalysis*, VCH, Weinheim, **1996**,

p. 1009; (d) H. C. Kolb, K. B. Sharpless in M. Beller, C. Bolm (Eds.), *Transition Metals for Organic Synthesis* Vol. 2, VCH-Wiley, Weinheim, **1998**, p. 219; (e) I. E. Marko, J. S. Svendsen in E. N. Jacobsen, A. Pfaltz, H. Yamamoto (Eds.), *Comprehensive Asymmetric Catalysis II*, Springer, Berlin, **1999**, p. 713.

[5] P. Dupau, R. Epple, A. A. Thomas, V. V. Fokin, K. B. Sharpless *Adv. Synth. Catal.* **2002**, *344*, 421.

[6] G. M. Mehltretter, C. Döbler, U. Sundermeier, M. Beller *Tetrahedron Lett.* **2000**, *41*, 8083.

[7] (a) S. G. Hentges, K. B. Sharpless *J. Am. Chem. Soc.* **1980**, *192*, 4263; (b) K. B. Sharpless, W. Amberg, Y. L. Bennani, G. A. Crispino, J. Hartung, K.-S. Jeong, H.-L. Kwong, K. Morikawa, Z.-M. Wang, D. Xu, X.-L. Zhang *J. Org. Chem.* **1992**, *57*, 2768.

[8] (a) Z.-M. Wang, H. C. Kolb, K. B. Sharpless *J. Org. Chem.* **1994**, *59*, 5104; (b) F. G. Fang, S. Xie, M. W. Lowery *J. Org. Chem.* **1994**, *59*, 6142; (c) D. P. Curran, S.-B. Ko *J. Org. Chem.* **1994**, *59*, 6139; (d) E. J. Corey, A. Guzman-Perez, M. C. Noe *J. Am. Chem. Soc.* **1994**, *116*, 12109; (e) J. Corey, A. Guzman-Perez, M. C. Noe *J. Am. Chem. Soc.* **1995**, *117*, 10805; (f) M. Nambu, J. D. White *Chem. Commmun.* **1996**, 1619; (g) E. J. Corey, M. C. Noe, A. Y. Ting *Tetrahedron Lett.* **1996**, *37*, 1735; (h) G. Li, H.-T. Chang, K. B. Sharpless *Angew. Chem., Int. Ed. Engl.* **1996**, *35*, 451; (i) K. Mori, H. Takikawa, Y. Nishimura, H. Horikiri *Liebigs Ann./Recueil* **1997**, 327; (j) B. M. Trost, T. L. Calkins, C. G. Bochet *Angew. Chem., Int. Ed. Engl.* **1997**, *36*, 2632; (k) S. C. Sinha, A. Sinha, S. C. Sinha, E. Keinan *J. Am. Chem. Soc.* **1998**, *120*, 4017; (l) A. J. Fisher, F. Kerrigan *Synth. Commun.* **1998**, *28*, 2959; (m) J. M. Harris, M. D. Keranen, G. A. O'Doherty *J. Org. Chem.* **1999**, *64*, 2982; (n) H. Takahata, M. Kubota, N. Ikota *J. Org. Chem.* **1999**, *64*, 8594; (o) M. Quitschalle, M. Kalesse *Tetrahedron Lett.* **1999**, *40*, 7765; (p) F. J. Aladro, F. M. Guerra, F. J. Moreno-Dorado, J. M. Bustamante, Z. D. Jorge, G. M. Massanet *Tetrahedron Lett.* **2000**, *41*, 3209; (q) J. Liang, E. D. Moher, R. E. Moore, D. W. Hoard *J. Org. Chem.* **2000**, *65*, 3143; (r) X. D. Zhou, F. Cai, W. S. Zhou *Tetrahedron Lett.* **2001**, *42*, 2537; (s) D. P. G. Hamon, K. L. Tuck, H. S. Christie *Tetrahedron* **2001**, *57*, 9499; (t) B. M. Choudary, N. S. Chowdari, K. Jyothi, N. S. Kumar, M. L. Kantam *Chem. Commun.* **2002**, 586; (u) P. Y. Hayes, W. Kitching *J. Am. Chem. Soc.* **2002**, *124*, 9718; (v) S. Chandrasekhar, T. Ramachandar, M. V. Reddy *Synthesis* **2002**, 1867; (w) P. R. Andreana, J. S. McLellan, Y. C. Chen, P. G. Wang *Org. Lett.* **2002**, *4*, 3875.

[9] K. A. Hofmann *Chem. Ber.* **1912**, *45*, 3329.

[10] (a) N. A. Milas, S. Sussmann *J. Am. Chem. Soc.* **1936**, *58*, 1302; (b) N. A. Milas, J.-H. Trepagnier, J. T. Nolan, M. I. Iliopulos *J. Am. Chem. Soc.* **1959**, *81*, 4730.

[11] K. B. Sharpless, K. Akashi *J. Am. Chem. Soc.* **1976**, *98*, 1986.

[12] (a) W. P. Schneider, A. V. McIntosh, (Upjohn) US-2.769.824 (1956) *Chem. Abstr.* **1957**, *51*, 8822e; (b) V. Van Rheenen, R. C. Kelly, D. Y. Cha *Tetrahedron Lett.* **1976**, *17*, 1973; (c) R. Ray, D. S. Matteson *Tetrahedron Lett.* **1980**, *21*, 449.

[13] M. Minamoto, K. Yamamoto, J. Tsuji *J. Org. Chem.* **1990**, *55*, 766.

[14] M. P. Singh, H. S. Singh, A. K. Arya, A. K. Singh, A. K. Sisodia *Indian J. Chem.* **1975**, *13*, 112.

[15] L. Ahlgren, L. Sutin *Org. Process Res. Dev.* **1997**, *1*, 425.

[16] (a) K. Bergstad, S. Y. Jonsson, J.-E. Bäckvall *J. Am. Chem. Soc.* **1999**, *121*, 10424; (b) S. Y. Jonsson, K. Färnegardh, J.-E. Bäckvall *J. Am. Chem. Soc.* **2001**, *123*, 1365.

[17] S. Y. Jonsson, H. Adolfsson, J.-E. Bäckvall *Org. Lett.* **2001**, *3*, 3463.

[18] A. H. Ell, S. Y. Jonsson, A. Börje, H. Adolfsson, J.-E. Bäckvall *Tetrahedron Lett.* **2001**, *42*, 2569.

[19] S. Y. Jonsson, H. Adolfsson, J.-E. Bäckvall *Chem. Eur. J.* **2003**, *9*, 2783.

[20] (a) R. W. Cummins, FMC-Corporation New York, US-Patent 3488394, **1970**;

(b) R. W. CUMMINS, FMC-Corporation New York, US-Patent 3846478, **1974**.

[21] G. M. MEHLTRETTER, S. BHOR, M. KLAWONN, C. DÖBLER, U. SUNDERMEIER, M. ECKERT, H.-C. MILITZER, M. BELLER *Synthesis* **2003**, *2*, 295.

[22] E. N. JACOBSEN, I. MARKO, W. S. MUNGALL, G. SCHRÖDER, K. B. SHARPLESS *J. Am. Chem. Soc.* **1988**, *110*, 1968.

[23] M. BELLER, A. ZAPF, W. MÄGERLEIN *Chem. Eng. Technol.* **2001**, *24*, 575.

[24] J. S. M. WAI, I. MARKÓ, J. S. SVENDSEN, M. G. FINN, E. N. JACOBSEN, K. B. SHARPLESS *J. Am. Chem. Soc.* **1989**, *111*, 1123.

[25] (a) J. F. CAIRNS, H. L. ROBERTS, *J. Chem. Soc. C* **1968**, 640; (b) CELANESE CORP., GB-1,028,940 (**1966**), *Chem. Abstr.* **1966**, *65*, 3064 f.; (c) R. S. MYERS, R. C. MICHAELSON, R. G. AUSTIN, (Exxon Corp.) US-4496779 (**1984**), *Chem. Abstr.* **1984**, *101*, P191362k.

[26] (a) A. KRIEF, C. COLAUX-CASTILLO *Tetrahedron Lett.* **1999**, *40*, 4189; (b) A. KRIEF, C. DELMOTTE, C. COLAUX-CASTILLO *Pure Appl. Chem.* **2000**, *72*, 1709.

[27] (a) A. KRIEF, C. COLAUX-CASTILLO, *Synlett* **2001**, 501; A. KRIEF, C. COLAUX-CASTILLO *Pure Appl. Chem.* **2002**, *74*, 107.

[28] A. KRIEF, A. DESTREE, V. DURISOTTI, N. MOREAU, C. SMAL, C. COLAUX-CASTILLO *Chem. Commun.* **2002**, 558.

[29] C. DÖBLER, G. MEHLTRETTER, M. BELLER *Angew. Chem., Int. Ed. Engl.* **1999**, *38*, 3026; (b) C. DÖBLER, G. MEHLTRETTER, U. SUNDERMEIER, M. BELLER *J. Am. Chem. Soc.* **2000**, *122*, 10289–10297.

[30] K. WEISSERMEL, H.-J. ARPE *Industrielle Organische Chemie*, 5th edn., Wiley-VCH, Weinheim, **1998**.

[31] C. DÖBLER, G. MEHLTRETTER, U. SUNDERMEIER, M. BELLER *J. Organomet. Chem.* **2001**, *621*, 70.

[32] Y. L. BENNANI, K. P. M. VANHESSCHE, K. B. SHARPLESS *Tetrahedron Asymmetry* **1994**, *5*, 1473.

[33] A. SEVEREYNS, D. E. DE VOS, P. A. JACOBS *Top. Catal.* **2002**, *19*, 125.

[34] C. CAINELLI, M. CONTENTO, F. MANESCALCHI, L. PLESSI *Synthesis* **1989**, 45.

[35] (a) B. M. KIM, K. B. SHARPLESS *Tetrahedron Lett.* **1990**, *31*, 3003; (b) B. B. LOHRAY, A. THOMAS, P. CHITTARI, J. R. AHUJA, P. K. DHAL *Tetrahedron Lett.* **1992**, *33*, 5453; (c) H. HAN, K. D. JANDA *J. Am. Chem. Soc.* **1996**, *118*, 7632; (d) D. J. GRAVERT, K. D. JANDA *Chem. Rev.* **1997**, *97*, 489; (e) C. BOLM, A. GERLACH *Angew. Chem., Int. Ed. Engl.* **1997**, *36*, 741; (f) C. BOLM, A. GERLACH *Eur. J. Org. Chem.* **1998**, 21; (g) P. SALVADORI, D. PINI, A. PETRI *Synlett* **1999**, 1181; (h) A. PETRI, D. PINI, S. RAPACCINI, P. SALVADORI *Chirality* **1999**, *11*, 745; (i) A. MANDOLI, D. PINI, A. AGOSTINI, P. SALVADORI *Tetrahedron: Asymmetry* **2000**, *11*, 4039; (j) C. BOLM, A. MAISCHAK *Synlett* **2001**, 93; (k) I. MOTORINA, C. M. CRUDDEN *Org. Lett.* **2001**, *3*, 2325; (l) H. M. LEE, S.-W. KIM, T. HYEON, B. M. KIM *Tetrahedron: Asymmetry* **2001**, *12*, 1537; (m) Y.-Q. KUANG, S.-Y. ZHANG, R. JIANG, L.-L. WIE *Tetrahedron Lett.* **2002**, *43*, 3669.

[36] (a) W. A. HERRMANN, G. WEICHSELBAUMER EP 0593425B1, **1994**; (b) W. A. HERRMANN, R. M. KRATZER, J. BLÜMEL, H. B. FRIEDRICH, R. W. FISCHER, D. C. APPERLEY, J. MINK, O. BERKESI *J. Mol. Catal. A* **1997**, *120*, 197.

[37] S. NAGAYAMA, M. ENDO, S. KOBAYASHI *J. Org. Chem.* **1998**, *63*, 609.

[38] S. KOBAYASHI, M. ENDO, S. NAGAYAMA *J. Am. Chem. Soc.* **1999**, *121*, 11229.

[39] S. KOBAYASHI, T. ISHIDA, R. AKIYAMA *Org. Lett.* **2001**, *3*, 2649.

[40] A. SEVEREYNS, D. E. DE VOS, L. FIERMANS, F. VERPOORT, P. J. GROBET, P. A. JACOBS *Angew. Chem.* **2001**, *113*, 606.

[41] (a) M. A. ANDERSSON, R. EPPLE, V. V. FOKIN, K. B. SHARPLESS *Angew. Chem.* **2002**, *114*, 490; (b) H. ADOLFSSON, F. STALFORS presented at the 221st National Am. Chem. Soc. Meeting **2001**.

[42] B. M. CHOUDARY, N. S. CHOWDARI, M. L. KANTAM, K. V. RAGHAVAN *J. Am. Chem. Soc.* **2001**, *123*, 9220.

[43] B. M. CHOUDARY, N. S. CHOWDARI, K. JYOTHI, M. L. KANTAM *J. Am. Chem. Soc.* **2002**, *124*, 5341.

[44] B. M. CHOUDARY, N. S. CHOWDARI, K. JYOTHI, S. MADHI, M. L. KANTAM *Adv. Synth. Catal.* **2002**, *344*, 503.

[45] Reviews: (a) T. WELTON *Chem. Rev.* **1999**, *99*, 2071; (b) P. WASSERSCHEID, W. KEIM *Angew. Chem., Int. Ed. Engl.* **2000**, *39*, 3772; (c) R. SHELDON *Chem.*

Commun. **2001**, 2399; (d) S. T. Handy Chem. Eur. J. **2003**, 9, 2938.

[46] (a) J. L. Reynolds, K. R. Erdner, P. B. Jones Org. Lett. **2002**, 4, 917; (b) I. A. Ansari, R. Gree Org. Lett. **2002**, 4, 1507; (c) K. G. Mayo, E. H. Nearhoof, J. J. Kiddle Org. Lett. **2002**, 4, 567; (d) T. Fukuyama, M. Shinmen, S. Nishitani, M. Sato, I. Ryu Org. Lett. **2002**, 4, 1691; (e) D. Semeril, H. Olivier-Bourbigou, C. Bruneau, P. H. Dixneuf Chem. Commun. **2002**, 146; (f) C. S. Consorti, G. Ebeling, J. Dupont Tetrahedron Lett. **2002**, 43 753; (g) S. J. Nara, J. R. Harjani, M. M. Salunkhe Tetrahedron Lett. **2002**, 43, 2979.

[47] (a) R. Yanada, Y. Takemoto Tetrahedron Lett. **2002**, 43, 6849; (b) Q. Yao Org. Lett. **2002**, 4, 2197.

[48] C. E. Song, D. Jung, E. J. Roh, S. Lee, D. Y. Chi Chem. Commun. **2002**, 3038.

[49] L. C. Branco, C. A. M. Afonso Chem. Commun. **2002**, 3036.

2
Transition Metal-catalyzed Epoxidation of Alkenes
Hans Adolfsson

2.1
Introduction

The formation of epoxides via metal-catalyzed oxidation of alkenes represents the most elegant and environmentally friendly route for the production of this compound class [1, 2]. This is of particular importance, considering that the conservation and management of resources should be the main focus of interest when novel chemical processes are developed. Thus, the innovation and improvement of catalytic epoxidation methods where molecular oxygen or hydrogen peroxide are employed as terminal oxidants is highly desirable. However, one of today's industrial routes for the formation of simple epoxides (e.g., propylene oxide) is the *chlorohydrin* process, where alkenes are reacted with chlorine in the presence of sodium hydroxide (Scheme 2.1) [3]. At present this process produces 2.01 ton NaCl and 0.102 ton 1,2-dichloropropane as byproducts per ton of propylene oxide. These significant amounts of waste are certainly not acceptable in the long run, and efforts aimed at replacing such chemical plants with "greener" epoxidation processes are under way. When it comes to the production of fine chemicals, non-catalyzed processes with traditional oxidants (e.g., peroxyacetic acid and *meta*-chloroperoxybenzoic acid) are often used. In these cases, however, transition metal-based systems using hydrogen peroxide as the terminal oxidant demonstrate several advantages. The scope and focus of this chapter will be to highlight some novel approaches to transition metal-catalyzed formation of epoxides by means of alkene oxidation using environmentally benign oxidants.

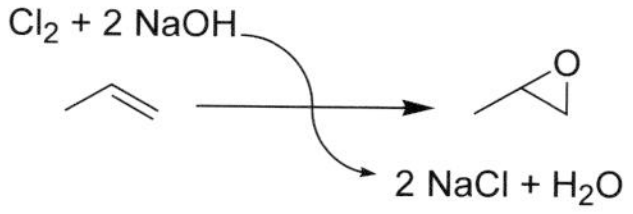

Scheme 2.1

$Cl_2 + 2\ NaOH \longrightarrow$ propene $\longrightarrow$ propylene oxide + $2\ NaCl + H_2O$

Modern Oxidation Methods. Edited by Jan-Erling Bäckvall
Copyright © 2004 WILEY-VCH Verlag GmbH & Co. KGaA, Weinheim
ISBN: 3-527-30642-0

2.2
Choice of Oxidant for Selective Epoxidation

There are several terminal oxidants available for the transition metal-catalyzed epoxidation of alkenes (Table 2.1). Typical oxidants compatible with a majority of metal-based epoxidation systems are various alkyl hydroperoxides, hypochlorite or iodosylbenzene. A problem associated with these oxidants is their low active oxygen content (Table 2.1). Considering the nature of the waste produced, there are further drawbacks using these oxidants. Hence, from an environmental and economical point of view, molecular oxygen should be the preferred oxidant, considering its high active oxygen content and that no waste products or only water is formed. One of the major limitations, however, using molecular oxygen as the terminal oxidant for the formation of epoxides is the poor product selectivity obtained in these processes [4]. In combination with the limited number of catalysts available for direct activation of molecular oxygen, this effectively restricts the use of this oxidant. On the other hand, hydrogen peroxide displays much better properties as the terminal oxidant. The active oxygen content of H_2O_2 is about as high as for typical applications of molecular oxygen in epoxidations (since a reductor is required in almost all cases), and the waste produced by employing this oxidant is plain water. As in the case of molecular oxygen, the epoxide selectivity using H_2O_2 can sometimes be relatively poor, although recent developments have led to transition metal-based protocols where excellent reactivity and epoxide selectivity can be obtained [5]. The various oxidation systems available for the selective epoxidation of alkenes using transition metal catalysts and hydrogen peroxide will be covered in the following sections.

Tab. 2.1 Oxidants used in transition metal-catalyzed epoxidations, and their active oxygen content

Oxidant	Active oxygen content (wt.%)	Waste product
Oxygen (O_2)	100	Nothing or H_2O
Oxygen (O_2)/reductor	50	H_2O
H_2O_2	47	H_2O
NaOCl	21.6	NaCl
CH_3CO_3H	21.1	CH_3CO_2H
^tBuOOH (TBHP)	17.8	^tBuOH
$KHSO_5$	10.5	$KHSO_4$
BTSP[a]	9	hexamethyldisiloxane
PhIO	7.3	PhI

[a] Bistrimethylsilyl peroxide.

2.3
Epoxidations of Alkenes Catalyzed by Early Transition Metals

High-valent early transition metals such as titanium(IV) and vanadium(V) have been shown to efficiently catalyze the epoxidation of alkenes. The preferred oxidants using these catalysts are various alkyl hydroperoxides, typically *tert*-butylhydroperoxide (TBHP) or ethylbenzene hydroperoxide (EBHP). One of the routes for the industrial production of propylene oxide is based on a heterogeneous Ti^{IV}/SiO_2 catalyst, which employs EBHP as the terminal oxidant [6].

The Sharpless-Katsuki asymmetric epoxidation (AE) protocol for the enantioselective formation of epoxides from allylic alcohols was a milestone in asymmetric catalysis [7]. This classical asymmetric transformation uses TBHP as the terminal oxidant, and the reaction has been widely used in various synthetic applications. There are several excellent reviews covering the scope and utility of the AE reaction [8]. On the other hand, the use of hydrogen peroxide as oxidant in combination with early transition metal catalysts (Ti and V) is rather limited. The reason for the poor reactivity can be traced to the severe inhibition of the metal complexes by strongly coordinating ligands such as alcohols and in particular water. The development of the heterogeneous titanium(IV)-silicate catalyst (TS-1) by chemists at Enichem represented a breakthrough for reactions performed with hydrogen peroxide [9]. This hydrophobic molecular sieve demonstrated excellent properties (i.e., high catalytic activity and selectivity) for the epoxidation of small linear alkenes in methanol. The substrates are adsorbed into the micropores of the TS-1 catalyst, which efficiently prevents the inhibition by water as observed using the Ti^{IV}/SiO_2 catalyst. After the epoxidation reaction, the TS-1 catalyst can easily be separated and reused. To extend the scope of this epoxidation method and thereby allow for the oxidation of a wider range of substrates, several different titanium containing silicate zeolites have been prepared. Consequently, the scope has been improved somewhat but the best epoxidation results using titanium silicates as catalysts are obtained with smaller, non-branched substrates.

2.4
Molybdenum and Tungsten-catalyzed Epoxidations

Epoxidation systems based on molybdenum and tungsten catalysts have been studied extensively for more than 40 years. The typical catalysts, Mo^{VI}-oxo or W^{VI}-oxo species do, however, behave quite differently depending on whether anionic or neutral complexes are employed. Whereas the former catalysts, especially the use of tungstates under phase-transfer conditions, are able to efficiently activate aqueous hydrogen peroxide for the formation of epoxides, neutral molybdenum or tungsten complexes give a lower selectivity with hydrogen peroxide. A better selectivity with the latter catalysts is often achieved using organic hydroperoxides (e.g., *tert*-butyl hydroperoxide) as terminal oxidants [10, 11].

2.4.1
Homogeneous Catalysts – Hydrogen Peroxide as the Terminal Oxidant

Payne and Williams reported in 1959 on the selective epoxidation of maleic, fumaric and crotonic acids using a catalytic amount of sodium tungstate (2 mol%) in combination with aqueous hydrogen peroxide as the terminal oxidant [12]. The key to success was careful control of the pH (4–5.5) in the reaction media. These electron-deficient substrates were notoriously difficult to oxidize selectively using the standard techniques (peroxy acid reagents) available at the time. Previous attempts to use sodium tungstate and hydrogen peroxide led to the isolation of the corresponding diols due to rapid hydrolysis of the intermediate epoxides. Significant improvements to this catalytic system were introduced by Venturello and coworkers [13, 14]. They found that the addition of phosphoric acid and the introduction of quaternary ammonium salts as PTC-reagents considerably increased the scope of the reaction. The active tungstate catalysts are often generated *in situ*, although catalytically active peroxo-complexes such as $(n\text{-hexyl}_4N)_3\{PO_4[W(O)(O_2)_2]_4\}$ have been isolated and characterized (Scheme 2.2) [15].

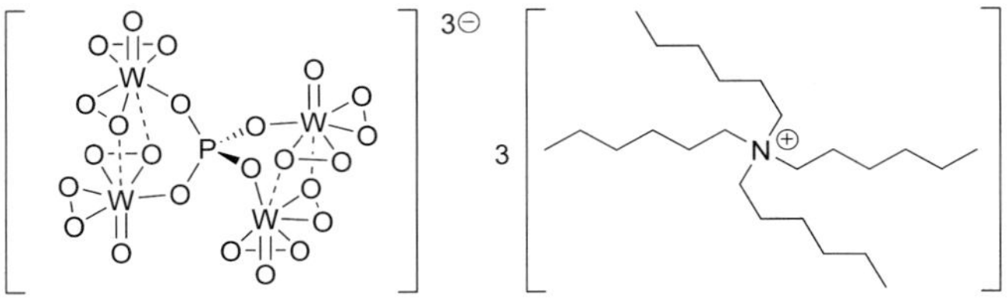

Scheme 2.2 The Venturello $(n\text{-hexyl}_4N)_3\{PO_4[W(O)(O_2)_2]_4\}$ catalyst

In recent work, Noyori and coworkers established conditions for the selective epoxidation of aliphatic terminal alkenes either in toluene, or using a completely solvent-free reaction setup [16, 17]. One of the disadvantages with the previous systems was the use of chlorinated solvents. The conditions established by Noyori, however, provided an overall "greener" epoxidation process since the reactions were performed efficiently in non-chlorinated solvents. In this reaction, sodium tungstate (2 mol%), (aminomethyl)phosphonic acid and methyltri-n-octylammonium bisulfate (1 mol% of each) were employed as catalysts for the epoxidation using aqueous hydrogen peroxide (30%) as the terminal oxidant. The epoxidation of various terminal alkenes using the above-mentioned conditions (90 °C, no solvent added) gave high yields for a number of substrates (Table 2.2). The work-up procedure was exceptionally simple, since the product epoxides could be distilled directly from the reaction mixture. The use of appropriate additives turned out to be crucial to a successful outcome of these epoxide-forming reactions.

When the (aminomethyl)phosphonic acid was replaced by other phosphonic acids or simply by phosphoric acid, significantly lower conversions were obtained. The nature of the phase-transfer reagent was further established as an important para-

Tab. 2.2 Epoxidation of terminal alkenes using the Noyori system

$$R\diagup\hspace{-0.3em}\diagdown \xrightarrow[\substack{Na_2WO_4 \cdot 2H_2O\ (2\ mol\%) \\ [CH_3(n\text{-}C_8H_{17})_3N]HSO_4\ (1\ mol\%) \\ NH_2CH_2PO_3H\ (1\ mol\%)}]{1.5\ equiv.\ H_2O_2\ (30\%\ aq)} R\diagup\hspace{-0.3em}\triangle\hspace{-0.3em}O$$

Entry	Alkene	Time (h)	Conversion (%)	Yield (%)
1	1-octene	2	89	86
2	1-decene	2	94	93
3[a]	1-decene	4	99	99
4[a]	allyl octyl ether	2	81	64
5[a]	styrene	3	70	2

[a] 20 mmol alkene in 4 mL toluene.

meter. The use of ammonium bisulfate (HSO_4^-) was superior to the corresponding chloride or hydroxide salts. The size, and hence the lipophilicity of the ammonium ion was important, since tetra-n-butyl- or tetra-n-hexyl ammonium bisulfate were inferior to phase-transfer agents containing larger alkyl groups. The epoxidation system was later extended to encompass other substrates, such as simple alkenes with different substitution patterns, and to alkenes containing various functionalities (alcohols, ethers, ketones and esters).

A major limitation of this method is the low pH under which the reactions are performed. This led to substantially lower yields in reactions with substrate progenitors of acid sensitive epoxides, where competing ring-opening processes effectively reduced the usefulness of the protocol. As an example, the oxidation of styrene led to 70% conversion after 3 h at 70 °C, although the observed yield for styrene oxide was only 2% (Table 2.2, entry 5).

The epoxidation method developed by Noyori, has subsequently been applied to the direct formation of dicarboxylic acids from alkenes [18]. Cyclohexene was oxidized to adipic acid in 93% yield using the tungstate, ammonium bisulfate system and 4 equiv. of hydrogen peroxide. The selectivity problem associated with the Noyori protocol was to a certain degree circumvented by the improvements introduced by Jacobs and coworkers [19]. To the standard catalytic mixture were added additional amounts of (aminomethyl)phosphonic acid and Na_2WO_4 and the pH of the reaction media was adjusted to 4.2–5 with aqueous NaOH. These changes allowed for the formation of epoxides from α-pinene, 1-phenyl-1-cyclohexene, and indene, in high conversions and good selectivity (Scheme 2.3).

Another highly efficient tungsten-based system for the epoxidation of alkenes was recently introduced by Mizuno and coworkers [20]. The tetrabutylammonium salt of a Keggin-type silicodecatungstate $[\gamma\text{-}SiW_{10}O_{34}(H_2O)_2]^{4-}$ (Scheme 2.4) was found to catalyze the epoxidation of various alkene substrates using aqueous hydrogen peroxide as the terminal oxidant. The characteristics of this system are very high epoxide selectivity (99%) and excellent efficiency in the use of the terminal oxidant (99%). Terminal- as well as di-and tri-substituted alkenes were all epoxidized in high yields

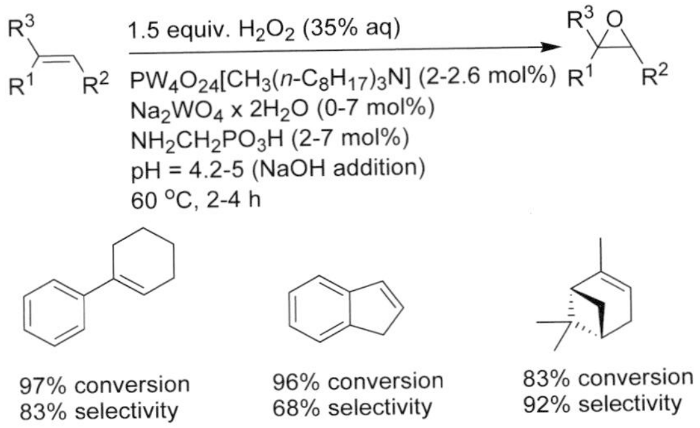

2 Transition Metal-catalyzed Epoxidation of Alkenes

Scheme 2.3

R³R¹C=CR² → R³R¹C(–O–)CR²

1.5 equiv. H_2O_2 (35% aq)
$PW_4O_{24}[CH_3(n-C_8H_{17})_3N]$ (2–2.6 mol%)
$Na_2WO_4 \times 2H_2O$ (0–7 mol%)
$NH_2CH_2PO_3H$ (2–7 mol%)
pH = 4.2–5 (NaOH addition)
60 °C, 2–4 h

97% conversion
83% selectivity

96% conversion
68% selectivity

83% conversion
92% selectivity

within reasonably short reaction times using 0.16 mol% catalyst (1.6 mol% in tungsten, Scheme 2.4). The X-ray structure of the catalyst precursor revealed 10 tungsten atoms connected to a central SiO_4 unit. *In situ* infrared spectroscopy of the reaction mixture during the epoxidation reaction indicated high structural stability of the catalyst. Furthermore, it was demonstrated that the catalyst can be recovered and reused up to 5 times without loss of activity or selectivity (epoxidation of cyclooctene). Interestingly, the often encountered problem with hydrogen peroxide decomposition was negligible using this catalyst. The efficient use of hydrogen peroxide (99%) combined with the high selectivity and productivity in propylene epoxidation opens up industrial applications.

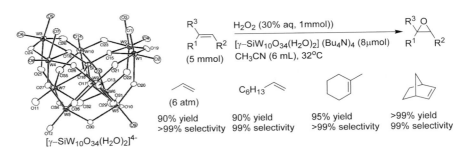

$[\gamma\text{-}SiW_{10}O_{34}(H_2O)_2]^{4-}$

R³R¹C=CR² (5 mmol) + H_2O_2 (30% aq, 1 mmol) → R³R¹C(–O–)CR²

$[\gamma\text{-}SiW_{10}O_{34}(H_2O)_2]$ $(Bu_4N)_4$ (8 μmol)
CH_3CN (6 mL), 32 °C

(6 atm) C_6H_{13}

90% yield   90% yield   95% yield   >99% yield
>99% selectivity 99% selectivity   >99% selectivity 99% selectivity

Scheme 2.4

The use of molybdenum catalysts in combination with hydrogen peroxide is not as common as for tungsten catalysts. There are, however, a number of examples where molybdates have been employed for the activation of hydrogen peroxide. A catalytic amount of sodium molybdate in combination with mono-dentate ligands (e.g., hexa-alkyl phosphorus triamides or pyridine-N-oxides), and sulfuric acid allowed for the epoxidation of simple linear or cyclic alkenes [21]. The selectivity obtained using this method was quite low, and significant amounts of diols were formed, even though highly concentrated hydrogen peroxide (>70%) was employed.

More recently, Sundermeyer and coworkers reported on the use of long-chain trialkylamine oxides, trialkylphosphane oxides or trialkylarsane oxides as mono-dentate ligands for neutral molybdenum peroxo complexes [22]. These compounds were employed as catalysts for the epoxidation of 1-octene and cyclooctene with aqueous hydrogen peroxide (30%), under biphasic conditions ($CHCl_3$). The epoxide products were obtained in high yields with good selectivity. The high selectivity achieved using this method was ascribed to the high solubility of the product in the organic phase, thus protecting the epoxide from hydrolysis. This protocol has not been employed for the formation of hydrolytically sensitive epoxides and the generality of the method can thus be questioned.

2.4.2
Heterogeneous Catalysts

One problem associated with the above described peroxotungstate catalyzed epoxidation system, is the separation of the catalyst after the completed reaction. To overcome this obstacle, efforts to prepare heterogeneous tungstate catalysts have been conducted. De Vos and coworkers employed W-catalysts derived from sodium tungstate and layered double hydroxides (LDH – coprecipitated $MgCl_2$, $AlCl_3$ and NaOH) for the epoxidation of simple alkenes and allyl alcohols with aqueous hydrogen peroxide [23]. They found that depending on the nature of the catalyst (either hydrophilic or hydrophobic catalysts were used), different reactivities and selectivities were obtained for non-polar and polar alkenes, respectively. The hydrophilic LDH-WO_4 catalyst was particularly effective for the epoxidation of allyl and homo-allyl alcohols, whereas the hydrophobic catalyst (containing *p*-toluensulfonate) showed better reactivity with non-functionalized substrates.

Gelbard and coworkers have reported on the immobilization of tungsten-catalysts using polymer-supported phosphine oxide, phosphonamide, phosphoramide and phosphotriamide ligands [24]. Employing these heterogeneous catalysts together with hydrogen peroxide for the epoxidation of cyclohexene resulted in moderate to good conversion of the substrate, although in most cases low epoxide selectivity was observed. A significantly more selective heterogeneous catalyst was obtained by Jacobs and coworkers upon treatment of the macroreticular ion-exchange resin Amberlite IRA-900 with an ammonium salt of the Venturello anion $\{PO_4[WO(O_2)_2]_4\}^{3-}$ [25]. The catalyst formed was used for the epoxidation of a number of terpenes, and high yields and good selectivity of the corresponding epoxides were achieved.

In a different strategy, siliceous mesoporous MCM-41 based catalysts were prepared. Quaternary ammonium salts and alkyl phosphoramides, respectively, were grafted onto MCM-41 and the material obtained was treated with tungstic acid for the preparation of heterogeneous tungstate catalysts. The catalysts were employed in the epoxidation of simple cyclic alkenes with aqueous hydrogen peroxide (35%) as the terminal oxidant, however conversion and selectivity for the epoxide formed was rather low. In the case of cyclohexene, the selectivity could be improved by the addition of pyridine. The low tungsten leaching (<2%) is certainly advantageous using these catalysts.

A particularly interesting system for the epoxidation of propylene to propylene oxide, working under pseudo-heterogeneous conditions, was reported by Zuwei and

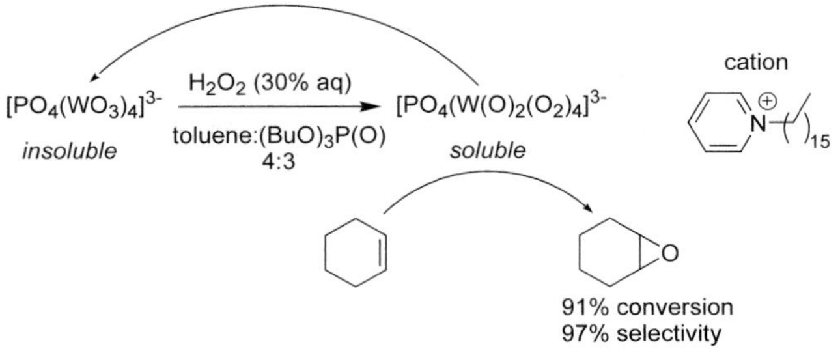

Scheme 2.5

coworkers [26]. The catalyst, which was based on the Venturello anion combined with long-chain alkylpyridinium cations, showed unique solubility properties. In the presence of hydrogen peroxide the catalyst was fully soluble in the solvent, a 4:3 mixture of toluene and tributylphosphate, but when no more oxidant remained, the tungsten catalyst precipitated and could simply be removed from the reaction mixture (Scheme 2.5). Furthermore, this epoxidation system was combined with the 2-ethylanthraquinone (EAQ)/2-ethylanthrahydroquinone (EAHQ) process for hydrogen peroxide formation (Scheme 2.6), and good conversion and selectivity were obtained for propylene oxide in three consecutive cycles. The catalyst was recovered by centrifugation in between every cycle, and used directly in the next reaction.

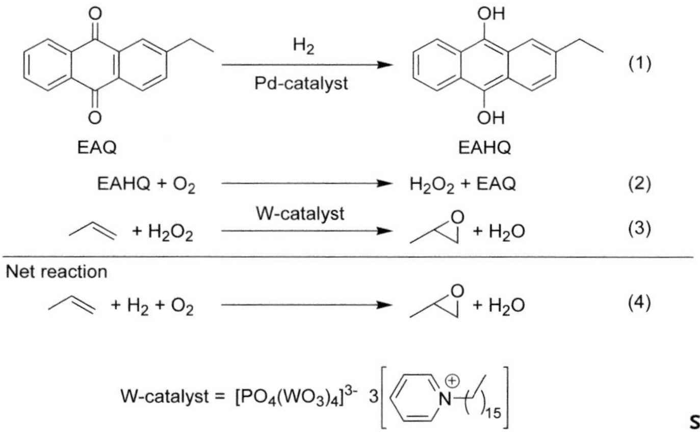

Scheme 2.6

2.5
Manganese-catalyzed Epoxidations

Historically, the interest in using manganese complexes as catalysts for the epoxidation of alkenes comes from biologically relevant oxidative manganese porphyrins. The terminal oxidants compatible with manganese porphyrins were initially restricted to iodosylbenzene, sodium hypochlorite, alkyl peroxides and hydroperoxides,

2.5 Manganese-catalyzed Epoxidations

N-oxides, KHSO$_5$ and oxaziridines. Molecular oxygen can also be used in the presence of an electron source. The use of hydrogen peroxide often results in oxidative decomposition of the catalyst due to the potency of this oxidant. However, the introduction of chlorinated porphyrins (**1**) (Scheme 2.7) allowed for hydrogen peroxide to be used as the terminal oxidant [27]. These catalysts, discovered by Mansuy and coworkers, were demonstrated to resist decomposition, and when used together with imidazole or imidazolium carboxylates as additives, efficient epoxidation of alkenes were achieved (Table 2.3, entries 1 and 2).

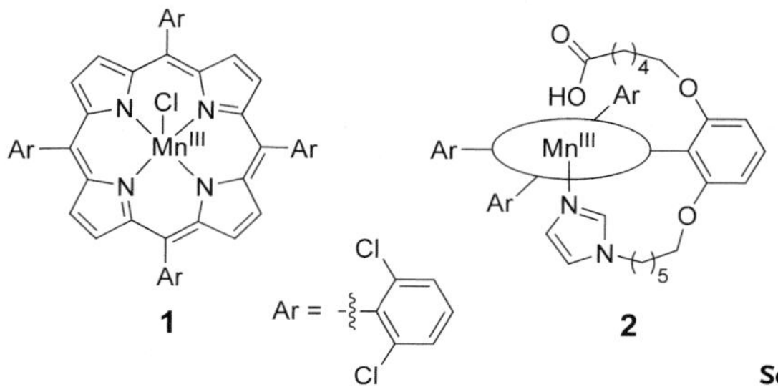

Scheme 2.7

Tab. 2.3 Manganese-porphyrin catalyzed epoxidation of *cis*-cyclooctene using aqueous H$_2$O$_2$ (30%)

Entry	Catalyst	Additive	Temp. (°C)	Time (min)	Yield (%)
1	**1** 2.5 mol%	imidazole (0.6 equiv.)	20	45	90
2	**1** 0.5 mol%	N-hexyl-imidazole (0.5 mol%) benzoic acid (0.5 mol%)	0	15	100
3	**2** 0.1 mol%	–	0	3	100

The observation that imidazoles and carboxylic acids significantly improved the epoxidation reaction led to the development of Mn-porphyrin complexes containing these groups covalently linked to the porphyrin platform as attached pendant arms (**2**) [28]. When these catalysts were employed in the epoxidation of simple alkenes with hydrogen peroxide, enhanced oxidation rates in combination with perfect product selectivity was obtained (Table 2.3, entry 3). In contrast to epoxidations catalyzed by other metals, the Mn-porphyrin system yields products with scrambled stereochemistry. For example, the epoxidation of *cis*-stilbene using Mn(TPP)Cl (TPP = tetraphenylporphyrin) and iodosylbenzene, generated *cis*- and *trans*-stilbene oxide in a ratio of 35:65. The low stereospecificity was improved using heterocyclic additives, such as pyridines or imidazoles. The epoxidation system using hydrogen peroxide as the terminal oxidant, was reported to be stereospecific for *cis*-alkenes, whereas *trans*-alkenes are poor substrates with these catalysts.

A breakthrough for manganese epoxidation catalysts came at the beginning of the 1990s when the groups of Jacobsen and Katsuki simultaneously discovered that chiral Mn-salen complexes (**3**) catalyzed the enantioselective formation of epoxides [29–31]. The discovery that simple non-chiral Mn-salen complexes could be used as catalysts for alkene epoxidation had already been established about 5 years earlier, and the typical terminal oxidants used with these catalysts closely resemble those of the porphyrin systems [32]. In contrast to the titanium-catalyzed asymmetric epoxidation discovered by Sharpless, the Mn-salen system does not require pre-coordination of the alkene substrate to the catalyst, hence unfunctionalized alkenes could efficiently and selectively be oxidized. The enantioselectivity was shown to be highly sensitive towards the substitution pattern of the alkene substrate. Excellent selectivity (>90% ee) was obtained for aryl- or alkynyl-substituted terminal-, *cis*-di-substituted- and tri-substituted alkenes, whereas *trans*-di-substituted alkenes were epoxidized with low rates and low ee (<40%). The typical oxidant used in Mn-salen asymmetric epoxidations is NaOCl, however, recent work by the groups of Berkessel and Katsuki have opened up the possibility of hydrogen peroxide being employed [33, 34]. Berkessel found that imidazole additives were crucial for the formation of the active oxo-manganese intermediates, and an Mn-catalyst (**4**) based on a salen ligand incorporating a pendant imidazole was used for the asymmetric epoxidation using aqueous H_2O_2. Yields and enantioselectivity did not, however, reach the levels obtained when other oxidants were used. In the work of Katsuki, imidazole was present as an additive in the reaction mixture containing a sterically hindered Mn-salen catalyst (**5**) (Scheme 2.8). In this way, high enantioselectivity could be obtained, although the catalytic activity was not as effective, and the epoxides were formed in low yields.

Considerably better ee values and yields were obtained when ammonium acetate (20 mol%) was used as an additive with the Jacobsen-catalyst (**3**) [35]. A major problem with the use of hydrogen peroxide in the Mn-salen catalyzed reactions is associated with catalyst deactivation due to the presence of water. Anhydrous hydrogen peroxide, either in the form of the urea/H_2O_2 adduct or in the triphenylphosphine oxide/H_2O_2 adduct, have been employed to circumvent this problem [36, 37]. Although epoxide yield and enantioselectivity are in the range of what can be obtained using NaOCl, the catalyst loading is significantly higher, and the removal of urea or Ph_3PO constitute an additional problem.

Apart from porphyrin and salen catalysts, manganese complexes of N-alkylated 1,4,7-triazacyclononane (e.g., TMTACN, **6**) have been found to catalyze the epoxida-

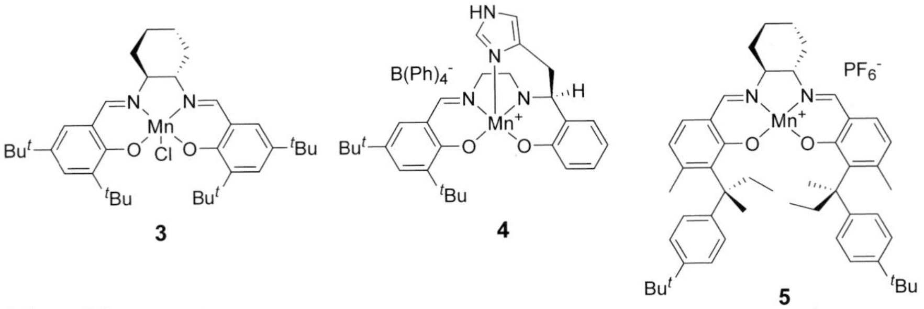

Scheme 2.8

2.5 Manganese-catalyzed Epoxidations

tion of alkenes efficiently in the presence of acid additives (typically oxalic, ascorbic or squaric acid) and hydrogen peroxide [38–40]. Reactions performed without any acid present required a huge excess (ca. 100 equiv.) of hydrogen peroxide for efficient epoxidation. The rather difficult preparation of the TACN ligands has led to an increased activity in order to find alternative ligands with similar coordinating properties. In this respect, pyridyl-amine ligands represent an interesting alternative. Feringa and coworkers found that the dinuclear manganese complex **8** (Scheme 2.9), prepared from the tetra-pyridyl ligand **7**, was an efficient catalyst for the epoxidation of simple alkenes [41]. Only 0.1 mol% of the catalyst (**8**) was required for high level of conversion (87%) of cyclohexene into its corresponding epoxide. An excess of aqueous hydrogen peroxide (8 equiv.) was used due to the usual problem of peroxide decomposition in the presence of manganese complexes.

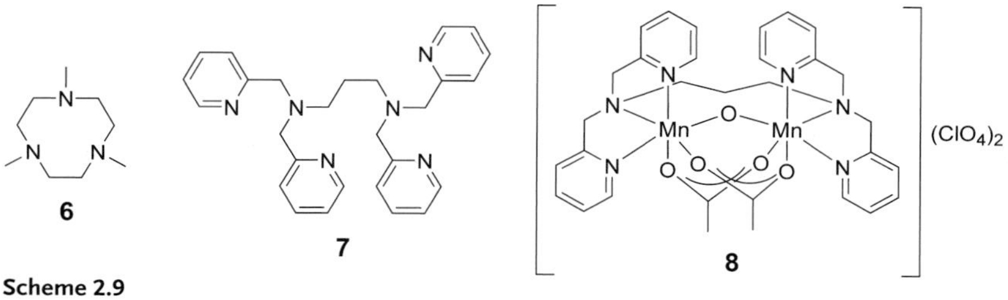

Scheme 2.9

In a recent screening of various metal salts, Lane and Burgess found that simple manganese(II) and -(III) salts catalyzed the formation of epoxides in DMF (N,N'-dimethylformamide) or ^{t}BuOH, using aqueous hydrogen peroxide (Scheme 2.10) [42]. It was further established that the addition of bicarbonate was of importance for the epoxidation reaction.

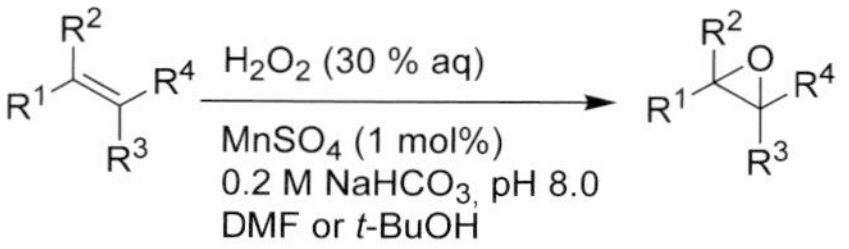

Scheme 2.10

Using spectroscopic methods, it was established that peroxymonocarbonate (HCO_4^-) is formed on mixing hydrogen peroxide and bicarbonate [43]. In the absence of the metal-catalyst, the oxidizing power of the peroxymonocarbonate formed *in situ* with respect to its reaction with alkenes was demonstrated to be moderate. In the initial reaction setup, this $MnSO_4$-catalyzed epoxidation required a considerable excess of hydrogen peroxide (10 equiv.) for efficient formation of the epoxide. Considering the scope of the reaction, it was found that electron-rich substrates such as di-, tri- and tetra-substituted alkenes were giving moderate to good yields of their corresponding epoxides. Styrene and styrene derivatives were also demonstrated to react

Tab. 2.4 Manganese sulfate catalyzed epoxidation of alkenes using aqueous H_2O_2 (30%)[a]

Alkene	No additive		Salicylic acid (4 mol%)	
	Equiv. H_2O_2	Yield	Equiv. H_2O_2	Yield
 cyclohexene	10	99	2.8	96
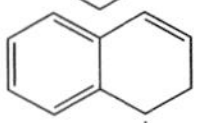 dihydronaphthalene	10	87	5	97[b]
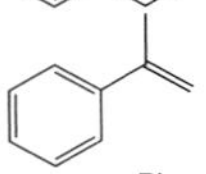 1-phenyl-dihydronaphthalene	10	96	5	95[b]
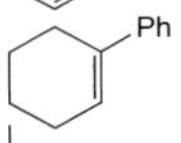 1-phenylcyclohexene (Ph)	10	95	5	95[b]
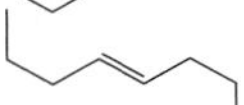 trans-alkene	25	60	25	75
 cis-4-octene	25	54	25	75
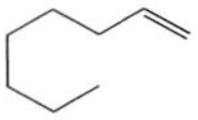 1-octene	25	0	25	0

[a] Conditions according to Scheme 2.5. [b] Isolated yields.

smoothly, whereas mono-alkyl substituted substrates were completely unreactive under these conditions. The basic reaction medium used was very beneficial for product protection, hence, acid sensitive epoxides were formed in good yields. Different additives were screened in order to improve this epoxidation system, and it was found that the addition of sodium acetate was beneficial for reactions performed in ^{t}BuOH. Similarly, the addition of salicylic acid improved the outcome of the reaction performed in DMF. The use of these additives efficiently reduced the number of hydrogen peroxide equivalents necessary for a productive epoxidation (Table 2.4). The reaction is not completely stereospecific, since the epoxidation of cis-4-octene yielded a cis/trans mixture of the product (1:1.45 without additive and 1:1.1 in the presence of 4 mol% salicylic acid).

The use of the ionic liquid [bmim][BF$_4$] further improved the Burgess epoxidation system [44]. Chan and coworkers found that replacing sodium bicarbonate with tetramethylammonium bicarbonate and performing the reaction in [bmim][BF$_4$] allowed for efficient epoxidation of a number of different alkenes, including substrates leading to acid labile epoxides [e.g., dihydronaphthalene (99% yield) and 1-phenylcyclohexene (80% yield)].

2.6
Rhenium-catalyzed Epoxidations

The use of rhenium-based systems for the epoxidation of alkenes has increased considerable during the last 10 years [45, 46]. In 1989, Jørgensen stated that "the catalytic activity of rhenium in epoxidation reactions is low". The very same year, a few

2.6 Rhenium-catalyzed Epoxidations

patents were released describing the use of porphyrin complexes containing rhenium as catalysts for the production of epoxides. The first major breakthrough, however, came in 1991 when Herrmann introduced methyltrioxorhenium (MTO, **9**, Scheme 2.11) as a powerful catalyst for alkene epoxidation, using hydrogen peroxide as the terminal oxidant [47]. This organometallic rhenium compound, formed in tiny amounts in the reaction between $(CH_3)_4ReO$ and air, was first detected by Beattie and Jones in 1979 [48].

A more reliable method for the preparation of MTO was introduced by Herrmann and coworkers in 1988 [49]. In this process, dirhenium heptoxide, Re_2O_7, was allowed to react with tetramethyltin, forming MTO and an equimolar amount of tin perrhenate. The maximum yield in this reaction was only 50% relative to the initial rhenium, and in order to improve this procedure, a more efficient route towards MTO was developed. Hence, treatment of Re_2O_7 with trifluoroacetic anhydride in acetonitrile generated $CF_3CO_2ReO_3$ quantitatively, which upon further reaction with $MeSn(Bu)_3$ gave MTO in high yield (95%) [50, 51]. The main advantage when using this route, apart from an efficient use of the rhenium source, is the replacement of the rather unpleasant tetramethyltin reagent with the more easily accessible alkyl-$Sn(Bu)_3$. This procedure is also compatible with the formation of other $RReO_3$ compounds, for example ethyltrioxorhenium (ETO).

An additional route towards MTO proceeds via the treatment of perrhenates with trialkylsilyl chloride to generate $ClReO_3$, followed by reaction with $(CH_3)_4Sn$ to form MTO in almost quantitative yield [52]. Today, there is a whole range of organorhenium oxides available, and they can be considered as one of the best examined classes of organometallic compounds [53, 54]. From a catalytic point of view, however, MTO is one of few organorhenium oxides that have been shown to effectively act as a catalyst in epoxidation reactions. Regarding the physical properties of organorhenium oxides, MTO shows the greatest thermal stability (decomposing at > 300 °C), apart from the catalytically inert 18e (η^5-C_5Me_5)ReO_3 complex Furthermore, the high solubility of MTO in virtually any solvent from pentane to water makes this compound particular attractive for catalytic applications.

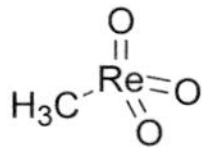

H_3C-ReO_3

9   Scheme 2.11

The 14e compound MTO readily forms coordination complexes of the type MTO-L and MTO-L_2 with anionic and uncharged Lewis bases [55]. These yellow adducts are typically 5- or 6-coordinate complexes and the Re-L system is highly labile. Apart from their fast hydrolysis in wet solvents, MTO-L adducts are much less thermally stable than MTO itself. For instance, the pyridine adduct of MTO decomposes even at room temperature. In solution, methyltrioxorhenium displays high stability in acidic aqueous media, albeit under increased hydroxide concentration its decomposition is strongly accelerated [56, 57]. Thus, under basic aqueous conditions MTO is decomposing according to Scheme 2.12:

$$CH_3ReO_3 + H_2O \longrightarrow CH_4 + ReO_4^- + H^+ \qquad \text{Scheme 2.12}$$

This decomposition is, however, rather slow and does not influence the use of MTO in catalysis to any greater extent.

For catalytic applications, perhaps the most important feature of MTO is its behavior in activating hydrogen peroxide. Upon treatment of MTO with hydrogen peroxide there is a rapid equilibrium taking place according to Scheme 2.13.

MTO reacts with hydrogen peroxide to form a mono-peroxo complex (**A**) which undergoes further reaction to yield a bis-peroxorhenium complex (**B**). The formation of the peroxo complexes is evident from the appearance of an intensive yellow color of the solution. Both peroxo complexes (**A** and **B**) have been detected by their methyl resonances using ^{1}H and ^{13}C NMR spectroscopy. Furthermore, the structure of the bis-peroxo complex **B** has been determined by crystallography [58]. In solution, **B** is the most abundant species in the equilibrium, suggesting that this is the thermodynamically most stable peroxo complex. The coordination of a water molecule to **B** has been established by NMR spectroscopy, however no such coordination has been observed for **A**, indicating either no coordinated water or high lability of such a

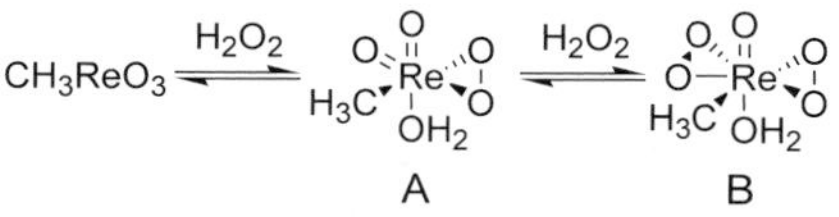

$$CH_3ReO_3 \xrightleftharpoons{H_2O_2} \underset{A}{\text{[mono-peroxo Re complex]}} \xrightleftharpoons{H_2O_2} \underset{B}{\text{[bis-peroxo Re complex]}} \qquad \text{Scheme 2.13}$$

ligand. The protons of the coordinated water molecule in **B** are highly acidic, and this has important implications for the epoxidation reaction (see below). As regards catalytic activity, however, it has been demonstrated that both complexes are active as oxygen-transfer species. Whereas decomposition of the MTO catalyst under basic conditions is often negligible, the presence of hydrogen peroxide completely changes the situation. The combination of basic media and H_2O_2 rapidly induces an irreversible decomposition of MTO according to Scheme 2.14, and this deleterious side reaction is usually a great problem in the catalytic system.

$$CH_3ReO_3 \xrightarrow[\text{Pyridine}]{H_2O_2} HOReO_3 \cdot 2py + CH_3OH \qquad \text{Scheme 2.14}$$

In this oxidative degradation, MTO is decomposing into catalytically inert perrhenate and methanol. The decomposition reaction is accelerated at higher pH, presumably through the reaction between the more potent nucleophile HO_2^- and MTO. The decomposition of MTO occurring under basic conditions is rather problematic, since the selectivity for epoxide formation certainly profits from the use of non-acidic conditions.

2.6.1
MTO as an Epoxidation Catalyst – Original Findings

The rapid formation of peroxo-complexes in the reaction between MTO and hydrogen peroxide makes this organometallic compound useful as an oxidation catalyst. In the original report on alkene epoxidation using MTO, Herrmann and coworkers employed a prepared solution of hydrogen peroxide in *tert*-butanol as the terminal oxidant. This solution was prepared by mixing *tert*-butanol and aqueous hydrogen peroxide followed by the addition of anhydrous $MgSO_4$. After filtration, this essentially water-free solution of hydrogen peroxide was used in the epoxidation reactions. It was further reported that MTO, or rather its peroxo-complexes were stable for weeks in this solution if kept at low temperatures (below 0 °C). As seen above, later studies by Espenson revealed the instability of MTO in hydrogen peroxide solutions. Epoxidation of various alkenes using 0.1–1 mol% of MTO and the $H_2O_2/^tBuOH$ solution generally resulted in high conversion into epoxide, but a significant amount of *trans*-1,2-diol was often formed via ring opening of the epoxide. The reason for using "anhydrous" hydrogen peroxide was of course an attempt to avoid the latter side-reaction; however, since hydrogen peroxide generates water upon reaction with MTO it was impossible to work under strictly water-free conditions. The ring opening process can either be catalyzed directly by MTO, due to the intrinsic metal Lewis acidity, or simply by protonation of the epoxide. To overcome this problem, Herrmann used an excess of amines (e.g., 4,4'-dimethyl-2,2'-bipyridine, quinine and cinchonine) which would coordinate to the metal and thus suppress the ring opening process [59]. This resulted in better selectivity for the epoxide, at the expense of decreased, or in some cases completely inhibited, catalytic activity. In an attempt to overcome the problems with low selectivity for epoxide formation and the decreased catalytic activity obtained using amine additives, Adam introduced the urea/hydrogen peroxide (UHP) adduct as the terminal oxidant for the MTO-catalyzed system [60]. This resulted in substantially better selectivity for several olefins, although substrates leading to highly acid-sensitive epoxides still suffered from deleterious ring opening reactions.

2.6.2
The Influence of Heterocyclic Additives

The second major discovery for the use of MTO as an epoxidation catalyst came in 1996, when Sharpless and coworkers reported on the use of sub-stoichiometric amounts of pyridine as co-catalysts in the system [61]. The switch of solvent from *tert*-butanol to dichloromethane, and the introduction of 12 mol% of pyridine allowed for the synthesis of even very sensitive epoxides using aqueous hydrogen peroxide as the terminal oxidant. A significant rate-acceleration was also observed for the epoxidation reaction performed in the presence of pyridine. This discovery was the first example of an efficient MTO-based system for epoxidation under neutral-to-basic conditions. Under these conditions the detrimental acid-induced decomposition of the epoxide is effectively avoided. Employing this novel system, a variety of alkene-substrates were converted into their corresponding epoxides in high yields and with high epoxide selectivity (Scheme 2.15 and Table 2.5).

The increased rate of epoxidation observed using pyridine as an additive has been studied by Espenson and Wang and was to a certain degree explained as an accelerated formation of peroxorhenium species in the presence of pyridine [62]. A stabilization of the rhenium-catalyst through pyridine coordination was also detected, although the excess of pyridine required in the protocol unfortunately led to increased catalyst deactivation. As can be seen above, MTO is stable under acidic conditions but at high pH an accelerated decomposition of the catalyst into perrhenate and methanol occurs. The Brønsted basicity of pyridine leads to increased amounts of HO_2^- which speeds up the formation of the peroxo-complexes and the decomposi-

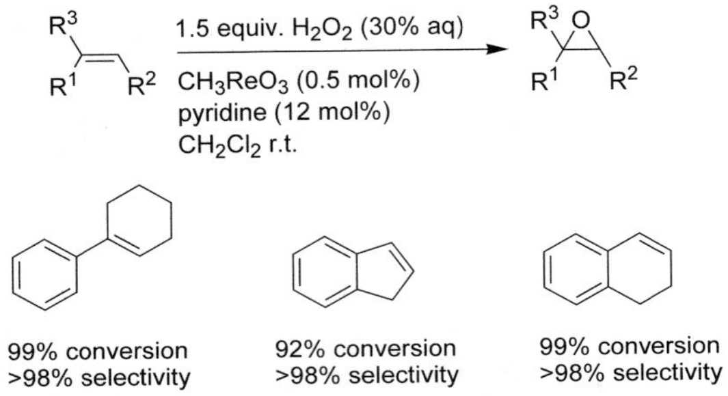

Scheme 2.15

99% conversion, >98% selectivity

92% conversion, >98% selectivity

99% conversion, >98% selectivity

Tab. 2.5 MTO-catalyzed epoxidation of alkenes using H_2O_2 [a]

Alkene	No additive[b]	Pyridine[c]	3-Cyanopyridine[c]	Pyrazole[c]
 cyclohexene	90 (5)	96 (6)		
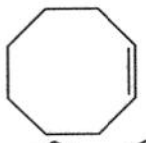 cyclooctene	100 (2)[c]	99 (2)		89 (0.02)
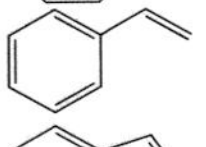 styrene		84 (16)	96 (5)[c]	96 (5)
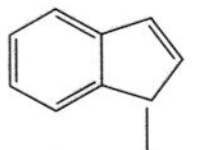 indene	48 (37)	96 (5)		
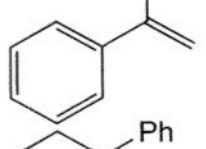 α-methylstyrene		82 (6)	74 (1.5)[d]	93 (1.5)
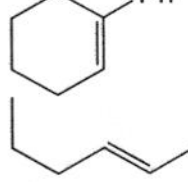 1-phenylcyclohexene		98 (1)	96 (1)[d]	95 (1)
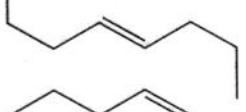 1-octene	95 (2)	91 (24)	97 (12)	
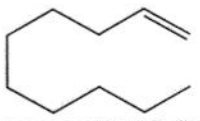 trans-alkene	75 (72)	82 (48)	99 (14)	99 (14)

[a] Yield % (reaction time h). [b] Anhydrous H_2O_2 in ^tBuOH. [c] Aqueous H_2O_2 (30%). [d] Pyridine and 3-cyanopyridine (6 mol% of each).

2.6 Rhenium-catalyzed Epoxidations

tion of the catalyst. Hence, the addition of pyridine to the epoxidation system led to certain improvements regarding rate and selectivity for epoxide formation, at the expense of catalyst lifetime. This turned out to be a minor problem for highly reactive substrates such as tetra-, tri- and *cis*-di-substituted alkenes, since these compounds are converted into epoxides at a rate significantly higher than the rate for catalyst decomposition. Less electron-rich substrates such as terminal alkenes, however, react slower with electrophilic oxygen-transfer agents, and require longer reaction times to reach acceptable conversions. Using the pyridine (12 mol%) conditions did not fully convert either 1-decene or styrene, even after prolonged reaction times.

A major improvement regarding epoxidation of terminal alkenes was achieved upon exchanging pyridine for its less basic analogue 3-cyanopyridine (pK_a pyridine = 5.4; 3-cyanopyridine = 1.9) [63]. This improvement turned out to be general for a number of different terminal alkenes, regardless of the existence of steric hindrance in the α-position of the alkene or whether other functional groups were present in the substrate (Scheme 2.16).

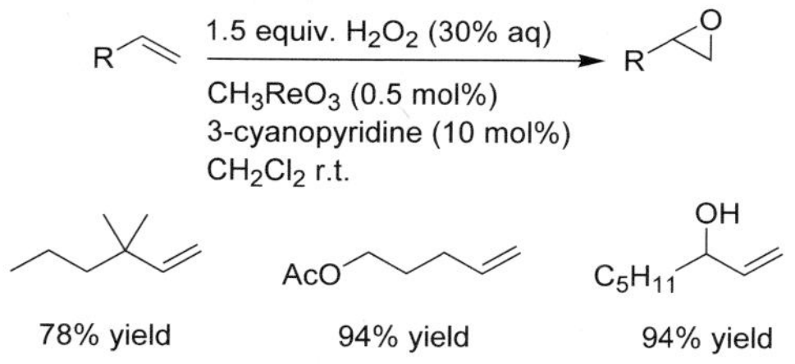

Scheme 2.16

Terminal alkenes leading to acid-labile epoxides were, however, not efficiently protected using this procedure. This problem was solved by using a cocktail consisting of 3-cyanopyridine and pyridine (5–6 mol% of each additive) in the epoxidation reaction. The additive 3-cyanopyridine was also successfully employed in epoxidation of *trans*-di-substituted alkenes, a problematic substance class using the parent pyridine system [64]. In these reactions, the amount of the MTO catalyst could be reduced down to 0.2–0.3 mol% with only 1–2 mol% of 3-cyanopyridine added. Again, acid sensitive epoxides were obtained using a mixture of 3-cyanopyridine and the parent pyridine. It should be pointed out that the pyridine additives do undergo oxidation reactions forming the corresponding pyridine-*N*-oxides [65]. This will of course effectively decrease the amount of additive present in the reaction mixture. In fact, as pointed out by Espenson, the use of a pyridinium salt (mixture of pyridine and, e.g., acetic acid) can be more effective in protecting the additive from *N*-oxidation [62, 66]. This can be beneficial for slow reacting substrates, where *N*-oxidation would compete with alkene epoxidation. The Herrmann group introduced an improvement to the Sharpless system by employing pyrazole as an additive [67]. Compared with pyridine, pyrazole is a less basic heterocycle (pK_a = 2.5) and does not undergo *N*-oxidation by the MTO/H_2O_2 system. Furthermore, employing pyrazole as the additive allowed for the formation of certain acid sensitive epoxides. With respect to which additive to choose, pyrazole is perhaps the most effective for the majority of alkenes, although for certain acid labile compounds, pyridine would be the preferred additive (Table 2.5) [68].

2.6.3
The Role of the Additive

The use of various heterocyclic additives in the MTO-catalyzed epoxidation has been demonstrated to be of great importance for substrate conversion, as well as for the product selectivity. Regarding the selectivity, the role of the additive is obviously to protect the product epoxides from deleterious, acid-catalyzed (Brønsted or Lewis acid) ring opening reactions. This can be achieved by direct coordination of the heterocyclic additive to the rhenium metal, thereby significantly decreasing the Lewis acidity of the metal. Also, the basic nature of the additives will increase the pH of the reaction media.

Concerning the accelerating effects observed when pyridine or pyrazole is added to the MTO-system, a number of different suggestions have been made. One likely explanation is that the additives do serve as phase-transfer agents. Hence, when MTO is added to an aqueous H_2O_2 solution, an immediate formation of the peroxo-complexes **A** and **B** (cf. Scheme 2.13) occurs, which is visualized by the intense bright yellow color of the solution. If a non-miscible organic solvent is added, the yellow color is still present in the aqueous layer, but addition of pyridine to this mixture results in an instantaneous transfer of the peroxo-complexes into the organic phase. The transportation of the active oxidants into the organic layer would thus favor the epoxidation reaction, since the alkene concentration is significantly higher in this phase (Scheme 2.17). Additionally, the rate at which MTO is converted into **A** and **B** is accelerated when basic heterocycles are added. This has been attributed to the Brønsted-basicity of the additives, which increases the amount of peroxide anion present in the reaction mixture. A higher concentration of HO_2^- is, however, detrimental to the MTO-catalyst, but the coordination of a Lewis base to the metal seems to have a positive effect in protecting the catalyst from decomposition.

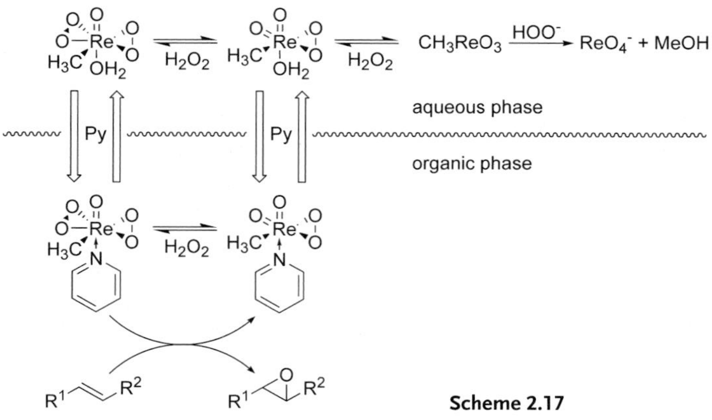

Scheme 2.17

2.6.4
Other Oxidants

While aqueous hydrogen peroxide is certainly the most practical oxidant for MTO-catalyzed epoxidations, the use of other terminal oxidants can sometimes be advantageous. As mentioned above the urea-hydrogen peroxide adduct has been employed in alkene epoxidations. The anhydrous conditions obtained using UHP improved the system by decreasing the amount of diol formed in the reaction. The absence of significant amounts of water further helped in preserving the active catalyst from decomposition. A disadvantage, however, is the poor solubility of UHP in many organic solvents, which makes these reactions heterogeneous.

Another interesting terminal oxidant which has been applied in MTO-catalyzed epoxidations is sodium percarbonate (SPC) [69]. The fundamental structure of SPC is composed of hydrogen peroxide encapsulated through hydrogen bonding in a matrix of sodium carbonate [70]. It slowly decomposes in water, and in organic solvents, to release hydrogen peroxide. The safety aspects associated with employing this oxidant are reflected in its common use as an additive to household washing detergents and toothpaste. When this "solid form" of hydrogen peroxide was employed in MTO-catalyzed (1 mol%) oxidation of a wide range alkenes, good yields of the corresponding epoxides were obtained. An essential requirement for a successful outcome of the reaction was the addition of an equimolar amount (with respect to the oxidant) of trifluoroacetic acid (TFA). In the absence of acid or with the addition of acetic acid, no or poor reactivity was observed. The role of the acid in this heterogeneous system is to facilitate the slow release of hydrogen peroxide. Despite the presence of acid, even hydrolytically sensitive epoxides were formed in high yields. This can be explained by an efficient buffering of the system by $NaHCO_3$ and CO_2, formed in the reaction between TFA and SPC. The initial pH was measured to be 2.5, but after 15 min a constant pH of 10.5 was established, ensuring protection of acid-sensitive products.

Bis(trimethylsilyl) peroxide (BTSP) represents another form of "anhydrous" hydrogen peroxide [71]. The use of strict anhydrous conditions in MTO-catalyzed alkene epoxidations would efficiently eliminate problems with catalyst deactivation and product decomposition due to ring opening reactions. BTSP, which is the di-silylated form of hydrogen peroxide, has been used in various organic transformations [72]. On reaction, BTSP is converted into hexamethyldisiloxane, thereby assuring anhydrous conditions. In initial experiments, MTO showed little or no reactivity towards BTSP under stoichiometric conditions [73]. This was very surprising, considering the high reactivity observed for BTSP compared with hydrogen peroxide in the oxidation of sulfides to sulfoxides [74]. The addition of one equivalent of water to the MTO/BTSP mixture, however, rapidly facilitated the generation of the active peroxocomplexes. This was explained by hydrolytic formation of H_2O_2 from BTSP in the presence of MTO (Scheme 2.18). In fact, other proton sources proved to be equally effective in promoting this hydrolysis. Thus, under strict water-free conditions no epoxidation occurred when the MTO/BTSP system was used. The addition of trace

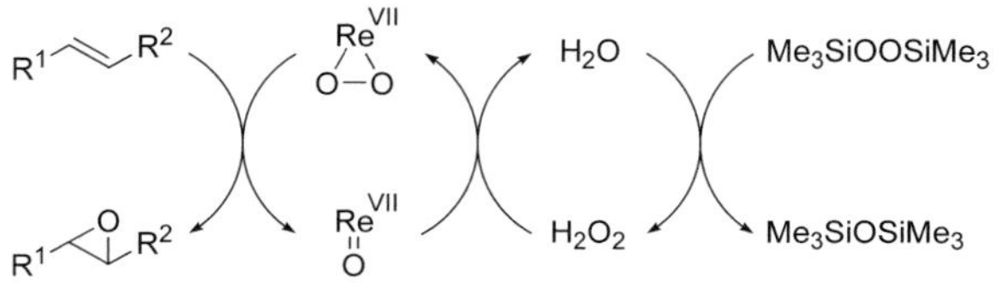

Scheme 2.18

amounts of a proton source triggered the activation of BTSP and the formation of epoxides was observed.

Under optimum conditions, MTO (0.5 mol%), water (5 mol%) and 1.5 equiv. of BTSP were used for efficient epoxide formation. The discovery of these essentially water-free epoxidation conditions led to another interesting breakthrough, namely the use of inorganic oxorhenium compounds as catalyst precursors [75]. The catalytic activity of rhenium compounds such as Re_2O_7, $ReO_3(OH)$ and ReO_3 in oxidation reactions with aqueous hydrogen peroxide as the terminal oxidant is typically very poor. Attempts to form epoxides using catalytic Re_2O_7 in 1,4-dioxane with H_2O_2 (60%) at elevated temperatures (90 °C) mainly yielded 1,2-diols [76]. However, when hydrogen peroxide was replaced by BTSP in the presence of a catalytic amount of a proton source, any of the inorganic rhenium oxides Re_2O_7, $ReO_3(OH)$ or ReO_3 were equally effective as MTO in alkene epoxidations. In fact, the use of ReO_3 proved to be highly practical, since this compound is hydrolytically stable as opposed to Re_2O_7. There are several benefits associated with these epoxidation conditions. The amount of BTSP used in the reaction can easily be monitored using gas chromatography. Furthermore, the simple work-up procedure associated with this protocol is very appealing, since evaporation of the solvent (typically dichloromethane) and the formed hexamethyldisiloxane yield the epoxide.

Tab. 2.6 MTO-catalyzed epoxidation of alkenes with anhydrous H_2O_2, or in fluorous solvents[a]

Alkene	UHP[b]	SPC[c]	BTSP[d]	UHP[e] ionic liquid	H_2O_2[f] CF_3CH_2OH	H_2O_2[g] $(CF_3)_2CHOH$
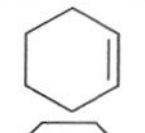 cyclohexene	97 (18)			99 (8)	99 (0.5)	
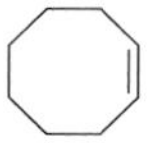 cyclooctene		94 (2)		95 (8)	99 (1)	93 (1)
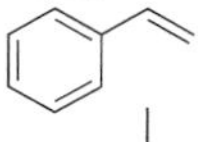 styrene	44 (19)	96 (12)		95 (8)	82 (2)	
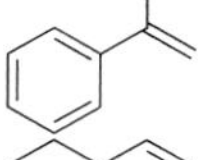 α-methylstyrene	55 (21)[h]	91 (3)				
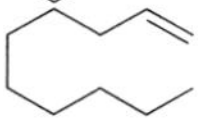 1-decene		94 (15)	94 (14)	46 (72)	97 (21)	88 (24)[i]

[a] Yield % (reaction time h). [b] 1 mol% MTO. [c] 1 mol% MTO, 12 mol% pyrazole. [d] 0.5 mol% Re_2O_7. [e] 2 mol% MTO. [f] 0.1 mol% MTO, 10 mol% pyrazole, 60% H_2O_2. [g] 0.1 mol% MTO, 10 mol% pyrazole, 30% H_2O_2. [h] Additional 26% of the diol was formed. [i] 1-dodecene was used as substrate.

2.6.5
Solvents/Media

The high solubility of the MTO catalyst in almost any solvent opens up the options for a broad spectrum of reaction media to choose from when performing epoxidations. The most commonly used solvent, however, is still dichloromethane. From an environmental point of view this is certainly not the most appropriate solvent in large scale epoxidations. Interesting solvent effects for the MTO-catalyzed epoxidation were reported by Sheldon and coworkers, who performed the reaction in trifluoroethanol [77]. The change from dichoromethane to the fluorinated alcohol allowed for a further reduction of the catalyst loading down to 0.1 mol%, even for terminal alkene substrates. It should be pointed out that this protocol does require 60% aqueous hydrogen peroxide for efficient epoxidations.

Bégué and coworkers recently reported on an improvement of this method by performing the epoxidation reaction in hexafluoro-2-propanol [78]. They found that the activity of hydrogen peroxide was significantly increased in this fluorous alcohol, as compared with trifluoroethanol, which allowed for the use of 30% aqueous H_2O_2. Interestingly, the nature of the substrate and the choice of additive turned out to have important consequences for the lifetime of the catalyst. Cyclic di-substituted alkenes were efficiently epoxidized using 0.1 mol% of MTO and 10 mol% pyrazole as the catalytic mixture; however, for tri-substituted substrates, the use of the additive 2,2'-bipyridine turned out to be crucial for a high conversion (Scheme 2.19). The use of pyrazole in the latter case proved to be highly deleterious for the catalyst, as indicated by the loss of the yellow color of the reaction solution. This observation is certainly contradictory, since more basic additives normally decrease the lifetime of the catalyst. The fact that full conversion of long-chain terminal alkenes was obtained after 24 h using pyrazole as the additive, and the observation that the catalyst was still active after this period of time, is very surprising considering the outcome with more functionalized substrates. To increase conversion for substrates which showed poor solubility in hexafluoro-2-propanol, trifluoromethylbenzene was added as a co-solvent. In this way, 1-dodecene was converted into its corresponding epoxide in high yield.

The use of non-volatile ionic liquids as environmentally benign solvents has received significant attention in recent years. Abu-Omar and coworkers developed an

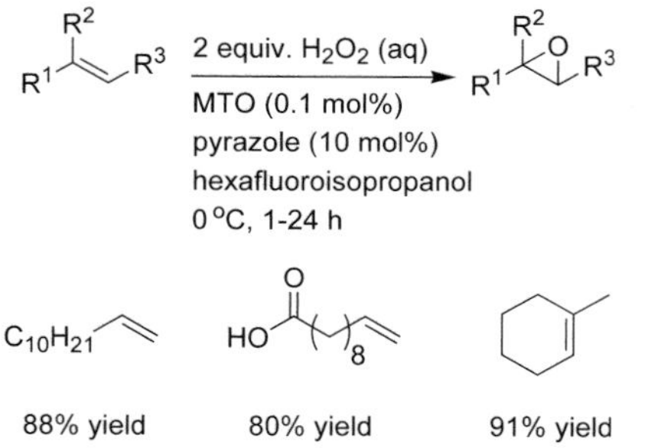

Scheme 2.19

efficient MTO-catalyzed epoxidation protocol using 1-ethyl-3-methylimidazolium tetrafluoroborate, [emim]BF$_4$, as the solvent and urea-hydrogen peroxide (UHP) as the terminal oxidant [79, 80]. A major advantage of this system is the high solubility of UHP, MTO and its peroxo-complexes, making the reaction media completely homogeneous. Employing these essentially water-free conditions, high conversions and good epoxide-selectivity were obtained for the epoxidation of variously substituted alkenes. Replacing UHP with aqueous hydrogen peroxide for the epoxidation of 1-phenylcyclohexene resulted in a poor yield of this acid sensitive epoxide, and instead the formation of the corresponding diol was obtained. A disadvantage of this system as compared with other MTO-protocols is the high catalyst loading (2 mol%) required for efficient epoxide formation.

2.6.6
Solid Support

The immobilization of catalysts or catalyst precursors by solid supports in order to simplify reaction procedures and to increase the stability of the catalyst is a common technique to render homogeneous systems heterogeneous. The MTO catalyst can be transferred into polymeric material in a number of different ways. When a water solution of MTO is heated for several hours (ca. 70 °C), the formation of a golden colored polymeric material occurs. The composition of this organometallic polymer is [H$_{0.5}$(CH$_3$)$_{0.92}$ReO$_3$]. This polymeric form of MTO is non-volatile, stable to air and moisture, and insoluble in all non-coordinating solvents. It can be used as a catalyst precursor for epoxidation of alkenes, since it is soluble in hydrogen peroxide, where it reacts to form the peroxo-rhenium species. Of course, the "heterogeneous" property of this material is lost on usage, but from a storage perspective, the polymeric MTO offers some advantages. An immobilization of MTO can, however, easily be obtained by the addition of a polymeric material containing Lewis basic groups with the ability to coordinate to the rhenium center. A number of different approaches have been reported. Herrmann and coworkers described the use of poly(vinylpyridines) as the organic support, but the resulting MTO-polymer complex showed low catalytic activity. A serious drawback with this supported catalyst was the oxidation of the polymeric backbone leading to loss of the rhenium catalyst.

In a recent improvement to this approach, poly(4-vinylpyridine) and poly(4-vinylpyridine) N-oxides were used as the catalyst carrier [81]. The MTO-catalyst obtained from 25% cross-linked poly(4-vinylpyridine) proved to efficiently catalyze the formation of even hydrolytically sensitive epoxides in the presence of aqueous hydrogen peroxide (30%). This catalyst could be recycled up to 5 times without any significant loss of activity. Attempts have been made to immobilize MTO with the use of either microencapsulation techniques, including sol–gel techniques, for the formation of silica bound rhenium compounds, and the attachment of MTO on silica tethered with polyethers. These approaches have provided catalysts with good activity using aqueous hydrogen peroxide as the terminal oxidant [83]. In the latter case, high selectivity for epoxide formation was also obtained for very sensitive substrates (e.g., indene).

An alternative approach to immobilization of the catalyst on a solid support is to perform the MTO-catalyzed epoxidation reactions in the presence of NaY zeolites. This technique has been employed by Malek and Ozin, and later by Bein et al., who used highly activated zeolites for the preparation of NaY/MTO using vacuum sublimation [84, 85]. More recently, Adam and coworkers found a significantly simpler approach towards this catalyst. The active catalyst was formed by mixing unactivated NaY zeolite with hydrogen peroxide (85%) in the presence of MTO and the substrate alkene [86]. Using this catalytic mixture, various alkenes were transformed into their corresponding epoxides, without the formation of diols (typical diol formation was <5%). The MTO catalyst is positioned inside the 12 Å supercages of the zeolite-Y, hence the role of the zeolite is to act as an absorbent for the catalyst and to provide heterogeneous microscopic reaction vessels for the reaction. The supernatant liquid was demonstrated to be catalytically inactive, even if Lewis bases (pyridine) were present. The high selectivity for epoxide formation was attributed to inhibition of the Lewis acid mediated hydrolysis of the product by means of steric hindrance.

Recently, Omar Bouh and Espenson reported that MTO supported on niobia catalyzed the epoxidation of various fatty oils using UHP as the terminal oxidant [87]. Oleic acid, elaidic acid, linoleic acid and linolenic acid were all epoxidized in high yields (80–100%) within less than 2 h. Furthermore, it was demonstrated that the catalyst could be recovered and reused without loss of activity.

2.6.7
Asymmetric Epoxidations Using MTO

The MTO-based epoxidation system offers a particularly effective and practical route for the formation of racemic epoxides. Attempts to prepare chiral MTO-complexes and to employ them in the catalytic epoxidation have so far been scarce and the few existing reports are unfortunately quite discouraging. In the epoxidation of cis-β-methylstyrene with MTO and hydrogen peroxide, in the presence of the additive (S)-1-(N,N-dimethyl)phenylethylamine, an enantiomeric excess of 86% of the product has been claimed [88]. The epoxides from other substrates such as styrene and 1-octene were obtained in significantly lower enantioselectivity (13% ee). Furthermore, the MTO-catalyzed epoxidation of 1-methylcyclohexene with L-prolineamide, (+)-2-aminomethylpyrrolidine or (R)-1-phenylethylamine as additives was reported to yield the product in low yield and enantioselectivity (up to 20% ee) [89]. A significant amount of the diol was formed in these reactions. Hence, a general protocol for the enantioselective formation of epoxides using rhenium catalysts is still lacking. There would certainly be a breakthrough if such a system could be developed, considering the efficiency of the MTO-catalyzed epoxidation reactions using hydrogen peroxide as the terminal oxidant.

2.7
Iron-catalyzed Epoxidations

The use of iron salts and complexes for alkene epoxidation is in many respects similar to that of manganese catalysts. Thus, iron porphyrins can be used as epoxidation catalysts, but often conversion and selectivity are inferior to what is obtained with its manganese counterpart. The possibilities of using hydrogen peroxide efficiently as the terminal oxidant are limited due to the rapid decomposition of the oxidant catalyzed by iron. Traylor and coworkers, however, found conditions where a polyfluorinated Fe(TPP)-catalyst (10) (Scheme 2.20) was employed in the epoxidation of cyclooctene to yield the corresponding epoxide in high yield (Scheme 2.21) [90]. High catalyst loading (5 mol%) and slow addition of the oxidant were required, which certainly limits the usefulness of this procedure.

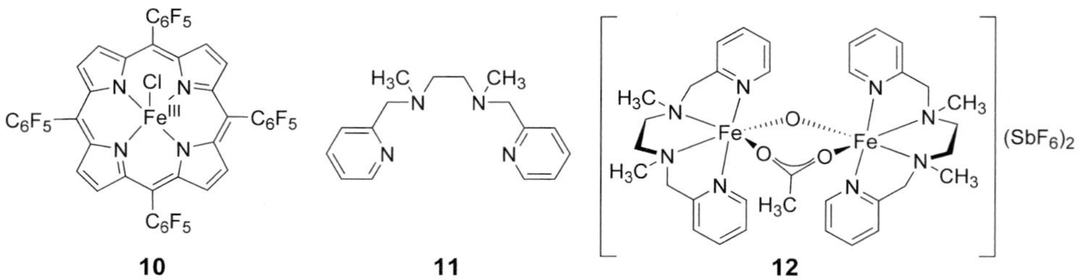

Scheme 2.20

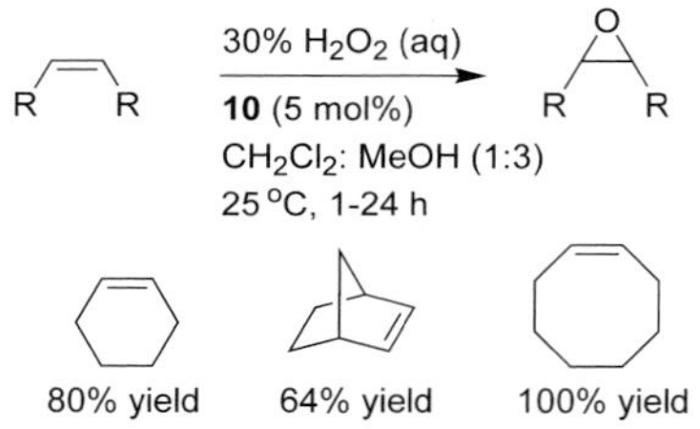

Scheme 2.21

Recently, a number of iron complexes with biomimetic non-heme ligands were introduced as catalysts for alkane hydroxylation, alkene epoxidation and dihydroxylation. These complexes were demonstrated to activate hydrogen peroxide without the formation of free hydroxyl radicals, a feature commonly observed in iron oxidation chemistry. A particularly efficient catalytic system for selective epoxidation of alkenes was developed by Jacobsen and coworkers [91]. In this protocol, a tetradentate ligand [BPMEN = N,N'-dimethyl-N,N'-bis(2-pyridylmethyl)-diaminoethane, 11] was combined with an iron(II) precursor and acetic acid to yield a self-assembled μ-oxo, carboxylate-bridged diiron(III) complex (12). This dimeric iron complex, resembling the active site found in the hydroxylase methane monooxygenase (MMO), was demonstrated to epoxidize alkenes efficiently in the presence of aqueous hydrogen peroxide (50%). This catalyst turned out to be particularly active for the epoxidation of terminal alkenes, which are normally the most difficult substrates to oxidize. Thus, 1-dodecene was transformed into its corresponding epoxide in 90% yield after 5 min using 3 mol% of the catalyst. This system was also effective for the epoxidation of other simple non-terminal alkenes, such as cyclooctene and *trans*-5-decene (Scheme 2.22).

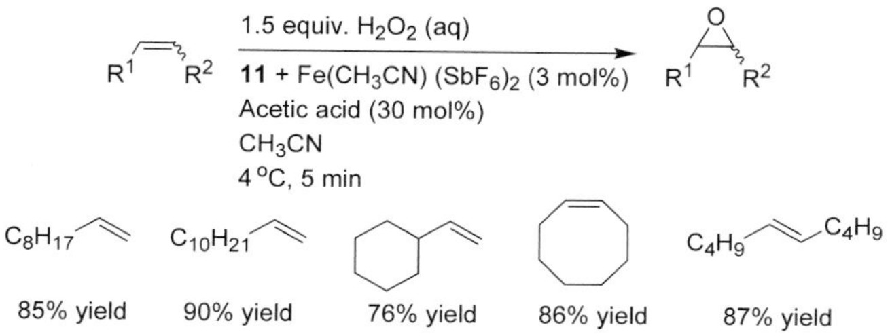

Scheme 2.22

Que and coworkers reported on a similar monomeric iron-complex, formed with the BPMEN ligand but excluding acetic acid [92]. This complex was able to epoxidize cyclooctene in reasonably good yield (75%), but at the same time a small amount of the *cis*-diol (9%) was formed. The latter feature observed with this class of complexes has been further studied and more selective catalysts have been prepared. Even though poor conversion is often obtained with the current catalysts, this method represents an interesting alternative to other *cis*-dihydroxylation systems [93, 94]. Using similar chiral ligands based on 1,2-diaminocyclohexane resulted in complexes which were able to catalyze the formation of epoxides in low yields and in low enantioselectivity (0–12% ee). The simultaneous formation of *cis*-diols was occurring with significantly better enantioselectivity (up to 82% ee), however, these products were also obtained in low yields.

Using high throughput screening techniques, Francis and Jacobsen discovered a novel iron-based protocol for the preparation of enantiomerically enriched epoxides [95]. In this system, chiral complexes prepared from polymer-supported peptide-like ligands and iron(II) chloride were evaluated as catalysts for the epoxidation of *trans*-β-methylstyrene employing aqueous hydrogen peroxide (30%) as the terminal oxidant. The best polymer-supported catalysts yielded the corresponding epoxide in up to 78% conversion with enantioselectivity ranging from 15 to 20% ee. Employing a homogeneous catalyst derived from this combinatorial study, *trans*-β-methylstyrene was epoxidized in 48% ee after 1 h (100% conversion, 5 mol% catalyst, 1.25 equiv. 50% hydrogen peroxide in ^{t}BuOH) (Scheme 2.23).

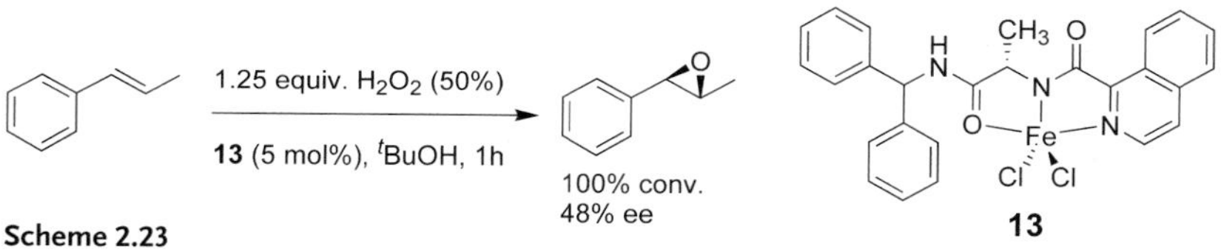

Scheme 2.23

2.8
Concluding Remarks

The epoxidation of alkenes using transition-metal based catalysts is certainly a well studied reaction. There are, however, only a few really good and general systems working with environmentally benign oxidants (i.e., aqueous hydrogen peroxide). A comparison of the efficiencies obtained with catalysts described in this chapter is presented in Table 2.7.

Tab. 2.7 Transition metal-catalyzed epoxidation of alkenes using H_2O_2 as terminal oxidant

Catalyst	S/C	Solvent	Temp. (°C)	1-Alkene[g] yield (%)/TOF (h^{-1})	Cyclooctene yield (%)/TOF (h^{-1})	Ref.
Ti[a]	100	MeOH	25	74/108[h]	–	[9]
W[b]	50/500	toluene	90	91/12	98/122	[17]
Mo[c]	100/200	CH_2Cl_2	60	96/4	100/100	[22]
Mn[d]	100	DMF	25	–	67/4	[43]
Re[e]	200	CH_2Cl_2	25	99/14	89/8900	[67]
Re[e]	1000	CF_3CH_2OH	25	97/48	99/990	[77]
Fe[f]	33	CH_3CN	4	85/337	86/341	[91]

[a] TS-1. [b] Na_2WO_4, $NH_2CH_2PO_3H_2$, $(C_8H_{17})_3NCH_3^+HSO_4^-$. [c] $MoO_5(OAs(C_{12}H_{25})_3$. [d] $MnSO_4$.
[e] MTO, pyrazole. [f] 12. [g] 1-decene. [h] 1-octene.

It is evident from the content of this chapter that there are advantages and limitations with almost all available epoxidation systems. The environmentally attractive TS-1 system is highly efficient, but restricted to linear substrates. The various tungsten systems available efficiently produce epoxides from simple substrates, but acid sensitive products undergo further reactions, thus effectively reducing the selectivity of the process. MTO is a highly active epoxidation catalyst, and when combined with heterocyclic additives, even hydrolytically sensitive products are obtained in good yield and selectivity. Most of the MTO-catalyzed reactions are, however, performed in chlorinated or fluorinated solvents. As regards asymmetric processes using hydrogen peroxide as the terminal oxidant, there are only a few reported systems that produce epoxides with some enantioselectivity. The iron-based catalyst developed by Francis and Jacobsen is a promising candidate, and further developments along these lines may produce more selective systems.

In conclusion, there is still room for further improvements in the field of selective alkene epoxidation using environmentally benign oxidants and solvents.

References

[1] R. A. Sheldon in *Applied Homogeneous Catalysis with Organometallic Compounds*, 2nd edn., *1* (Eds.: B. Cornils, W. A. Herrmann), Wiley-VCH, Weinheim, **2002**, pp. 412–427.

[2] K. A. Jørgensen, *Chem. Rev.*, **1989**, *89*, 431.

[3] D. Kahlich, K. Wiechern, J. Lindner in *Ullmann's Encyclopedia of Industrial Chemistry*, 5th edn., Vol. A22 (Eds.: B. Elvers, S. Hawkins, W. Russey, G. Schultz), VCH, Weinheim, **1993**, pp. 239–260.

[4] J. R. Monnier, *Applied Catalysis A: General*, **2001**, *221*, 73.

[5] B. S. Lane, K. Burgess, *Chem. Rev.*, **2003**, *103*, 2457.

[6] R. A. Sheldon in *Aspects of Homogeneous Catalysis*, Vol. 4 (Ed.: R. Ugo) Reidel, Dordrecht, **1981**, pp. 3–70.

[7] T. Katsuki, K. B. Sharpless, *J. Am. Chem. Soc.*, **1980**, *102*, 5974.

[8] For a recent comprehensive review, see: T. Katsuki in *Comprehensive Asymmetric Catalysis* II (Eds.: E. N. Jacobsen, A. Pfaltz, H. Yamamoto), Springer, Heidelberg, **1999**, pp. 621–648.

[9] B. Notari, *Catal. Today*, **1993**, *18*, 163.

[10] K. B. Sharpless, T. R. Verhoeven, *Aldrichim. Acta*, **1979**, *12*, 63.

[11] W. R. Thiel in *Transition Metals for Organic Synthesis* 2 (Eds.: M. Beller, C. Bolm), Wiley-VCH, Weinheim, **1998**, pp. 290–300.

[12] G. B. Payne, P. H. Williams, *J. Org. Chem.*, **1959**, *24*, 54.

[13] C. Venturello, E. Alneri, M. Ricci, *J. Org. Chem.*, **1983**, *48*, 3831.

[14] C. Venturello, R. D'Aloisio, *J. Org. Chem.*, **1988**, *53*, 1553.

[15] C. Venturello, R. D'Aloisio, J. C. J. Bart, M. Ricci, *J. Mol. Catal.*, **1985**, *32*, 107.

[16] K. Sato, M. Aoki, M. Ogawa, T. Hashimoto, R. Noyori, *J. Org. Chem.*, **1996**, *61*, 8310.

[17] K. Sato, M. Aoki, M. Ogawa, T. Hashimoto, D. Paynella, R. Noyori, *Bull. Chem. Soc. Jpn.*, **1997**, *70*, 905.

[18] K. Sato, M. Aoki, R. Noyori, *Science*, **1998**, *281*, 1646.

[19] A. L. Villa de P., B. F. Sels, D. E. De Vos, P. A. Jacobs, *J. Org. Chem.*, **1999**, *64*, 7267.

[20] K. Kamata, K. Yonehara, Y. Sumida, K. Yamaguchi, S. Hikichi, N. Mizuno, *Science*, **2003**, *300*, 964.

[21] O. Bortolini, F. Di Furia, G. Modena, R. Seraglia, *J. Org. Chem.*, **1985**, *50*, 2688.

[22] G. Wahl, D. Kleinhenz, A. Schorm, J. Sundermeyer, R. Stowasser, C. Rummey, G. Bringmann, C. Fikkert, W. Kiefer. *Chem. Eur. J.*, **1999**, *5*, 3237.

[23] B. F. Sels, D. E. De Vos, P. A. Jacobs, *Tetrahedron Lett.*, **1996**, *37*, 8557.

[24] For a comprehensive summary, see: G. Gelbard, *C.R. Chim.*, **2000**, *3*, 757.

[25] D. Hoegaerts, B. F. Sels, D. E. De Vos, F. Verpoort, P. A. Jacobs, *Catal. Today*, **2000**, *60*, 209.

[26] X. Zuwei, Z. Ning, S. Yu, L. Kunlan, *Science*, **2001**, *292*, 1139.

[27] P. Battioni, J.-P. Renaud, J. F. Bartoli, M. Reina-Artiles, M. Fort, D. Mansuy, *J. Am. Chem. Soc.*, **1988**, *110*, 8462.

[28] P. L. Anelli, L. Banfi, F. Legramandi, F. Montanari, G. Pozzi, S. Quici, *J. Chem. Soc., Perkin Trans. 1*, **1993**, 1345.

[29] W. Zhang, J. L. Loebach, S. R. Wilson, E. N. Jacobsen, *J. Am. Chem. Soc.*, **1990**, *112*, 2801.

[30] R. Irie, K. Noda, Y. Ito, N. Matsumoto, T. Katsuki, *Tetrahedron Lett.*, **1990**, *31*, 7345.

[31] E. N. Jacobsen, M. H. Wu in *Comprehensive Asymmetric Catalysis* II (Eds.: E. N. Jacobsen, A. Pfaltz, H. Yamamoto), Springer, Heidelberg, **1999**, pp. 649–677.

[32] K. Srinivasan, P. Michaud, J. K. Kochi, *J. Am. Chem. Soc.*, **1986**, *108*, 2309.

[33] A. Berkessel, M. Frauenkron, T. Schwenkreis, A. Steinmetz, G. Baum, D. Fenske, *J. Mol. Catal. A: Chem.*, **1996**, *113*, 321.

[34] R. Irie, N. Hosoya, T. Katsuki, *Synlett*, **1994**, 255.

[35] P. Pietikäinen, *Tetrahedron*, **1998**, *54*, 4319.
[36] P. Pietikäinen, *J. Mol. Catal. A: Chem.*, **2001**, *165*, 73.
[37] R. I. Kureshy, N. H. Khan, S. H. R. Abdi, S. T. Patel, R. V. Jasra, *Tetrahedron: Asymmetry*, **2001**, *12*, 433.
[38] R. Hage, J. E. Iburg, J. Kerschner, J. H. Koek, E. L. M. Lempers, R. J. Martens, U. S. Racheria, S. W. Russell, T. Swarthoff, M. R. P. van Vliet, J. B. Warnaar, L. van der Wolf, B. Krijnen, *Nature*, **1994**, *369*, 637.
[39] D. E. De Vos, B. F. Sels, M. Reynaers, Y. V. S. Rao, P. A. Jacobs, *Tetrahedron Lett.*, **1998**, *39*, 3221.
[40] A. Berkessel, C. A. Sklorz, *Tetrahedron Lett.*, **1999**, *40*, 7965.
[41] J. Brinksma, R. Hage, J. Kerschner, B. L. Feringa, *Chem. Commun.*, **2000**, 537.
[42] B. S. Lane, K. Burgess, *J. Am. Chem. Soc.*, **2001**, *123*, 2933.
[43] B. S. Lane, M. Vogt, V. J. DeRose, K. Burgess, *J. Am. Chem. Soc.*, **2002**, *124*, 11946.
[44] K.-H. Tong, K.-Y. Wong, T. H. Chan, *Org. Lett.*, **2003**, *5*, 3423.
[45] F. E. Kühn, W. A. Herrmann, *Chemtracts – Org. Chem.*, **2001**, *14*, 59.
[46] G. S. Owens, A. Arias, M. M. Abu-Omar, *Catalysis Today*, **2000**, *55*, 317. Please note: this review contains several serious errors in the chapter dealing with MTO-catalyzed epoxidations (i.e., Tables 5 and 6).
[47] W. A: Herrmann, R. W. Fischer, D. W. Marz, *Angew. Chem., Int. Ed. Engl.*, **1991**, *30*, 1638.
[48] I. R. Beattie, P. J. Jones, *Inorg. Chem.*, **1979**, *18*, 2318.
[49] W. A. Herrmann, J. G. Kuchler, J. K. Felixberger, E. Herdtweck, W. Wagner, *Angew. Chem., Int. Ed. Engl.*, **1988**, *27*, 394.
[50] W. A. Herrmann, F. E. Kühn, R. W. Fischer, W. R. Thiel, C. C. Romão, *Inorg. Chem.*, **1992**, *31*, 4431.
[51] W. A. Herrmann, W. R. Thiel, F. E. Kühn, R. W. Fischer, M. Kleine, E. Herdtweck, W. Scherer, J. Mink, *Inorg. Chem.*, **1993**, *32*, 5188.
[52] W. A. Herrmann, R. Kratzer, R. W. Fischer, *Angew. Chem., Int. Ed. Engl.*, **1997**, *36*, 2652.
[53] W. A. Herrmann, F. E. Kühn, *Acc. Chem. Res.*, **1997**, *30*, 169.
[54] C. C. Romão, F. E. Kühn, W. A. Herrmann, *Chem. Rev.*, **1997**, *97*, 3197.
[55] J. H. Espenson, *Chem. Commun.*, **1999**, 479.
[56] M. M. Abu-Omar, P. J. Hansen, J. H. Espenson, *J. Am. Chem. Soc.*, **1996**, *118*, 4966.
[57] G. Laurenczy, F. Lukács, R. Roulet, W. A. Herrmann, R. W. Fischer, *Organometallics*, **1996**, *15*, 848.
[58] W. A. Herrmann, R. W. Fischer, W. Scherer, M. U. Rauch, *Angew. Chem., Int. Ed. Engl.*, **1993**, *32*, 1157.
[59] W. A. Herrmann, R. W. Fischer, m. U. Rauch, W. Scherer, *J. Mol. Catal.*, **1994**, *86*, 243.
[60] W. Adam, C. M. Mitchell, *Angew. Chem., Int. Ed. Engl.*, **1996**, *35*, 533.
[61] J. Rudolph, K. L. Reddy, J. P. Chiang, K. B. Sharpless, *J. Am. Chem. Soc.*, **1997**, *119*, 6189.
[62] W.-D. Wang, J. H. Espenson, *J. Am. Chem. Soc.*, **1998**, *120*, 11335.
[63] C. Copéret, H. Adolfsson, K. B. Sharpless, *Chem. Commun.*, **1997**, 1565.
[64] H. Adolfsson, C. Copéret, J. P. Chiang, A. K. Yudin, *J. Org. Chem.*, **2000**, *65*, 8651.
[65] C. Copéret, H. Adolfsson, T.-A. V. Khuong, A. K. Yudin, K. B. Sharpless, *J. Org. Chem.*, **1998**, *63*, 1740.
[66] H. Adolfsson, K. B. Sharpless, *unpublished results*.
[67] W. A. Herrmann, R. M. Kratzer, H. Ding, W. R. Thiel, H. Glas, *J. Organomet. Chem.*, **1998**, *555*, 293.
[68] H. Adolfsson, A. Converso, K. B. Sharpless, *Tetrahedron Lett.*, **1999**, *40*, 3991.
[69] A. R. Vaino, *J. Org. Chem.*, **2000**, *65*, 4210.
[70] A. McKillop, W. R. Sanderson, *Tetrahedron*, **1995**, *51*, 6145.
[71] W. P. Jackson, *Synlett*, **1990**, 536.
[72] C. Jost, G. Wahl, D. Kleinhenz, J. Sundermeyer in *Peroxide Chemistry*, (Ed. W. Adam), Wiley-VCH, Weinheim, **2000**, pp. 341–364.
[73] A. K. Yudin, K. B. Sharpless, *J. Am. Chem. Soc.*, **1997**, *119*, 11536.

[74] R. Curci, R. Mello, L. Troisi, *Tetrahedron*, **1986**, *42*, 877.
[75] A. K. Yudin, J. P. Chiang, H. Adolfsson, C. Copéret, *J. Org. Chem.*, **2001**, *66*, 4713.
[76] S. Warwel, M. Rüsch den Klaas, M. Sojka, *J. Chem. Soc., Chem. Commun.*, **1991**, 1578.
[77] M. C. A. van Vliet, I. W. C. E. Arends, R. A. Sheldon, *Chem. Commun.*, **1999**, 821.
[78] J. Iskra, D. Bonnet-Delpon, J.-P. Bégué, *Tetrahedron Lett.*, **2002**, *43*, 1001.
[79] G. S. Owens, M. M. Abu-Omar, *Chem. Commun.*, **2000**, 1165.
[80] G. S. Owens, A. Durazo, M. M. Abu-Omar, *Chem. Eur. J.*, **2002**, *8*, 3053.
[81] R. Saladino, V. Neri, A. R. Pelliccia, R. Caminiti, C, Sadun, *J. Org. Chem.*, **2002**, *67*, 1323.
[82] K. Dallmann, R. Buffon, *Catal. Commun.*, **2000**, *1*, 9.
[83] R. Neumann, T.-J. Wang, *Chem. Commun.*, **1997**, 1915.
[84] A. Malek, G. Ozin, *Adv. Mater.*, **1995**, *7*, 160.
[85] T. Bein, C. Huber, K. Moller, C.-G. Wu, L, Xu, *Chem. Mater.*, **1997**, *9*, 2252.
[86] W. Adam, C. R. Saha-Möller, O. Weichold, *J. Org. Chem.*, **2000**, *65*, 2897.
[87] A. Omar Bouh, J. H. Espenson, *J. Mol. Catal. A: Chem.*, **2003**, *200*, 43.
[88] C. E. Tucker, K. G. Davenport, **1997**, Hoechst Celanese Corporation US Patent 5618958.
[89] M. J. Sabater, M. E. Domine, A. Corma, *J. Catal.*, **2002**, *210*, 192.
[90] T. G. Traylor, S. Tsuchiya, Y.-S. Byun, C. Kim, *J. Am. Chem. Soc.*, **1993**, *115*, 2775.
[91] M. C. White, A. G. Doyle, E. N. Jacobsen, *J. Am. Chem. Soc.*, **2001**, *123*, 7194.
[92] K. Chen, L. Que, Jr., *Chem. Commun.*, **1999**, 1375.
[93] M. Costas, A. K. Tipton, K. Chen, D.-H. Jo, L. Que, Jr., *J. Am. Chem. Soc.*, **2001**, *123*, 6722.
[94] K. Chen, M. Costas, J. Kim, A. K. Tipton, L. Que, Jr., *J. Am. Chem. Soc.*, **2002**, *124*, 3026.
[95] M. B. Francis, E. N. Jacobsen, *Angew. Chem., Int. Ed. Engl.*, **1999**, *38*, 937.

3
Organocatalytic Oxidation. Ketone-catalyzed Asymmetric Epoxidation of Olefins

YIAN SHI

3.1
Introduction

Epoxides are very versatile intermediates, and asymmetric epoxidation of olefins is an effective approach to the synthesis of enantiomerically enriched epoxides [1–3]. Great success has been achieved for the epoxidation of allylic alcohols [1], the metal-catalyzed epoxidation of unfunctionalized olefins (particularly conjugated cis- and trisubstituted) [2], and the nucleophilic epoxidation of electron-deficient olefins [3]. In recent years, chiral dioxiranes have been shown to be powerful agents for asymmetric epoxidation of olefins. Dioxiranes can be isolated or generated *in situ* from Oxone (potassium peroxymonosulfate) and ketones (Scheme 3.1) [4, 5]. When the dioxirane is used *in situ*, the corresponding ketone is regenerated upon epoxidation. Therefore, in principle, a catalytic amount of ketone can be used. When a chiral ketone is used, asymmetric epoxidation should also be possible [6]. Extensive studies have been carried out in this area since the first chiral ketone was reported by Curci in 1984 [7]. This chapter describes some of the recent progress in this area.

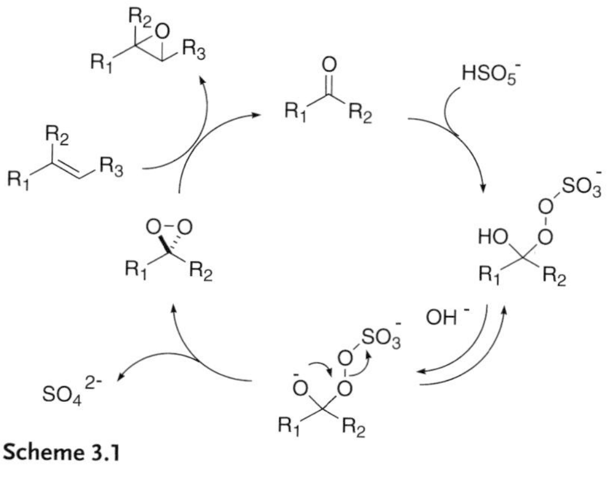

Scheme 3.1

Modern Oxidation Methods. Edited by Jan-Erling Bäckvall
Copyright © 2004 WILEY-VCH Verlag GmbH & Co. KGaA, Weinheim
ISBN: 3-527-30642-0

3.2
Early Ketones

In 1984 [7], Curci and coworkers first reported the asymmetric epoxidation of *trans*-β-methylstyrene and 1-methylcyclohexene using (+)-isopinocamphone (**1**) and (S)-(+)-3-phenylbutan-2-one (**2**) as the catalyst in CH_2Cl_2–H_2O at pH 7–8 (Scheme 3.2). As shown in Scheme 3.3, up to 12.5% ee was obtained, and the amount of ketone could be reduced to as little as 20 mol% without reducing the ee, which demonstrated the possibility of asymmetric induction by a chiral ketone.

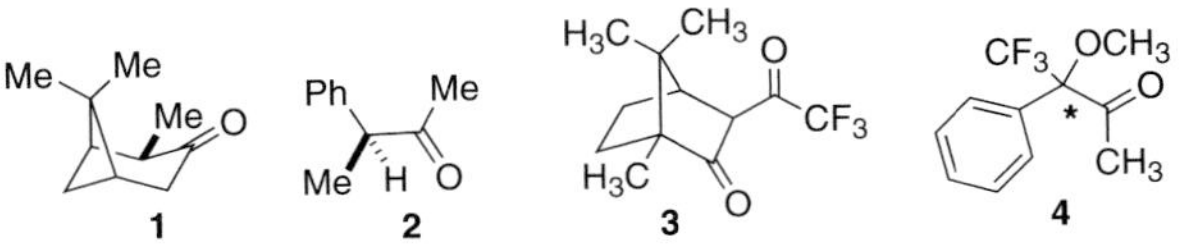

Scheme 3.2

1 (1.0 equiv.), 60% yield*, 12.5% ee
1 (0.2 equiv.), 68% yield**, 11.2% ee
2 (1.0 equiv.), 85% yield**, 9.5% ee

1 (1.0 equiv.) 90% yield**, 10.4% ee
1 (0.2 equiv.) 85% yield**, 10.2% ee
2 (0.5 equiv.) 92% yield**, 12% ee

(* isolated yield; ** GC yield based on the substrate reacted)

Scheme 3.3 Epoxidation of olefins with ketones **1** and **2**

Epoxidations with ketones **1** and **2** were somewhat sluggish, thus requiring relatively high catalyst loading and long reaction time to achieve high conversion. To further increase the reaction rate of the epoxidation, ketones **3** and **4**, which contain a trifluoromethyl group, were subsequently chosen for studies since ketones with electron-withdrawing substituents are usually more reactive for epoxidation [4]. Indeed, the epoxidations with ketones **3** and **4** were found to be much faster [8]. As shown in Scheme 3.4, high conversions could be achieved with 0.8–1.2 equiv. ketone at 2–5 °C within 17–48 h, and the ketones could be recovered from the reaction with little loss (2–5%). Up to 20% ee was obtained for *trans*-2-octene. The activation of ketones with electron withdrawing groups was also illustrated by Marples and coworkers in 1995 in their studies of asymmetric epoxidation with fluorinated 1-tetralones and 1-indanones (**5–8**) (Scheme 3.5) although no enantioselectivity was obtained for the epoxidation with these ketones [9].

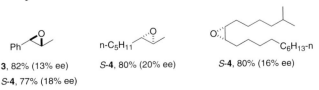

3, 82% (13% ee)   S-**4**, 80% (20% ee)   S-**4**, 80% (16% ee)
S-**4**, 77% (18% ee)

Scheme 3.4 Epoxidation of olefins with ketones **3** and **4** (0.8–1.2 equiv.)

3.3
C$_2$ Symmetric Binaphthyl-based and Related Ketones

Elegant binaphthyl derived chiral ketones were first reported by Yang and coworkers in 1996 (Scheme 3.6) [10–12]. In this type of ketone catalyst, C$_2$ symmetry was introduced to limit the competing reaction modes of the dioxirane, and a remote binaphthalene unit was used as the chiral control element. Substituents at the α-carbon to the carbonyl were avoided to eliminate the potential problems of racemization of chiral centers and steric hindrance at the α-carbon. The C$_2$ symmetric, 11-membered chiral ketone 9a derived from 1,1′-binaphthyl-2,2′-dicarboxylic acid, was initially investigated for the epoxidation (Scheme 3.7) [10]. The unhindered carbonyl and the presence of electron withdrawing groups at the α-carbon made ketone 9a a very reactive catalyst, providing high conversion for epoxidation with as little as 10 mol% catalyst in a few hours at pH 7–7.5 (Scheme 3.7). Running the epoxidation in a homogeneous solvent system (CH$_3$CN–H$_2$O) [5i, 13] could also enhance the reaction efficiency by facilitating the dioxirane–olefin interaction. The enantioselectivity of the epoxidation was found to be dependent upon the size of the *para*-substituents on *trans*-stilbenes, and the ee values of the resulting epoxides increased from 47% to 87% as the size of the substituents increased from H to Ph (Scheme 3.7) [10, 12]. Ketone 9a was found to be stable under the reaction conditions and could be recovered in >80% yield.

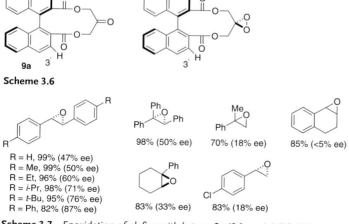

Scheme 3.7 Epoxidation of olefins with ketone 9a (0.1 equiv.) [10, 12]

3 Organocatalytic Oxidation. Ketone-catalyzed Asymmetric Epoxidation of Olefins

The stereodifferentiation for the epoxidation is largely dependent upon the steric interaction between the ketone catalyst and the reacting olefin. As revealed in the X-ray structure of ketone 9a [10–12], H-3 and H-3' are likely to be the interacting points between the ketone catalyst and the reacting olefin. It appears that increasing the steric bulkiness at the 3 and 3' positions would lead to a stronger steric interaction, and therefore higher enantioselectivity. Thus ketones 9b–k were designed and prepared by replacing the hydrogens at the 3 and 3' positions with larger groups [11, 12]. As the substituents became larger going from H to Cl to Br to I, the enantioselectivity first increased and then decreased, suggesting that an appropriate size of substituent is required for optimal selectivities (Scheme 3.8).

9a X = H, 91% (47% ee)
9b X = Cl, 95% (76% ee)
9c X = Br, 92% (75% ee)
9d X = I, 90% (32% ee)
9e X = Me, 93% (56% ee)
9f X = CH$_2$OCH$_3$, 92% (66% ee)
9g X = Ph, 50% (55% ee)

9h X = TMS, 44% ee
9i X = 95% (71% ee)
9j X = 90% (77% ee)
9k X = 91% (75% ee)

Scheme 3.8 Epoxidation of *trans*-stilbene with ketones 9a–k (0.1 equiv.)

As shown in Scheme 3.9, *para*-substituted *trans*-stilbenes were found to be effective substrates for ketone 9, and the enantioselectivity of the epoxidation varied with the size of the substituents on the olefins. The ee values of the epoxide product increased (84 to 95% ee for 9i) as the substituents became larger (from H to Me to Et to *i*-Pr to *t*-Bu). On the other hand, little effect on enantioselectivity was observed by the *meta*-substituents on the phenyl group of the stilbene.

R = H	R = Me	R = Et	R = *i*-Pr	R = *t*-Bu
9b, 76% ee (rt)	80% ee (rt)	85% ee (rt)	85% ee (rt)	91% ee (rt)
9c, 75% ee (rt)	85% ee (rt)	88% ee (rt)	90% ee (rt)	93% ee (rt)
9c, 80% ee (0 °C)	88% ee (0 °C)	92% ee (0 °C)	92% ee (0 °C)	95% ee (0 °C)
9i, 71% ee (rt)	84% ee (rt)	82% ee (rt)	88% ee (rt)	90% ee (rt)
9i, 84% ee (0 °C)	88% ee (0 °C)	91% ee (0 °C)	91% ee (0 °C)	95% ee (0 °C)

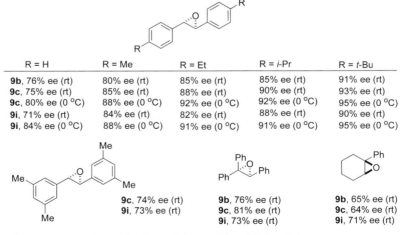

9c, 74% ee (rt)
9i, 73% ee (rt)

9b, 76% ee (rt)
9c, 81% ee (rt)
9i, 73% ee (rt)

9b, 65% ee (rt)
9c, 64% ee (rt)
9i, 71% ee (rt)

Scheme 3.9 Epoxidation of olefins with ketones 9b, c, i (0.1 equiv.)

3.3 C_2 Symmetric Binaphthyl-based and Related Ketones

Further studies on the effect of chiral elements showed that ketone **10** derived from 6,6'-dinitro-2,2'-diphenic acid gave ee values similar to ketone **9** (Scheme 3.10) [12]. Epoxidations with ketones **9** and **10** have recently been extended to cinnamates (Scheme 3.11) [14], and various efforts have also been made to further improve the synthesis of ketones **9a, b** by Seki and coworkers [15].

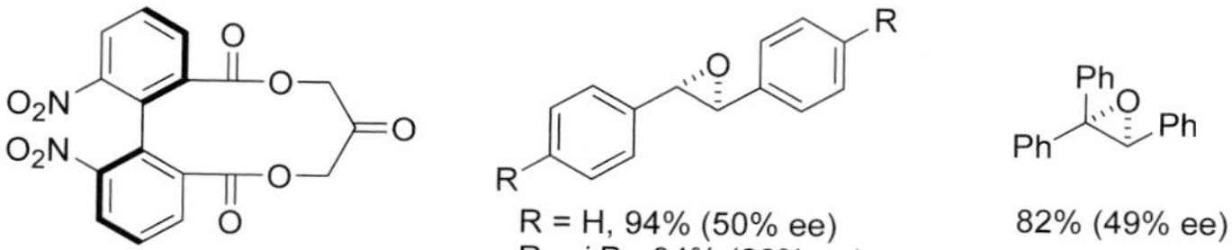

R = H, 94% (50% ee)
R = i-Pr, 94% (66% ee)
R = t-Bu, 91% (77% ee)

82% (49% ee)

Scheme 3.10 Epoxidation of olefins with ketones **10** (0.1 equiv.)

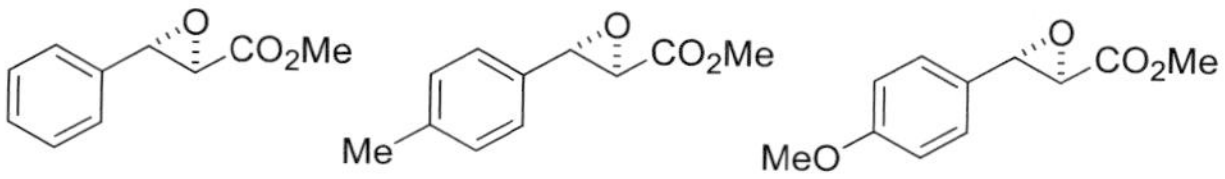

9a, 75% (74% ee)   9a, 95% (72% ee)   9a, 92% (80% ee)
                                         9b, 74% (85% ee)
                                         10, 86% (68% ee)

Scheme 3.11 Epoxidation of cinnamates with ketones **9a, b**, and **10** (0.05 equiv.)

The epoxidations of stilbene with the 11-membered ether and sulfonylamide ketones **11** and **12** were investigated by Tomioka and coworkers (Scheme 3.12) [16]. Relatively high yields were obtained for the epoxide, but the enantioselectivity was rather low. A number of other C_2 symmetric ether linked chiral ketones were also investigated for the asymmetric epoxidation. In 1997, Song and coworkers reported that the replacement of the ester groups of ketone **9** with ether groups (ketone **13a**) lowered both reactivity and enantioselectivity (Scheme 3.13) [17, 18]. The lower reactivity of ketone **13a** compared with ketone **9** could be due to the weaker electron-withdrawing ability of the ether compared with the ester. When ketones **13b** and **13c** were used for the epoxidation, 24% ee and 2% ee were obtained for stilbene, respectively [17b]. Up to 59% ee was obtained for ketone **14**, which uses simple phenyl groups as the chiral control element (Scheme 3.13) [17a].

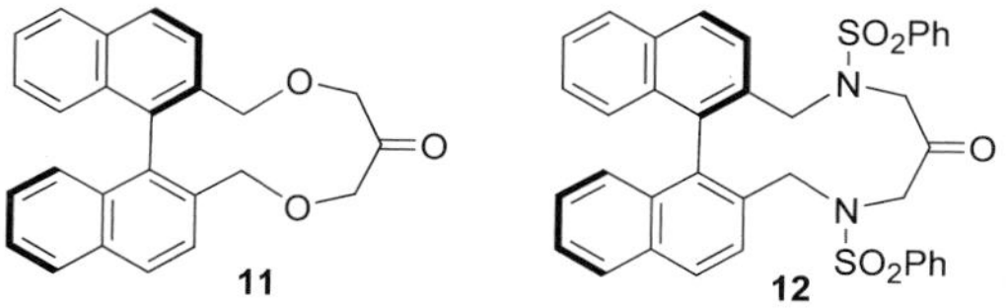

Scheme 3.12

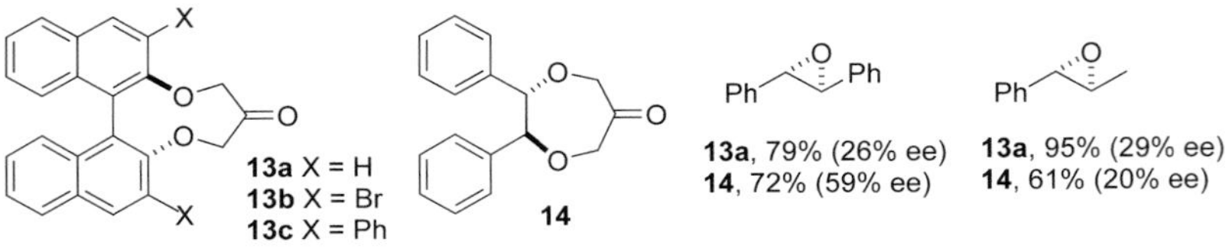

3 Organocatalytic Oxidation. Ketone-catalyzed Asymmetric Epoxidation of Olefins

Scheme 3.13 Epoxidation of olefins with ketones **13** and **14** (1.0 equiv.) [17a]

13a X = H
13b X = Br
13c X = Ph

13a, 79% (26% ee)
14, 72% (59% ee)

13a, 95% (29% ee)
14, 61% (20% ee)

In 1997, Adam and coworkers reported another variation of C_2 symmetric ether linked ketones using mannitol (**15**) and tartaric acid (**16**) as chiral backbones (Scheme 3.14) [19]. Up to 81% ee was obtained. In 2001, Tomioka and coworkers reported epoxidations with C_2 symmetric 7-membered sulfonylamide ketones **17** and **18** (Scheme 3.15) [16, 20]. Up to 30% ee was obtained for stilbene with **17b**. In their subsequent studies, higher ee values were obtained with tricyclic ketone **19** and bicyclic ketone **20** (Scheme 3.16) [21]. For the epoxidation of 1-phenylcyclohexene, a quantitative yield and 83% ee were obtained with a catalytic amount (20 mol%) of ketone **19**.

In 1999 and 2002, Denmark and coworkers reported several 7-membered carbocyclic biaryl chiral ketones (**22a–d**) (Scheme 3.17) [6a, 22]. For these ketones the chirality is closer to the reacting carbonyl compared with 11-membered ketone **9**, which could further enhance the stereodifferentiation for the epoxidation. It was found that the epoxidation rate was greatly accelerated by fluorine substitution at the α-carbon

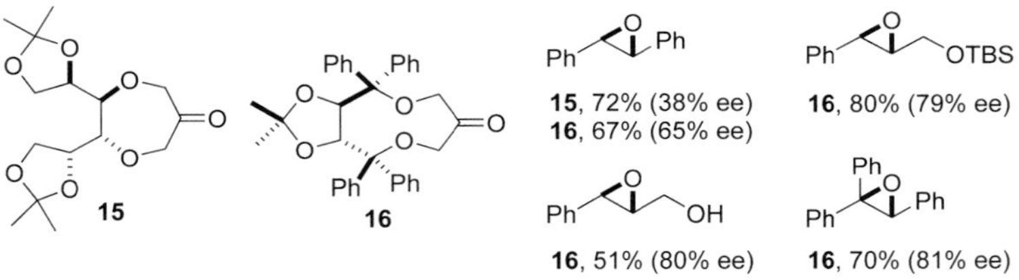

15, 72% (38% ee)
16, 67% (65% ee)

16, 80% (79% ee)

16, 51% (80% ee)

16, 70% (81% ee)

Scheme 3.14 Epoxidation of olefins with ketones **15** and **16** (0.5–2.0 equiv.)

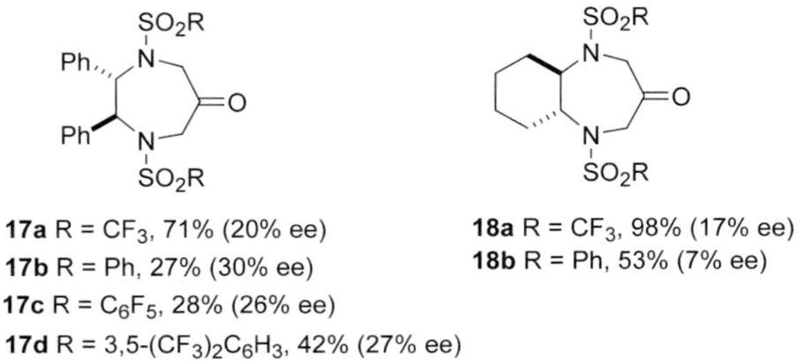

17a R = CF$_3$, 71% (20% ee)
17b R = Ph, 27% (30% ee)
17c R = C$_6$F$_5$, 28% (26% ee)
17d R = 3,5-(CF$_3$)$_2$C$_6$H$_3$, 42% (27% ee)

18a R = CF$_3$, 98% (17% ee)
18b R = Ph, 53% (7% ee)

Scheme 3.15 Epoxidation of stilbene with ketones **17** and **18** (1.0 equiv.)

3.3 C₂ Symmetric Binaphthyl-based and Related Ketones

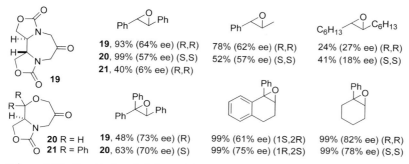

Scheme 3.16 Epoxidation of olefins with ketones **19–21** (1.0 equiv.)

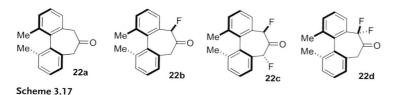

Scheme 3.17

[23, 24]. High reactivity and enantioselectivity were obtained with difluoroketones **22c** and **22d** (Scheme 3.18). Up to 94% ee was obtained for stilbene. In 2002, related fluorinated binaphthyl ketones were also reported by Behar (Scheme 3.19) [25]. Among these ketones, α,α′-difluoroketone **23c** gave the best results for the epoxidation of *trans*-β-methylstyrene (100% yield, 86% ee).

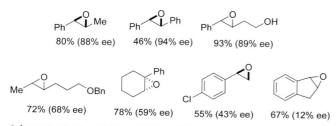

Scheme 3.18 Epoxidation of olefins with ketone **22c** (0.3 equiv.) [22]

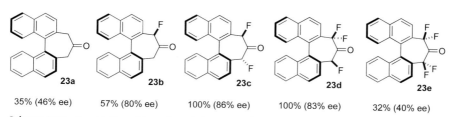

Scheme 3.19 Epoxidation of *trans*-β-methylstyrene with ketone **23** (0.1 equiv.)

3.4
Ammonium Ketones

In their elegant studies on reaction parameters for the ketone-catalyzed epoxidation, Denmark and coworkers showed that 4-oxopiperidinium salt **24** was an effective catalyst (Scheme 3.20) [5h]. In this ketone, the ammonium ion acts not only as an electron-withdrawing group to inductively activate the carbonyl and to suppress Baeyer-Villiger oxidation, but also as a phase transfer mediator to facilitate the partitioning of the ketone and its dioxirane between the organic and aqueous phases. The phase transfer ability of **24** can be adjusted by varying the alkyl groups on the nitrogen. Based on this work, a number of chiral ammonium ketones were investigated for the epoxidation (Scheme 3.20) [26]. In the initial studies with ammonium ketones **25** and **26** [5h, 5l, 6a], low reactivity was observed probably due to the steric congestion near the carbonyl. Ketone **26** gave 34% ee for trans-β-methylstyrene and 58% ee for 1-phenylcyclohexene. To further activate the carbonyl and suppress the Baeyer-Villiger oxidation, bis(ammonium) ketones **27–30** were evaluated for the epoxidation. These ketones were found to be effective catalysts. For example, >95% conversion was obtained with 10 mol% of **29** and **30** for the epoxidation of trans-β-methylstyrene. Up to 40% ee was obtained with ketone **28**.

The high flexibility of the 7-membered ring of ketones **27–30** could be one of the contributing factors to the low enantioselectivities. Therefore, a more rigid tropinone based ammonium ketone **31** was then investigated (Scheme 3.21) [6a, 22, 23]. The

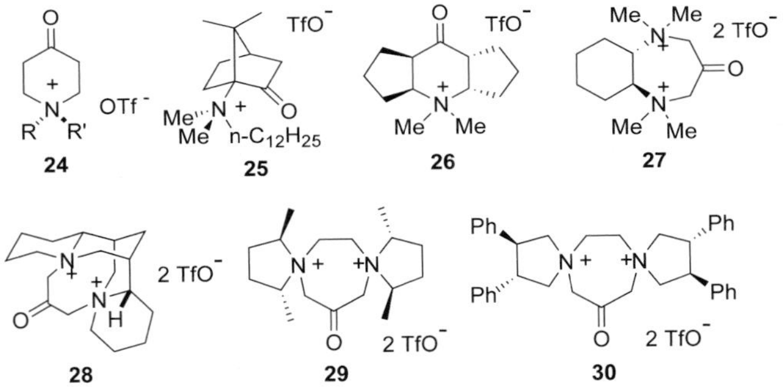

Scheme 3.20

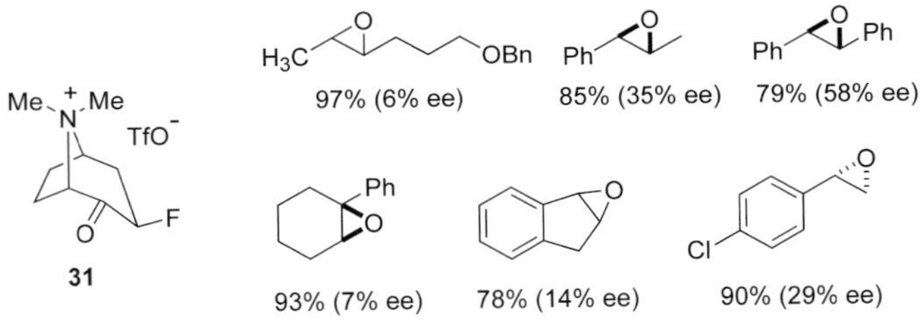

Scheme 3.21 Epoxidation of olefins with ketone **31** (0.1 equiv.) [22]

fluorine was introduced to further activate the carbonyl, and it was found to be highly reactive. Up to 58% ee was obtained for *trans*-stilbene.

3.5
Bicyclo[3.2.1]octan-3-ones

In 1998, Armstrong and coworkers reported the uncharged tropinone-based ketone **32** (Scheme 3.22) [27, 28]. A combination of the bridgehead nitrogen at the β-position and the fluorine atom at the α-position made this ketone a highly reactive catalyst, yielding up to 83% ee for phenylstilbene (Scheme 3.22). Subsequent studies showed that the enantioselectivity could be further increased by replacing the fluorine of **32** with an acetate and/or replacing the bridgehead nitrogen with an oxygen (Scheme 3.23) [28–30]. Up to 98% ee_{max} [31] was obtained for phenylstilbene with ketone **35**. The α-fluorotropinone was also immobilized on silica materials. Similar enantioselectivity was obtained with the supported catalyst compared with the non-supported ketone catalyst [32].

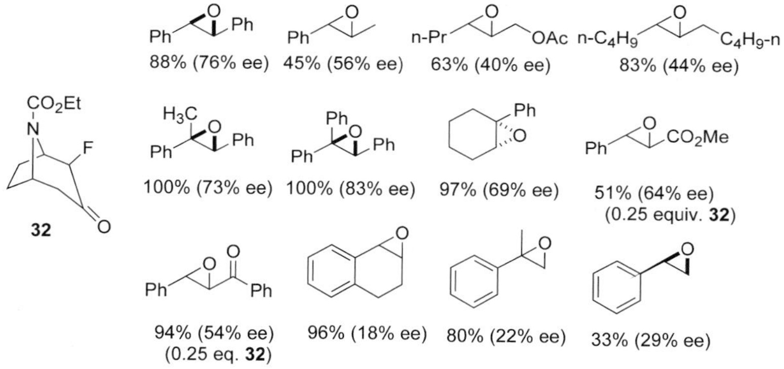

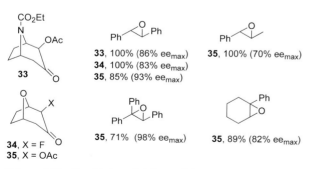

Scheme 3.22 Epoxidation of olefins with ketone **32** (0.1 equiv.) [28]

Scheme 3.23 Epoxidation of olefins with ketone **33–35** (0.2 equiv.)

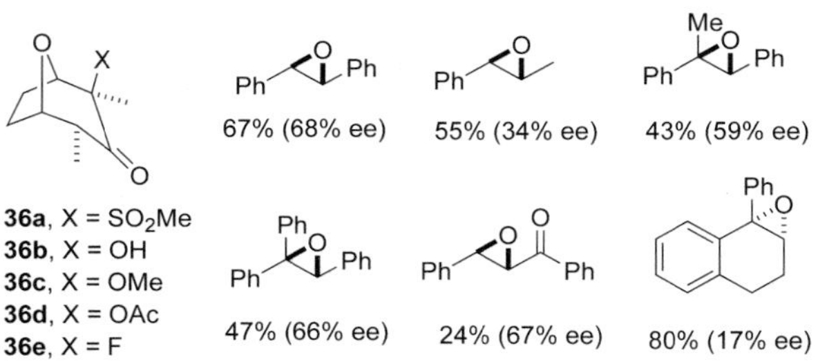

Scheme 3.24 Epoxidation of olefins with ketone **36e** (0.3 equiv.)

36a, X = SO₂Me
36b, X = OH
36c, X = OMe
36d, X = OAc
36e, X = F

67% (68% ee)
55% (34% ee)
43% (59% ee)
47% (66% ee)
24% (67% ee)
80% (17% ee)

A number of 2-substituted-2,4-dimethyl-8-oxabicyclo[3.2.1]octan-3-ones **36** were also investigated for epoxidation by Klein and coworkers (Scheme 3.24) [33]. Among these, the fluoro ketone **36e** was found to be most reactive. The epoxidation of several olefins was evaluated with **36e**, and up to 68% ee was obtained for stilbene (Scheme 3.24).

3.6
Carbohydrate Based and Related Ketones

In 1996, a fructose-derived ketone (**39**) was reported to be a highly effective epoxidation catalyst for a wide range of olefins (Scheme 3.25) [34]. The synthesis of ketone **39** can be readily achieved in two steps from D-fructose by ketalization and oxidation [34–37]. The synthesis of the enantiomer of ketone **39** can be performed similarly from L-fructose, which can be prepared from readily available L-sorbose based on a literature procedure [35, 38]. Similar enantioselectivities were observed for the epoxidation with ketone **ent-39** prepared in this way.

Ketone **39** is one member of a broad class of ketones designed based on the following general considerations (Scheme 3.26): (1) placement of the stereogenic centers close to

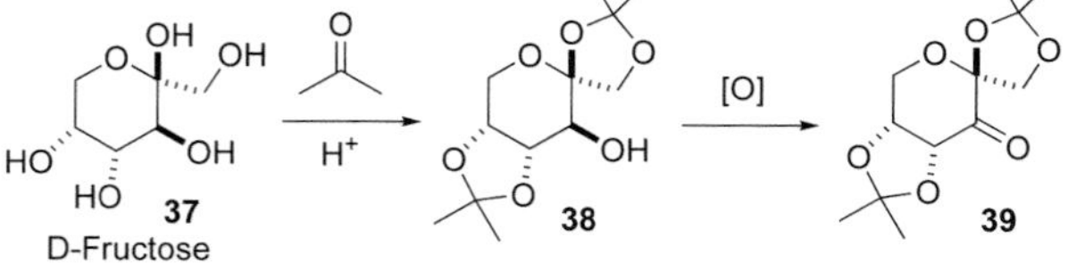

Scheme 3.25

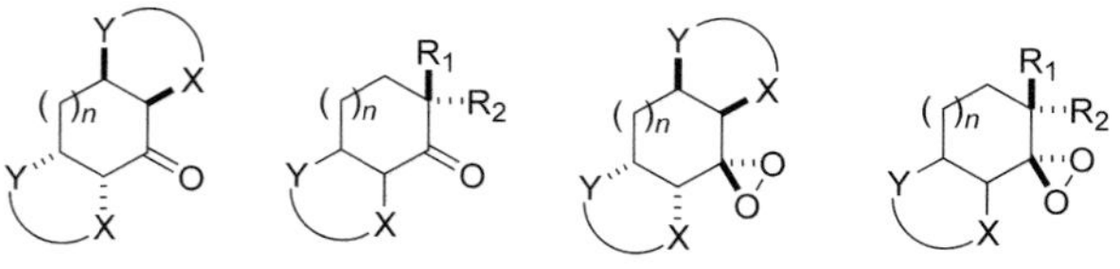

Scheme 3.26

3.6 Carbohydrate Based and Related Ketones

the reacting carbonyl to have an effective stereochemical interaction between substrate and catalyst; (2) use of fused ring(s) α to the carbonyl group to minimize the epimerization of the stereogenic centers; (3) control of the approach of an olefin to the reacting dioxirane by sterically blocking one face or by a C_2 or pseudo C_2 symmetric element; (4) introduction of inductively withdrawing substituents to activate the carbonyl.

Controlling the reaction pH is often crucial for the epoxidation with dioxiranes generated *in situ* [5a, 5h]. Earlier ketone-mediated epoxidations were usually performed at pH 7–8 [5], since Oxone rapidly autodecomposes at high pH [39, 40], thus decreasing the epoxidation efficiency. Therefore, the epoxidation with ketone **39** was initially performed at pH 7–8. While high enantioselectivies (>90% ee) were obtained for a variety of *trans*-disubstituted and trisubstituted olefins [34], ketone **39** decomposed very rapidly at this pH and an excess amount of ketone was required for good conversion of the substrate. The Baeyer-Villiger reaction resulting from intermediate **40** was assumed to be one of the possible decomposition pathways for ketone **39**, although the corresponding lactones **43** and **44** had not been isolated, presumably due to their facile hydrolysis under the reaction conditions (Scheme 3.27). It was surmised that raising the reaction pH could favor the formation of anion **41** and subsequent formation of the desired dioxirane **42**, thus reducing the competition of the undesired Baeyer-Villiger oxidation. It was also hoped that ketone **39** could react with Oxone fast enough to override the rapid autodecomposition of Oxone at high pH.

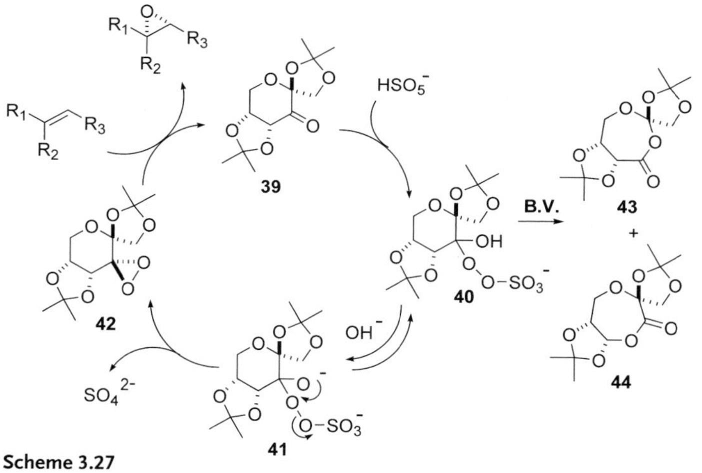

Scheme 3.27

Based on these assumptions, the epoxidation of *trans*-β-methylstyrene was then performed under a different reaction pH [35, 41]. As shown in Figure 3.1, the reaction pH displayed a profound impact on the substrate conversion, and a higher pH was indeed beneficial to the catalyst efficiency, with more than a 10-fold increase in conversion from a lower pH (7–8) to a higher pH (>10). This dramatic pH effect led to a catalytic asymmetric epoxidation process, consequently enhancing the potential of ketone **39** for practical use. Typically, the epoxidation is performed at a pH of

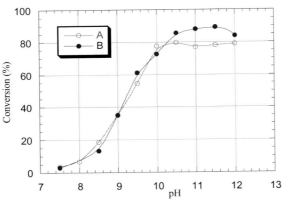

Fig. 3.1 Plot of the conversion of trans-β-methylstyrene against pH using ketone **39** (0.2 equiv.) in H_2O–CH_3CN (1:1.5, v/v) (**A**) or H_2O–CH_3CN–DMM (2:1:2, v/v) (**B**) [35, 41]

around 10.5 by adding either K_2CO_3 or KOH as the reaction proceeds. Maintaining a steady pH throughout the reaction is very important for the epoxidation.

Comparative studies showed greater conversions were also obtained with acetone and trifluoroacetone at higher pH (Figures 3.2 and 3.3) [5m, 35, 42]. For example, the conversion increased from ~5% to 80% when the pH was changed from 7.5 to 10 for trans-β-methylstyrene when 5 mol% of CF_3COCH_3 was used as the catalyst (Figure 3.3). Higher reaction pH could enhance the nucleophilicity of Oxone, thus increasing Oxone's reactivity towards acetone and trifluoroacetone. Therefore, the increased epoxidation efficiency at higher pH for ketone catalyst **39** is not only due to the reduction of the Baeyer-Villiger reaction, but also a result of increased reaction between ketone **39** and Oxone. A clearer mechanistic understanding awaits further study.

The generality of this asymmetric epoxidation was subsequently explored with a variety of olefins with a catalytic amount of ketone **39** (typically 20–30 mol%). Some of the epoxidation results are summarized in Schemes 3.28–3.34. High enantioselectivities can be obtained for a wide variety of unfunctionalized trans- and trisubstituted olefins (Schemes 3.28 and 3.29) [35]. Significantly, high ee can be obtained with trans-7-tetradecene, indicating that this epoxidation is quite general for simple

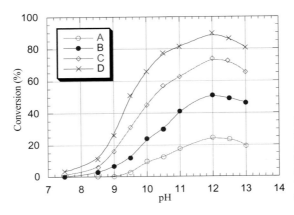

Fig. 3.2 Plot of the conversion of trans-β-methylstyrene against pH using acetone (3 equiv.) as catalyst in H_2O–CH_3CN (1:1.5, v/v). Samples were taken at 0.5 h (**A**), 1.0 h (**B**), 1.5 h (**C**), and 2.0 h (**D**) to determine the conversion [35]

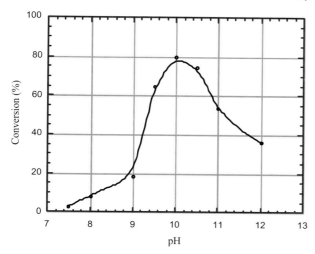

Fig. 3.3 Plot of the conversion of *trans*-β-methylstyrene against pH using CF$_3$COCH$_3$ (0.05 equiv.) as catalyst [42]

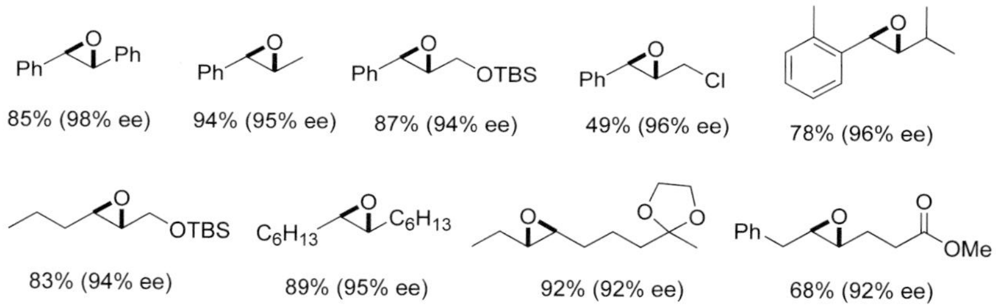

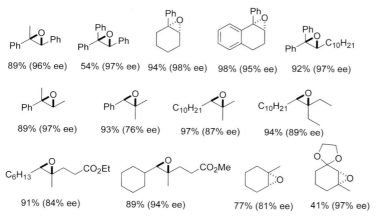

Scheme 3.28 Epoxidation of *trans*-disubstituted olefins with ketone **39** [35]

Scheme 3.29 Epoxidation of trisubstituted olefins with ketone **39** [35]

trans-olefins. A variety of 2,2-disubstituted vinyl silanes can also be epoxidized with high enantioselectivity (Scheme 3.30), and the resulting epoxide may be desilylated using TBAF to provide 1,1-disubstituted terminal epoxides with high enantioselectivity [43]. Hydroxyalkenes are also effective substrates (Scheme 3.31) [44], which is complementary to the Sharpless asymmetric epoxidation since high ee values can also be obtained for homoallylic and bishomoallylic alcohols. Vinyl epoxides can be obtained with high ee values by the regio- and enantioselective epoxidation of conju-

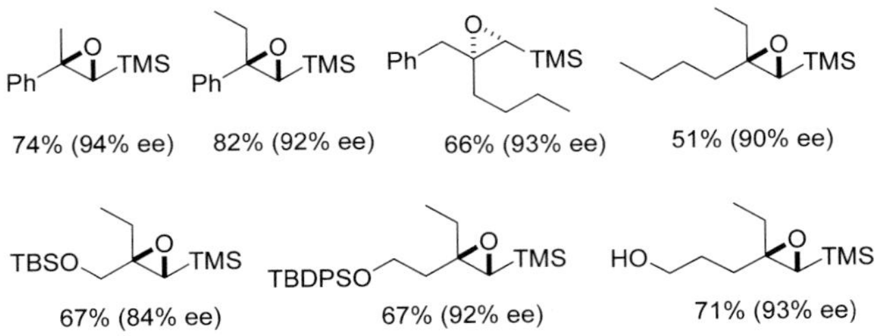

Scheme 3.30 Epoxidation of 2,2-disubstituted vinylsilanes with ketone **39** [43]

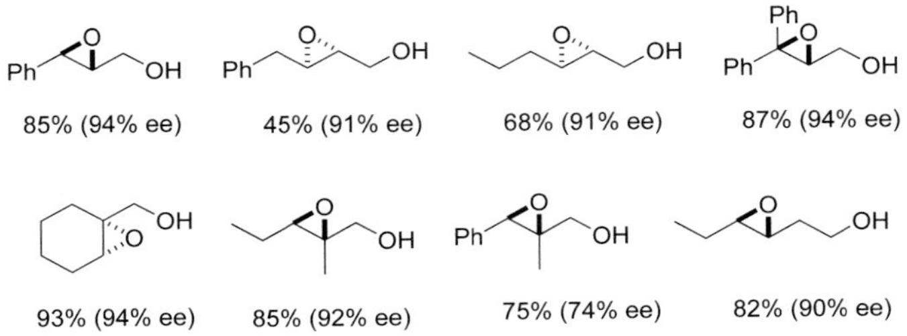

Scheme 3.31 Epoxidation of hydroxyalkenes with ketone **39** [44]

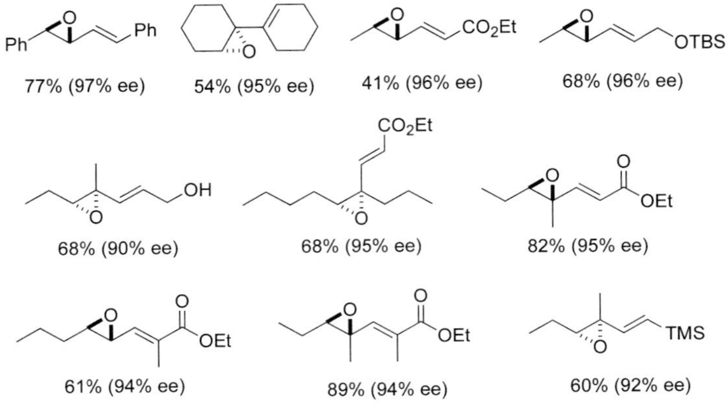

Scheme 3.32 Epoxidation of conjugated dienes with ketone **39** [45]

gated dienes (Scheme 3.32) [45]. Upon epoxidation of one olefin, the remaining olefin is inductively deactivated by the first epoxide, thus monoepoxides can be formed predominately if the amount of catalyst is properly controlled. For unsymmetrical dienes, the regioselectivity can be controlled by using steric and/or electronic effects. Conjugated enynes can be highly chemo- and enantioselectively epoxidized to produce chiral propargyl epoxides (Scheme 3.33) [46, 47].

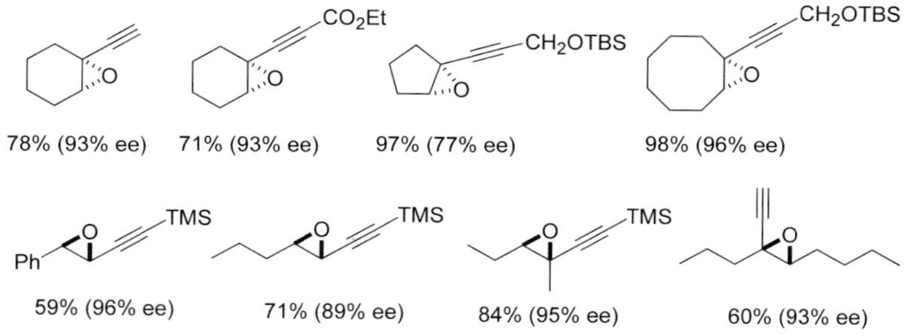

CO2Et
CH2OTBS
CH2OTBS
78% (93% ee)
71% (93% ee)
97% (77% ee)
98% (96% ee)
TMS
TMS
TMS
Ph
59% (96% ee)
71% (89% ee)
84% (95% ee)
60% (93% ee)

Scheme 3.33 Epoxidation of enynes with ketone **39** [46, 47]

Some silyl enol ethers could also be epoxidized to give enantiomerically enriched α-hydroxy ketones (Scheme 3.34) [48, 49]. However, some α-hydroxy ketones are prone to racemization or dimerization. In addition, some α-hydroxy ketones formed during the reaction could also act as a catalyst for the epoxidation, thus affecting the overall enantioselectivity. Therefore, silyl enol ethers are generally less effective substrates than enol esters under the current reaction conditions. When enol esters are epoxidized, the corresponding enol ester epoxides are obtained with high ee (Scheme 3.34) [49, 50]. From the resulting chiral enol ester epoxides, optically active α-hydroxy or α-acyloxy ketones can be obtained by hydrolysis or stereoselective rearrangement. As illustrated in Scheme 3.35 [50, 51], one enantiomer of an epoxide can be converted into either enantiomer of the α-acyloxy ketone by judicious choice of reaction conditions [49–52].

Ph
OH
BzO
BzO
BzO
80% (90% ee)
82% (93% ee)
79% (80% ee)
87% (91% ee)
BzO
BzO
AcO
AcO
Ph
Ph
Ph
82% (95% ee)
92% (88% ee)
66% (91% ee)
46% (91% ee)

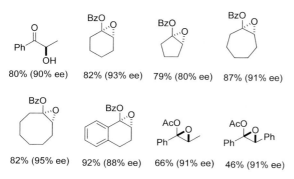

Scheme 3.34 Epoxidation of silyl enol ethers and esters with ketone **39** [49, 50]

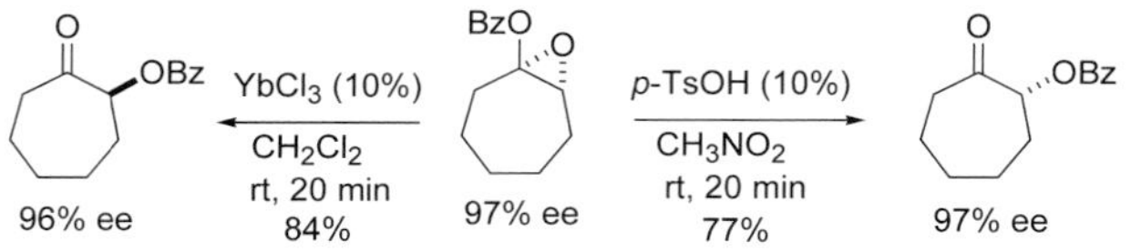

Scheme 3.35

Understanding the transition state of the dioxirane-mediated epoxidation is extremely important for predicting the stereochemical outcome of the reaction and designing new ketone catalysts. The two extreme transition state geometries (spiro and planar) are shown in Figure 3.4 [4c, 4d, 11, 12, 24, 34, 35, 53–58]. Baumstark and coworkers found that *cis*-hexenes were 7–9-fold more reactive than the corresponding *trans*-hexenes while using dimethyldioxirane, and proposed that a spiro transition state was consistent with the observed reactivity difference between *cis*- and *trans*-hexenes [53, 54]. The spiro transition state has also been shown to be the optimal transition state for oxygen atom transfer from dimethyldioxirane to ethylene from computational studies [24, 55–58]. The favoring of a spiro transition state over a planar one could be due to the stabilizing interaction between the non-bonding orbital (lone pair) of the dioxirane oxygen and the π^* orbital of the alkene in the spiro transition state (stereoelectronic origin) [55–58]. Such stabilizing orbital interactions are not geometrically feasible in the planar transition state (Figure 3.4).

The stereochemistry of epoxides generated by chiral dioxiranes provides the opportunity to further address the transition state. The dioxirane derived from 39 has two diastereomeric oxygens. The equatorial oxygen is likely to be sterically more accessible for the epoxidation. Our studies show that while the epoxidation of *trans*- and trisubstitued olefins with ketone 39 proceeds mainly through spiro **A**, planar **B** is also competing (Figure 3.5) [34, 35, 43–47, 49, 50, 59]. Spiro **A** and planar **B** give the opposite stereochemistry for the epoxide product, thus their competition will affect the ee obtained for epoxidation. Studies have shown that the extent of involvement of

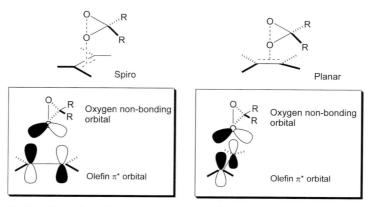

Fig. 3.4 The spiro and planar transition states for the dioxirane epoxidation of olefins

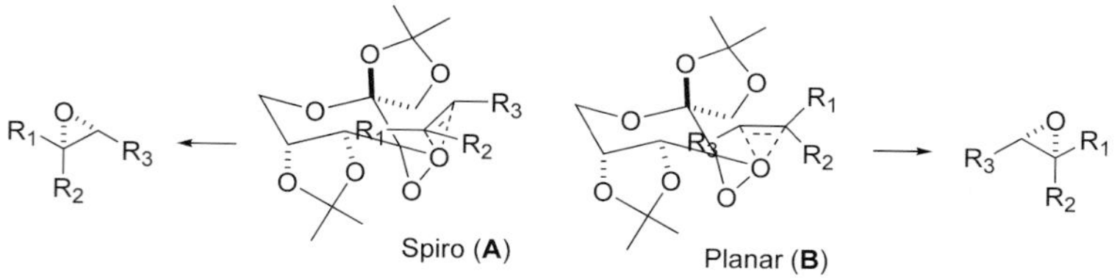

Fig. 3.5 The spiro and planar transition states for the epoxidation with ketone **39**

the planar transition state is dependent on the substituents on the olefins [35]. Generally speaking, higher enantioselectivity can be obtained by decreasing the size of R_1 (favoring spiro **A**) and/or increasing the size of R_3 (disfavoring planar **B**) [60].

The transition state model suggests that certain racemic olefins might be able to be kinetically resolved. Studies showed that a number of 1,6 and 1,3-disubstituted cyclohexenes could indeed efficiently be resolved with ketone **39** (Scheme 3.36) [61]. A rationalization for the kinetic resolution of 1,6-disubstituted cyclohexene using ketone **39** is shown in Scheme 3.37. Spiro **C** and **D** represent the major transition states for the epoxidation of each enantiomer of the racemic olefin. The destabilizing steric interaction between R_2 and one of the dioxirane oxygens in spiro **D** disfavors this transition state, thus the epoxidation of the corresponding enantiomer proceeds at a lower rate. This kinetic resolution not only provides a valuable route to preparing certain chiral intermediates but also further validates the transition state model.

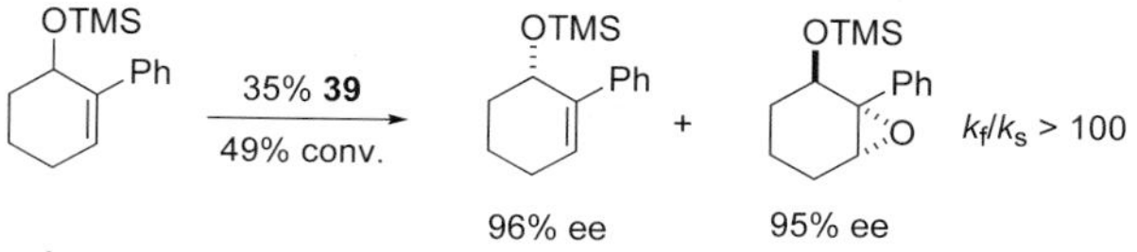

Scheme 3.36

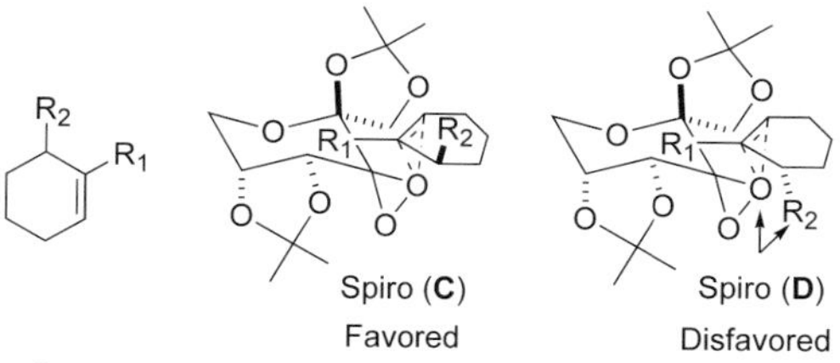

Scheme 3.37

In almost every case, the dioxirane is generated using potassium peroxymonosulfate ($KHSO_5$) as oxidant (Scheme 3.1) [62–64]. The effectiveness of potassium peroxymonosulfate as oxidant is probably due to the fact that the sulfate moiety is a good leaving group, which facilitates the formation of the dioxirane. It is of particular interest whether oxidants with poorer leaving groups than sulfate are capable of generating dioxiranes. Hydrogen peroxide (H_2O_2) is among the highly desirable oxidants since it has a high active oxygen content and its reduction product is water [65]. In 1999, it was reported that indeed H_2O_2 could be used as a primary oxidant in combination with a nitrile for the epoxidation with the fructose-derived ketone **39** (Scheme 3.38) [66–68]. Among various nitriles tested, CH_3CN and CH_3CH_2CN proved to be the most effective for the epoxidation [67]. Under these conditions, peroxyimidic acid **45** (an analogous intermediate of Payne oxidation [69]) is likely to be the active oxidant that reacts with the ketone to form the dioxirane. High yields and ee have been obtained for a variety of olefins with this RCN-H_2O_2 oxidant (Scheme 3.39) [67]. This epoxidation system proceeds under mild conditions, and less solvents and salts are involved in the reaction. In addition to ketone **39**, some other ketones can be effective catalysts using the RCN-H_2O_2 system. For example, a variety of olefins can be efficiently epoxidized with 10–30% trifluoroacetone [42].

Fructose-derived ketone **39** is readily available, and is a highly general and enantioselective catalyst for the epoxidation of *trans*- and trisubstituted olefins. Its utilization in synthesis has been reported by other researchers [70]. For example, recently Corey and coworkers reported that (*R*)-2,3-dihydroxy-2,3-dihydrosqualene (**47**) was enantio-

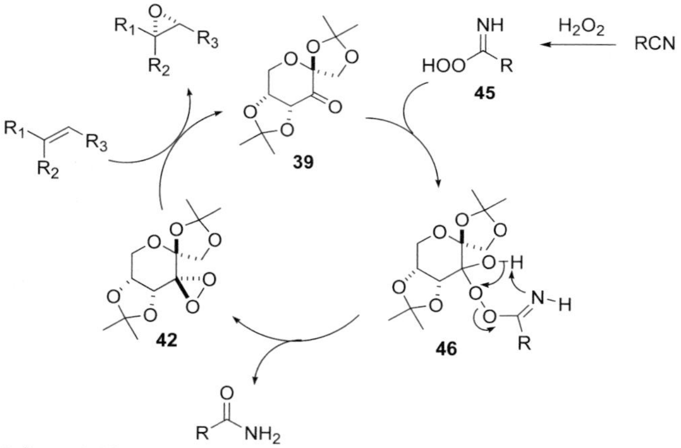

R1
R3
R2
O
NH
HOO
R
45
H2O2
RCN
39
42
46
N-H
O-H
NH2

Scheme 3.38

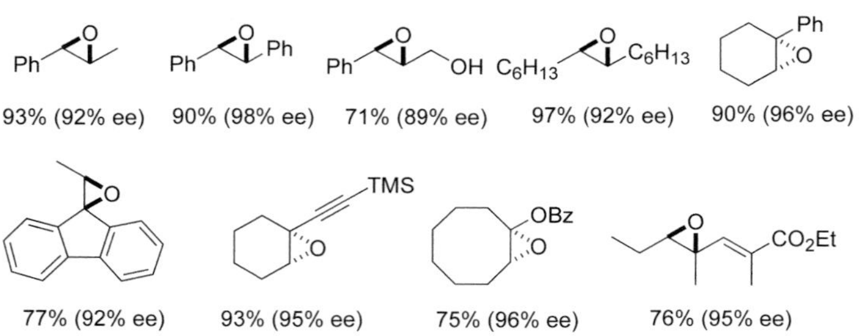

Ph
93% (92% ee)
90% (98% ee)
OH
71% (89% ee)
C6H13
97% (92% ee)
90% (96% ee)
TMS
OBz
CO2Et
77% (92% ee)
93% (95% ee)
75% (96% ee)
76% (95% ee)

Scheme 3.39 Epoxidation of olefins with ketone **39** (0.1–0.3 equiv.) using CH_3CN-H_2O_2

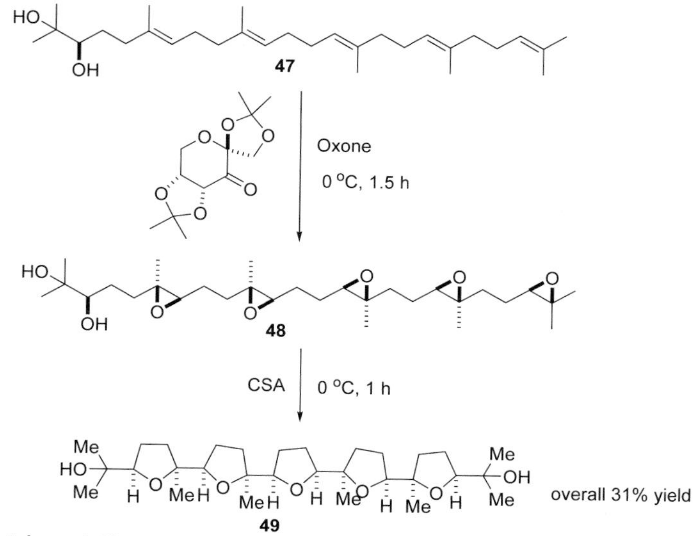

3.6 Carbohydrate Based and Related Ketones | 69

Scheme 3.40

selectively epoxidized with ketone **39** to give pentaepoxide **48**, which was subsequently converted into pentaoxacyclic compound **49** in 31% overall yield (Scheme 3.40) [70d].

To further probe and understand the structural requirements for the chiral ketone catalyzed epoxidation, a variety of ketone catalysts were prepared from various carbohydrates (such as arabinose, glucose, fructose, mannose, and sorbose) and investigated [71]. The size of the groups attached to the ketals is important. Generally speaking, the smaller the group, the higher the reactivity and selectivity (**39** vs. **50**). The rigid 5-membered spiro ketal of **39** was structurally better than the 6-membered ketal of **51** and the acyclic groups of **52** and **53** for both reactivity and enantioselectivity [72]. The carbocyclic analogue **54** gave lower conversion and ee than **39**, suggesting that the oxygen of the pyranose ring is beneficial to catalysis [73]. Studies also showed that the 5-membered ketones were poorer catalysts, largely

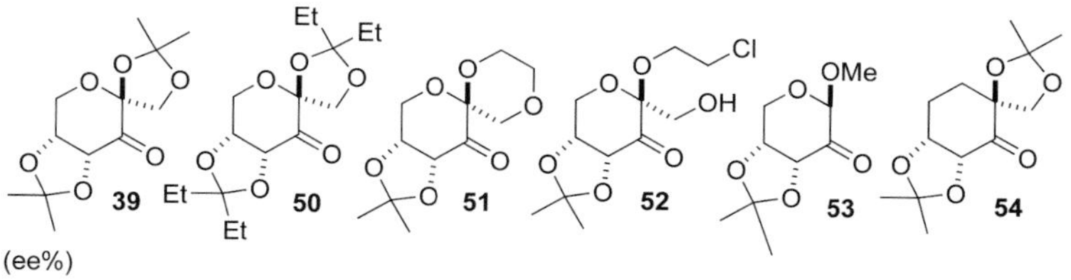

conv.% (ee%)	39	50	51	52	53	54
Ph–CH=CH$_2$	93 (92)	32 (86)	44 (61)	8 (65)	15 (59)	61 (87)
Ph–CH=CH–Ph	75 (97)	16 (96)	34 (90)	2 (nd)	10 (88)	10 (88)

Scheme 3.41 Epoxidation of olefins with ketones **39** and **50–54** (0.3 equiv.)

due to their facile Baeyer-Villiger oxidative decomposition caused by the 5-membered ring strain [71].

To further understand how structure affects the stability and reactivity of the ketone, and to search for more robust ketone catalysts, ketone **55** was prepared and investigated (Scheme 3.42) [74]. It was hoped that the replacement of the ketal of **39** with a more electron withdrawing oxazolidinone would reduce the undesired Baeyer-Villiger oxidation (Scheme 3.27), thus providing a more stable ketone catalyst. Ketone **55** was indeed found to be highly active. The catalyst loading can be reduced to 5 mol% and even 1 mol% in some cases.

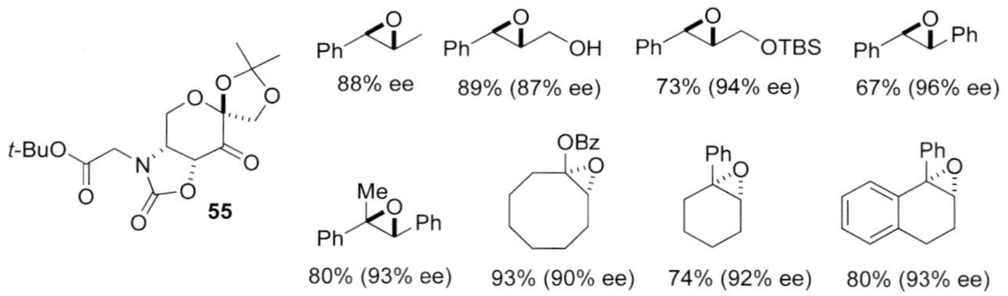

Scheme 3.42 Epoxidation of olefins with ketone **55** (0.01–0.05 equiv.)

Being electrophilic reagents, dioxiranes epoxidize electron-deficient olefins sluggishly. An effective catalyst for this class of olefin requires high structural stringency for being both highly active and enantioselective. Ketone **56**, readily available from **39**, was found to be effective for the epoxidation of α,β-unsaturated esters [75]. High ee results and good yields can be obtained for a variety of α,β-unsaturated esters using 20–30 mol% ketone **56** (Scheme 3.43).

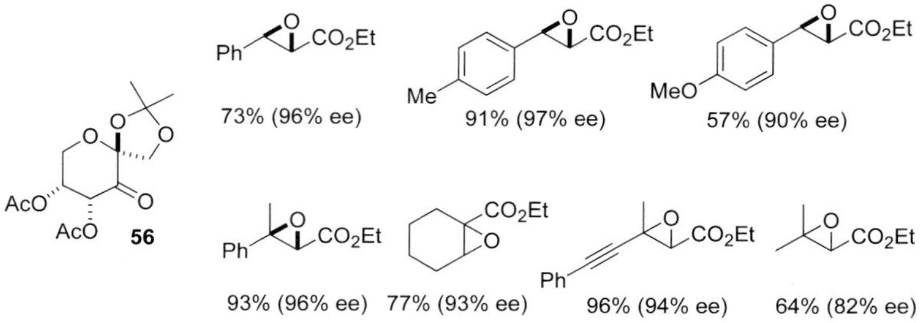

Scheme 3.43 Epoxidation of α,β-unsaturated esters with ketone **56** (0.2–0.3 equiv.)

In efforts to expand the scope of the ketone catalyzed epoxidation, glucose-derived ketone **57** was reported to be an effective catalyst for the epoxidation of *cis*-olefins in 2000 (Scheme 3.44). High ee can be obtained for a number of both acyclic and cyclic olefins (Scheme 3.44) [76–79]. The epoxidation is stereospecific and no isomeriza-

3.6 Carbohydrate Based and Related Ketones

tion has been observed in the epoxidation of acyclic systems. In addition, ketone **57** provides encouragingly high ee for certain terminal olefins [77, 79, 80]. The studies suggest that the stereodifferentiation for the epoxidation of *cis-* and terminal olefins with ketone **57** probably involves electronic interactions. It appears that there is an attractive interaction between the R_π group of the olefin and the oxazolidinone moiety of the ketone catalyst in the transition state (Scheme 3.45). As a result, spiro transition state **E** overrides the competing spiro **F**, yielding high enantioselectivity [76, 77, 79, 80]. A precise mechanistic understanding of the origin of the enantioselectivity with ketone **57** awaits further investigation. The scope for the substrate for the epoxidation of **57** is expected to be further expanded in the future.

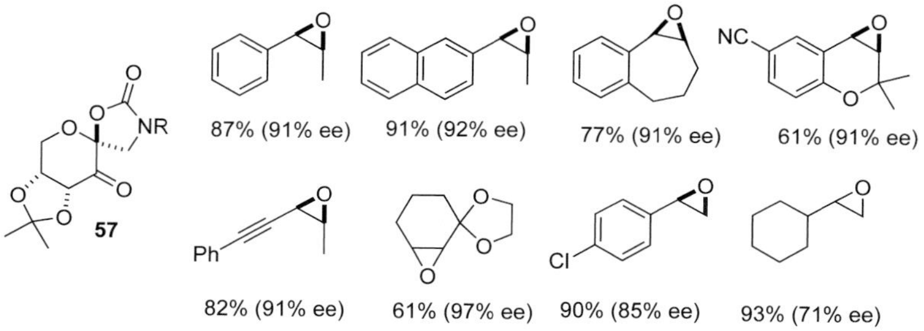

Scheme 3.44 Epoxidation of olefins with ketone **57** (0.15–0.30 equiv.) [76, 80]

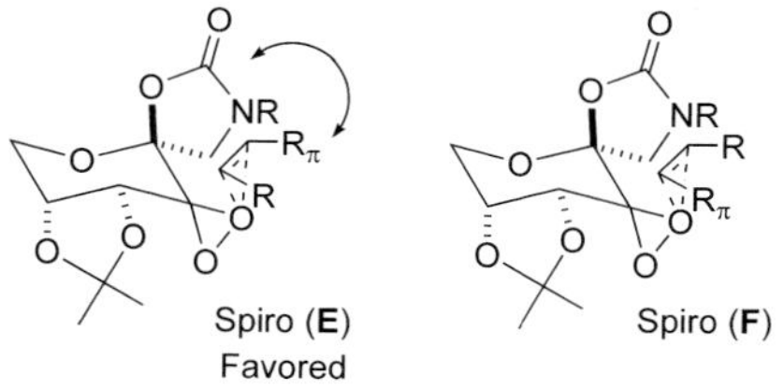

Scheme 3.45

In 2002, Shing and coworkers reported three glucose derived ketones **58–60** as epoxidation catalysts (Scheme 3.46) [81]. Ketone **58** was found to be more effective than **59** and **60**, and up to 71% ee was obtained for stilbene with this ketone. In 2003, Shing and coworkers reported their studies on epoxidation with the L-arabinose derived ketones **61–64** (Scheme 3.47) [72]. Up to 90% ee was obtained for stilbene with **64**. Further studies showed that a higher yield of the epoxidation could be obtained with the ester substituted ketones **65–67**, and up to 68% ee was obtained for phenylstilbene (Scheme 3.48) [72].

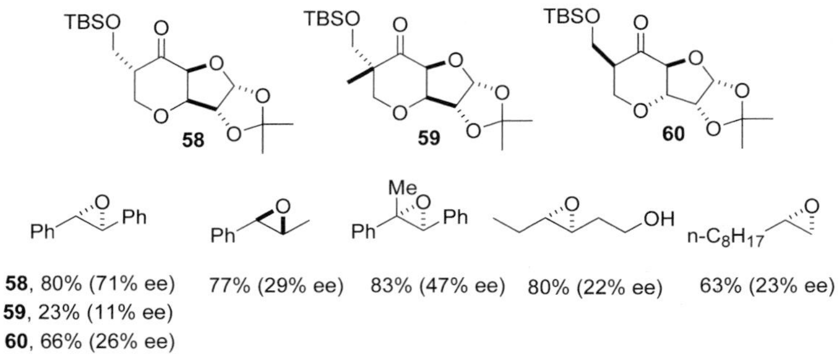
TBSO
58
59
60
Me
Ph
OH
n-C8H17
58, 80% (71% ee)
59, 23% (11% ee)
60, 66% (26% ee)
77% (29% ee)
83% (47% ee)
80% (22% ee)
63% (23% ee)

Scheme 3.46 Epoxidation of olefins with ketones **58–60** (0.1 equiv.)

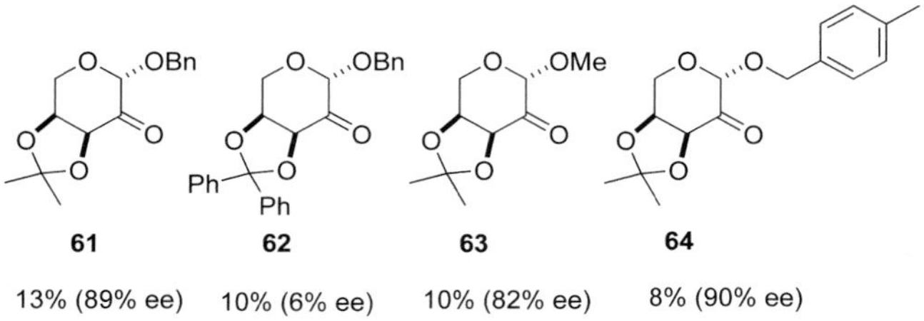
OBn
OMe
Ph
61
62
63
64
13% (89% ee)
10% (6% ee)
10% (82% ee)
8% (90% ee)

Scheme 3.47 Epoxidation of stilbene with ketones **61–64** (0.3 equiv.)

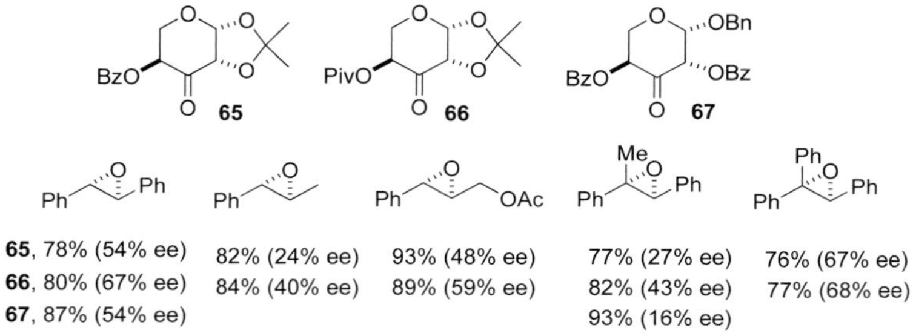
BzO
PivO
OBn
OBz
65
66
67
Me
Ph
OAc
65, 78% (54% ee)
66, 80% (67% ee)
67, 87% (54% ee)
82% (24% ee)
84% (40% ee)
93% (48% ee)
89% (59% ee)
77% (27% ee)
82% (43% ee)
93% (16% ee)
76% (67% ee)
77% (68% ee)

Scheme 3.48 Epoxidation of olefins with ketones **65–67** (0.1 equiv.)

3.7
Carbocyclic Ketones

Ketones **39** and **57** utilize a fused ring and a quaternary carbon α to the carbonyl group to place the stereogenic centers close to the reacting carbonyl and to minimize potential epimerization of the chiral elements (Scheme 3.49). Related ketones containing two fused rings at each side of the carbonyl group were also investigated (Scheme 3.49). In 1997, the pseudo C_2 symmetric ketone **68**, prepared from quinic acid, was reported as a member of this class of ketones [82, 83]. Some examples of epoxidation with ketone **68** (R = CMe_2OH) are shown in Scheme 3.50. Ketone **68** is a

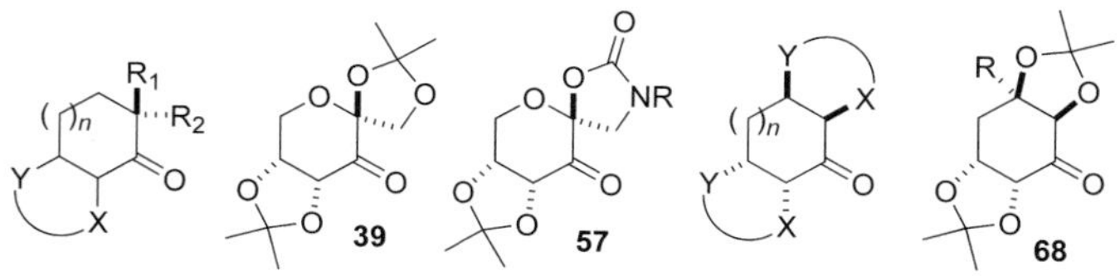

3.7 Carbocyclic Ketones

Scheme 3.49

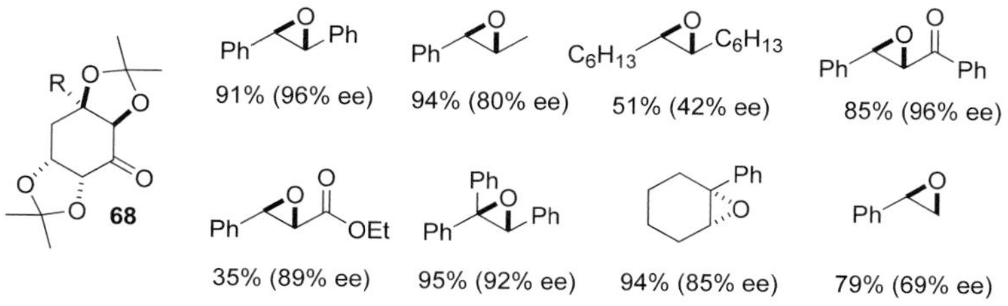

Scheme 3.50 Epoxidation of olefins with ketone **68** (R = CMe$_2$OH) (0.05–0.1 equiv.)

very active catalyst, and certain electron deficient olefins can also be epoxidized, indicating that the dioxirane is very electrophilic. Generally speaking, ketone **68** is less enantioselective than **39** for the epoxidation of *trans-* and trisubstituted olefins.

Two C_2-symmetric 5-membered ring ketones **69** and **70** were reported by Armstrong and coworkers (Scheme 3.51) [84, 85]. Studies with these ketones showed that the 5-membered ring is more prone to Baeyer-Villiger oxidation, and the activation of the carbonyl by electron-withdrawing substituents is important for the epoxidation.

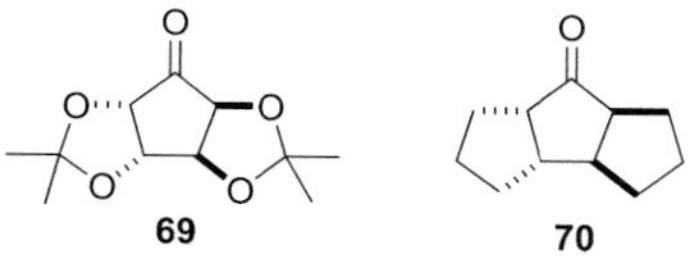

Scheme 3.51

In ketones **39**, **57**, and **68**, the ketals or oxazolidinone are placed at α-positions of the carbonyl. Studies with ketones **71–73** showed that moving the ketal from the α- to β-positions lowered the enantioselectivity for the epoxidation, suggesting that placing the stereogenic centers close to the carbonyl is important for an efficient stereochemical communication between the substrate and the catalyst (Scheme 3.52) [73, 86]. Adam and coworkers also reported their studies on ketones **71** and **74**, and up to 87% ee was obtained with **71** (Scheme 3.53) [87].

In 1998, Yang and coworkers reported a ketone with a quaternary carbon at the α-position at one side and a substituent at the β-position of the other side of the carbonyl group (**75**) (Scheme 3.54) [88]. Studies on a series of *meta-* and *para-*substituted *trans-*stilbenes with **75b** showed that the ee of the epoxide varied with the substituent

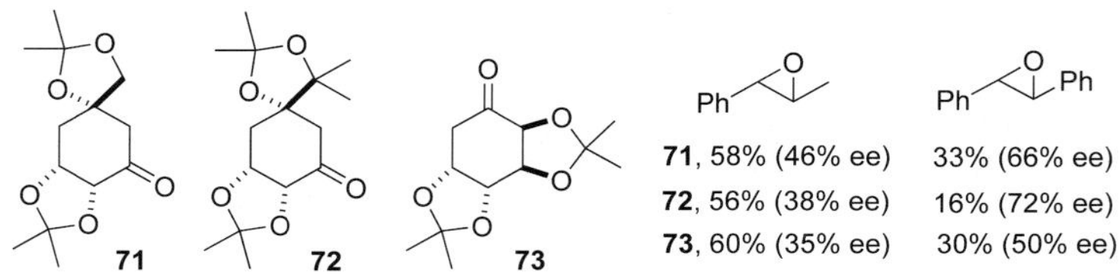

Scheme 3.52 Epoxidation of olefins with ketones **71–73** (0.3–0.5 equiv.) [73, 86]

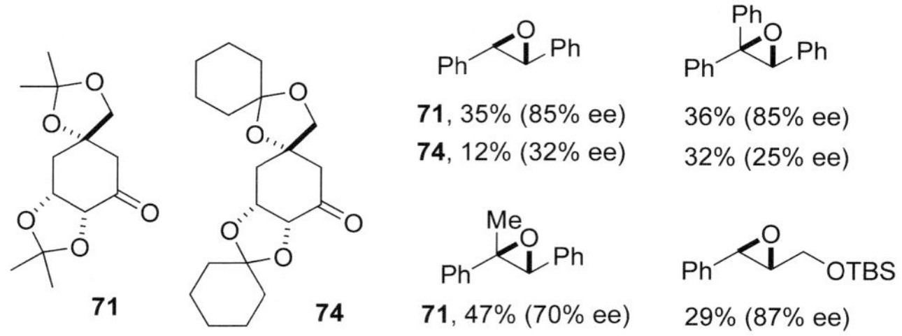

Ketones **71**, **72**, **73**.

Ph–epoxide–Me: **71**, 58% (46% ee); **72**, 56% (38% ee); **73**, 60% (35% ee)

Ph–epoxide–Ph: 33% (66% ee); 16% (72% ee); 30% (50% ee)

Scheme 3.53 Epoxidation of olefins with ketones **71** and **74** (1.0–3.0 equiv.) [87]

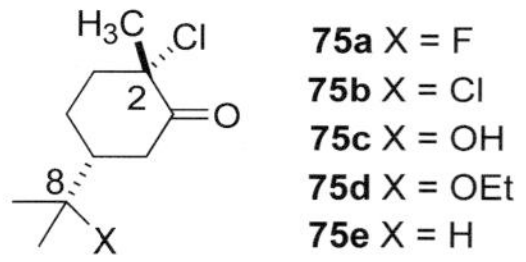

Ketones **71**, **74**.

Ph–epoxide–Ph: **71**, 35% (85% ee); **74**, 12% (32% ee)

Ph(Ph)–epoxide–Ph: 36% (85% ee); 32% (25% ee)

Ph–epoxide(Me)–Ph: **71**, 47% (70% ee)

Ph–epoxide–OTBS: 29% (87% ee)

Scheme 3.54

Ketone **75**: H$_3$C, Cl at C2; X at C8.
- **75a** X = F
- **75b** X = Cl
- **75c** X = OH
- **75d** X = OEt
- **75e** X = H

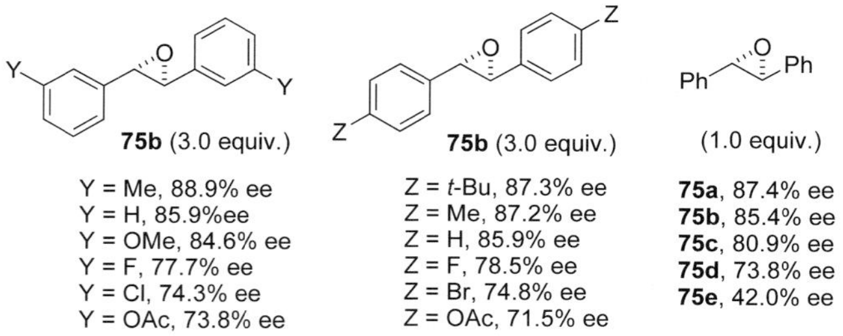

Scheme 3.55 Epoxidation of stilbenes with ketone **75**

Y-substituted stilbene epoxide, **75b** (3.0 equiv.):
- Y = Me, 88.9% ee
- Y = H, 85.9% ee
- Y = OMe, 84.6% ee
- Y = F, 77.7% ee
- Y = Cl, 74.3% ee
- Y = OAc, 73.8% ee

Z-substituted stilbene epoxide, **75b** (3.0 equiv.):
- Z = t-Bu, 87.3% ee
- Z = Me, 87.2% ee
- Z = H, 85.9% ee
- Z = F, 78.5% ee
- Z = Br, 74.8% ee
- Z = OAc, 71.5% ee

Ph–epoxide–Ph (1.0 equiv.):
- **75a**, 87.4% ee
- **75b**, 85.4% ee
- **75c**, 80.9% ee
- **75d**, 73.8% ee
- **75e**, 42.0% ee

on the phenyl group of the olefin (Scheme 3.55). This observed ee difference was attributed to the n–π electronic repulsion effect between the Cl atom of **75b** and the phenyl group rather than a steric interaction. The epoxidation of stilbene with **75** showed that the substituent at C$_8$ could also significantly influence the ee through the electrostatic interaction between of the polar C–X bond and the phenyl group of the stilbene. Recently, Solladié-Cavallo and coworkers reported their studies on the fluoro ketone **76** [89–92]. Up to 90% ee was obtained for stilbene [92]. The increased ee resulting from the change of substituents at C-5 in ketones **76b–d** was attributed to the reduction of the axial approach of the olefin towards the dioxirane, thus increasing the enantioselectivity [92].

In 2001, Bortolini, Fogagnolo, and coworkers reported the epoxidation of cinnamic acid derivatives using bile acid based ketones [93, 94]. As shown in Scheme 3.57, up to 95% ee was obtained for *p*-methylcinnamic acid with ketone **77a** [94]. Studies with various ketone analogues showed that the epoxidation could be significantly influenced by the substituents at C-7 and C-12 of the bile acid [94].

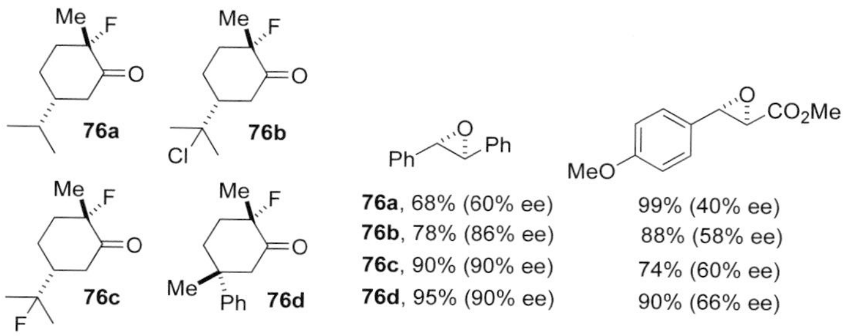
Me
F
O
Cl
F
Ph
76a
76b
76c
76d
Ph
Ph
MeO
CO_2Me
76a, 68% (60% ee)
76b, 78% (86% ee)
76c, 90% (90% ee)
76d, 95% (90% ee)
99% (40% ee)
88% (58% ee)
74% (60% ee)
90% (66% ee)

Scheme 3.56 Epoxidation of olefins with ketones **76a–d** (0.1–0.3 equiv.) [92]

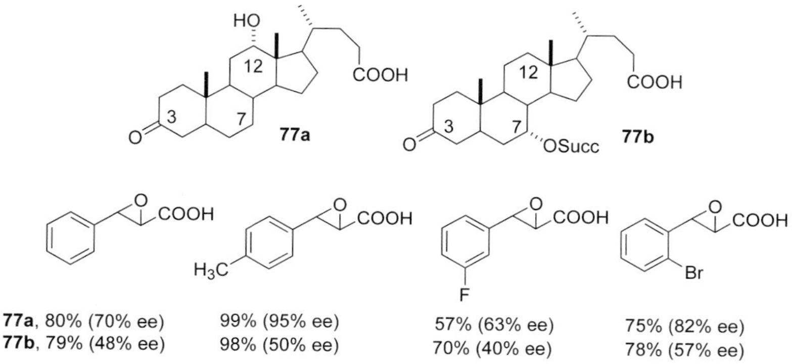
HO
12
3
7
COOH
77a
OSucc
77b
COOH
H_3C
F
Br
77a, 80% (70% ee)
77b, 79% (48% ee)
99% (95% ee)
98% (50% ee)
57% (63% ee)
70% (40% ee)
75% (82% ee)
78% (57% ee)

Scheme 3.57 Epoxidation of olefins with ketone **77** (1.0 equiv.) [94]

3.8
Ketones with an Attached Chiral Moiety

The reacting carbonyl and the stereogenic centers are usually contained in a cyclic structure for most of the ketones discussed above. Several ketones in which the carbonyl and the chiral moiety are combined by a non-cyclic structure have also been studied for epoxidation.

In 1999, Armstrong and coworkers reported trifluoromethyl ketone **78**, using the chiral oxazolidinone as the chiral control element. Up to 34% ee was obtained (Scheme 3.58) [85]. Carnell and coworkers found that *N,N*-dialkylalloxans such as **79** were very reactive epoxidation catalysts and can be recovered without decomposition (Scheme 3.59) [95]. Unfortunately, no asymmetric induction was achieved on epoxidation of *trans*-stilbene with ketone **80**, presumably due to the fact that the chiral center was not close to the reacting carbonyl. In 2003, Wong and coworkers reported a β-cyclodextrin-modified ketoester **81** as an epoxidation catalyst (Scheme 3.60) [96]. Up to 40% ee was obtained with 4-chlorostyrene. In 2003, Zhao and coworkers reported the epoxidation studies with fructose-derived ketone **82** and aldehyde **83**. Up to 94% ee was obtained for stilbene with aldehyde **83b** (Scheme 3.61) [97].

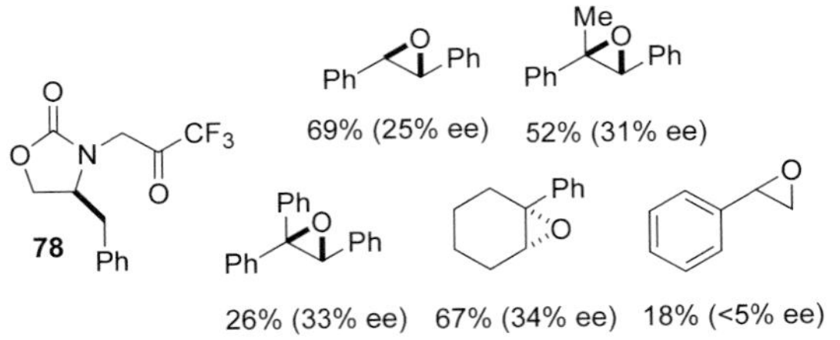

Scheme 3.58 Epoxidation of olefins with ketone **78** (3.0 equiv.)

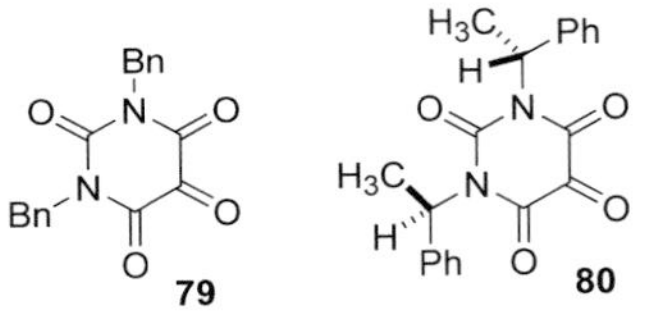

Scheme 3.59

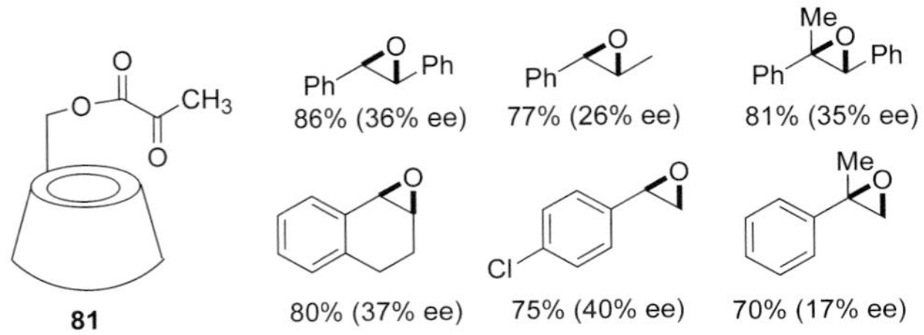

Scheme 3.60 Epoxidation of olefins with ketone **81**

3.9
Conclusion

Discovering highly enantioselective ketone catalysts for asymmetric epoxidation has proven to be a challenging process. As shown in Scheme 3.62, quite a few processes are competing with the catalytic cycle of the ketone mediated epoxidation, including racemization of chiral control elements, excessive hydration of the carbonyl, facile

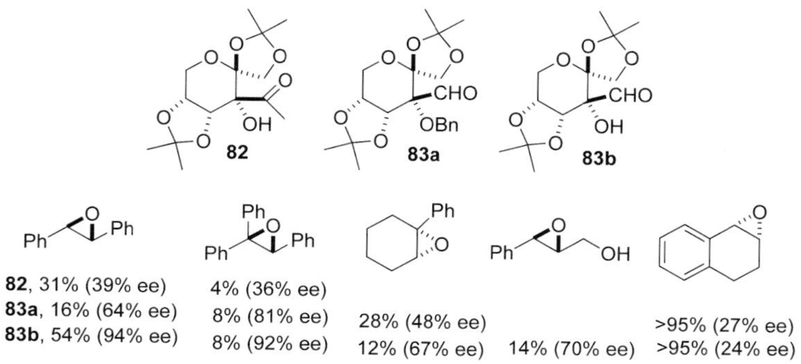

82
83a
83b
CHO
OBn
OH
Ph
82, 31% (39% ee)
83a, 16% (64% ee)
83b, 54% (94% ee)
4% (36% ee)
8% (81% ee)
8% (92% ee)
28% (48% ee)
12% (67% ee)
14% (70% ee)
>95% (27% ee)
>95% (24% ee)

Scheme 3.61 Epoxidation of olefins with ketone **82** and aldehyde **83** (3.0 equiv.)

HSO4-
HSO5-
k
d
e
f
a
g
i
j
decomposition
c
b
B.V.
h
I
II
III
IV
V
VI
VII
SO4 2- + O2
SO5 2-
SO4 2- + HSO4- + O2

Scheme 3.62 Pathways **a–k**: (**a**) nucleophilic addition of the ketone by peroxymonosulfate; (**b**) formation of the oxy-anion intermediate; (**c**) formation of the dioxirane; (**d**) epoxidation of an olefin by the dioxirane; (**e**) epimerization of the stereogenic centers of the ketone; (**f**) hydration of the ketone; (**g**) self-decomposition of peroxymonosulfate; (**h**) Baeyer-Villiger oxidation; (**i**) consumption of the dioxirane by peroxymonosulfate; (**j**) self-decomposition of the dioxirane. (**k**) racemic epoxidation of the olefin by peroxymonosulfate itself

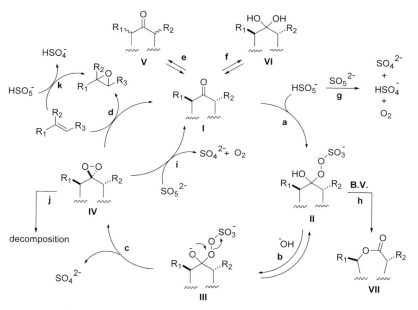

self-decomposition of the oxidant, undesired Baeyer-Villiger oxidation, decomposition of the dioxirane, consumption of the dioxirane by oxidant, background epoxidation by oxidant itself, etc. [98]. Achieving the desired outcome requires delicately balancing the sterics and electronics of the chiral control elements around the carbonyl group, which puts high structural stringency on chiral ketone catalysts in order for them to be highly reactive and enantioselective.

During the past few years, a variety of chiral ketones have been investigated in a number of laboratories, and significant progress has been made in the field. Chiral dioxiranes have been shown to be very effective epoxidation reagents for a wide variety of *trans*- and trisubstituted olefins, and have great promise for *cis*- and terminal olefins. With further efforts, the chiral ketone-catalyzed asymmetric epoxidation has the potential to become a practical and predictable epoxidation process with broad substrate scope. The mechanistic understanding gained thus far will certainly provide useful information for the future development of this field.

Acknowledgments

The author is grateful to Christopher Burke, Zackary Crane, David Goeddel, Dr. Jiang Long, and other research group members for their assistance during the preparation of this manuscript.

References

[1] For recent reviews on highly enantioselective epoxidation of allylic alcohols, see: (a) R.A. Johnson, K.B. Sharpless In *Catalytic Asymmetric Synthesis*; I. Ojima Ed.; VCH: New York, **1993**, pp. 103–158; (b) T. Katsuki, V.S. Martin *Org. React.* **1996**, *48*, 1–299; (c) R.A. Johnson, K.B. Sharpless In *Catalytic Asymmetric Synthesis*; I. Ojima Ed.; VCH: New York, **2000**, pp. 231–280.

[2] For recent reviews on metal catalyzed highly enantioselective epoxidation of unfunctionalized olefins, see: (a) E.N. Jacobsen In *Catalytic Asymmetric Synthesis*; I. Ojima Ed.; VCH: New York, **1993**, pp. 159–202; (b) J.P. Collman, X. Zhang, V.J. Lee, E.S. Uffelman, J.I. Brauman *Science* **1993**, *261*, 1404–1411; (c) T. Mukaiyama *Aldrichim. Acta* **1996**, *29*, 59–76; (d) T. Katsuki In *Catalytic Asymmetric Synthesis*; I. Ojima Ed.; VCH: New York, **2000**, pp. 287–325.

[3] For recent reviews on asymmetric epoxidation of electron-deficient olefins, see: (a) M.J. Porter, J. Skidmore *Chem. Commun.* **2000**, 1215–1225; (b) C. Lauret, S.M. Roberts *Aldrichim. Acta* **2002**, *35*, 47–51; (c) T. Nemoto, T. Ohshima, M. Shibasaki *J. Synth. Org. Chem. Jpn.* **2002**, *60*, 94–105.

[4] For general leading references on dioxiranes see: (a) R.W. Murray *Chem. Rev.* **1989**, *89*, 1187–1201; (b) W. Adam, R. Curci, J.O. Edwards *Acc. Chem. Res.* **1989**, *22*, 205–211; (c) R. Curci, A. Dinoi, M.F. Rubino *Pure Appl. Chem.* **1995**, *67*, 811–822; (d) W. Adam, A.K. Smerz *Bull. Soc. Chim. Belg.* **1996**, *105*, 581–599; (e) W. Adam, C.R. Saha-Möller, C-G. Zhao *Org. React.* **2002**, *61*, 219–516.

[5] For examples of *in situ* generation of dioxiranes see: (a) J.O. Edwards, R.H. Pater, R. Curci, F. Di Furia *Photochem. Photobiol.* **1979**, *30*, 63–70; (b) R. Curci, M. Fiorentino, L. Troisi, J.O. Ed-

WARDS, R.H. PATER *J. Org. Chem.* **1980**, *45*, 4758–4760; (c) A.R. GALLOPO, J.O. EDWARDS *J. Org. Chem.* **1981**, *46*, 1684–1688; (d) G. CICALA, R. CURCI, M. FIORENTINO,.O. LARICCHIUTA *J. Org. Chem.* **1982**, *47*, 2670–2673; (e) P.F. COREY, F.E. WARD *J. Org. Chem.* **1986**, *51*, 1925–1926; (f) W. ADAM, L. HADJIARAPOGLOU, A. SMERZ *Chem. Ber.* **1991**, *124*, 227–232; (g) M. KURIHARA, S. ITO, N. TSUTSUMI, N. MIYATA *Tetrahedron Lett.* **1994**, *35*, 1577–1580; (h) S.E. DENMARK, D.C. FORBES, D.S. HAYS, J.S. DEPUE, R.G. WILDE *J. Org. Chem.* **1995**, *60*, 1391–1407; (i) D. YANG, M-K. WONG, Y-C. YIP *J. Org. Chem.* **1995**, *60*, 3887–3889; (j) S.E. DENMARK, Z. WU *J. Org. Chem.* **1997**, *62*, 8964–8965; (k) T.R. BOEHLOW, P.C. BUXTON, E.L. GROCOCK, B.A. MARPLES, V.L. WADDINGTON *Tetrahedron Lett.* **1998**, *39*, 1839–1842; (l) S.E. DENMARK, Z. WU *J. Org. Chem.* **1998**, *63*, 2810–2811; (m) M. FROHN, Z.-X. WANG, Y. SHI *J. Org. Chem.* **1998**, *63*, 6425–6426; (n) D. YANG, Y.-C. YIP, G.-S. JIAO, M.-K. WONG *J. Org. Chem.* **1998**, *63*, 8952–8956; (o) D. YANG, Y.-C. YIP, M.-W. TANG, M.-K. WONG, K.-K. CHEUNG *J. Org. Chem.* **1998**, *63*, 9888–9894.

[6] For recent reviews on chiral ketone catalyzed asymmetric epoxidation see: (a) S.E. DENMARK, Z. WU *Synlett* **1999**, 847–859; (b) M. FROHN, Y. SHI *Synthesis* **2000**, 1979–2000; (c) Y. SHI *J. Synth. Org. Chem. Jpn.* **2002**, *60*, 342–349.

[7] R. CURCI, M. FIORENTINO, M. R. SERIO *J. Chem. Soc., Chem. Commun.* **1984**, 155–156.

[8] R. CURCI, L. D'ACCOLTI, M. FIORENTINO, A. ROSA *Tetrahedron Lett.* **1995**, *36*, 5831–5834.

[9] D.S. BROWN, B.A. MARPLES, P. SMITH, L. WALTON *Tetrahedron* **1995**, *51*, 3587–3606.

[10] D. YANG, Y.-C. YIP, M.-W. TANG, M.-K. WONG, J.-H. ZHENG, K.-K. CHEUNG *J. Am. Chem. Soc.* **1996**, *118*, 491–492.

[11] D. YANG, X.-C. WANG, M.-K. WONG, Y.-C. YIP, M.-W. TANG, *J. Am. Chem. Soc.* **1996**, *118*, 11311–11312.

[12] D. YANG, M.-K. WONG, Y.-C. YIP, X.-C. WANG, M.-W. TANG, J.-H. ZHENG, K.-K. CHEUNG *J. Am. Chem. Soc.* **1998**, *120*, 5943–5952.

[13] For a related iminium-catalyzed epoxidation under homogenous conditions (CH_3CN-H_2O) with Oxone-$NaHCO_3$ see: L. BOHE, G. HANQUET, M. LUSINCHI, X. LUSINCHI *Tetrahedron Lett.* **1993**, *34*, 7271–7274.

[14] (a) M. SEKI, T. FURUTANI, R. IMASHIRO, T. KURODA, T. YAMANAKA, N. HARADA, H. ARAKAWA, M. KUSAMA, T. HASHIYAMA *Tetrahedron Lett.* **2001**, *42*, 8201–8205; (b) T. FURUTANI, R. IMASHIRO, M. HATSUDA, M. SEKI *J. Org. Chem.* **2002**, *67*, 4599–4601.

[15] (a) T. FURUTANI, M. HATSUDA, R. IMASHIRO, M. SEKI *Tetrahedron: Asymmetry* **1999**, *10*, 4763–4768; (b) M. SEKI, T. FURUTANI, M. HATSUDA, R. IMASHIRO *Tetrahedron Lett.* **2000**, *41*, 2149–2152; (c) T. KURODA, R. IMASHIRO, M. SEKI *J. Org. Chem.* **2000**, *65*, 4213–4216; (d) M. SEKI, S.-I. YAMADA, T. KURODA, R. IMASHIRO, T. SHIMIZU *Synthesis* **2000**, 1677–1680; (e) M. HATSUDA, H. HIRAMATSU, S.-I. YAMADA, T. SHIMIZU, M. SEKI *J. Org. Chem.* **2001**, *66*, 4437–4439; (f) T. FURUTANI, M. HATSUDA, T. SHIMIZU, M. SEKI *Biosci. Biotechnol. Biochem.* **2001**, *65*, 180–184.

[16] K. MATSUMOTO, K. TOMIOKA *Chem. Pharm. Bull.* **2001**, *49*, 1653–1657.

[17] (a) C.E. SONG, Y.H. KIM, K.C. LEE, S.-G. LEE, B.W. JIN *Tetrahedron: Asymmetry* **1997**, *8*, 2921–2926; (b) Y.H. KIM, K.C. LEE, D.Y. CHI, S.-G. LEE, C.E. SONG *Bull. Korean Chem. Soc.* **1999**, *20*, 831–834.

[18] For a related study on ketone **13a** see: ref. [12].

[19] W. ADAM, C.-G. ZHAO *Tetrahedron: Asymmetry* **1997**, *8*, 3995–3998.

[20] K. MATSUMOTO, K. TOMIOKA *Heterocycles* **2001**, *54*, 615–617.

[21] K. MATSUMOTO, K. TOMIOKA *Tetrahedron Lett.* **2002**, *43*, 631–633.

[22] S.E. DENMARK, H. MATSUHASHI *J. Org. Chem.* **2002**, *67*, 3479–3486.

[23] For the stereoelectronic effect of α-fluorine substituents on the 2-fluorocyclohexanone catalyzed epoxidation see: S.E. DENMARK, Z. WU, C.M. CRUDDEN, H. MATSUHASHI *J. Org. Chem.* **1997**, *62*, 8288–8289.

[24] For a calculation study on stereoelectronics of the transition state for fluorina-

ted dioxirane mediated epoxidation see: A. Armstrong, I. Washington, K.N. Houk *J. Am. Chem. Soc.* **2000**, *122*, 6297–6298.

[25] C.J. Stearman, V. Behar *Tetrahedron Lett.* **2002**, *43*, 1943–1946.

[26] For a detailed discussion on this class of ketones see: ref. [6a].

[27] A. Armstrong, B.R. Hayter *Chem. Commun.* **1998**, 621–622.

[28] A. Armstrong, G. Ahmed, B. Dominguez-Fernandez, B.R. Hayter, J.S. Wailes *J. Org. Chem.* **2002**, *67*, 8610–8617.

[29] A. Armstrong, B.R. Hayter, W.O. Moss, J.R. Reeves, J.S. Wailes *Tetrahedron: Asymmetry* **2000**, *11*, 2057–2061.

[30] A. Armstrong, W.O. Moss, J.R. Reeves *Tetrahedron: Asymmetry* **2001**, *12*, 2779–2781.

[31] Ketones 33–35 were obtained with ~80% ee. The ee$_{max}$ is an extrapolated value based on enantiomerically pure ketone catalysts.

[32] G. Sartori, A. Armstrong, R. Maggi, A. Mazzacani, R. Sartorio, F. Bigi, B. Dominguez-Fernandez *J. Org. Chem.* **2003**, *68*, 3232–3237.

[33] S. Klein, S.M. Roberts *J. Chem. Soc., Perkin Trans. 1* **2002**, 2686–2691.

[34] Y. Tu, Z.-X. Wang, Y. Shi *J. Am. Chem. Soc.* **1996**, *118*, 9806–9807.

[35] Z.-X. Wang, Y. Tu, M. Frohn, J.-R. Zhang, Y. Shi *J. Am. Chem. Soc.* **1997**, *119*, 11224–11235.

[36] S. Mio, Y. Kumagawa, S. Sugai *Tetrahedron* **1991**, *47*, 2133–2144.

[37] Y. Tu, M. Frohn, Z.-X. Wang, Y. Shi *Org. Synth.* **2003**, *80*, 1–8.

[38] C.-C. Chen, R.L. Whistler *Carbohydr. Res.* **1988**, *175*, 265–271.

[39] D.L. Ball, J.O. Edwards *J. Am. Chem. Soc.* **1956**, *78*, 1125–1129.

[40] R.E. Montgomery *J. Am. Chem. Soc.* **1974**, *96*, 7820–7821.

[41] Z.-X. Wang, Y. Tu, M. Frohn, Y. Shi *J. Org. Chem.* **1997**, *62*, 2328–2329.

[42] L. Shu, Y. Shi *J. Org. Chem.* **2000**, *65*, 8807–8810.

[43] J.D. Warren, Y. Shi *J. Org. Chem.* **1999**, *64*, 7675–7677.

[44] Z.-X. Wang, Y. Shi *J. Org. Chem.* **1998**, *63*, 3099–3104.

[45] M. Frohn, M. Dalkiewicz, Y. Tu, Z.-X. Wang, Y. Shi *J. Org. Chem.* **1998**, *63*, 2948–2953.

[46] G.-A. Cao, Z.-X. Wang, Y. Tu, Y. Shi *Tetrahedron Lett.* **1998**, *39*, 4425–4428.

[47] Z.-X. Wang, G.-A. Cao, Y. Shi *J. Org. Chem.* **1999**, *64*, 7646–7650.

[48] W. Adam, R.T. Fell, C.R. Saha-Moller, C.-G. Zhao *Tetrahedron: Asymmetry* **1998**, *9*, 397–401.

[49] Y. Zhu, Y. Tu, H. Yu, Y. Shi *Tetrahedron Lett.* **1998**, *39*, 7819–7822.

[50] Y. Zhu, L. Shu, Y. Tu, Y. Shi *J. Org. Chem.* **2001**, *66*, 1818–1826.

[51] Y. Zhu, K.L. Manske, Y. Shi *J. Am. Chem. Soc.* **1999**, *121*, 4080–4081.

[52] X. Feng, L. Shu, Y. Shi *J. Am. Chem. Soc.* **1999**, *121*, 11002–11003.

[53] A.L. Baumstark, C.J. McCloskey *Tetrahedron Lett.* **1987**, *28*, 3311–3314.

[54] A.L. Baumstark, P.C. Vasquez *J. Org. Chem.* **1988**, *53*, 3437–3439.

[55] R.D. Bach, J.L. Andres, A.L. Owensby, H.B. Schlegel, J.J.W. McDouall *J. Am. Chem. Soc.* **1992**, *114*, 7207–7217.

[56] K.N. Houk, J. Liu, N.C. DeMello, K.R. Condroski *J. Am. Chem. Soc.* **1997**, *119*, 10147–10152.

[57] C. Jenson, J. Liu, K.N. Houk, W.L. Jorgensen *J. Am. Chem. Soc.* **1997**, *119*, 12982–12983.

[58] D.V. Deubel *J. Org. Chem.* **2001**, *66*, 3790–3796.

[59] In their studies, Yang and coworkers also showed that a spiro transition state was favored for the ketone 9 catalyzed epoxidation, see refs. [11, 12].

[60] Based on the above analysis, it is conceivable that planar transition state **B** could become the major reaction mode if a large R$_1$ group is chosen to strongly discourage spiro **A** and a small R$_3$ group is chosen to strongly encourage planar **B**. One such example has been observed. The epoxidation of (Z)-3,3-dimethyl-1-phenyl-2-trimethylsiloxy-1-butene with 39 led to the formation of (S)-3,3-dimethyl-1-hydroxy-1-phenyl-2-butanone in 43% ee (see ref. [48]). The S-configuration of the product suggested that a planar transition state is favored.

[61] M. Frohn, X. Zhou, J.-R. Zhang, Y.

TANG, Y. SHI *J. Am. Chem. Soc.* **1999**, *121*, 7718–7719.

[62] Oxone (2KHSO$_5$ · KHSO$_4$ · K$_2$SO$_4$) is currently the common source of potassium peroxymonosulfate (KHSO$_5$).

[63] As close analogues of potassium peroxymonosulfate, arenesulfonic peracids generated from (arenesulfonyl)imidazole-H$_2$O$_2$-NaOH have also been shown to react with acetone and trifluoroacetone to generate dioxiranes as illustrated by ^{18}O-labeling experiments see: M. SCHULZ, S. LIEBSCH, R. KLUGE, W. ADAM *J. Org. Chem.* **1997**, *62*, 188–193.

[64] It has been reported that some dioxiranes can also be generated when a ketone reacts with an oxidant such as CH$_3$COOOH, HOF, and ONOO$^-$: (a) for CH$_3$COOOH see: R.W. MURRAY, V. RAMACHANDRAN *Photochem. Photobiol.* **1979**, *30*, 187–189; (b) for HOF see: S. ROZEN, Y. BAREKET, M. KOL *Tetrahedron* **1993**, *49*, 8169–8178; (c) for ONOO$^-$ see: D. YANG, Y.-C. TANG, J. CHEN, X.-C. WANG, M.D. BARTBERGER, K.N. HOUK, L. OLSON *J. Am. Chem. Soc.* **1999**, *121*, 11976–11983.

[65] For general references on hydrogen peroxide see: (a) G. STRUKUL, *Catalytic Oxidations with Hydrogen Peroxide as Oxidant*, Kluwer Academic Publishers, **1992**; (b) G. GRIGOROPOULOU, J.H. CLARK, J.A. ELINGS *Green Chem.* **2003**, *5*, 1–7.

[66] L. SHU, Y. SHI. *Tetrahedron Lett.* **1999**, *40*, 8721–8724.

[67] L. SHU, Y. SHI *Tetrahedron* **2001**, *57*, 5213–5218.

[68] Z.-X. WANG, L. SHU, M. FROHN, Y. TU, Y. SHI *Org. Synth.* **2003**, *80*, 9–17

[69] For leading references on epoxidation using H$_2$O$_2$ and RCN see: (a) G.B. PAYNE, P.H. DEMING, P.H. WILLIAMS *J. Org. Chem.* **1961**, *26*, 659–663; (b) G.B. PAYNE *Tetrahedron* **1962**, *18*, 763–765; (c) J.E. MCISAAC JR., R.E. BALL, E.J. BEHRMAN *J. Org. Chem.* **1971**, *36*, 3048–3050; (d) R.D. BACH, J.W. KNIGHT, *Org. Synth.* **1981**, *60*, 63; (e) L.A. ARIAS, S. ADKINS, C.J. NAGEL, R.D. BACH *J. Org. Chem.* **1983**, *48*, 888–890.

[70] For recent applications using ketone **39** see: (a) G. BLUET, J.-M. CAMPAGNE *Synlett* **2000**, 221–222; (b) T. TOKIWANO, K. FUJIWARA, A. MURAI *Synlett*, **2000**, 335–338; (c) H. HIOKI, C. KANEHARA, Y. OHNISHI, Y. UMEMORI, H. SAKAI, S. YOSHIO, M. MATSUSHITA, M. KODAMA *Angew. Chem., Int. Ed. Engl.* **2000**, *39*, 2552–2554; (d) Z. XIONG, E.J. COREY *J. Am. Chem. Soc.* **2000**, *122*, 4831–4832; (e) Z. XIONG, E.J. COREY *J. Am. Chem. Soc.* **2000**, *122*, 9328–9329; (f) F.E. MCDONALD, X. WANG, B. DO, K.I. HARDCASTLE *Org. Lett.* **2000**, *2*, 2917–2919; (g) Y. MORIMOTO, T. IWAI, T. KINOSHITA, *Tetrahedron Lett.* **2001**, *42*, 6307–6309; (h) N.R. GUZ, P. LORENZ, F.R. STERMITZ *Tetrahedron Lett.* **2001**, *42*, 6491–6494; (i) K-H. SHEN, S.-F. LUSH, T.-L. CHEN, R.-S. LIU *J. Org. Chem.* **2001**, *66*, 8106–8111; (j) F.E. MCDONALD, S. WEI *Org. Lett.* **2002**, *4*, 593–595; (k) F.E. MCDONALD, F. BRAVO, X. WANG, X. WEI, M. TOGANOH, J.R. RODRIGUEZ, B. DO, W.A. NEIWERT, K.I. HARDCASTLE *J. Org. Chem.* **2002**, *67*, 2515–2523; (l) D.W. HOARD, E.D. MOHER, M.J. MARTINELLI, B.H. NORMAN *Org. Lett.* **2002**, *4*, 1813–1815; (m) V.S. KUMAR, D.L. AUBELE, P.E. FLOREANCIG *Org. Lett.* **2002**, *4*, 2489–2492; (n) Y. MORIMOTO, M. TAKAISHI, T. IWAI, T. KINOSHITA, H. JACOBS *Tetrahedron Lett.* **2002**, *43*, 5849–5852; (o) B. OLOFSSON, P. SOMFAI *J. Org. Chem.* **2002**, *67*, 8574–8583; (p) K.-H. ALTMANN, G. BOLD, G. CARAVATTI, D. DENNI, A. FLÖRSHEIMER, A. SCHMIDT, G. RIHS, M. WARTMANN *Helv. Chim. Acta* **2002**, *85*, 4086–4110; (q) R.J. MADHUSHAW, C.-L. LI, H.-L. SU, C.-C. HU, S.-F. LUSH, R.-S. LIU *J. Org. Chem.* **2003**, *68*, 1872–1877; (r) B. OLOFSSON, P. SOMFAI *J. Org. Chem.* **2003**, *68*, 2514–2517; (s) F. BRAVO, F.E. MCDONALD, W.A. NEIWERT, B. DO, K.I. HARDCASTLE *Org. Lett.* **2003**, *5*, 2123–2126; (t) T.P. HEFFRON, T.F. JAMISON *Org. Lett.* **2003**, *5*, 2339–2342.

[71] Y. TU, Z.-X. WANG, M. FROHN, M. HE, H. YU, Y. TANG, Y. SHI *J. Org. Chem.* **1998**, *63*, 8475–8485.

[72] For a recent study on related analogues of arabinose derived ketone **53**, see:

T.K.M. Shing, Y.C. Leung, K.W. Yeung *Tetrahedron* **2003**, *59*, 2159–2168.

[73] Z.-X. Wang, S.M. Miller, O.P. Anderson, Y. Shi *J. Org. Chem.* **2001**, *66*, 521–530.

[74] H. Tian, X. She, Y. Shi *Org. Lett.* **2001**, *3*, 715–718.

[75] X. Wu, X. She, Y. Shi *J. Am. Chem. Soc.* **2002**, *124*, 8792–8793.

[76] H. Tian, X. She, L. Shu, H. Yu, Y. Shi *J. Am. Chem. Soc.* **2000**, *122*, 11551–11552.

[77] H. Tian, X. She, H. Yu, L. Shu, Y. Shi *J. Org. Chem.* **2002**, *67*, 2435–2446.

[78] L. Shu, Y.-M. Shen, C. Burke, D. Goeddel, Y. Shi *J. Org. Chem.* **2003**, *68*, 4963–4965.

[79] L. Shu, P. Wang, Y. Gan, Y. Shi *Org. Lett.* **2003**, *5*, 293–296.

[80] H. Tian, X. She, J. Xu, Y. Shi *Org. Lett.* **2001**, *3*, 1929–1931.

[81] T.K.M. Shing, G.Y.C. Leung *Tetrahedron* **2002**, *58*, 7545–7552.

[82] Z.-X. Wang, Y. Shi *J. Org. Chem.* **1997**, *62*, 8622–8623.

[83] Z.-X. Wang, S.M. Miller, O.P. Anderson, Y. Shi *J. Org. Chem.* **1999**, *64*, 6443–6458.

[84] A. Armstrong, B.R. Hayter *Tetrahedron: Asymmetry* **1997**, *8*, 1677–1684.

[85] A. Armstrong, B.R. Hayter *Tetrahedron* **1999**, *55*, 11119–11126.

[86] Z.-X. Wang, Y. Shi, unpublished results.

[87] W. Adam, C.R. Saha-Moller, C.-G. Zhao *Tetrahedron: Asymmetry* **1999**, *10*, 2749–2755.

[88] D. Yang, Y.-C. Yip, J. Chen, K.-K. Cheung *J. Am. Chem. Soc.* **1998**, *120*, 7659–7660.

[89] A. Solladié-Cavallo, L. Bouérat *Tetrahedon: Asymmetry* **2000**, *11*, 935–941.

[90] A. Solladié-Cavallo, L. Jierry, L. Bouérat, P. Taillasson *Tetrahedron: Asymmetry* **2001**, *12*, 883–891.

[91] A. Solladié-Cavallo, L. Bouérat *Org. Lett.* **2000**, *2*, 3531–3534.

[92] A. Solladié-Cavallo, L. Bouérat, L. Jierry *Eur. J. Org. Chem.* **2001**, 4557–4560.

[93] O. Bortolini, M. Fogagnolo, G. Fantin, S. Maietti, A. Medici *Tetrahedron: Asymmetry* **2001**, *12*, 1113–1115.

[94] O. Bortolini, G. Fantin, M. Fogagnolo, R. Forlani, S. Maietti, P. Pedrini *J. Org. Chem.* **2002**, *67*, 5802–5806.

[95] A.J. Carnell, R.A.W. Johnstone, C.C. Parsy, W.R. Sanderson *Tetrahedron Lett.* **1999**, *40*, 8029–8032

[96] W.-K. Chan, W.-Y. Yu, C.-M. Che, M.-K. Wong *J. Org. Chem.* **2003**, *68*, 6576–6582.

[97] G. Bez, C.-G. Zhao *Tetrahedron Lett.* **2003**, *44*, 7403–7406.

[98] For detailed discussions on these pathways see: ref. [6 b] and references cited therein.

# 4
Modern Oxidation of Alcohols Using Environmentally Benign Oxidants

I. W. C. E. ARENDS and R. A. SHELDON

4.1
Introduction

The oxidation of primary and secondary alcohols to the corresponding carbonyl compounds plays a central role in organic synthesis [1]. However, standard organic textbooks [2] still recommend classical oxidation methods using stoichiometric quantities of inorganic oxidants, notably chromium(VI) reagents [3] or ruthenium or manganese salts [4], which are highly toxic and environmentally polluting. Other classic non-green methods are based on the use of high valent iodine compounds (notably the Dess Martin reagent) or involve the stoichiometric use of DMSO (dimethyl sulfoxide) (Swern oxidation) [4]. However the state-of-the-art in alcohol oxidation nowadays is far better. Numerous catalytic methods are now known which can be used to oxidize alcohols using either O_2 or H_2O_2 as the oxidant. These oxidants are to be preferred because they are inexpensive and produce water as the sole byproduct. In this chapter we will focus on the use of metal catalysts to mediate the selective oxidation of alcohols using O_2 or H_2O_2 as the primary oxidant. Predominantly homogeneous catalysts are described, but where relevant heterogeneous catalysts (mainly ruthenium) will be covered as well. For an excellent review on heterogeneous oxidation of alcohols and carbohydrates we refer to the publications by Gallezot et al. [5]. Before turning to metal-mediated oxidation of alcohols we will first describe the recent developments in catalytic oxoammonium-mediated oxidation of alcohols.

4.2
Oxoammonium-based Oxidation of Alcohols – TEMPO as Catalyst

A very useful and frequently applied method in the fine chemical industry to convert alcohols into the corresponding carbonyl compounds is the use of oxoammonium salts as oxidants as denoted in Scheme 4.1 [6]. These are very selective oxidants for alcohols, which operate under mild conditions and tolerate a large variety of functional groups. The oxidation proceeds in acidic as well as alkaline media.

Modern Oxidation Methods. Edited by Jan-Erling Bäckvall
Copyright © 2004 WILEY-VCH Verlag GmbH & Co. KGaA, Weinheim
ISBN: 3-527-30642-0

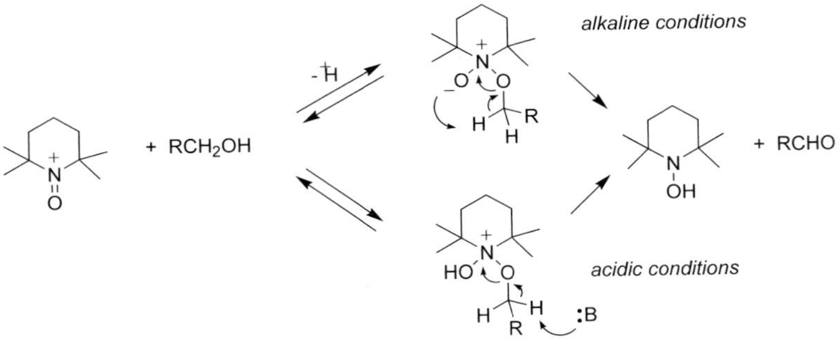

Scheme 4.1 Mechanism for the oxoammonium catalyzed oxidation of alcohols

The oxoammonium is generated *in situ* from its precursor, TEMPO (2,2′,6,6′-tetramethylpiperidine-*N*-oxyl) (or derivatives thereof), which is used in catalytic quantities (see Scheme 4.2). Various oxidants can be applied as the final oxidant [7–12]. In particular, the TEMPO–bleach protocol using bromide as the co-catalyst introduced by Anelli et al. is finding wide application in organic synthesis [7]. TEMPO is used at levels as low as 1 mol% relative to the substrate and full conversion of substrates can commonly be achieved within 30 min.

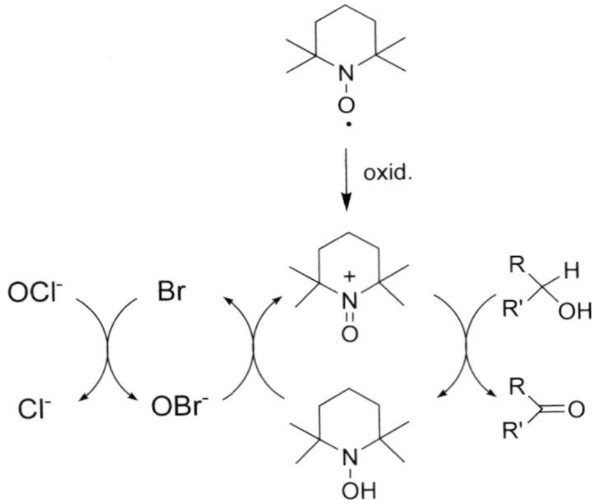

Scheme 4.2 TEMPO catalyzed oxidation of alcohols using hypochlorite as the oxidant

The major drawbacks of this method are the use of NaOCl as the oxidant, the need for the addition of bromine ions and the necessity to use chlorinated solvents. Recently a great deal of effort has been devoted towards a greener oxoammonium-based method, by, for example, replacing TEMPO by heterogeneous variations or the replacement of NaOCl by a combination of metal as the co-catalyst and molecular oxygen as the oxidant. Examples of heterogeneous variants of TEMPO are anchoring

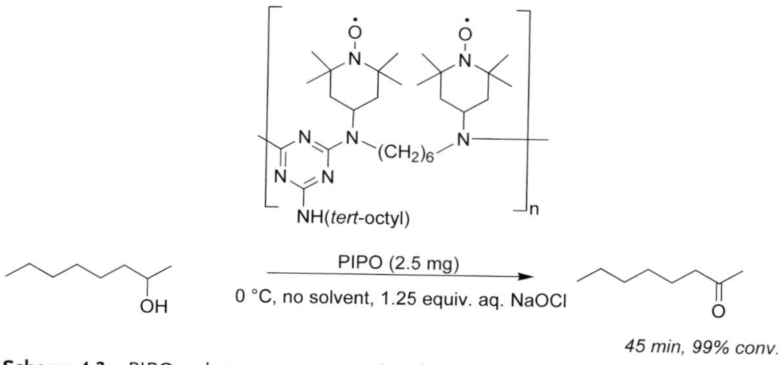

4.2 Oxoammonium-based Oxidation of Alcohols – TEMPO as Catalyst

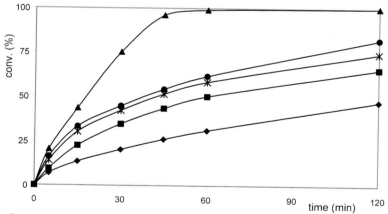

Scheme 4.3 PIPO as heterogeneous catalyst for alcohol oxidation

TEMPO to solid supports such as silica [13, 14] and the mesoporous silica, MCM-41 [15] or by entrapping TEMPO in sol-gel [16]. Alternatively, in our group we developed an oligomeric TEMPO (Scheme 4.3), derived from Chimassorb 944 [17].

This new polymer immobilized TEMPO, which we refer to as PIPO (Polymer Immobilized Piperidinyl Oxyl), proved to be a very effective catalyst for the oxidation of alcohols with hypochlorite [17] (Scheme 4.3). Under the standard conditions PIPO is dissolved in the dichloromethane layer. In contrast, in the absence of solvent PIPO was a very effective recyclable heterogeneous catalyst. Furthermore, the enhanced activity of PIPO compared with TEMPO made the use of a bromide co-catalyst redundant. Hence, the use of PIPO, in an amount equivalent to 1 mol% of nitroxyl radical,

Scheme 4.4 Bleach oxidation of octan-2-ol under chlorinated hydrocarbon solvent- and bromide-free conditions using 1 mol% of nitroxyl catalyst: (▲) PIPO (3.19 mmol g^{-1}; amine linker) [17]; (●) MCM-41 TEMPO (0.60 mmol g^{-1}; ether linker) [15]; (*) SiO$_2$ TEMPO (0.87 mmol g^{-1}, amine linker) [13]; (■) SiO$_2$ TEMPO (0.40 mmol g^{-1}, amide linker) [14]; (◆) TEMPO provided an effective (heterogeneous) catalytic method for the oxidation of a variety of alcohols with 1.25 equiv. of 0.35 M NaOCl (pH 9.1) in a bromide- and chlorinated hydrocarbon solvent-free medium (Scheme 4.4). Under these environmentally benign conditions, PIPO was superior to the already mentioned heterogeneous TEMPO systems and homogeneous TEMPO [18]. In the solvent-free system primary alcohols, such as octan-1-ol, gave low selectivities to the corresponding aldehyde owing to over-oxidation to the carboxylic acid. This problem was circumvented by using n-hexane as the solvent, in which PIPO is insoluble, affording an increase in aldehyde selectivity from 50 to 94%.

Recently two papers were published that dealt with the bleach free TEMPO-catalyzed oxidation. In one approach a heteropolyacid, which is a known redox catalyst, was able to generate oxoammonium ions *in situ* with 2 atm of molecular oxygen at 100 °C [19]. In the other approach, a combination of manganese and cobalt (5 mol%) was able to generate oxoammonium ions under acidic conditions at 40 °C [20]. Results for both methods are compared in Table 4.1. Although these conditions are still prone to improvement, both processes use molecular oxygen as the ultimate oxidant, are chlorine free and therefore are valuable examples of progress in this area. Later on in this chapter we will discuss examples where the use of TEMPO in combination with an Ru or Cu catalyst results in even higher active catalytic systems. The mechanism however in these cases is metal-based instead of oxoammonium-based and is therefore listed in the appropriate section. Another approach to generate oxoammonium ions *in situ* is an enzymatic one. Laccase, which is an abundant highly potent redox enzyme, is capable of oxidizing TEMPO to the oxoammonium ion [21]. Although this method still requires large amounts of TEMPO (30 mol%) and long reaction times (24 h), it demonstrates that a combination of laccase and TEMPO is able to catalyze the aerobic oxidation of alcohols (see Table 4.1).

Tab. 4.1 Aerobic oxoammonium based oxidation of alcohols

Substrate	Aldehyde or ketone yield [a]		
	2 mol% Mn(NO$_3$)$_2$ 2 mol% Co(NO$_3$)$_2$ 10 mol% TEMPO acetic acid, 40 °C 1 atm O$_2$ [b]	1 mol% H$_5$[PMo$_{10}$V$_2$O$_{40}$] 3 mol% TEMPO acetone, 100 °C 2 atm O$_2$ [c]	Laccase (3U mL^{-1}) water pH 4.5 30 mol% TEMPO 25 °C, 1 atm O$_2$ [d]
n-C$_6$H$_{13}$-CH$_2$OH	97% (6 h)		
n-C$_7$H$_{15}$-CH$_2$OH		98% (18 h)	
n-C$_9$H$_{19}$-CH$_2$OH			15% (24 h)
n-C$_7$H$_{15}$-CH(CH$_3$)OH	100% (5 h)		
n-C$_6$H$_{13}$-CH(CH$_3$)OH		96% (18 h)	
PhCH$_2$OH	98% (10 h) [e]	100% (6 h)	92% (24 h)
PhCH(CH$_3$)OH	98% (6 h) [e]		
cis-C$_3$H$_7$-CH=CH-CH$_2$OH		100% (10 h)	
Ph-CH=CH-CH$_2$OH	99% (3 h)		94% (24 h)

[a] GLC yields.   [b] Minisci and coworkers [20].   [c] Neumann and coworkers [19].   [d] Fabbrini et al. [21].
[e] Reaction performed at 20 °C with air.

4.3
Metal-mediated Oxidation of Alcohols – Mechanism

Noble metal salts, e.g., of Pd^{II} or Pt^{II}, undergo reduction by primary and secondary alcohols in homogeneous solution. Indeed, the ability of alcohols to reduce Pd^{II} has already been described in 1828 by Berzelius who showed that K_2PdCl_4 was reduced to palladium metal in an aqueous ethanolic solution [22]. The reaction involves a β-hydride elimination from an alkoxymetal intermediate and is a commonly used method for the preparation of noble metal hydrides [Eq. (1)]. In the presence of dioxygen this leads to catalytic oxidative dehydrogenation of the alcohol, e.g., with palladium salts [23–27].

$$M-O-C(H)(C) \longrightarrow MH + C=O \qquad (1)$$

The aerobic oxidation of alcohols catalyzed by low-valent late transition metal ions, particularly those of Group VIII elements, involves an oxidative dehydrogenation mechanism. In the catalytic cycle (see Scheme 4.5) a hydridometal species, formed by β-hydride elimination from an alkoxymetal intermediate, is reoxidized by dioxygen, presumably via insertion of O_2 into the M–H bond with the formation of H_2O_2. Alternatively, an alkoxymetal species can decompose to a proton and the reduced form of the catalyst (see Scheme 4.5), either directly or via the intermediacy of a hydridometal intermediate. These reactions are promoted by bases as co-catalysts, which presumably facilitate the formation of an alkoxymetal intermediate and/or β-hydride elimination. Examples of metal ions that operate via this pathway are Pd^{II}, Ru^{II} and Rh^{III}.

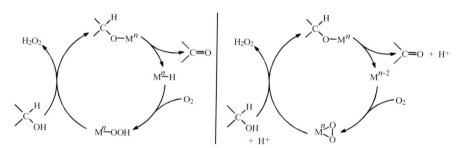

Scheme 4.5 Hydridometal pathways for alcohol oxidation

Metal-catalyzed oxidations of alcohols with peroxide reagents can be conveniently divided into two categories, involving peroxometal and oxometal species, respectively, as the active oxidant (Scheme 4.6). In the peroxometal pathway the metal ion remains in the same oxidation state throughout the catalytic cycle and no stoichiometric oxidation is observed in the absence of the peroxide. In contrast, oxometal pathways involve a two-electron change in oxidation state of the metal ion and a stoichiometric oxida-

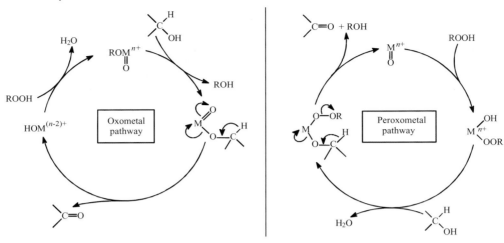

Scheme 4.6 Oxometal *versus* peroxometal pathways in metal catalyzed alcohol oxidations

tion is observed, with the oxidized form of the catalyst, in the absence of, e.g., H_2O_2. Indeed, this is a test for distinguishing between the two pathways.

Peroxometal pathways are typically observed with early transition metal ions with a d^0 configuration, e.g., Mo^{VI}, W^{VI}, Ti^{IV} and Re^{VII}, which are relatively weak oxidants. Oxometal pathways are characteristic of late transition elements and first row transition elements, e.g., Cr^{VI}, Mn^V, Os^{VIII}, Ru^{VI} and Ru^{VIII}, which are strong oxidants in high oxidation states. Some metals can operate *via* both pathways depending, *inter alia*, on the substrate, e.g., vanadium(V) operates *via* a peroxometal pathway in olefin epoxidations but an oxometal pathway is involved in alcohol oxidations [28].

In some cases, notably ruthenium, the aerobic oxidation of alcohols is catalyzed by both low- and high-valent forms of the metal (see later). In the former case the reaction involves (see Scheme 4.5) the formation of a hydridometal species (or its equivalent) while the latter involves an oxometal intermediate (see Scheme 4.6), which is regenerated by reaction of the reduced form of the catalyst with dioxygen instead of a peroxide. It is difficult to distinguish between the two and one should bear in mind, therefore, that aerobic oxidations with high-valent oxometal catalysts could involve the formation of low-valent species, even the (colloidal) metal, as the actual catalyst.

4.4
Ruthenium-catalyzed Oxidations with O_2

Ruthenium compounds are widely used as catalysts in organic synthesis [29, 30] and have been studied extensively as catalysts for the aerobic oxidation of alcohols [31]. In 1978, Mares and coworkers [32] reported that $RuCl_3 \cdot nH_2O$ catalyzes the aerobic oxidation of secondary alcohols into the corresponding ketones, albeit in modest yields. In 1981, Matsumoto and Ito showed that $RuCl_3$ and $RuCl_2(Ph_3P)_3$ catalyze the aerobic oxidation of activated allylic and benzylic alcohols under mild conditions [33], e.g., the oxidation of retinol to retinal could be performed at 25 °C (57% yield was obtained after 48 h). Aliphatic primary and secondary alcohols were more efficiently oxidized using trinuclear ruthenium carboxylates, $Ru_3O(O_2CR)_6L_n$ ($L = H_2O$, Ph_3P) as the catalysts [34]. With lower aliphatic alcohols, e.g., 1-propanol, 2-propanol and 1-butanol, activities were ca. 10 times higher than with $RuCl_3$ and $RuCl_2(Ph_3P)_3$. Recently somewhat higher activities were reached using $RuCl_2PPh_3$ as the catalyst with ionic liquids as solvents (Scheme 4.7). These solvents have been tested as environmentally friendly solvents for a large variety of reactions [35]. In this particular case tetramethylammonium hydroxide and aliquat® 336 (tricaprylylmethylammonium chloride) were used as the solvent and rapid conversion of benzyl alcohol was observed [36]. Moreover the tetramethylammonium hydroxide/$RuCl_2(PPh_3)_3$ could be reused after extraction of the product.

Ruthenium compounds are widely used as catalysts for hydrogen transfer reac-

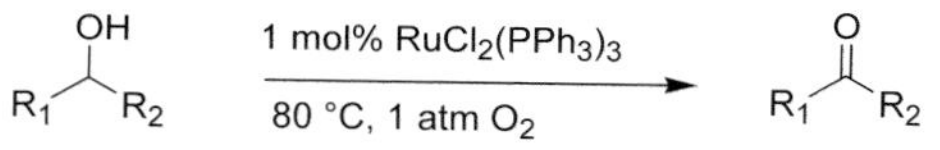

solvent:	tetramethylammonium hydroxide	aliquat
R_1=$PhCH_2$, R_2=H	*91% conv. (5 h)*	*58% conv. (5 h)*
R_1, R_2, = c-C_7H_{14}	*61% conv. (11 h)*	*92% conv. (11 h)*
R_1=C_6H_{13}, R_2=CH_3	*43% conv. (25 h)*	*81% conv. (25 h)*

Scheme 4.7 Aerobic Ru-catalyzed oxidation in ionic liquids

tions. These systems can be readily adapted to the aerobic oxidation of alcohols by employing dioxygen, in combination with a hydrogen acceptor as a cocatalyst, in a multistep process. For example, Bäckvall and coworkers [37] used low-valent ruthenium complexes in combination with a benzoquinone and a cobalt–Schiff's base complex. The coupled catalytic cycle is shown in Scheme 4.8. A low-valent ruthenium complex reacts with the alcohol to afford the aldehyde or ketone product and a ruthenium dihydride. The latter undergoes hydrogen transfer to the benzoquinone to give hydroquinone with concomitant regeneration of the ruthenium catalyst. The cobalt–Schiff's base complex catalyzes the subsequent aerobic oxidation of the hydroquinone to benzoquinone to complete the catalytic cycle. Optimization of the electron-rich quinone, combined with the so-called "Shvo" Ru-catalyst, led to one of the fastest catalytic systems reported for the oxidation of secondary alcohols [37 c]. The reaction conditions and results for selected alcohols are reported in Table 4.2.

The regeneration of the benzoquinone can also be achieved with dioxygen in the absence of the cobalt co-catalyst. Thus, Ishii and coworkers [38] showed that a combination of $RuCl_2(Ph_3P)_3$, hydroquinone and dioxygen, in $PhCF_3$ as solvent, oxidized primary aliphatic, allylic and benzylic alcohols to the corresponding aldehydes in quantitative yields [Eq. (2)].

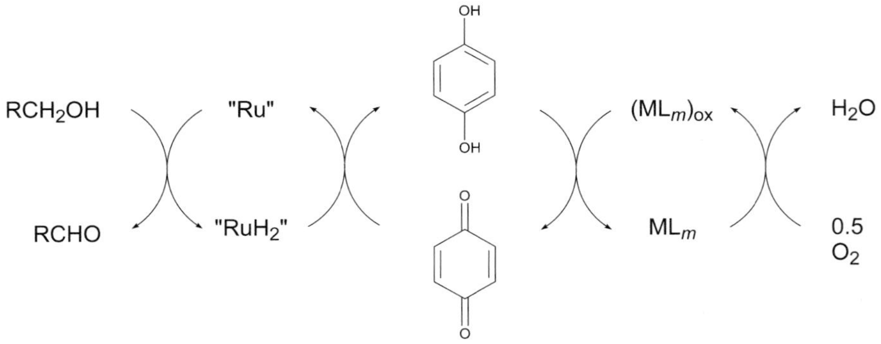
RCH2OH
RCHO
"Ru"
"RuH2"
OH
OH
O
O
(MLm)ox
MLm
H2O
0.5 O2

Scheme 4.8 Ruthenium catalyst in combination with a hydrogen acceptor for aerobic oxidation

Tab. 4.2 Ruthenium/quinone/Co-salen catalyzed aerobic oxidation of secondary alcohols [a]

Substrate	*Time (h)*	*Isolated yield (%)*
n-C_6H_{13}-$CH(CH_3)$-OH	1	92
Cyclohexanol	1	92
Cyclododecanol	1.5	86
$PhCH(CH_3)$-OH	1	89
L-menthol	2	80

[a] According to ref. [37c]. Reaction conditions: 1 mmol substrate, 1 mL toluene, 100 °C, 1 atm air; employing 0.5 mol% $[(C_4Ph_4COHOCC_4Ph_4)(\mu\text{-}H)(CO)_4Ru_2]$, 20 mol% 2,6-dimethoxy-1,4-benzoquinone, and 2 mol% bis(salicylideniminato-3-propyl)methylamino-cobalt(II) as catalysts.

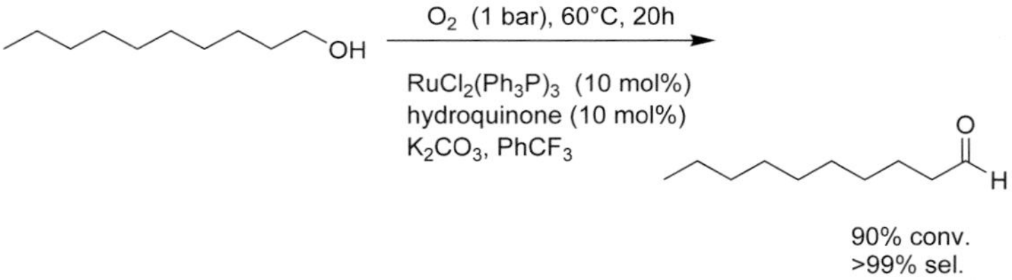
OH
O2 (1 bar), 60°C, 20h
RuCl2(Ph3P)3 (10 mol%)
hydroquinone (10 mol%)
K2CO3, PhCF3
O
H
90% conv.
>99% sel.

(2)

A combination of $RuCl_2(Ph_3P)_3$ and the stable nitroxyl radical, 2,2′,6,6′-tetramethylpiperidine-*N*-oxyl (TEMPO) is a remarkably effective catalyst for the aerobic oxidation of a variety of primary and secondary alcohols, giving the corresponding aldehydes and ketones, respectively, in >99% selectivity [39]. The best results were obtained using 1 mol% of $RuCl_2(Ph_3P)_3$ and 3 mol% of TEMPO [Eq. (3)].

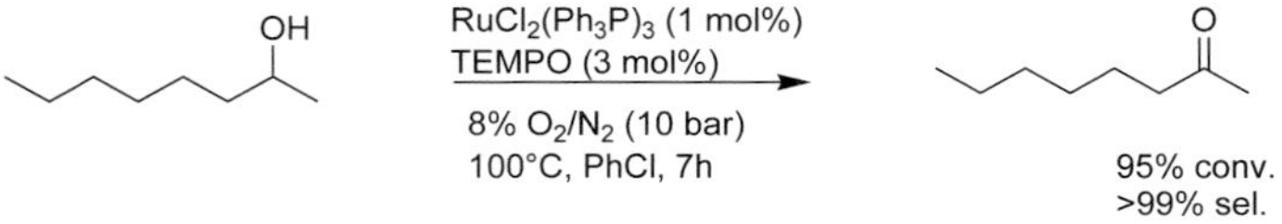
OH
RuCl2(Ph3P)3 (1 mol%)
TEMPO (3 mol%)
8% O2/N2 (10 bar)
100°C, PhCl, 7h
O
95% conv.
>99% sel.

(3)

Tab. 4.3 Ruthenium-TEMPO catalyzed oxidation of primary and secondary alcohols to the corresponding aldehyde using molecular oxygen [a]

Substrate	S/C ratio [b]	Time (h)	Conv. (%) [c]
n-$C_7H_{15}CH_2OH$	50	7	85
n-$C_6H_{13}CH(CH_3)OH$	100	7	98
Adamantan-2-ol	100	7	92
Cyclooctanol	100	7	92
$(CH_3)_2C=CHCH_2OH$	67	7	96
$(CH_3)_2C=CH(CH_2)_2CH(CH_3)=CHCH_2OH$ [d]	67	7	91
$PhCH_2OH$ [e]	200	2.5	>99
$(4-NO_2)PhCH_2OH$ [e]	200	6	97
$PhCH(CH_3)-OH$	100	4	>99

[a] 15 mmol substrate, 30 mL chlorobenzene, $RuCl_2(PPh_3)_3$/TEMPO ratio of 1/3, 10 mL min^{-1} O_2/N_2 (8/92; v/v), $P = 10$ bar, $T = 100\,°C$. [b] Substrate/Ru ratio. [c] Conversion of substrate, selectivity to aldehyde or ketone >99%. [d] Geraniol. [e] 1 atm O_2.

The results obtained in the oxidation of representative primary and secondary aliphatic alcohols and allylic and benzylic alcohols using this system are shown in Table 4.3.

Primary alcohols give the corresponding aldehydes in high selectivity, e.g., 1-octanol affords 1-octanal in >99% selectivity. Over-oxidation to the corresponding carboxylic acid, normally a rather facile process, is completely suppressed in the presence of a catalytic amount of TEMPO. For example, attempted oxidation of octanal under the reaction conditions, in the presence of 3 mol% TEMPO, gave no reaction in one week. In contrast, in the absence of TEMPO octanal was completely converted into octanoic acid within 1 h under the same conditions. These results are consistent with over-oxidation of aldehydes occurring *via* a free radical autoxidation mechanism. TEMPO suppresses this reaction by efficiently scavenging free radical intermediates resulting in the termination of free radical chains, i.e., it acts as an antioxidant. Allylic alcohols were selectively converted into the corresponding unsaturated aldehydes in high yields. No formation of the isomeric saturated ketones *via* intramolecular hydrogen transfer, which is known to be promoted by ruthenium phosphine complexes [40], was observed.

Although, in separate experiments, secondary alcohols are oxidized faster than primary ones, in competition experiments the Ru/TEMPO system displayed a preference for primary over secondary alcohols. This can be explained by assuming that initial complex formation between the alcohol and the ruthenium precedes rate-limiting hydrogen transfer and determines substrate specificity, i.e., complex formation with a primary alcohol is favored over a secondary one.

An oxidative hydrogenation mechanism, analogous to that proposed by Bäckvall for the Ru/quinone system (see above), can be envisaged for the Ru/TEMPO system (see Scheme 4.9).

The intermediate hydridoruthenium species is most probably $RuH_2(Ph_3P)_3$ as was observed in $RuCl_2(Ph_3P)_3$-catalyzed hydrogen transfer reactions [41]. The obser-

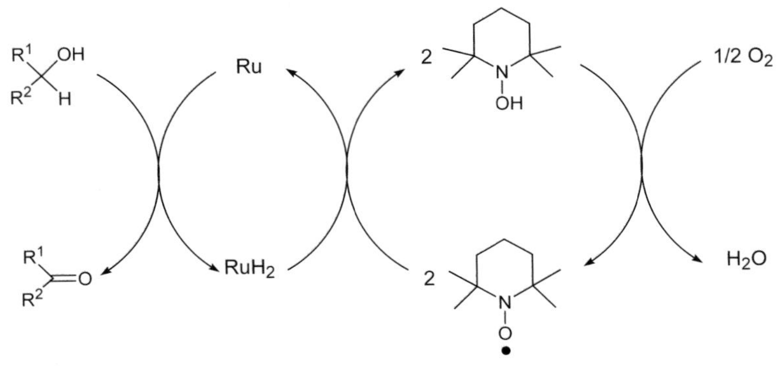

Scheme 4.9 Ruthenium/TEMPO catalyzed aerobic oxidation of alcohols

vation that $RuH_2(Ph_3P)_4$ exhibits the same activity as $RuCl_2(Ph_3P)_3$ in the Ru/TEMPO catalyzed aerobic oxidation of 2-octanol is consistent with this notion. The TEMPO acts as a hydrogen transfer mediator by promoting the regeneration of the ruthenium catalyst, *via* oxidation of the ruthenium hydride, resulting in the concomitant formation of the corresponding hydroxylamine, TEMPOH. The latter then undergoes rapid reoxidation to TEMPO, by molecular oxygen, to complete the catalytic cycle (see Scheme 4.9).

A linear increase in the rate of 2-octanol oxidation was observed with increasing TEMPO concentration in the range 0–4 mol% but above 4 mol% further addition of TEMPO had a negligible effect on the rate. Analogous results were observed by Bäckvall and coworkers [42] in the Ru/benzoquinone system and were attributed to a change in the rate-limiting step. Hence, by analogy, we propose that at relatively low TEMPO/Ru ratios (up to 4:1) reoxidation of the ruthenium hydride species is the slowest step while at high ratios dehydrogenation of the alcohol becomes rate-limiting.

Under an inert atmosphere $RuCl_2(Ph_3P)_3$ catalyzes the stoichiometric oxidation of 2-octanol by TEMPO, to give 2-octanone and the corresponding piperidine, TEMPH, in a stoichiometry of 3:2 as denoted in [Eq. (4)] [39].

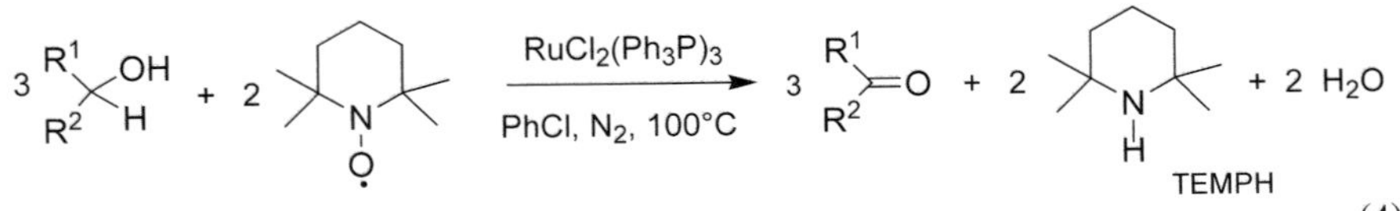

$$(4)$$

This result can be explained by assuming that the initially formed TEMPOH (see above) undergoes disproportionation to TEMPH and the oxoammonium cation [Eq. (5)]. Reduction of the latter by the alcohol affords another molecule of TEMPOH and this leads, ultimately, to the formation of the ketone and TEMPH in the observed stoichiometry of 3:2. The observation that attempts to prepare TEMPOH [43] under an inert atmosphere always resulted in the formation of TEMPH is consistent with this hypothesis.

(5)

Based on the results discussed above the detailed catalytic cycle depicted in Scheme 4.10 is proposed for the Ru/TEMPO catalyzed aerobic oxidation of alcohols.

The alcohol oxidations discussed above involve, as a key step, the oxidative dehydrogenation of the alcohol to form low-valent hydridoruthenium intermediates. On the other hand, high-valent oxoruthenium species are also able to dehydrogenate alcohols via an oxometal mechanism (see Scheme 4.6). It has long been known that ruthenium tetroxide, generated by reaction of ruthenium dioxide with periodate, smoothly oxidizes a variety of alcohols to the corresponding carbonyl compounds [44].

Griffith and coworkers [45] reported the synthesis of the organic soluble tetra-n-butylammoniumperruthenate (TBAP), $n\text{-}Bu_4N^+RuO_4^-$, in 1985. They later found that tetra-n-propylammoniumperruthenate (TPAP), $n\text{-}Pr_4N^+RuO_4^-$, is even easier to prepare, from RuO_4 and $n\text{-}Pr_4NOH$ in water [46, 47]. TBAB and TPAP are air-stable, non-volatile and soluble in a wide range of organic solvents. Griffith and Ley [48, 49] subsequently showed that TPAP is an excellent catalyst for the selective oxidation of

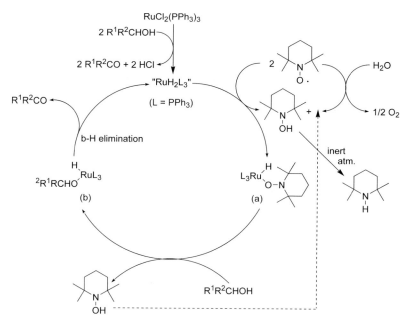

Scheme 4.10 Proposed mechanism for the ruthenium/TEMPO catalyzed oxidation of alcohols

a wide variety of alcohols using N-methylmorpholine-N-oxide (NMO) as the stoichiometric oxidant [Eq. (6)].

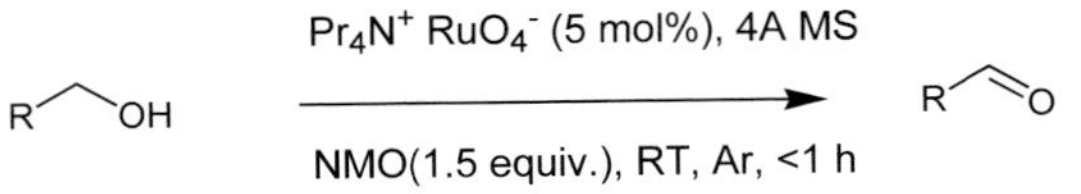

$$R\text{-OH} \xrightarrow[\text{NMO(1.5 equiv.), RT, Ar, <1 h}]{Pr_4N^+ RuO_4^- (5\ mol\%),\ 4A\ MS} R\text{-CHO} \quad (6)$$

More recently, the groups of Ley [50] and Marko [51] independently showed that TPAP is able to catalyze the oxidation of alcohols using dioxygen as the stoichiometric oxidant. In particular, polymer supported perruthenate (PSP), prepared by anion exchange of $KRuO_4$ with a basic anion exchange resin (Amberlyst A-26), has emerged as a versatile catalyst for the aerobic oxidation [Eq. (7)] of alcohols [52]. However the activity was ca. 4 times lower than homogeneous TPAP, and this catalyst could not be recycled, which was attributed to oxidative degradation of the polystyrene support. PSP displays a marked preference for primary *versus* secondary alcohol functionalities [52]. The problem of deactivation was also prominent for the homogeneous TPAP oxidation, which explains the high (10 mol%) loading of catalyst required.

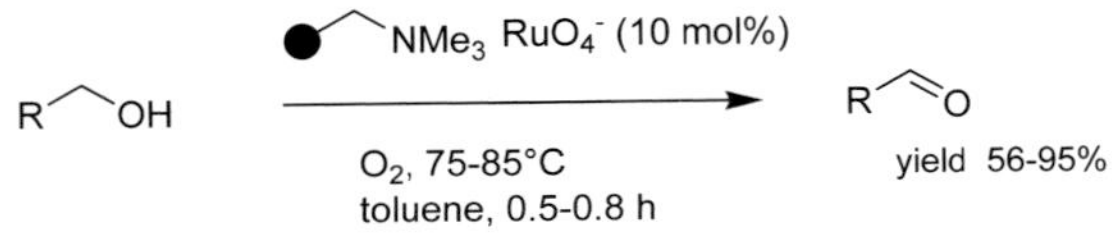

$$R\text{-OH} \xrightarrow[\substack{O_2,\ 75\text{-}85°C \\ \text{toluene, 0.5-0.8 h}}]{\bullet\text{-NMe}_3\ RuO_4^-\ (10\ mol\%)} R\text{-CHO} \quad \text{yield 56-95\%} \quad (7)$$

Examples illustrating the scope of TPAP-catalyzed aerobic oxidation of primary and secondary alcohols to the corresponding aldehydes are shown in Table 4.4.

Recently two heterogeneous TPAP-catalysts were developed, which could be recycled successfully and displayed no leaching: In the first example the tetraalkylam-

Tab. 4.4 Perruthenate catalyzed oxidation of primary and secondary alcohols to aldehydes using molecular oxygen

Substrate	Toluene, 75–85 °C 10 mol% polymer supported perruthenate (PSP)[b]	Carbonyl yield[a] Toluene, 70–80 °C, 4 Å MS, 5 mol% tetrapropyl-ammoniumperruthenate (TPAP)[c]	Toluene, 75 °C, 10 mol% TPAP doped sol-gel ormosil[d]
$C_7H_{15}CH_2OH$	91% (8 h)		70% (7 h)
$C_9H_{19}CH_2OH$		73% (0.5 h)[e]	
$C_9H_{19}CH(CH_3)\text{-}OH$		88% (0.5 h)	
$(H_3C)_2N(CH_2)_2CH_2OH$	>95% (8 h)		
$PhCH_2OH$	>95% (0.5 h)		100% (0.75 h)
$(4\text{-}Cl)PhCH_2OH$		81% (0.5 h)	
$Ph\text{-}CH=CHCH_2OH$	>95% (1 h)	70% (0.5 h)	90% (5 h)

[a] Yields at 100% conversion. [b] Ley and coworkers [52]. [c] Marko et al. [51]. [d] Pagliaro and Ciriminna [54]. [e] 94% conversion, no molecular sieves were added.

monium perruthenate was tethered to the internal surface of mesoporous silica (MCM-41) and was shown [53] to catalyze the selective aerobic oxidation of primary and secondary allylic and benzylic alcohols (Scheme 4.11). Surprisingly, both cyclohexanol and cyclohexenol were unreactive although these substrates can easily be accommodated in the pores of MCM-41. No mechanistic interpretation for this surprising observation was offered by the authors.

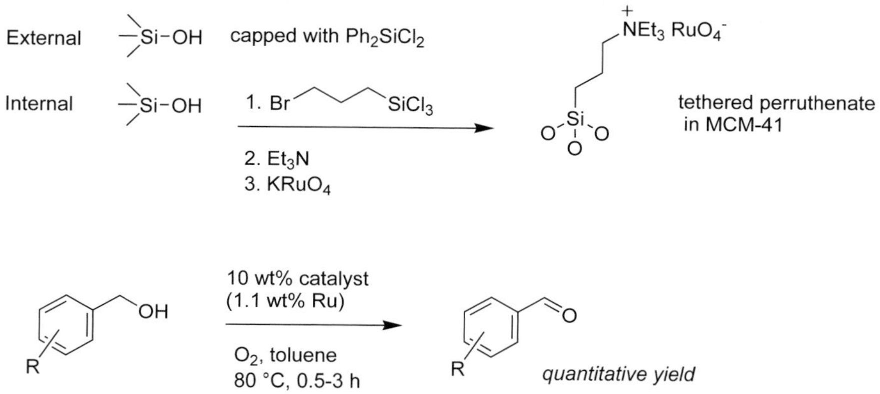

Scheme 4.11 Aerobic alcohol oxidation catalyzed by perruthenate tethered to the internal surface of MCM-41

The second example involves straightforward doping of methyl modified silica, denoted as ormosil, with tetrapropylammonium perruthenate *via* the sol-gel process [54] (see Table 4.4). A serious disadvantage of this system is the low-turnover frequency (1.0 and 1.8 h^{-1}) observed) for primary aliphatic alcohol and allylic alcohol, respectively.

Sparse attention has been paid to the mechanism of perruthenate-catalyzed alcohol oxidations [55]. Although TPAP can act as a three-electron oxidant (Ru^{VII} → Ru^{IV}) the fact that it selectively oxidizes cyclobutanol to cyclobutanone and *tert*-butyl phenylmethanol to the corresponding ketone, militates against free radical intermediates and is consistent with a heterolytic, two-electron oxidation [55, 56]. Presumably, the key step involved β-hydride elimination from a high-valent, e.g., alkoxyruthenium (VII) intermediate followed by reoxidation of the lower valent ruthenium by dioxygen. However, as shown in Scheme 4.12, if this involved the Ru^{VII}/Ru^{V} couple the reoxidation would require the close proximity of two ruthenium centers, which would seem unlikely in a polymer-supported catalyst. A plausible alternative, which can occur at an isolated ruthenium center involves the oxidation of a second molecule of alcohol, resulting in the reduction of ruthenium(V) to ruthenium(III), followed by reoxidation of the latter to ruthenium(VII) by dioxygen (see Scheme 4.12).

More detailed mechanistic studies are obviously necessary in order to elucidate the details of this fascinating reaction. It is worth noting, in this context, that the reaction of TPAP with 2-propanol was found to be autocatalytic, possibly due to the for-

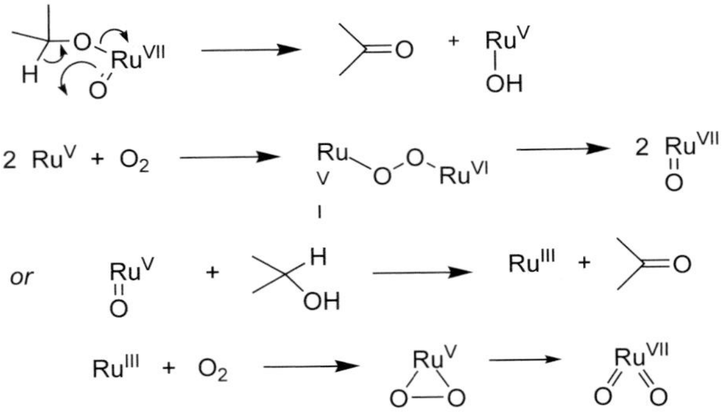

Scheme 4.12 Proposed catalytic cycle for reoxidation of perruthenate in the oxidation of alcohols

mation of colloidal RuO_2 [57]. Another possible alternative is one involving the initial formation of oxoruthenium(VI), followed by cycling between ruthenium(VI), ruthenium(IV) and possibly ruthenium(II).

We note, in this context, that James and coworkers [58] showed that a *trans*-dioxoruthenium(VI) complex of meso-tetrakismesitylporphyrin dianion (tmp) oxidizes isopropanol, in a stoichiometric reaction, with concomitant formation of a dialkoxyruthenium(IV) tmp complex [Eq. (8)].

$$Ru^{VI}(tmp)O_2 + 3\,Me_2CHOH \longrightarrow Ru^{IV}(tmp)(OCHMe_2)_2 + Me_2CO + 2\,H_2O \quad (8)$$

The oxoruthenium(VI) complex was prepared by exposing a benzene solution of *trans*-Ru^{II} (tmp)(MeCN)$_2$ to air at 20 °C. Addition of isopropanol to the resulting solution, in the absence of air, afforded the dialkoxyruthenium(IV) complex, in quantitative yield, within 24 h. In the presence of air, benzene solutions of the dioxoruthenium(VI) or the dialkoxyruthenium(IV) complex effected catalytic oxidation of isopropanol at room temperature, albeit with a modest rate (1.5 catalytic turnovers per day). Interestingly, with the dialkoxyruthenium(IV) complex, catalytic oxidation was observed with air but not with dry oxygen, suggesting that hydrolysis to an oxoruthenium(IV) complex is necessary for a catalytic cycle.

Other ruthenium-based catalysts for the aerobic oxidation of alcohols have been described where it is not clear if they involve oxidative dehydrogenation by low-valent ruthenium, to give hydridoruthenium intermediates, or by high-valent oxoruthenium. For example, both RuO_2 and 5% Ru-on-charcoal catalyze the aerobic oxidation of activated alcohols such as allylic alcohols [59] and α-ketols [60], e.g., Eq. (9).

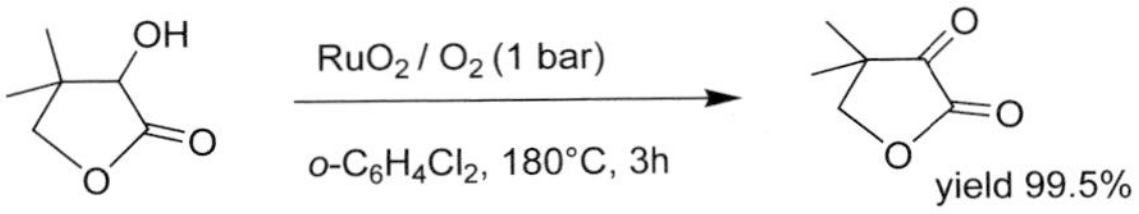

Kagan and coworkers [61] have described the use of ruthenium supported on ceria, CeO_2, as a catalyst for the aerobic oxidation of alcohols. Primary and secondary alcohols are oxidized to the corresponding aldehydes (carboxylic acids) and ketones, respectively, at elevated temperatures (>140 °C). Surprisingly, allylic alcohols, such as geraniol, and some cyclic alcohols, e.g., menthol, are unreactive. The former result suggests that low-valent ruthenium species are possibly involved and that coordination of ruthenium to the double bond inhibits alcohol oxidation.

Recently Mizuno and Yamaguchi [62] reported ruthenium on alumina to be a powerful and recyclable catalyst for selective alcohol oxidation. This method displayed a large substrate scope [see Eq. (10) and Table 4.5] and tolerates the presence of sulfur and nitrogen groups. Only primary aliphatic alcohols required the addition of hydroquinone. Turnover frequencies in the range of 4 h^{-1} (for secondary allylic alcohols) to 18 h^{-1} (for 2-octanol) were obtained in trifluorotoluene, while in the solvent-free oxidation at 150 °C a TOF of 300 h^{-1} was observed for 2-octanol.

$$R_1R_2CH(OH) + 0.5\,O_2 \xrightarrow[\text{PhCF}_3,\ 83\,°C,\ 1\ \text{atm}\ O_2]{2.5\ \text{mol\%}\ Ru(OH)_3\ \text{on}\ Al_2O_3} R_1R_2C=O + H_2O \quad (10)$$

The catalyst consists of highly dispersed $Ru(OH)_3$ on the surface of γ-Al_2O_3. Based, inter alia, on the fact that this catalyst is also capable of performing a transfer hydrogenation using 2-propanol as the hydrogen donor, it was concluded that the mechanism of this reaction proceeds via a hydridometal pathway.

Ruthenium-exchanged hydrotalcites were shown by Kaneda and coworkers [63], to be heterogeneous catalysts for the aerobic oxidation of reactive allylic and benzylic alcohols. Hydrotalcites are layered anionic clays consisting of a cationic Brucite layer with anions (hydroxide or carbonate) situated in the interlayer region. Various cations can be introduced in the Brucite layer by ion exchange. For example, ruthenium-exchanged hydrotalcite with the formula $Mg_6Al_2Ru_{0.5}(OH)_{16}CO_3$, was prepared by treating an aqueous solution of $RuCl_3 \cdot 3H_2O$, $MgCl_2 \cdot 6H_2O$ and

Tab. 4.5 $Ru(OH)_3$-Al_2O_3 catalyzed oxidation of primary and secondary alcohols to the corresponding aldehydes and ketones using O_2 [a]

Substrate	Time (h)	Conversion (%)	Selectivity (%)
n-$C_6H_{13}CH(CH_3)OH$	2	91	>99
Cyclooctanol	6	81	>99
n-$C_7H_{15}CH_2OH$ [b]	4	87	98
$PhCH(CH_3)OH$	1	>99	>99
$(CH_3)_2C=CH(CH_2)_2CH(CH_3)=CHCH_2OH$ [c]	6	89	97
$PhCH_2OH$	1	>99	>99
$(4$-$NO_2)PhCH_2OH$	3	97	>99

[a] According to ref. [62]; 2.5 mol% Ru/Al_2O_3, $PhCF_3$ as solvent, 83 °C, 1 atm O_2; conversion and yields determined by GLC. [b] 5 mol% Ru/Al_2O_3 and 5 mol% hydroquinone (to suppress over-oxidation) were used. [c] Geraniol

AlCl$_3$ · H$_2$O with a solution of NaOH and Na$_2$CO$_3$ followed by heating at 60 °C for 18 h [63]. The resulting slurry was cooled to room temperature, filtered, washed with water and dried at 110 °C for 12 h. The resulting ruthenium–hydrotalcite showed the highest activity amongst a series of hydrotalcites exchanged with, e.g., Fe, Ni, Mn, V and Cr.

Subsequently, the same group showed that the activity of the ruthenium–hydrotalcite was significantly enhanced by the introduction of cobalt(II), in addition to ruthenium(III), in the Brucite layer [64]. For example, cinnamyl alcohol underwent complete conversion in 40 min in toluene at 60 °C, in the presence of Ru/Co-HT, compared with 31% conversion under the same conditions with Ru–HT. A secondary aliphatic alcohol, 2-octanol, was smoothly converted into the corresponding ketone but primary aliphatic alcohols, e.g., 1-octanol, exhibited extremely low activity. The authors suggested that the introduction of cobalt induced the formation of higher oxidation states of ruthenium, e.g., Ru^{IV} to Ru^{VI}, leading to a more active oxidation catalyst. However, on the basis of the reported results it is not possible to rule out low-valent ruthenium species as the active catalyst in a hydridometal pathway. The results obtained in the oxidation of representative alcohols with Ru–HT and Ru–Co–HT are compared in Table 4.6.

Tab. 4.6 Oxidation of various alcohols to their corresponding aldehydes or ketones with Ru-hydrotalcites using molecular oxygen[a]

Substrate	Ru-Mg-Al-CO$_3$-HT[b]		Ru-Co-Al-CO$_3$-HT[c]	
	Time	Yield (%)	Time	Yield (%)
PhCH=CHCH$_2$OH	8 h	95[d]	40 min	94
PhCH$_2$OH	8 h	95[d]	1 h	96
4-ClPhCH$_2$OH	8 h	61[e]	1.5 h	95
PhCH(CH$_3$)OH	18 h	100	1.5 h	100
n-C$_6$H$_{13}$CH(CH$_3$)OH	–	–	2 h	97
(CH$_3$)$_2$C=CH(CH$_2$)$_2$CH(CH$_3$)CH$_2$OH[f]	–	–	12 h	71[g]

[a] 2 mmol substrate, 0.3 g hydrotalcite ($\approx$14 mol%), in toluene, 60 °C, 1 bar O$_2$. Conversion 100%.
[b] See ref. [63]. [c] See ref. [64]. [d] Conversion 98%. [e] Conversion 64%. [f] Geraniol. [g] Conversion 89%.

In 2000, Kaneda et al. synthesized a ruthenium-based hydroxyapatite catalyst, with the formula (RuCl)$_{10}$(PO$_4$)$_6$(OH)$_2$ [65]. This catalyst could also be recycled and displayed a reasonable substrate scope in the aerobic alcohol oxidations [e.g., Eq. (11)]. Turnover frequencies reported in this case were generally somewhat lower, in the order of 1 h^{-1} for 2-octanol, to 12 h^{-1} for benzyl alcohol. The fact that distinct Ru-Cl species are present at the surface points in the direction of a hydridometal mechanism.

The same group recently reported the use of a ferrite spinel catalyst ($MnFe_2O_4$), where the iron was partially substituted with Ru and Cu, i.e., $MnFe_{1.5}Ru_{0.35}Cu_{0.15}O_4$ for the room temperature oxidation of alcohols [66]. However, 20 mol% catalyst (based on ruthenium) was necessary to accomplish even the oxidation of benzyl alcohol. For primary and secondary aliphatic alcohols turnover frequencies of 2 h^{-1} and 3.5 h^{-1}, respectively, were the maximum rates achieved.

Another class of ruthenium catalysts, which has attracted considerable interest due to their inherent stability under oxidative conditions, is the polyoxometalates [67]. Recently, Yamaguchi and Mizuni [68] reported that a mono-ruthenium-substituted silicotungstate, synthesized by the reaction of the lacunary polyoxometalate $[SiW_{11}O_{39}]^{8-}$ with Ru^{3+} in an organic solvent, acts as an efficient heterogeneous catalyst with high turnover frequencies for the aerobic oxidation of alcohols (see Table 4.7). Among the solvents used 2-butyl acetate was the most effective and this Ru-heteropolyanion could be recycled. The low loading used resulted in very long reaction times of >2 d (see Table 4.7).

Tab. 4.7 Ru substituted-polyoxometalates as catalysts for the oxidation of alcohols

Substrate	Mono-Ru-silicotungstate[a]		Over-exchanged Ru/$SiW_{11}O_{39}$/THA[b]	
	Time (h)	Conv. (Sel.) (%)[c]	Time (h)	Conv. (Sel.) (%)
n-$C_7H_{15}CH(CH_3)OH$	120	90 (88)		
Cyclohexanol	48	67 (81)	48	44 (38)
n-$C_6H_{13}CH(CH_3)OH$	48	14 (44)[d]		99 (92)
Ph-CH=CH_2-CH_2OH			2	100 (96)
$PhCH_2OH$	120	36 (65)[e]		100 (96)

[a] 0.05 mol% $[TBA]_4H[SiW_{11}Ru(H_2O)O_{39}]\cdot 2H_2O$, isobutyl acetate as solvent, 110 °C, 1 atm O_2 see ref. [68]. [b] 5.8 mol% (on Ru) $(THA)_x(Ru_7Oy)SiW_{11}O_{39}$, a homogeneous catalyst, prepared by ex-changing lacunary $K_8[SiW_{11}O_{39}]$ with basic $RuCl_3$, followed by precipitation in organic solution: temperature 80 °C, solvent PhCl see ref. [69]. [c] Selectivity towards aldehyde or ketone. [d] 30% acid was also formed. [e] 10% benzoic acid was also formed.

An example of a homogeneous Ru-heteropolyanion derivative is also shown in Table 4.7. In this case the same lacunary silicotungstate was over-exchanged with a basic $RuCl_3$ solution. The resulting solution was precipitated in organic solution using $[(C_6H_{13})_4N]HSO_4$ [69] and elemental analysis showed that 7 Ru molecules were present per molecule of $[SiW_{11}O_{39}]^{8-}$. This led us to postulate a structure comprising Ru-oxide clusters stabilized by the HPA. This material displayed better results than Ru–HPA molecules which were prepared according to previous publications and subjected to the conditions in Table 4.7 [70, 71]

In contrast to the above mentioned reactions, which involve either oxoruthenium or ruthenium-hydride species as intermediates, free radical reactions can also be promoted by ruthenium. The aerobic oxidation of alcohols proceeds smoothly at room temperature in the presence of one equivalent of an aldehyde, e.g., acetaldehyde, and a catalyst comprising a 1:1 mixture of $RuCl_3 \cdot nH_2O$ and $Co(OAc)_2$, in ethyl acetate [Eq. (12)] [72].

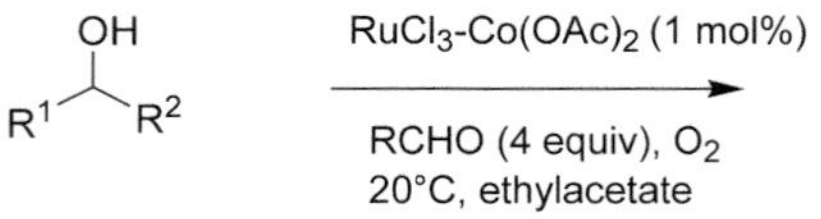

$$R^1R^2CH(OH) \xrightarrow[20°C, \text{ethylacetate}]{\text{RuCl}_3\text{-Co(OAc)}_2 \text{ (1 mol\%)}, \text{RCHO (4 equiv), O}_2} R^1C(O)R^2 \quad (12)$$

Representative examples are shown in Table 4.8. The results were rationalized by assuming that the corresponding percarboxylic acid is formed by cobalt-mediated free radical autoxidation of the aldehyde. Subsequent reaction of ruthenium(III) with the peracid affords oxoruthenium(V) carboxylate which is the active oxidant. Compared with the aerobic oxidations discussed above the method suffers from the drawback that one equivalent of a carboxylic acid is formed as a coproduct.

Tab. 4.8 Oxidation of alcohols with a ruthenium/cobalt/aldehyde catalytic system using molecular oxygen[a]

Substrate	Product	Isolated yield (%)
$n\text{-}C_3H_7CH(OH)\text{-}n\text{-}C_4H_9$	$n\text{-}C_3H_7C(=O)\text{-}n\text{-}C_4H_9$	89
Cyclooctanol	cyclooctanone	95
L-menthol	L-menthone	91
$PhCH(CH_3)OH$	$PhC(O)CH_3$	94
$n\text{-}C_7H_{15}CH_2OH$	$n\text{-}C_7H_{15}COOH$	96

[a] According to ref. [72]. Reaction conditions: 10 mmol alcohol, 0.10 mmol $RuCl_3 \cdot nH_2O$, 0.10 mmol $Co(OAc).4H_2O$, EtAc, 40 mmol acetaldehyde added over 1.5 h, 20 °C, 1 bar O_2.

4.5
Palladium-catalyzed Oxidations with O$_2$

Palladium(II) is also capable of mediating the oxidation of alcohols via the hydrido-metal pathway shown in Scheme 4.5. Blackburn and Schwarz first reported [73] the $PdCl_2$–NaOAc-catalyzed aerobic oxidation of alcohols in 1977. However, activities were very low, with turnover frequencies of the order of 1 h^{-1}. Subsequently, much effort has been devoted to finding synthetically useful methods for the palladium-catalyzed aerobic oxidation of alcohols. For example, the giant palladium cluster, $Pd_{561}phen_{60}(OAc)_{180}$ [74], was shown to catalyze the aerobic oxidation of primary allylic alcohols to the corresponding α,β-unsaturated aldehydes [Eq. (13)] [75].

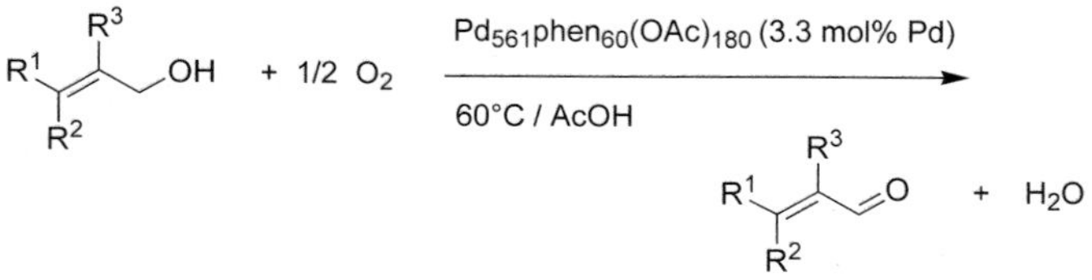

$$R^1R^2C=C(R^3)CH_2OH + 1/2\, O_2 \xrightarrow[60°C / \text{AcOH}]{Pd_{561}phen_{60}(OAc)_{180} \text{ (3.3 mol\% Pd)}} R^1R^2C=C(R^3)CHO + H_2O \quad (13)$$

In 1998, Peterson and Larock showed that $Pd(OAc)_2$, in combination with $NaHCO_3$ as a base in DMSO as the solvent, catalyzed the aerobic oxidation of primary and secondary allylic and benzylic alcohols to the corresponding aldehydes and ketones, respectively, in fairly good yields [76]. In both cases, ethylene carbonate and DMSO acted both as the solvent as well as the ligand necessary for a smooth reoxidation [77]. Similarly, $PdCl_2$, in combination with sodium carbonate and a tetraalkylammonium salt, Adogen 464, as a phase transfer catalyst, catalyzed the aerobic oxidation of alcohols, e.g., 1,4- and 1,5-diols afforded the corresponding lactones [Eq. (14)] [78, 79].

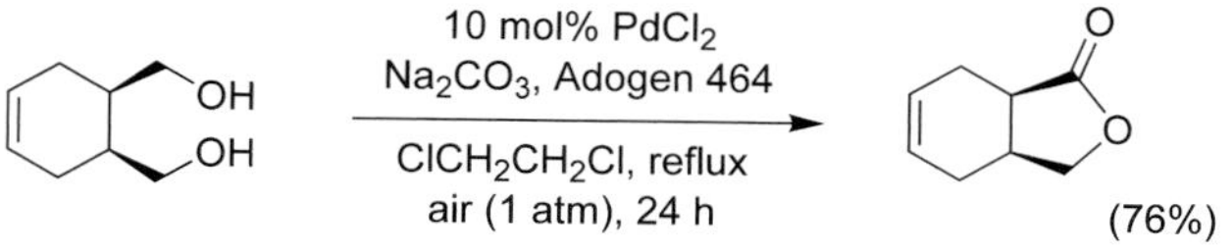

$$\text{cyclohexane-1,2-diol} \xrightarrow[\text{ClCH}_2\text{CH}_2\text{Cl, reflux}]{\substack{10\ \text{mol}\%\ PdCl_2 \\ Na_2CO_3,\ \text{Adogen 464} \\ \text{air (1 atm), 24 h}}} \text{lactone (76\%)} \quad (14)$$

However, these methods suffer from low activities and/or narrow scope. Uemura and coworkers [80, 81] reported an improved procedure involving the use of $Pd(OAc)_2$ (5 mol%) in combination with pyridine (20 mol%) and 3 Å molecular sieves (500 mg per mmol of substrate) in toluene at 80 °C. This system smoothly catalyzed the aerobic oxidation of primary and secondary aliphatic alcohols to the corresponding aldehydes and ketones, respectively, in addition to benzylic and allylic alcohols. Representative examples are summarized in Table 4.9. 1,4- and 1,5-Diols afforded the corresponding lactones. This approach could also be employed under fluorous biphasic conditions [82].

Tab. 4.9 Pd^{II}-catalyzed oxidation of various alcohols to their corresponding ketones or aldehydes in the presence of pyridine using molecular oxygen [a]

Substrate	Conversion after 2 h (%)	Yield aldehyde/ketone (%)
$PhCH_2OH$	100	100
$4\text{-}ClPhCH_2OH$	100	98
$n\text{-}C_{11}H_{23}CH_2OH$	97	93[b]
$n\text{-}C_{10}H_{21}CH(CH_3)OH$	98	97[b]
$PhCH=CHCH_2OH$	46	35[b]

[a] Data from ref. [81]. Reaction conditions: alcohol 1.0 mmol, 5 mol% $Pd(OAc)_2$, 20 mol% pyridine, 500 mg MS 3A, toluene 10 mL, 80 °C, 1 bar O_2, 2 h. [b] Isolated yield.

Recently Stahl et al. conducted mechanistic studies on both systems: the Pd/DMSO and the Pd/pyridine system [83, 84]. Kinetic studies revealed that in the Pd/pyridine system, the rate exhibits no dependence on the oxygen pressure, and kinetic isotope effect studies support turnover-limiting substrate oxidation. In contrast the Pd/DMSO system features turnover-limiting oxidation of palladium(0) (see Scheme 4.13). Moreover in the Pd/pyridine system, pyridine is very effective in oxidizing palladium(0) by molecular oxygen, but at the same time inhibits the rate of alcohol oxidation by palladium(II).

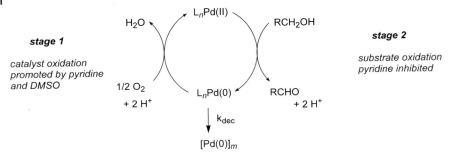

Scheme 4.13 Mechanistic insights to Pd/pyridine and Pd/DMSO systems

Although this methodology constitutes an improvement on those previously reported, turnover frequencies were still generally < 10 h^{-1} and, hence, there is considerable room for further improvement. Recent attempts to replace either pyridine by triethylamine [85], or Pd(OAc)$_2$ by palladacycles [86] all resulted in lower activities.

Recently, we described the use of a water-soluble palladium(II) complex of sulfonated bathophenanthroline as a stable, recyclable catalyst for the aerobic oxidation of alcohols in a two-phase aqueous–organic medium, e.g., in Eq. (15) [16, 87, 88]. Reactions were generally complete in 5 h at 100 °C/30 bar air with as little as 0.25 mol% catalyst. No organic solvent is required (unless the substrate is a solid) and the product ketone is easily recovered by phase separation. The catalyst is stable and remains in the aqueous phase, which can be recycled to the next batch.

$$\text{alcohol} \xrightarrow[\text{pH 11.5, 100°C, air (30 atm), 10 h}]{\text{Pd}^{II} \text{ (0.25 mol\%)}} \text{ketone} \quad \text{(yield 92\%)} \quad (15)$$

A wide range of alcohols were oxidized with TOFs ranging from 10 h^{-1} to 100 h^{-1}, depending on the solubility of the alcohol in water (since the reaction occurs in the aqueous phase the alcohol must be at least sparingly soluble in water). Thus, in a series of straight-chain secondary alcohols the TOF decreased from 100 h^{-1} to 13 h^{-1} on increasing the chain length from 1-pentanol to 1-nonanol. Representative examples of secondary alcohols that were smoothly oxidized using this system are collected in Table 4.10. The corresponding ketones were obtained in >99% selectivity in virtually all cases.

Primary alcohols afforded the corresponding carboxylic acids *via* further oxidation of the aldehyde intermediate, e.g., 1-hexanol afforded 1-hexanoic acid in 95% yield. It is important to note, however, that this was achieved without the requirement of 1 equiv. of base to neutralize the carboxylic acid product (which is the case with sup-

Tab. 4.10 Conversion of secondary alcohols into their corresponding ketones using a Pd^{II}-bathophenanthroline-complex in a two-phase system[a]

Substrate	Time (h)	Conversion (%)	Selectivity[b] (%)	Isolated yield (%)
n-$C_3H_7CH(CH_3)OH$	5	100	100	90
n-$C_4H_9CH(CH_3)OH$	10	100	100	90
Cyclopentanol	5	100	100	90
$PhCH(CH_3)OH$	10	90	100	85
$CH_3CH=CHCH(CH_3)OH$	10	95	83[c]	79
n-$C_4H_9OCH_2CH(CH_3)OH$	10	100	100	92

[a] Reaction conditions 20 mmol alcohol, 0.05 mmol; $PhenS*Pd(OAc)_2$, 1 mmol NaOAc, 100 °C, 30 bar air. [b] Selectivity to ketone, determined by gas chromatography with an external standard. [c] Ether (17%) was formed.

ported noble metal catalysts) [5]. In contrast, when 1 mol% TEMPO (4 equiv. per Pd) was added the aldehyde was obtained in high yield, e.g., 1-hexanol afforded 1-hexanal in 97% yield. Some representative examples of primary alcohol oxidations using this system are shown in Table 4.11. The TEMPO was previously shown to suppress the autoxidation of aldehydes to the carboxylic acids (see earlier).

Compared with existing systems for the aerobic oxidation of alcohols, the Pd–bathophenanthroline system is among the fastest catalytic systems reported today, requires no solvent and product/catalyst isolation involves simple phase separation. The system has broad scope but is not successful with all alcohols. Some examples of unreactive alcohols are shown in Scheme 4.14. Low reactivity was generally observed with alcohols containing functional groups which could strongly coordinate to the palladium.

The reaction is half-order in palladium and first-order in the alcohol substrate, when measured with a water soluble alcohol to eliminate the complication of mass transfer [88]. A possible mechanism is illustrated in Scheme 4.15. The resting catalyst is a dimeric complex containing bridging hydroxyl groups. Reaction with the alcohol in the presence of a base, added as a cocatalyst (NaOAc) or free ligand, affords a mono-

Tab. 4.11 Conversion of primary alcohols into their corresponding aldehydes or acids using a Pd^{II}–bathophenanthroline complex in a two-phase system[a]

Substrate	Product	Time (h)	Conv. (%)	Sel.[b] (%)	Isolated yield (%)
n-$C_4H_9CH_2OH$[c]	n-C_4H_9CHO	15	98	97[d]	90
n-$C_5H_9CH_2OH$	n-C_5H_9COOH	12	95	90[e]	80
$PhCH_2OH$	$PhCHO$	10	100	99.8[d]	93
$(CH_3)_2CH=CHCH_2OH$	$(CH_3)_2CH=CHCHO$	10	95	83[d]	79

[a] Reaction conditions 10 mmol alcohol, 0.05 mmol $PhenS*Pd(OAc)_2$, 1 mmol NaOAc, 100 °C, 30 bar air. [b] Selectivity to product, determined by gas chromatography with an external standard. [c] TEMPO(4 equiv. to Pd) was added. [d] Acid was formed as the major byproduct. [e] Hexanal and hexanoate were formed.

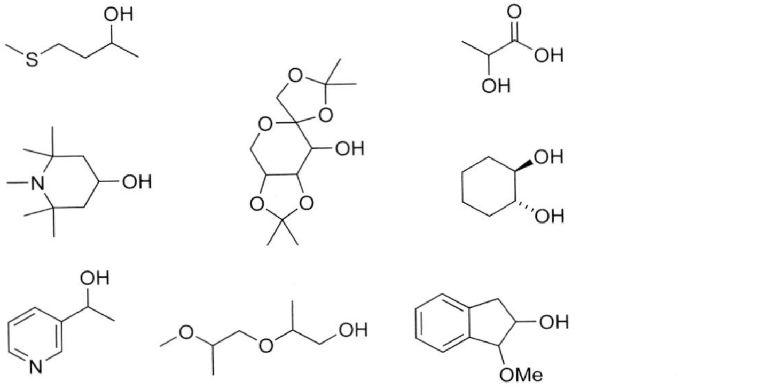

Scheme 4.14 Unreactive alcohols in the Pd–bathophenanthroline catalytic system

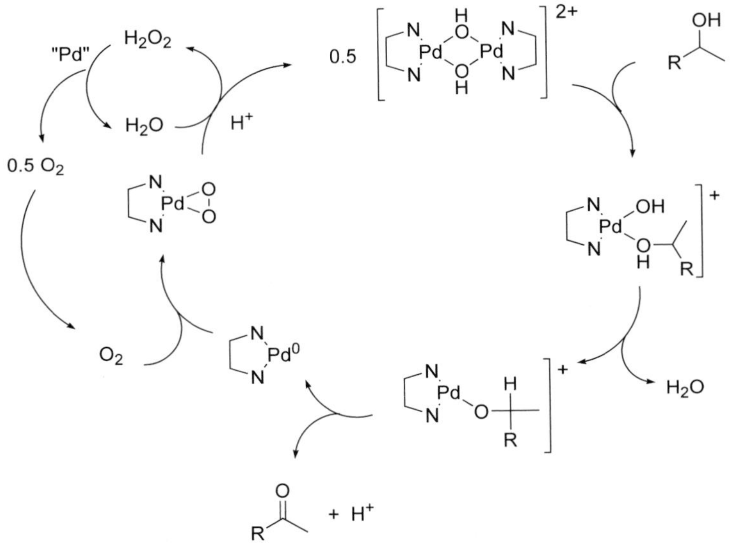

Scheme 4.15 Mechanism of Pd–bathophenanthroline catalyzed oxidation of alcohols

meric alkoxy palladium(II) intermediate which undergoes β-hydride elimination to give the carbonyl compound, water and a palladium(0) complex. Oxidative addition of dioxygen to the latter affords a palladium(II) η-peroxo complex, which can react with the alcohol substrate to regenerate the catalytic intermediate, presumably with concomitant formation of hydrogen peroxide as was observed in analogous systems [89].

According to the proposed mechanism the introduction of substituents at the 2- and 9-positions in the PhenS ligand would, as a result of steric hindrance (see Scheme 4.15), promote dissociation of the dimer and enhance the reactivity of the catalyst. This proved to be the case: introduction of methyl groups at the 2- and 9-position (the ligand is commercially available and is known as bathocuproin) tripled the activity in 2-hexanol oxidation [88].

It is worth noting, in this context, that palladium complexes of substituted phenanthrolines were recently shown [90] to catalyze the formation of hydrogen peroxide, by reaction of a primary or a secondary alcohol with dioxygen, in the presence of an acid cocatalyst, e.g., $C_7F_{15}CO_2H$, in a biphasic chlorobenzene/water medium at 70 °C and 5 bar. Turnover frequencies up to 220 h^{-1} were observed. The hydrogen peroxide is formed in the organic phase, via palladium catalyzed oxidation of the alcohol, but is subsequently extracted into the water phase where it is protected from decomposition by the palladium complex. The same catalyst system was also used for the production of hydrogen peroxide from a mixture of carbon monoxide, water and dioxygen, with turnover frequencies up to 600 h^{-1}, according to Eq. (16) [89].

$$O_2 + CO + H_2O \xrightarrow[\text{1,2,4 } C_6H_3Cl_3 \text{ / 2-methyl-2-butanol / } H_2O]{Pd^{II}(L)(C_7F_{15}COO)_2 \quad 70°C} H_2O_2 + CO_2 \quad (16)$$

$$\binom{65}{atm} \binom{6}{atm}$$

L = 2,9-dimethyl-4,7-diphenyl-1,10-phenanthroline

In the context of heterogeneous palladium catalysts, the previously mentioned Pd/C catalysts are commonly used for water-soluble substrates, i.e., carbohydrates [91]. Other examples of heterogeneous Pd catalysts are rare. Recently it was shown that in addition to ruthenium, palladium can also be introduced in the brucite-layer of the hydrotalcite [92]. As with Ru/Co–hydrotalcite (see above), apart from benzylic and allylic also aliphatic and cyclic alcohols are smoothly oxidized using this palladium–hydrotalcite. However a major shortcoming is the necessity of at least 5 mol% catalyst and the co-addition of 20–100 mol% pyridine.

4.6
Copper-catalyzed Oxidations with O_2

Copper would seem to be an appropriate choice of metal for the catalytic oxidation of alcohols with dioxygen since it comprises the catalytic center in a variety of enzymes, e.g., galactose oxidase, which catalyze this conversion in vivo [93, 94]. However, despite extensive efforts [95] synthetically useful copper-based systems have generally not been forthcoming. For instance, in the absence of other metals, CuCl in combination with 2,2'-bipyridine (bipy) as the base/ligand shows catalytic activity in the aerobic oxidation of alcohols. However, benzhydrol is the only suitable substrate and at least 1 equiv. of bipy (relative to substrate) is required to reach complete conversion. On the other hand, with ortho-phenanthroline as the ligand, $CuCl_2$ can catalyze the aerobic oxidation of a variety of primary and secondary alcohols to the corresponding carboxylic acids and ketones in alkaline media [95].

A special class of active copper-based aerobic oxidation systems comprises the biomimetic models of galactose oxidase, i.e., Cu^{II}-phenoxyl radical complexes, reported by Stack and Wieghardt [96–99]. Just like the enzyme itself, these monomeric Cu^{II} species are effective only with easily oxidized benzylic and allylic alcohols, simple primary and secondary aliphatic alcohols being largely unreactive. A good example of a biomimetic model of galactose oxidase is [Cu^{II}BSP], in which BSP stands for a salen-ligand with a binaphthyl backbone (Scheme 4.16). The rate determining step (RDS) of this interesting system was suggested to involve inner sphere one-electron transfer from the alkoxide ligand to Cu^{II} followed by hydrogen-transfer to the phenoxyl radical yielding Cu^{I}, phenol and the carbonyl product (Scheme 4.16) [100].

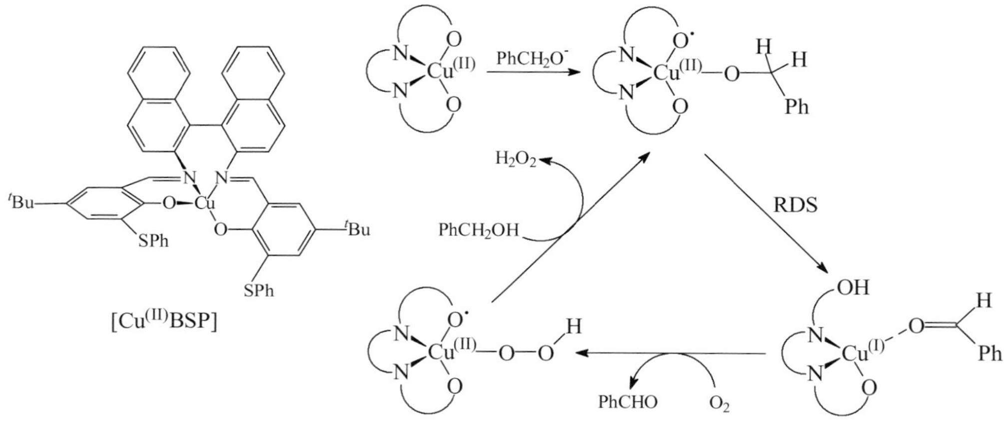

Scheme 4.16 [Cu^{II}BSP]-catalyzed aerobic oxidation of benzyl alcohol

More recently, Marko and coworkers [101, 102] reported that a combination of CuCl (5 mol%), phenanthroline (5 mol%) and di-*tert*-butylazodicarboxylate, DBAD (5 mol%), in the presence of 2 equiv. of K_2CO_3, catalyzes the aerobic oxidation of allylic and benzylic alcohols [Eq. (17)]. Primary aliphatic alcohols, e.g., 1-decanol, could be oxidized but required 10 mol% catalyst for smooth conversion.

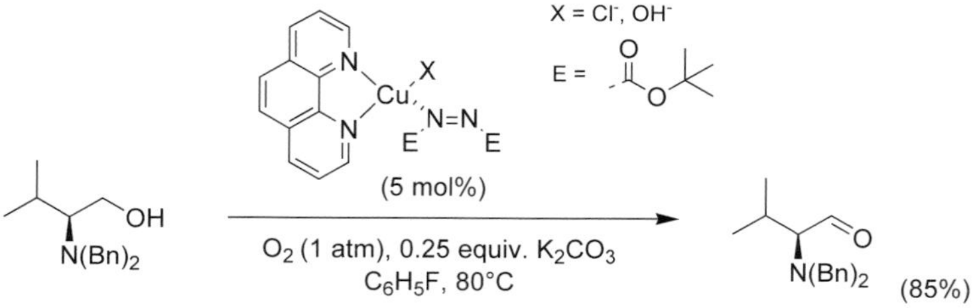

(17)

The nature of the copper counterion was critical, with chloride, acetate and triflate proving to be the most effective. Polar solvents such as acetonitrile inhibit the reaction whereas smooth oxidation takes place in apolar solvents such as toluene. An ad-

Tab. 4.12 Copper-catalyzed aerobic oxidation of alcohols to the corresponding aldehyde or ketone using DBAD and K_2CO_3 [a]

Substrate	Carbonyl yield[b] (%)
MeS-PhCH$_2$OH	81
Ph-CH=CHCH$_2$OH	89
(CH$_3$)$_2$C=CH(CH$_2$)$_2$CH(CH$_3$)CH$_2$OH[c]	71
C$_9$H$_{19}$CH$_2$OH	65
C$_9$H$_{19}$CH(CH$_3$)OH	88

[a] Table adapted from ref. [102]. Conditions: 5 mol% CuCl, 5 mol% phenanthroline, 5 mol% DBAD-H$_2$ (DBAD = dibutylazodicarboxylate), 2 equiv. K_2CO_3, gentle stream of O_2, solvent is toluene, 90 °C. After 1 h reaction was complete. [b] Isolated yields at 100% conversion. [c] Geraniol.

vantage of the system is that it tolerates a variety of functional groups (see Table 4.12 for examples). Serious drawbacks of the system are the low activity, the need for 2 equiv. of K_2CO_3 (relative to substrate) and the expensive DBAD as a co-catalyst. According to a later report [103] the amount of K_2CO_3 can be reduced to 0.25 equiv. by changing the solvent to fluorobenzene.

The active catalyst is heterogeneous, being adsorbed on the insoluble K_2CO_3 (filtration gave a filtrate devoid of activity). Besides fulfilling a role as a catalyst support the K_2CO_3 acts as a base and as a water scavenger. The mechanism illustrated in Scheme 4.17 was postulated to explain the observed results.

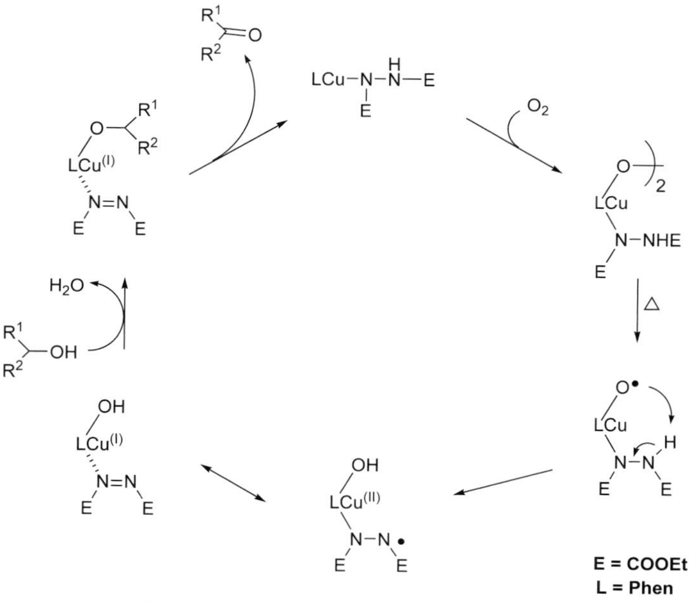

Scheme 4.17 Mechanism of CuCl.phen catalyzed oxidation of alcohols using DEAD-H$_2$ (diethylazo dicarboxylate) as an additive

Semmelhack et al. [104] reported that the combination of CuCl and 4-hydroxy TEMPO catalyzes the aerobic oxidation of alcohols. However, the scope was limited to active benzylic and allylic alcohols and activities were low (10 mol% of catalyst was needed for smooth reaction). They proposed that the copper catalyzes the reoxidation of TEMPO to the oxoammonium cation. Based on our results with the Ru/TEMPO system we doubted the validity of this mechanism. Hence, we subjected the Cu/TEMPO to the same mechanistic studies described above for the Ru/TEMPO system [105]. The results of stoichiometric experiments under anaerobic conditions, Hammett correlations and kinetic isotope effect studies showed a similar pattern to those with the Ru/TEMPO system, i.e., they are inconsistent with a mechanism involving an oxoammonium species as the active oxidant. Hence, we propose the mechanism shown in Scheme 4.18 for Cu/TEMPO-catalyzed aerobic oxidation of alcohols.

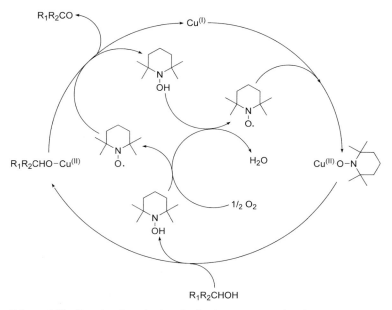

Scheme 4.18 Postulated mechanism for the Cu/TEMPO catalyzed oxidation of alcohols

We have shown, in stoichiometric experiments, that reaction of copper(I) with TEMPO affords a piperidinyloxyl copper(II) complex. Reaction of the latter with a molecule of alcohol afforded the alkoxycopper(II) complex and TEMPOH. Reaction of the alkoxycopper(II) complex with a second molecule of TEMPO gave the carbonyl compound, copper(I) and TEMPOH. This mechanism resembles that proposed for the aerobic oxidation of alcohols catalyzed by the copper-dependent enzyme, galactose oxidase, and mimics thereof. Finally, TEMPOH is reoxidized to TEMPO by oxygen. We have also shown that copper in combination with PIPO affords an active and recyclable catalyst for alcohol oxidation [18].

Recently, two improvements of the Cu/Tempo system were published, which are related to solvent innovation. In the first example, Knochel and coworkers [106] showed that $CuBr \cdot Me_2S$ with perfluoroalkyl substituted bipyridine as the ligand was capable of oxidizing a large variety of primary and secondary alcohols in a fluorous biphasic system of chlorobenzene and perfluorooctane [see Eq. (18) and Table 4.13]. In the second example Ansari and Gree [107] showed that the combination of CuCl and TEMPO can be used as a catalyst in 1-butyl-3-methylimidazolium hexafluorophosphate, an ionic liquid, as the solvent. However in this case turnover frequencies were still rather low even for benzylic alcohol (around $1.3\ h^{-1}$).

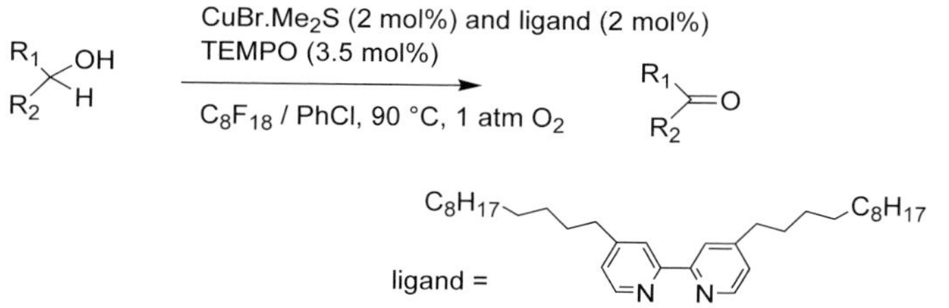

$$\begin{array}{c} R_1 \\ R_2 \end{array}\!\!\!\!\!\!\!\!\!\!\diagdown\!\!\!\!OH \quad \xrightarrow[C_8F_{18}\ /\ PhCl,\ 90\ °C,\ 1\ atm\ O_2]{CuBr \cdot Me_2S\ (2\ mol\%)\ and\ ligand\ (2\ mol\%),\ TEMPO\ (3.5\ mol\%)} \quad \begin{array}{c} R_1 \\ R_2 \end{array}\!\!\!\!\!\!\!\!\!=\!O \qquad (18)$$

ligand = C_8H_{17}—[bipyridine]—C_8H_{17}

Tab. 4.13 Copper-catalyzed aerobic oxidation of alcohols under fluorous biphasic conditions[a]

Substrate	Time (h)	Isolated yield (%)
$n\text{-}C_8H_{17}CH(CH_3)OH$	7–13	71
$n\text{-}C_9H_{19}CH_2OH$	7–13	73
$(2\text{-}Br)PhCH_2OH$	2–7	96
$PhCH=CH\text{-}CH_2OH$	2–7	79

[a] For conditions see reaction 18, ref [106].

Osborn and coworkers [108–110] reported that CuCl in combination with OsO_4 or Pr_4NRuO_4 (TPAP) catalyzes the aerobic oxidation of alcohols. The scope is rather limited, however, and the system would not appear to have any advantages over the earlier described ruthenium- and palladium-based systems. Similarly an $MoO_2(\text{a-cac})_2$–$Cu(NO_3)_2$ system [111] resulted in rather low activities and selectivities for the oxidation of primary activated and secondary alcohols.

4.7
Other Metals as Catalysts for Oxidation with O_2

In addition to ruthenium, other late and first-row transition elements are capable of dehydrogenating alcohols *via* an oxometal pathway. Some are used as catalysts, in combination with H_2O_2 or RO_2H, for the oxidative dehydrogenation of alcohols (see later). By analogy with ruthenium, one might expect that regeneration of the active oxidant with dioxygen would be possible. For example, one could easily envisage alcohol oxidation by oxovanadium(V) followed by reoxidation of the resulting vanadium(III) by dioxygen. However, scant attention appears to have been paid to such possibilities. The aerobic oxidation of 1-propanol to 1-propanal (94–99% selectivity), in the gas phase at 210 °C over a V_2O_5 catalyst modified with an alkaline earth metal oxide (10 mol%), was described [112] in 1979. However, to our knowledge vanadium-catalyzed aerobic oxidation of alcohols have not been further investigated, in the liquid or gas phase [113].

$Co(acac)_3$ in combination with N-hydroxyphthalimide (NHPI) as co-catalyst mediates the aerobic oxidation of primary and secondary alcohols, to the corresponding carboxylic acids and ketones, respectively, e.g., Eq. (19) [114].

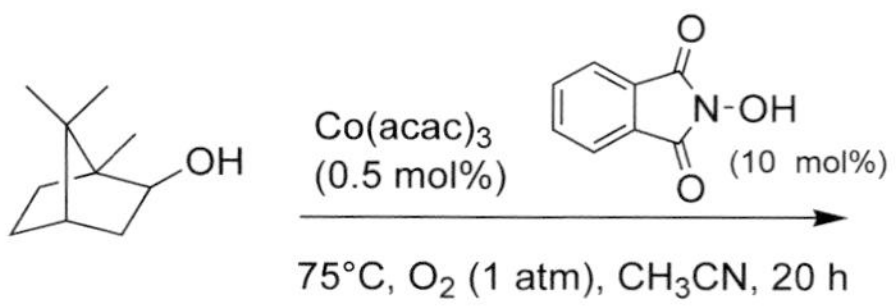

$$\text{substrate} \xrightarrow[\text{75°C, O}_2 \text{ (1 atm), CH}_3\text{CN, 20 h}]{Co(acac)_3 \text{ (0.5 mol\%)}, \text{N-OH (10 mol\%)}} \text{product} \quad (19)$$

By analogy with other oxidations mediated by the Co/NHPI catalyst studied by Ishii and coworkers [115, 116], Eq. (19) probably involves a free radical mechanism. We attribute the promoting effect of NHPI to its ability to efficiently scavenge alkylperoxy radicals, suppressing the rate of termination by combination of alkylperoxy radicals. The resulting PINO radical subsequently abstracts a hydrogen atom from the α-C–H bond of the alcohol to propagate the autoxidation chain [Eqs. (20–22)].

$$\text{PhthN-OH} + \text{R-OO}^\bullet \longrightarrow \text{PhthN-O}^\bullet + \text{R-OOH} \quad (20)$$

$$\text{PhthN-O}^\bullet + \text{R-H} \longrightarrow \text{PhthN-OH} + \text{R}^\bullet \quad (21)$$

$$\text{R}^\bullet + O_2 \longrightarrow \text{R-OO}^\bullet \quad (22)$$

Recently a nickel substituted hydrotalcite was reported as a catalyst for the aerobic oxidation of benzylic and allylic alcohols [117]. Analogous to cobalt, nickel is expected to catalyze oxidation *via* a free radical mechanism.

After their leading publication on the osmium-catalyzed dihydroxylation of olefins in the presence of dioxygen [118], Beller et al. [119] recently reported that alcohol oxidations could also be performed using the same conditions [see Eq. (23)]. The reactions were carried out in a buffered two-phase system with a constant pH of 10.4.

Under these conditions a remarkable catalyst productivity (TON up to 16600 for acetophenone) was observed. The pH value is critical in order to ensure the reoxidation of Os^{VI} to Os^{VIII}. The scope of this system seems to be limited to benzylic and secondary alcohols.

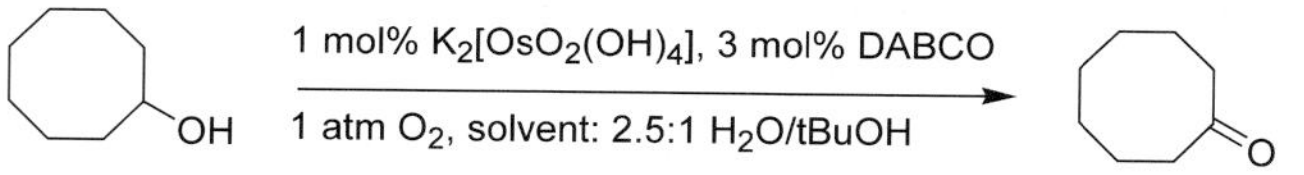

$$\text{cyclooctanol} \xrightarrow[\text{1 atm } O_2,\ \text{solvent: 2.5:1 } H_2O/tBuOH]{\text{1 mol\% } K_2[OsO_2(OH)_4],\ 3\ \text{mol\% DABCO}} \text{cyclooctanone} \quad (23)$$

A striking example of a heterogeneous and recyclable catalyst for aerobic alcohol oxidation is formed by manganese substituted octahedral molecular sieves [120]. In this case benzylic and allylic alcohols could be converted within 4 h. However 50 mol% of catalyst was needed to achieve this.

4.8
Catalytic Oxidation of Alcohols with Hydrogen Peroxide

In the aerobic oxidations discussed in the preceding sections the most effective catalysts tend to be late transition elements, e.g., Ru and Pd, which operate *via* oxometal or hydridometal mechanisms. In contrast, the most effective catalysts with H_2O_2 (or RO_2H) as the oxidant tend to be early transition metal ions with a d^0 configuration, e.g., Mo^{VI}, W^{VI} and Re^{VII}, which operate *via* peroxometal pathways. Ruthenium and palladium are generally not effective with H_2O_2 because they display high catalase activity, i.e., they catalyze rapid decomposition of H_2O_2. Early transition elements, on the other hand, are generally poor catalysts for H_2O_2 decomposition.

One of the few examples of ruthenium-based systems is the $RuCl_3 \cdot 3\ H_2O$/didecyldimethylammonium bromide combination reported by Sasson and coworkers [121]. This system catalyzes the selective oxidation of a variety of alcohols, at high (625:1) substrate:catalyst ratios, in an aqueous/organic biphasic system. However, 3–6 equiv. of H_2O_2 were required, reflecting the propensity of ruthenium for catalyzing the non-productive decomposition of H_2O_2.

Jacobsen et al. [122] showed, in 1979, that anionic molybdenum(VI) and tungsten(VI) peroxo complexes are effective oxidants for the stoichiometric oxidation of secondary alcohols to the corresponding ketones. Subsequently, Trost and Masuyama [123] showed that ammonium molybdate, $(NH_4)_6Mo_7O_{24} \cdot 4\ H_2O$ (10 mol%), is able to catalyze the selective oxidation of secondary alcohols, to the corresponding ketones, using hydrogen peroxide in the presence of tetrabutylammonium chloride and a stoichiometric amount of a base (K_2CO_3). It is worth noting that a more hindered alcohol moiety was oxidized more rapidly than a less hindered one, e.g., Eq. (24).

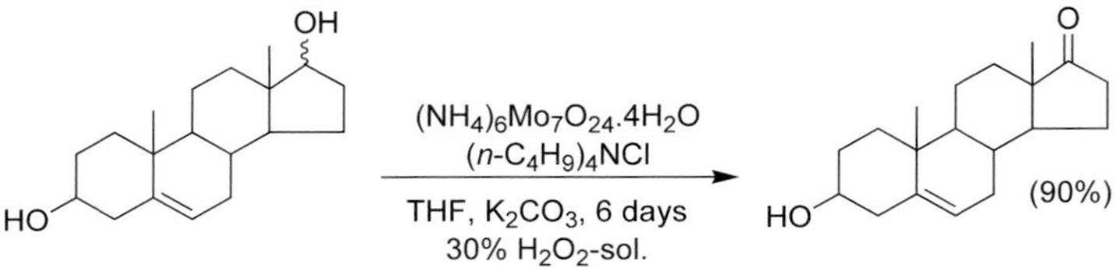

[steroid structure with OH] → [steroid structure with ketone] (24)

Reagents: $(NH_4)_6Mo_7O_{24} \cdot 4H_2O$, $(n\text{-}C_4H_9)_4NCl$, THF, K_2CO_3, 6 days, 30% H_2O_2-sol. (90%)

The above mentioned reactions were performed in a single phase using tetrahydrofuran as solvent. Subsequently, the group of Di Furia and Modena reported [124] the selective oxidation of alcohols with 70% aq. H_2O_2, using $Na_2MoO_4 \cdot 2 H_2O$ or $Na_2WO_4 \cdot 2 H_2O$ as the catalyst and methyltrioctylammonium chloride (Aliquat 336) as a phase transfer agent in a biphasic (dichloroethane–water) system.

More recently, Noyori and coworkers [125, 126] have achieved substantial improvements in the sodium tungstate-based, biphasic system by employing a phase transfer agent containing a lipophilic cation and bisulfate as the anion, e.g., $CH_3(n\text{-}C_8H_{17})_3NHSO_4$. This afforded a highly active catalytic system for the oxidation of alcohols using 1.1 equiv. of 30% aq. H_2O_2 in a solvent-free system. For example, 1-phenylethanol was converted into acetophenone with turnover numbers up to 180 000. As with all Mo- and W-based systems, the Noyori system shows a marked preference for secondary alcohols, e.g., Eq. (25).

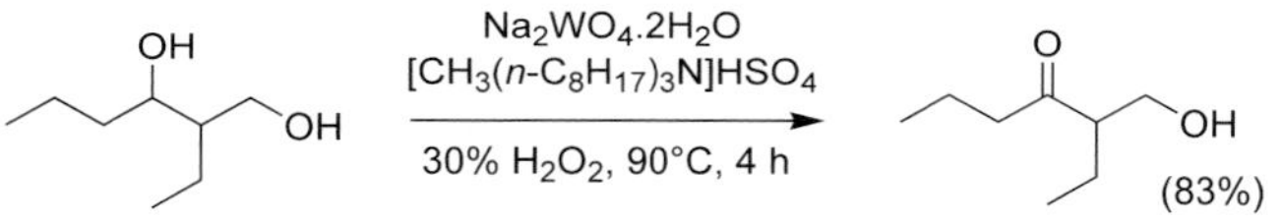

[secondary alcohol] → [ketone with OH] (25)

Reagents: $Na_2WO_4 \cdot 2H_2O$, $[CH_3(n\text{-}C_8H_{17})_3N]HSO_4$, 30% H_2O_2, 90°C, 4 h (83%)

Unsaturated alcohols generally undergo selective oxidation of the alcohol moiety [Eqs. (26) and (27)] but when an allylic alcohol contained a reactive trisubstituted double bond, selective epoxidation of the double bond was observed [Eq. (28)].

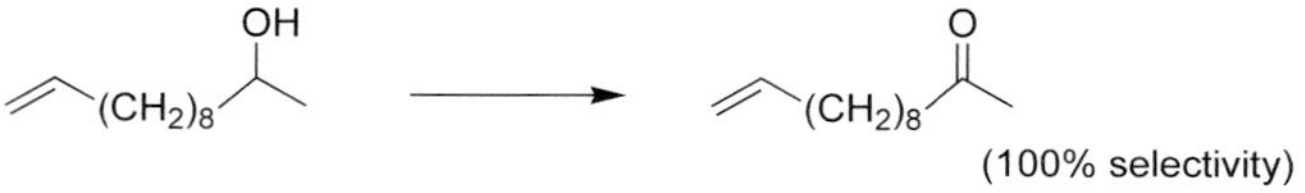

[allylic alcohol] → [allylic ketone] (100% selectivity) (26)

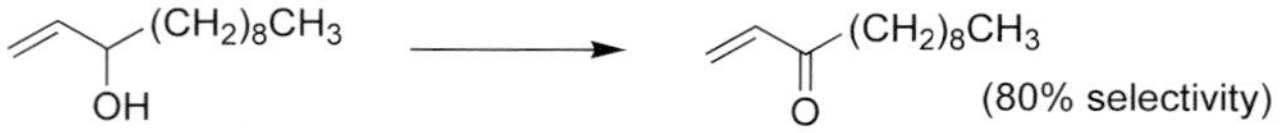

[allylic alcohol] → [ketone] (80% selectivity) (27)

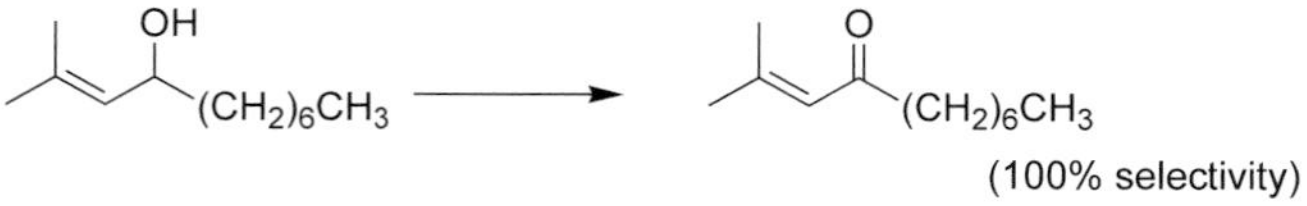

[trisubstituted allylic alcohol] → [epoxide product] (100% selectivity) (28)

Molybdenum- and tungsten-containing heteropolyanions are also effective catalysts for alcohol oxidations with H_2O_2 [127–129]. For example, $H_3PMo_{12}O_{40}$ or $H_3PW_{12}O_{40}$, in combination with cetylpyridinium chloride as a phase transfer agent, were shown by Ishii and coworkers [127] to be effective catalysts for alcohol oxidations with H_2O_2 in a biphasic, chloroform/water system.

Methyltrioxorhenium (MTO) also catalyzes the oxidation of alcohols with H_2O_2 via a peroxometal pathway [130, 131]. Primary benzylic and secondary aliphatic alcohols afforded the corresponding aldehydes and ketones, respectively, albeit using 2 equiv. of H_2O_2. In the presence of bromide ion the rate was increased by a factor 1 000. In this case the active oxidant may be hypobromite (HOBr), formed by MTO-catalyzed oxidation of bromide ion by H_2O_2.

Vanadium-pillared montmorillonite clay (V-PILC) [132] and a zeolite-encapsulated vanadium picolinate complex [133] were shown to catalyze alcohol oxidations with 30% aq. H_2O_2 and an H_2O_2–urea adduct, respectively. However, it seems highly likely that the observed catalysis is due to leached vanadium. Indeed, as we have noted elsewhere, heterogeneous catalysts based on Mo, W, Cr, V, etc. are highly susceptible to leaching by H_2O_2 or alkyl hydroperoxides [134]. Hence, in the absence of rigorous experimental proof, it is questionable whether the observed catalysis is heterogeneous in nature. In contrast, titanium silicalite (TS-1), an isomorphously substituted molecular sieve [135] is a truly heterogeneous catalyst for oxidations with 30% aq. H_2O_2, including the oxidation of alcohols [136].

A dinuclear manganese(IV) complex of trimethyl triazacyclononane (tmtacn) catalyzed the selective oxidation of reactive benzylic alcohols with hydrogen peroxide in acetone [137]. However, a large excess (up to 8 equiv.) of H_2O_2 was required, suggesting that there is substantial non-productive decomposition of the oxidant. Moreover, we note that the use of acetone as a solvent for oxidations with H_2O_2 is not recommended owing to the formation of explosion-sensitive peroxides. The exact nature of the catalytically active species in this system is rather obscure; for optimum activity it was necessary to pre-treat the complex with H_2O_2 in acetone. Presumably the active oxidant is a high-valent oxomanganese species but further studies are necessary to elucidate the mechanism.

4.9
Concluding Remarks

The economic importance of alcohol oxidations in the fine chemical industry will, in the future, continue to stimulate the quest for effective catalysts that utilize dioxygen or hydrogen peroxide as the primary oxidant. Although much progress has been made in recent years there is still room for further improvement with regard to catalyst activity and scope in organic synthesis. A better understanding of mechanistic details regarding the nature of the active intermediate and the rate-determining step would certainly facilitate this since many of these systems are poorly understood. It may even lead to the development of efficient methods for the enantioselective oxidation of chiral alcohols, e.g., the ruthenium-based system recently described by Katsuki and coworkers [138] and the palladium–chiral base systems reported by Sigman and Stolz and their coworkers [139, 140].

References

[1] S. V. Ley, J. Norman, W. P. Griffith, S. P. Marsden, *Synthesis* **1994**, 639.

[2] See Organic Textbooks, e.g., J. Clayden, N. Greeves, S. Warren, P. Wothers, *Organic Chemistry*, Oxford Univ. Press, New York **2001**; P. Y. Bruice, *Organic Chemistry*, 3rd edn., Prentice Hall, New Jersey **2001**.

[3] G. Cainelli, G. Cardillo, *Chromium Oxidations in Organic Chemistry*, Springer, Berlin, **1984**.

[4] R. A. Sheldon, J. K. Kochi, *Metal-catalysed Oxidations of Organic Compounds*, Academic Press, New York, **1981**.

[5] For recent reviews see: M. Besson, P. Gallezot, *Catal. Today* **2000**, *57*, 127; T. Mallat, A. Baiker, *Catal. Today* **1994**, *19*, 247; M. Besson, P. Gallezot, in *Fine Chemicals through Heterogeneous Catalysis*, R. A. Sheldon, H. van Bekkum, Eds., Wiley-VCH, Weinheim **2001**, p. 491.

[6] A. E. J. de Nooy, A. C. Besemer, H. van Bekkum, *Synthesis* **1996**, 1153 and references cited therein.

[7] P. L. Anelli, C. Biffi, F. Montanari and S. Quici, *J. Org. Chem.* **1987**, *52*, 2559.

[8] J. A. Cella, J. A. Kelley, E. F. Kenehan, *J. Org. Chem.* **1975**, *40*, 1860; S. D. Rychovsky, R. Vaidyanathan, *J. Org. Chem.* **1999**, *64*, 310.

[9] T. Inokuchi, S. Matsumoto, T. Nishiyama, S. Torii, *J. Org. Chem.* **1990**, *55*, 462.

[10] M. Zhao, J. Li, E. Mano, Z. Song, D. M. Tschaen, E. J. J. Grabowski, P. J. Reider, *J. Org. Chem.* **1999**, *64*, 2564.

[11] C.-J. Jenny, B. Lohri, M. Schlageter, Eur. Patent. **1997**, 0775684A1.

[12] C. Bolm, A. S. Magnus, J. P. Hildebrand, *Org. Lett.* **2000**, *2*, 1173.

[13] C. Bolm, T. Fey, *Chem. Commun.* **1999**, 1795; D. Brunel, F. Fajula, J. B. Nagy, B. Deroide, M. J. Verhoef, L. Veum, J. A. Peters, H. van Bekkum, *Appl. Catal. A: General* **2001**, *213*, 73.

[14] D. Brunel, P. Lentz, P. Sutra, B. Deroide, F. Fajula, J. B. Nagy, *Stud. Surf. Sci. Catal.* **1999**, *125*, 237.

[15] M. J. Verhoef, J.A. Peters, H. van Bekkum, *Stud. Surf. Sci. Catal.* **1999**, *125*, 465.

[16] R. Ciriminna, J. Blum, D. Avnir, M. Pagliaro, *Chem. Commun.* **2000**, 1441.

[17] A. Dijksman, I. W. C. E. Arends, R. A. Sheldon, *Chem. Commun.* **2000**, 271.

[18] A. Dijksman, I. W. C. E. Arends, R. A. Sheldon, *Synlett* **2001**, 102.

[19] R. B. Daniel, P. Alsters, R. Neumann, *J. Org. Chem.* **2001**, *66*, 8650.

[20] A. Cecchetto, F. Fontana, F. Minisci, F. Recupero, *Tetrahedron Lett.* **2001**, *42*, 6651.

[21] M. Fabbrini, C. Galli, P. Gentilli, D. Macchitella, *Tetrahedron Lett.* **2001**, *42*, 7551; F. d'Acunzo, P. Baiocco, M. Fabbrini, C. Galli, P. Gentilli, *Eur. J. Org. Chem.* **1995**, 4195.

[22] J. J. Berzelius, *Ann.* **1828**, *13*, 435.

[23] W. G. Lloyd, *J. Org. Chem.* **1967**, *32*, 2816.

[24] T. F. Blackburn, J. J. Schwarz, *J. Chem. Soc. Chem. Commun.* **1977**, 157.

[25] S. Aït-Mohand, J. J. Muzart, *J. Mol. Catal. A: Chemical* **1998**, *129*, 135.

[26] K. P. Peterson, R. C. J. Larock, *Org. Chem.* **1998**, *63*, 3185.

[27] T. Nishimura, T. Onoue, K. Ohe, S. Uemura, *Tetrahedron Lett.* **1998**, *39*, 6011; T. Nishimura, T. Onoue, K. Ohe, S. Uemura, *J. Org. Chem.* **1999**, *64*, 6750.

[28] R. A. Sheldon, *Top. Curr. Chem.* **1993**, *164*, 21.

[29] T. Naota, H. Takaya, S.-I. Murahashi, *Chem. Rev.* **1998**, *98*, 2599.

[30] S.-I. Murahashi, N. Komiya, in *Biomimetic Oxidations Catalyzed by Transition Metal Complexes*, B. Meunier, Ed., Imperial College Press, London, **2000**, pp. 563; see also E. Gore, *Plat. Met. Rev.* **1983**, *27*, 111.

[31] R. A. Sheldon, I. W. C. E. Arends, A. Dijksman, *Catal. Today* **2000**, *57*, 158.

[32] R. Tang, S. E. Diamond, N. Neary, F. J. Mares, *J. Chem. Soc., Chem. Commun.* **1978**, 562.

[33] M. Matsumoto, S. J. Ito, *J. Chem. Soc. Chem. Commun.* **1981**, 907.
[34] C. Bilgrien, S. Davis, R. S. Drago, *J. Am. Chem. Soc.* **1987**, *109*, 3786.
[35] R. A. Sheldon, *Chem. Commun.* **2001**, *23*, 2399.
[36] A. Wolfson, S. Wuyts, D. E. de Vos, I. F. J. Vancelecom, P. A. Jacobs, *Tetrahedron Lett.* **2002**, *43*, 8107.
[37] (a) J. E. Bäckvall, R. L. Chowdhury, U. Karlsson, *J. Chem. Soc., Chem. Commun.* **1991**, 473; (b) G.-Z. Wang, U. Andreasson, J. E. Bäckvall, *J. Chem. Soc., Chem. Commun.* **1994**, 1037; (c) G. Csjernyik, A. Ell, L. Fadini, B. Pugin, J. E. Backvall, *J. Org. Chem.* **2002**, *67*, 1657.
[38] A. Hanyu, E. Takezawa, S. Sakaguchi, Y. Ishii, *Tetrahedron Lett.* **1998**, *39*, 5557.
[39] A. Dijksman, I. W. C. E. Arends, R. A. Sheldon, *Chem. Commun.* **1999**, 1591; A. Dijksman, A. Marino-González, A. Mairata i Payeras, I. W. C. E. Arends, R. A. Sheldon, *J. Am. Chem. Soc.* **2001**, *123*, 6826. For a related study see T. Inokuchi, K. Nakagawa, S. Torii, *Tetrahedron Lett.* **1995**, *36*, 3223.
[40] J.-E. Bäckvall, U. Andreasson, *Tetrahedron Lett.* **1993**, *34*, 5459; B. M. Trost, R. J. Kulawiec, *Tetrahedron Lett.* **1991**, *32*, 3039.
[41] A. Aranyos, G. Csjernyik, K.J. Szabo, J.-E. Bäckvall, *Chem. Commun.* **1999**, 351.
[42] U. Karlson, G.-Z. Wang, J.-E. Bäckvall, *J. Org. Chem.* **1994**, *59*, 1196.
[43] C. M. Paleos, P. J. Dais, *J. Chem. Soc., Chem. Commun.* **1977**, 345.
[44] P. J. Beynon, P. M. Collins, D. Gardiner, W. G. Overend, *Carbohydr. Res.* **1968**, *6*, 431; see also H. B. Friedrich, *Plat. Met. Rev.* **1999**, *43*, 94.
[45] A. C. Dengel, R. A. Hudson, W. P. Griffith, *Trans. Met. Chem.* **1985**, *10*, 98.
[46] W. P. Griffith, S. V. Ley, G. P. Whitcombe, A. D. White, *J. Chem. Soc., Chem. Commun.* **1987**, 1625.
[47] A. C. Dengel, A. M. El-Hendawy, W. P. Griffith, *Trans. Met. Chem.* **1989**, *40*, 230.
[48] W. P. Griffith, S. V. Ley, *Aldrichim. Acta* **1990**, *23*, 13.

[49] S. V. Ley, J. Norman, W. P. Griffith, S. P. Marsen, *Synthesis* **1994**, 639.
[50] R. Lenz, S. V. Ley, *J. Chem. Soc., Perkin Trans. 1* **1997**, 3291.
[51] I. E. Marko, P. R. Giles, M. Tsukazaki, I. Chelle-Regnaut, C. J. Urch, S. M. Brown, *J. Am. Chem. Soc.* **1997**, *119*, 12661.
[52] B. Hinzen, R. Lenz, S. V. Ley, *Synthesis* **1998**, 977.
[53] A. Bleloch, B. F. G. Johnson, S. V. Ley, A. J. Price, D. S. Shepard, A. N. Thomas, *Chem. Commun.* **1999**, 1907.
[54] M. Pagliaro, R. Ciriminna, *Tetrahedron Lett.* **2001**, *42*, 4511.
[55] M. Hasan, M. Musawir, P. N. Davey, I. Kozhevnikov, *J. Mol. Catal. A: Chemical* **2002**, *180*, 77.
[56] J. Rocek, C.-S. Ng, *J. Am. Chem. Soc.* **1974**, *96*, 1522.
[57] D. G. Lee, Z. Wang, W. D. Chandler, *J. Org. Chem.* **1992**, *57*, 3276.
[58] S. Y. S. Cheng, N. Rajapakse, S. J. Rettig, B. R. James, *J. Chem. Soc., Chem. Commun.*, **1994**, 2669; see also N. Rajapakse, B. R. James, D. Dolphin, *Stud. Surf. Sci. Catal.*, **1990**, *55*, 109.
[59] M. Matsumoto, N. J. Watanabe, *Org. Chem.* **1984**, *49*, 3435.
[60] M. Matsumoto, S. Ito, *Synth. Commun.* **1984**, *14*, 697.
[61] F. Vocanson, Y. P. Guo, J. L. Namy, H. B. Kagan, *Synth. Commun.* **1998**, *28*, 2577.
[62] K. Yamaguchi, N. Mizuno, *Angew. Chem., Int. Ed. Engl.* **2002**, *41*, 4538.
[63] K. Kaneda, T. Yamashita, T. Matsushita, K. Ebitani, *J. Org. Chem.* **1998**, *63*, 1750.
[64] T. Matsushita, K. Ebitani, K. Kaneda, *Chem. Commun.* **1999**, 265.
[65] K. Yamaguchi, K. Mori, T. Mizugaki, K. Ebitani, K. Kaneda, *J. Am. Chem. Soc.* **2000**, *122*, 7144.
[66] H. B. Ji, K. Ebitani, T. Mizugaki, K. Kaneda, *Catal. Commun.* **2002**, *3*, 511.
[67] (a) R. Neumann, *Prog. Inorg. Chem.* **1998**, *47*, 317; (b) C. L. Hill, C. M. Prosser-McCartha, *Coord. Chem. Rev.* **1995**, *143*, 407; (c) M. T. Pope, A. Müller, *Angew. Chem., Int. Ed. Engl.* **1991**, *30*, 34.
[68] K. Yamaguchi, N. Mizuno, *New. J. Chem.* **2002**, *26*, 972.

[69] T. Nishimura, I. W. C. E. Arends, R. A. Sheldon, results to be published.
[70] (a) R. Neumann, C. Abu-Gnim, *J. Chem. Soc., Chem. Commun.* **1989**, 124; (b) R. Neumann, C. Abu-Gnim, *J. Am. Chem. Soc.* **1990**, *112*, 6025.
[71] M. Higashijima, *Chem. Lett.* **1999**, 1093.
[72] S.-I. Murahashi, T. Naota, N. Hirai, N. *J. Org. Chem.* **1993**, *58*, 7318.
[73] T. F. Blackburn, J. J. Schwartz, *J. Chem. Soc., Chem. Commun.* **1977**, 157.
[74] M. N. Vargaftik, V. P. Zagorodnikov, I. P. Storarov, I. I. Moiseev, *J. Mol. Catal.* **1989**, *53*, 315; see also I. I. Moiseev, M. N. Vargaftik, in *Catalysis by Di- and Polynuclear Metal Cluster Complexes*, R. D. Adams, F. A. Cotton, Eds., Wiley-VCH, Weinheim, **1998**, p. 395.
[75] K. Kaneda, M. Fujii, K. Morioka, *J. Org. Chem.* **1996**, *61*, 4502; K. Kaneda, Y. Fujie, K. Ebitani, *Tetrahedron Lett.* **1997**, *38*, 9023.
[76] K. P. Peterson, R. C. J. Larock, *Org. Chem.* **1998**, *63*, 3185.
[77] R. A. T. M. van Benthem, H. Hiemstra, P. W. N. M. van Leeuwen, J. W. Geus, W. N. Speckamp, *Angew. Chem.* **1995**, *107*, 500; *Angew. Chem., Int. Ed. Engl.* **1995**, *34*, 457.
[78] S. Ait-Mohand, F. Hénin, J. Muzart, *Tetrahedron Lett.* **1995**, *36*, 2473.
[79] S. Ait-Mohand, J. J. Muzart, *J. Mol. Catal. A: Chemical* **1998**, *129*, 135.
[80] T. Nishimura, T. Onoue, K. Ohe, S. Uemura, *Tetrahedron Lett.* **1998**, *39*, 6011.
[81] T. Nishimura, T. Onoue, K. Ohe, S. J. Uemura, *J. Org. Chem.* **1999**, *64*, 6750; T. Nishimura, K. Ohe. S. J. Uemura, *J. Am. Chem. Soc.* **1999**, *121*, 2645.
[82] T. Nishimura, Y. Maeda, N. Kakiuchi, S. Uemura, *J. Chem. Soc., Perkin Trans. 1* **2000**, 4301.
[83] B. A. Steinhoff, S. R. Fix, S. S. Stahl, *J. Am. Chem. Soc.* **2002**, *124*, 766.
[84] B. A. Steinhoff, S. S. Stahl, *Org. Lett.* **2002**, *4*, 4179–4181.
[85] M. J. Schultz, C. C. Park, M. S. Sigman, *Chem. Commun.* **2002**, 3034.
[86] K. Hallman, C. Moberg, *Adv. Synth. Catal.* **2001**, *343*, 260.
[87] G.-J. ten Brink, I. W. C. E. Arends, R. A. Sheldon, *Science* **2000**, *287*, 1636.
[88] G.-J. ten Brink, I. W. C. E. Arends, R. A. Sheldon, *Adv. Synth. Catal.* **2002**, *344*, 355.
[89] (A) D. Bianchi, R. Bortolo, R. D'Aloisio, M. Ricci, *Angew. Chem., Int. Ed. Engl.* **1999**, *38*, 706; (b) *J. Mol. Catal. A: Chemical* **1999**, *150*, 87.
[90] R. Bortolo, D. Bianchi, R. D'Aloisio, C. Querici, M. J. Ricci, *J. Mol. Catal. A: Chemical* **2000**, *153*, 25.
[91] For an example in toluene see C. Keresszegi, T. Burgi, T. Mallat, A. Baiker, *J. Catal.* **2002**, *211*, 244.
[92] T. Nishimura, N. Kakiuchi, M. Inoue, S. Uemura, *Chem. Commun.* **2000**, 1245; see also N. Kakiuchi, T. Nishimura, M. Inoue, S. Uemura, *Bull. Chem. Soc. Jpn.* **2001**, *74*, 165.
[93] N. Ito, S. E. V. Phillips, C. Stevens, Z. B. Ogel, M. J. McPherson, J. N. Keen, K. D. S. Yadav, P. F. Knowles, *Nature* **1991**, *350*, 87.
[94] K. Drauz, H. Waldmann, *Enzyme Catalysis in Organic Synthesis*, VCH, Weinheim, **1995**, Chapter 6.
[95] For example see: I. P. Skibida, A. M. Sakharov, *Catal. Today* **1996**, *27*, 187; A. M. Sakharov, I. P. Skibida, *J. Mol. Catal.* **1988**, *48*, 157; L. Feldberg, Y. L. Sasson, *J. Chem. Soc., Chem. Commun.* **1994**, 1807; P. Capdevielle, D. Sparfel, J. Baranne-Lafont, N. K. Cuong, D. Maumy, *J. Chem. Res. (S)* **1993**, 10; M. Munakata, S. Nishibayashi, S. Sakamoto, *J. Chem. Soc., Chem. Commun.* **1980**, 219; S. Bhaduri, N. Y. Sapre, *J. Chem. Soc., Dalton Trans.* **1981**, 2585; C. Jallabert, H. Rivière, *Tetrahedron Lett.* **1977**, 1215 and *J. Mol. Catal.* **1980**, *7*, 127; C. Jallabert, C. Lapinte, H. Rivière, *J. Mol. Catal.* **1986**, *14*, 75; C. Jallabert, H. Rivière, *Tetrahedron* **1980**, *36*, 1191.
[96] Y. Wang, J. L. DuBois, B. Hedman, K. O. Hodgson, T. D. P. Stack, *Science* **1998**, *279*, 537.
[97] P. Chauhuri, M. Hess, U. Flörke, K. Wieghardt, *Angew. Chem., Int. Ed. Engl.* **1998**, *37*, 2217.
[98] P. Chauhuri, M. Hess, T. Weyhermüller, K. Wieghardt, *Angew. Chem., Int. Ed. Engl.* **1998**, *38*, 1095.

[99] V. Mahadevan, R. J. M. Klein Gebbink, T. D. P. Stack, *Curr. Opin. Chem. Biol.* **2000**, *4*, 228.

[100] M. M. Whittaker, C. A. Ekberg, J. Peterson, M. S. Sendova, E. P. Day, J. W. Whittaker, *J. Mol. Catal. B: Enzymatic* **2000**, *8*, 3.

[101] I. E Marko, P. R. Giles, M. Tsukazaki, S. M. Brown, C. J. Urch, *Science* **1996**, *274*, 2044; I. E. Marko, M. Tsukazaki, P. R. Giles, S. M. Brown, C. J. Urch, *Angew. Chem., Int. Ed. Engl.* **1997**, *36*, 2208.

[102] I. E. Marko, P. R. Giles, M. Tsukazaki, I. Chellé-Regnaut, A. Gautier, S. M. Brown, C. J. Urch, *J. Org. Chem.* **1999**, *64*, 2433.

[103] I. E. Marko, A. Gautier, I. Chellé-Regnaut, P. R. Giles, M. Tsukazaki, C. J. Urch, S. M. Brown, *J. Org. Chem.* **1998**, *63*, 7576.

[104] M. F. Semmelhack, C. R. Schmid, D. A. Cortés, C. S. Chou, *J. Am. Chem. Soc.* **1984**, *106*, 3374.

[105] A. Dijksman, Thesis, Delft University of Technology, Delft, **2001**.

[106] B. Betzemeier, M. Cavazzine, S. Quici, P. Knochel, P. *Tetrahedron Lett.* **2000**, *41*, 4343.

[107] I. A. Ansari, R. Gree, *Org. Lett.* **2002**, *4*, 1507.

[108] K. S. Coleman, C. Y. Lorber, J. A. Osborn, *Eur. J. Inorg. Chem.* **1998**, 1673.

[109] K. S. Coleman, M. Coppe, C. Thomas, J. A. Osborn, *Tetrahedron Lett.* **1999**, *40*, 3723.

[110] For another example of Os/Cu catalyzed alcohol oxidation see J. Muldoon, S. N. Brown, *Org. Lett.* **2002**, *4*, 1043.

[111] C. Y. Lorber, S. P. Schmidt, J. A. Osborn, *Eur. J. Inorg. Chem.* **2000**, 655.

[112] Kh. M. Minachev, G. V Antoshin, D. G. Klissurski, N. K. Guin, N. Ts. Abadzhijeva, *React. Kinet. Catal. Lett.* **1979**, *10*, 163.

[113] But see M. Kirihara, Y. Ochiai, S. Takizawa, H. Takahata, H. Nemoto, *Chem. Commun.* **1999**, 1387.

[114] T. Iwahama, S. Sakaguchi, Y. Nishiyama, Y. Ishii, *Tetrahedron Lett.* **1995**, *36*, 6923.

[115] Y. Yoshino, Y. Hanyashi, T. Iwahama, S. Sakaguchi, Y. Ishii, *J. Org. Chem.* **1997**, *62*, 6810; S. Kato, T. Iwahama, S. Sakaguchi, Y. Ishii, *J. Org. Chem.* **1998**, *63*, 222; S. Sakaguchi, S. Kato, T. Iwahama, Y. Ishii, *Bull. Chem. Soc. Jpn.* **1988**, *71*, 1.

[116] See also F. Minisci, C. Punta, F. Recupero, F. Fontana, G. F. Pedulli, *Chem. Commun.* **2002**, *7*, 688.

[117] B. M. Choudary, M. Lakshmi Kantam, Ateeq Rahman, Ch. V. Reddy, K. K. Rao, *Angew. Chem., Int. Ed. Engl.* **2001**, *40*, 763.

[118] (a) C. Döbler, G. Mehltretter, M. Beller, *Angew. Chem., Int. Ed. Engl.* **1999**, *38*, 3026; (b) C. Döbler, G. Mehltretter, G. M. Sundermeier, M. J. Beller, *J. Am. Chem. Soc.* **2000**, *122*, 10289.

[119] C. Döbler, G. M. Mehltretter, U. Sundermeier, M. Eckert, H.-C. Militzer, M. Beller, *Tetrahedron Lett.* **2001**, *42*, 8447.

[120] Y.-C. Son, V. D. Makwana, A. R. Howell, S. L. Suib, *Angew. Chem., Int. Ed. Engl.* **2001**, *40*, 4280.

[121] G. Barak, J. Dakka, Y. Sasson, *J. Org. Chem.* **1988**, *53*, 3553.

[122] S. E. Jacobsen, D. A. Muccigrosso, F. Mares, *J. Org. Chem.* **1979**, *44*, 921; see also O. Bortolini, S. Campestrini, F. Di Furia, G. Modena, *J. Org. Chem.* **1987**, *52*, 5467.

[123] B. M. Trost, Y. Masuyama, *Tetrahedron Lett.* **1984**, *25*, 173.

[124] O. Bortolini, V. Conte, F. Di Furia, G. Modena, *J. Org. Chem.* **1986**, *51*, 2661.

[125] K. Sato, M. Aoki, J. Takagi, R. Noyori, *J. Am. Chem. Soc.* **1997**, *119*, 12386.

[126] K. Sato, J. Takagi, M. Aoki, R. Noyori, *Tetrahedron Lett.* **1998**, *39*, 7549.

[127] Y. Ishii, K. Yamawaki, T. Yoshida, T. Ura, M. Ogawa, M. J. *Org. Chem.* **1987**, *52*, 1868; Y. Ishii, K. Yamawaki, T. Ura, H. Yamada, T. Yoshida, M. Ogawa, *J. Org. Chem.* **1988**, *53*, 3587; K. Yamawaki, H. Nishihara, T. Yoshida, T. Ura, H. Yamada, Y. Ishii, M. Ogawa, *Synth. Commun.* **1988**, *18*, 869; K. Yamawaki, T. Yoshida, H. Nishihara, Y. Ishii, M. Ogawa, *Synth. Commun.* **1986**, *16*, 537.

[128] C. Venturello, M. Gambaro, *J. Org. Chem.* **1991**, *56*, 5924.

[129] R. Neumann, M. Gara, *J. Am. Chem. Soc.* **1995**, *117*, 5066.
[130] T. H. Zauche, J. H. Espenson, *Inorg. Chem.* **1995**, *37*, 6827.
[131] J. H. Espenson, Z. Zhu, T. H. Zauche, *J. Org. Chem.* **1991**, *64*, 1191
[132] B. M. Choudary, V. L. K. Vialli, *J. Chem. Soc., Chem. Commun.* **1990**, 1115.
[133] A. Kozlov, A. Kozlova, K. Asakura, Y. Iwasawa, *J. Mol. Catal. A: Chemical* **1999**, *137*, 223.
[134] R. A. Sheldon, M. Wallau, I. W. C. E. Arends, U. Schuchardt, *Acc. Chem. Res.* **1998**, *31*, 485.
[135] I. W. C. E. Arends, R. A. Sheldon, M. Wallau, U. Schuchardt, *Angew. Chem., Int. Ed. Engl.* **1997**, *36*, 1144.
[136] F. Maspero, U. Romano, *J. Catal.* **1994**, *146*, 476.
[137] C. Zondervan, R. Hage, B. L. Feringa, *Chem. Commun.* **1997**, 419; J. Brinksma, M. T. Rispens, R. H. Hage, B. L. Feringa, *Inorg. Chem. Acta* **2002**, *337*, 75.
[138] K. Masutani, T. Uchida, R. Irie, T. Katsuki, *Tetrahedron Lett.* **2000**, *41*, 5119.
[139] D. R. Jensen, J. S. Pugsley, M. S. Sigman, *J. Am. Chem. Soc.* **2001**, *123*, 7475.
[140] E. M. Ferreira, B. M. Stolz, *J. Am. Chem. Soc.* **2001**, *123*, 7725.

5
Aerobic Oxidations and Related Reactions Catalyzed by *N*-Hydroxyphthalimide

Yasutaka Ishii and Satoshi Sakaguchi

5.1
Introduction

In parallel with the development of the petrochemical industry, over 90% of organic chemicals are derived from petroleum. Among them, a wide range of oxygen-containing molecules including alcohols, aldehydes, ketones, epoxides and carboxylic acids have become necessary to supply the starting materials for producing, in particular, plastics and synthetic fiber materials for polyamides, polyesters and polycarbonates, etc. For instance, ethylene oxide, acrolein, acrylic acid, and methacrolein are produced by the vapor-phase partial oxidation of lower alkenes such as ethylene, propylene and butanes [1], while acetic acid, KA oil (a mixture of cyclohexanone and cyclohexanol), benzoic acid, terephthalic acid and phenol accompanied by acetone are manufactured by the liquid-phase catalytic oxidation of alkanes, such as butane, cyclohexane and alkylbenzenes, etc. [2]. Liquid-phase aerobic oxidation, which is generally referred to as autoxidation, is practiced extensively in industry worldwide, although the efficiency of this oxidation methodology is not necessarily too high [2a, 3]. As a result, nitric acid is still widely used as a useful oxidizing agent for manufacturing carboxylic acids such as adipic acid, nicotinic acid and pyromellitic acid, etc. [2a, 4]. Today, however, environmentally unacceptable traditional oxidation methods using high valent metal oxo complexes, halogens and nitric acid are being replaced by cleaner oxidation methods.

Although the partial aerobic oxidation of alkanes leading to alcohols and carbonyl compounds has considerable potential from both ecological and economical points of view, current oxidation technology is not totally feasible as it incurs extensive oxidative cleavage or concomitant combustion of alkanes. The most important liquid-phase oxidation methods include the transformation of *p*-xylene to terephthalic acid and cyclohexane to KA oil [2a]. However, the reaction conditions are often harsh, the reagent mixture is corrosive, and the reaction is often unselective. Therefore, it is apparent that selective transformation of hydrocarbons, especially saturated hydrocarbons such as alkanes, into valuable oxygenated compounds constitutes an extremely important area of contemporary industrial chemistry. Thus, considerable research effort has been made towards the development of the selective oxidation of alkanes

Modern Oxidation Methods. Edited by Jan-Erling Bäckvall
Copyright © 2004 WILEY-VCH Verlag GmbH & Co. KGaA, Weinheim
ISBN: 3-527-30642-0

with molecular oxygen, leading to alcohols, ketones and carboxylic acids [5, 6]. A number of catalytic oxidations of alkanes with dioxygen by transition metals in the presence of reducing agents such as aldehydes have appeared in the literature, but these oxidations may be limited to laboratory-scale synthesis [6]. In recent years, there has been a growing demand for the development of fundamentally new and environmentally benign catalytic systems for oxidation of hydrocarbons that are operative on an industrial scale under moderate conditions in the liquid-phase with a high degree of selectivity.

Recently, we have developed an innovative strategy for the catalytic carbon radical generation from hydrocarbons by a phthalimide N-oxyl (PINO) radical generated *in situ* from N-hydroxyphthalimide (NHPI) and molecular oxygen in the presence or absence of a cobalt ion under mild conditions. The carbon radicals derived from a variety of hydrocarbons under the influence of molecular oxygen lead to oxygenated products such as alcohols, ketones and carboxylic acids in good yields. In this chapter we present a novel methodology for the functionalizations of hydrocarbons, including oxygenation, nitration, sulfoxidation, epoxidation, carboxylation and oxyalkylation through the catalytic carbon radical generation. In particular, the NHPI catalyzed aerobic oxidations of alkanes, which are very important in industry worldwide, are described in detail.

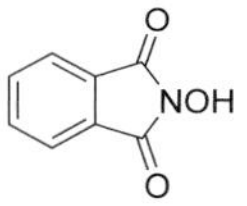

Scheme 5.1

5.2
NHPI-catalyzed Aerobic Oxidation

NHPI was first used as a catalyst by Grochowski et al. [7a] for the addition of ethers to diethylazodicarboxylate and as an efficient mediator by Masui and coworkers [7b, 7c] for the electrochemical oxidation of alcohols. There has been a patent taken out on the oxidation of allylic hydrogen of isoprenoid with dioxygen using NHPI in the presence of a radical initiator [7d]. In 1995, we found that NHPI serves as a carbon radical producing catalyst (CRPC) from hydrocarbons in the presence or absence of transition metals, such as Co and Mn ions, under dioxygen [8].

5.2.1
Alkane Oxidations with Dioxygen

Over the past several decades, a number of catalytic systems have been developed for the oxidation of alkanes with dioxygen in the presence of reducing agents, e.g., H_2, metals, aldehydes, etc., under mild conditions [9–13]. In 1981 Tabushi et al. reported the oxidation of adamantane to 1- and 2-adamantanols by the Mn^{III}porphyrin/Pt/H_2 system under a dioxygen atmosphere at room temperature [10]. Barton developed a family of systems, the so-called Gif systems, for aerobic oxidation and oxidative functionalization of alkanes under mild conditions using Fe and Zn as reductants [11a–c]. Alkane oxidation using aldehydes as reducing agents was reported by Murahashi and coworkers, who attempted the aerobic oxidation of cyclohexane and adamantane by ruthenium or iron catalysts in the presence of acetaldehyde [12a–c]. There have been several reports on the photo-oxidations of alkanes with O_2 catalyzed by polyoxotungstates [14], heteropolyoxometalates [15], and $FeCl_3$ [16]. Shul'pin et al. carried out the vanadium-catalyzed oxidation of alkanes to alcohols, ketones and hydroperoxides by O_2 in combination with H_2O_2 [17]. Lyon and Ellis reported that halogenated metalloporphyrin complexes are efficient catalysts for the oxidation of isobutane with dioxygen [18]. Mizuno and coworkers have shown that heteropolyanions containing Fe catalyze the aerobic oxidation of cyclohexane and adamantane into the corresponding alcohols and ketones [19]. A Ru^{III}–EDTA system [24], Ru-substituted polyoxometalate [21] and $[Co(NCMe)_4](PF_6)_2$ [22] have been reported to catalyze the aerobic oxidation of cyclohexane and adamantane. However, effective and selective methods for the catalytic oxygenation of alkanes with dioxygen still remain a major challenge.

Adipic acid, which is used as a raw material for nylon-6,6 and polyester, is the most important acid of all of the aliphatic dicarboxylic acids manufactured at present. The current production of adipic acid consists of a two-step oxidation process involving the aerobic oxidation of cyclohexane in the presence of a soluble Co catalyst at 150–170 °C to a KA oil and the nitric acid oxidation of the KA oil to adipic acid [2a, 23]. The drawbacks of this process are that the oxidation in the first step must be operated within 3–6% conversion of cyclohexane to maintain a high selectivity (80%) of the KA oil, and that the nitric acid oxidation generates a large amount of undesired global-warming nitrogen oxides, in particular N_2O. Therefore, the direct conversion of cyclohexane into adipic acid with molecular oxygen has long been sought after as a desirable and promising method in industrial chemistry worldwide. Tanaka succeeded in achieving conversion of cyclohexane to adipic acid under 30 atm of O_2 by the use of a higher concentration of Co^{III} acetate combined with acetaldehyde or cyclohexanone, which serves as promoter [25]. More recently, Noyori and coworkers reported the oxidation of cyclohexene to adipic acid with aqueous hydrogen peroxide by a polyoxometalate having a phase-transfer function as an alternative clean route [26].

The direct conversion of cyclohexane to adipic acid was successfully achieved by the use of a combined catalyst of NHPI with Co and Mn ions [27]. The oxidation of cyclohexane (**1**) in the presence of a catalytic amount of NHPI (10 mol%) and $Mn(acac)_2$ (1 mol%) under a dioxygen atmosphere (1 atm) in acetic acid at 100 °C for 20 h gave adipic acid (**4**) in 73% selectivity at 73% conversion [Eq. (1)]. This is the first example of a one-step oxidation of **1** to **4** under a normal pressure of dioxygen at a reasonably low reaction temperature with high conversion and selectivity. The oxidation using the $NHPI/Co(OAc)_2$ system in acetonitrile, on the other hand, gave rise to cyclohexanone (**2**) in good selectivity [Eq. (1)] [28]. The oxidation by the NHPI/Co system in acetonitrile provides an alternative direct route to cyclohexanone, although the autoxidation of cyclohexane leads to a mixture of KA oil consisting of cyclohexanol as a main product [2a, 23]. The present catalytic system can be extended to the oxidation of large-membered cycloalkanes to the corresponding dicarboxylic acids. Cyclooctane, cyclodecane and cyclododecane were oxidized to suberic acid, sebacic acid and dodecanedioic acid, respectively

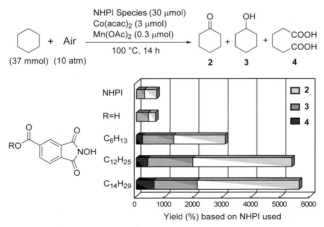

(1)

The aerobic oxidations of **1** must be carried out in an appropriate solvent such as acetic acid or acetonitrile due to the lower solubility of NHPI in non-polar solvents such as hydrocarbons. It is worth noting that the NHPI-catalyzed reaction of **1** could proceed without any solvent by the use of a lipophilic NHPI derivative. From a series of 4-alkyloxycarbonyl N-hydroxyphthalimides examined as lipophilic NHPI catalysts, 4-laulyloxycarbony N-hydroxyphthalimide was found to be an efficient catalyst for the aerobic oxidation of **1** under solvent-free conditions (Figure 5.1) [29].

Methane and ethane are the main components of natural gas, which has been used worldwide for a long time as a clean fuel. Since ethane comprises 5–10% of the natural gas, a vast amount of ethane is formed alongside the production of methane. If a new methodology for converting ethane into useful oxygen-containing compounds such as ethanol and acetic acid is developed, it would vastly contribute to the efficient use of the feedstock. Previously, Sen and coworkers reported on the liquid-phase oxidation of ethane to acetic acid and formic acid by Pd/C with H_2O_2

Fig. 5.1 Oxidation of **1** catalyzed by NHPI derivatives under solvent-free conditions

generated *in situ* from H_2, arising from a metal-catalyzed water gas shift reaction with CO and water, and O_2 [30]. Ethane can be converted into N,N-dimethylpropylamine by the reaction with N,N-trimethylamine N-oxide on using $Cu(OAc)_2$ as the catalyst [31]. Ethane oxidation with H_2O_2, catalyzed homogeneously by V-containing polyphosphomolybdates, has been carried out by Shul'pin and coworkers [32].

Catalytic aerobic oxidation of ethane to acetic acid was successfully performed through a catalytic radical process using NHPI derivatives combined with a Co^{II} salt in acetonitrile or propionic acid. Among the catalysts examined, N,N-dihydroxypyromellitimide (NDHPI) was found to be the best. For instance, when a mixture of ethane (20 atm) and air (20 atm) in acetonitrile was allowed to react in the presence of NDHPI (100 μmol) and $Co(OAc)_2$ (30 μmol) at 150 °C for 15 h, 830 μmol of acetic acid were obtained and the turnover number (TON) of NDHPI reached 8.3 [Eq. (2)]. In this reaction, other products such as ethanol or acetaldehyde were not detected at all. In the oxidation of ethane using NHPI as a catalyst under these conditions, the amount of NHPI used was twice that of the NDHPI, but the yield of acetic acid and the TON of the catalyst were 530 μmol and 2.7, respectively. The highest TON (15.3) was obtained when the reaction was carried out using NDPHI combined with $CoCl_2$ in propionic acid [33].

$$C_2H_6 + Air \xrightarrow[150\ °C]{Catalyst} CH_3COOH$$
20 atm   20 atm

Catalyst (μmol)	Solvent	Yield (μmol)	TON
NHPI/Co(OAc)$_2$ (200 / 30)	CH$_3$CN	530	2.7
NDHPI/Co(OAc)$_2$ (100 / 30)	CH$_3$CN	830	8.3
NDHPI/CoCl$_2$ (100 / 30)	CH$_3$CH$_2$COOH	1532	15.3

(2)

NDHPI

The autoxidation of isobutane is now mainly carried out to obtain *tert*-butyl hydroperoxide [34]. Halogenated metalloporphyrin complexes are reported to be efficient catalysts for the aerobic oxidation of isobutene [18, 35]. It was found that the oxidation of isobutane by air (10 atm) catalyzed by NHPI and $Co(OAc)_2$ in benzonitrile at 100 °C produced *tert*-butyl alcohol in high yield (81%) along with acetone (14%) [Eq. (3)] [36]. 2-Methylbutane was converted into the carbon–carbon bond cleaved products, acetone and acetic acid, rather than the alcohols, as principal products. These cleaved products seem to be formed *via* β-scission of an alkoxy radical derived from the decomposition of a hydroperoxide by Co ions. The extent of the β-scission is known to depend on the stability of the radicals released from the alkoxy radicals [37]. It is thought that the β-scission of a *tert*-butoxy radical to acetone and a methyl radical is more difficult than that of a 2-methylbutoxy radical to acetone and an ethyl radical. As a result, isobutane produces *tert*-butyl alcohol as the principal product, while 2-methylbutane affords mainly acetone and acetic acid.

$$\text{(isobutane)} + \text{Air} \xrightarrow[\text{PhCN, 100 °C}]{\text{NHPI (10 mol\%)} \atop \text{Co(OAc)}_2 \text{ (0.25 mol\%)}} \underset{81\%}{t\text{-BuOH}} + \underset{14\%}{\text{acetone}} + \underset{21\%}{t\text{-amyl alcohol}} + \underset{32\%}{\text{MEK}} + \underset{15\%}{\text{AcOH}} \quad (3)$$

(10 atm)

There have been a few reports on the catalytic hydroxylation of adamantane with dioxygen in the presence of aldehydes [12]. Mizuno and coworkers reported that the aerobic oxidation of adamantane by the PW_9Fe_2Ni heteropolyanion without any reducing agents gives 1-adamantanol and 2-adamantanone with 29% conversion [19a]. The NHPI-catalyzed aerobic oxidation of adamantane is accelerated considerably by adding a small amount of a Co salt [28, 38, 39]. Thus, the oxidation of adamantane (5) in the presence of NHPI (10 mol%) and Co(acac)$_2$ (0.5 mol%) in acetic acid under dioxygen (1 atm) for 6 h produced 1-adamantanol (6) (43%), 1,3-adamantanediol (8) (40%), and 2-adamantanone (7) (8%) [Eq. (4)]. The relative reactivity of the tertiary C–H bond to the secondary C–H bond in the oxidation by NHPI/Co^{II} was 31.1. This value is considerably higher than that attained by the conventional autoxidation (3.8–5.4). The preferential oxidation of the tertiary C–H bond over the secondary bond may be attributed to the electron-deficient character of PINO, which is a key radical species in the NHPI-catalyzed oxidation (*vide infra*).

$$\mathbf{5} + O_2 \xrightarrow[\text{AcOH, 75 °C} \atop \text{Conv. 93\%}]{\text{NHPI (10 mol\%)} \atop \text{Co(acac)}_2 \text{ (0.5 mol\%)}} \underset{\mathbf{6}\ 43\%}{\text{1-AdOH}} + \underset{\mathbf{7}\ 8\%}{\text{2-adamantanone}} + \underset{\mathbf{8}\ 40\%}{\text{1,3-diol}} \quad (4)$$

(1 atm)

It is important that the oxidation led to diol **8** in high selectivity, because **8** is rarely produced by conventional oxidation. Hirobe and coworkers obtained **8** in 25% yield by the oxidation of **5** using a Ru complex with 2,6-dichloropyridine *N*-oxide as the oxidant [40]. In the stepwise hydroxylation of **5** by the NHPI/Co(acac)$_2$ system, the diol **8** and triol **9** were obtained in high selectivity [Eqs. (5) and (6)]. These alcohols are now manufactured as important components of photoresistent polymer materials on an industrial scale by Daicel Chemical Industry Ltd.

$$\mathbf{6} + O_2 \xrightarrow[\text{AcOH, 75 °C} \atop \text{Conv. 95\%}]{\text{NHPI (10 mol\%)} \atop \text{Co(acac)}_2 \text{ (0.5 mol\%)}} \underset{\mathbf{8}\ 76\%}{\text{diol}} + \underset{\mathbf{9}\ 18\%}{\text{triol}} \quad (5)$$

(1 atm)

$$\mathbf{8} + O_2 \xrightarrow[\text{AcOH, 75 °C} \atop \text{Conv. 46\%}]{\text{NHPI (10 mol\%)} \atop \text{Co(acac)}_2 \text{ (0.5 mol\%)}} \underset{\mathbf{9}\ 85\%}{\text{triol}} \quad (6)$$

(1 atm)

5.2.2
Oxidation of Alkylarenes

5.2.2.1 Oxidation of Alkylbenzenes

Aerobic oxidation of alkylbenzenes is a promising subject in industrial chemistry. Many bulk chemicals such as terephthalic acid, phenol, benzoic acid, etc. are manufactured by homogeneous liquid-phase oxidations with O_2 [2, 41]. The largest-scale liquid-phase oxidation is the conversion of *p*-xylene into terephthalic acid which is chiefly used as polyethylene terephthalate polymer material [2a]. *m*-Xylene is also commercially oxidized to isophthalic acid. Benzoic acid derived from the oxidation of toluene is an important raw material in the production of various pharmaceuticals and pesticides. Commercially important cumene hydroperoxide and ethylbenzene hydroperoxide are also manufactured by the aerobic oxidation of isopropylbenzene and ethylbenzene, respectively [2a, 5a]. These oxidation processes are usually operated at higher temperatures and pressures of air. A great deal of effort has been made to develop the homogeneous oxidations of alkylbenzenes with better selectivity under milder conditions. The first successful oxidation of a variety of alkylbenzenes with O_2 by the use of NHPI as the catalyst under very mild conditions has been achieved.

Currently, the oxidation of toluene is practiced commercially in the presence of a catalytic amount of cobalt(II) 2-ethylhexanoate under a pressure of 10 atm of air at 140–190 °C [42]. The oxidation of toluene under normal pressure of dioxygen at room temperature is achieved by the use of a combined catalyst of NHPI and Co^{II} species. The fact that the toluene was oxidized with dioxygen through the catalytic process in high yield under ambient conditions is very important, from ecological and technical points of view, as a promising strategy in oxidation chemistry. As a typical example, the oxidation of toluene in the presence of NHPI (10 mol%) and Co(OAc)$_2$ (0.5 mol%) in acetic acid under an atmosphere of O_2 at 25 °C for 20 h afforded benzoic acid and benzaldehyde in 81% and 3% yields, respectively [Eq. (7)] [43]. This finding suggests that an efficient cleavage of a C–H bond with a bond dissociation energy (BDE) of 88 kcal mol^{-1} (corresponding to the BDE of toluene) is possible at room temperature by the use of NHPI catalyst. However, when Co^{III} was employed in place of Co^{II}, no reaction took place at all at room temperature.

$$\text{PhCH}_3 + O_2 \xrightarrow[\text{AcOH, 25 °C}]{\text{NHPI (10 mol\%)} \atop \text{Co(OAc)}_2 \text{ (0.5 mol\%)}} \text{PhCOOH} + \text{PhCHO} \quad (7)$$

(1 atm), Conv. 84%, 96%, 4%

Representative results for the NHPI-catalyzed aerobic oxidation of various alkylbenzenes in the presence of Co(OAc)$_2$ in acetic acid under ambient conditions are listed in Table 5.1. Both *p*- and *o*-xylenes are selectively oxidized to *p*- and *o*-toluic acids without the formation of dicarboxylic acids. *o*-Ethyltoluene undergoes selective oxidation to form a mixture of the corresponding alcohol and ketone in which the ethyl moiety was selectively functionalized. It is of interest to examine the effect of substituents on

Tab. 5.1 Aerobic oxidation of various alkylbenzenes at room temperature [a]

Run	Substrate	Time (h)	Conv. (%)	Products (Yield (%))
1 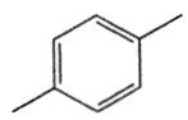 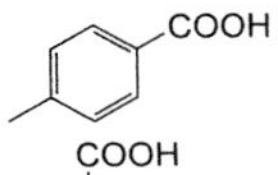	toluene	20	95	benzoic acid (85)
2	o-xylene	20	93	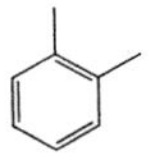 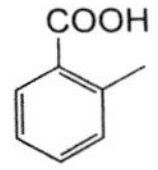 o-methylbenzoic acid (83)
3[b] 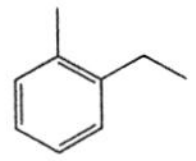 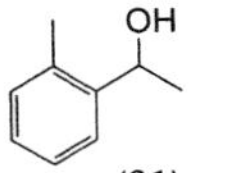 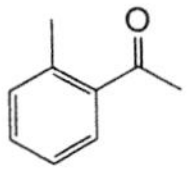	1,2,4-trimethylbenzene	20	82	(2-methylphenyl)methanol type OH product (21); ketone (37)
4	p-tBu-toluene	20	95	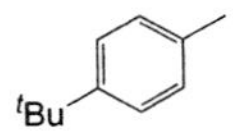 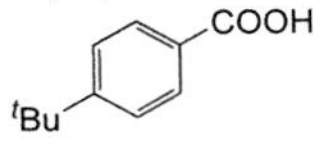 p-tBu-benzoic acid (91)
5[b] 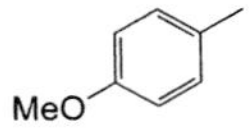 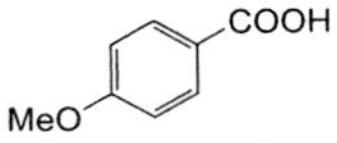	p-MeO-toluene	6	89	p-MeO-benzoic acid (80)
6	p-Cl-toluene	20	71	p-Cl-benzoic acid (67)
7	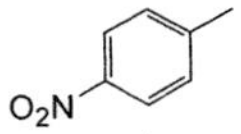 p-O$_2$N-toluene	20	No reaction	
8	1,2,4-trimethyl (isopropyl) substrate	12	>99	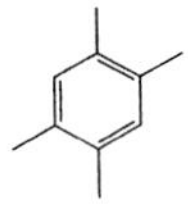 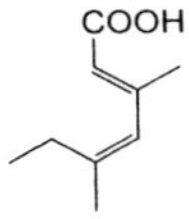 COOH product (93)

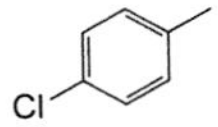

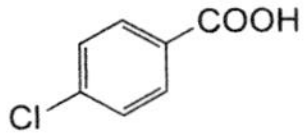

[a] Substrates (3 mmol) were allowed to react in the presence of NHPI (10 mol%) and Co(OAc)$_2$ (0.5 mol%) in AcOH (5 mL) under dioxygen (1 atm) at 25 °C. [b] CH$_3$CN was used as the solvent.

the aromatic ring in the oxidation of substituted toluenes. p-Methoxytoluene is oxidized more rapidly than the toluene itself, while p-chlorotoluene is oxidized at a relatively slow rate. An electron-donating substituent anchoring to toluene stabilizes the partial positive charge on the benzylic carbon atom in the transition state, for the abstraction of a benzylic hydrogen atom by PINO possessing an electrophilic character (Scheme 5.2) (vide infra) [44]. Therefore, the oxidation of toluenes having electron-donating groups by the NHPI catalyst is facilitated. Indeed, p-nitrotoluene substituted by a strong electron-withdrawing nitro group is not oxidized at all under these conditions. Recently, various substituted NHPI derivatives were prepared, and studied by Nolte and coworkers in the aerobic oxidation of ethylbenzene [45]. It was found that NHPI with an electron-withdrawing fluorine substituent increases the rate, while the NHPI substituted by a methoxy group decreases the oxidation rate.

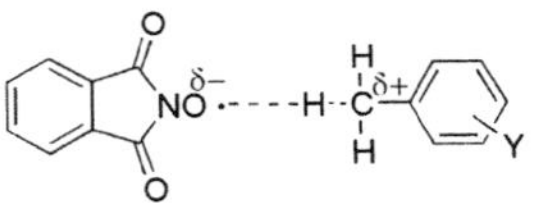

Scheme 5.2 Transition state for the reaction of PINO with substituted benzenes

A plausible reaction pathway for the aerobic oxidation of alkanes catalyzed by NHPI and Co^{II} is illustrated in Scheme 5.3. A labile dioxygen complex such as the superoxocobalt(III) or peroxocobalt(III) complexes is known to be formed by the complexation of Co^{II} with O$_2$. The *in situ* generation of PINO from NHPI by the action of the cobalt(III)–oxygen complex formed is a key step in the present oxidation. The next step involves the hydrogen atom abstraction from alkanes by PINO to form alkyl radicals. Trapping the resulting alkyl radicals by dioxygen provides peroxy radicals, which are eventually converted into oxygenated products through alkyl hydroperoxides. In fact, on exposing NHPI in benzonitrile containing a small amount of Co(OAc)$_2$ to dioxygen at 80 °C, an ESR signal attributed to PINO as a triplet signal having hyperfine splitting (hfs) by the nitrogen atom (g = 2.0074, A_N = 4.3 G) is observed (Figure 5.2). The g-value and hyperfine splitting constants observed here are consistent with those (g = 2.0073, A_N = 4.23 G) of PINO reported previously [46]. In addition, PINO is observed during the oxidation of toluene by the NHPI/Co^{II} system under ambient conditions [43]. Quite recently, Minisci, Pedulli and coworkers found, by means of ESR spectroscopy, that the BDE value of the O–H bond for NHPI is >86 kcal mol^{-1}. This suggests that PINO could abstract the benzylic hydrogen atom of toluene, whose BDE is 88 kcal mol^{-1} [47].

5.2.2.2 Synthesis of Terephthalic Acid

Terephthalic acid (TPA) as well as dimethyl terephthalate (DMT) have recently become important as raw materials for polyethylene terephthalate [48]. In 1999, ca. 17 million tons of TPA were manufactured worldwide and its production was esti-

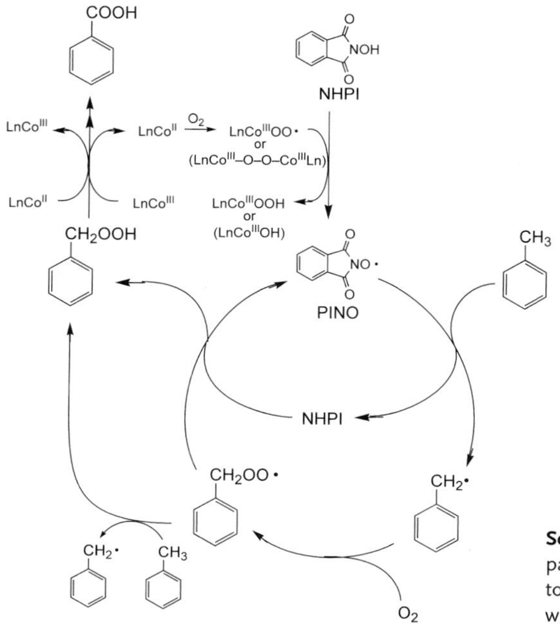

Scheme 5.3 A plausible reaction path for the aerobic oxidation of toluene catalyzed by NHPI combined with Co^{II}

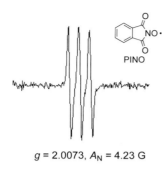

Fig. 5.2 ESR spectrum of PINO

$g = 2.0073$, $A_N = 4.23$ G

mated to be increasing at a minimum growth rate of 10% annually up to the year 2002. Until the 1980s, the following four-step process developed by Witten and modified by Hercules and Dynamit-Nobel (Witten-Hercules process) had been the main operation for the production of DMT [2a, 48]. The first step is the conversion of p-xylene (PX) into p-toluic acid (PTA). It then passes to an esterification step to form methyl p-toluate, which is subjected to further oxidation to monomethyl terephthalate, followed by esterification to DMT. From the 1990s, these processes were changed to the aerobic one-stage oxidation of PX to TPA by the combined use of cobalt and manganese salts, in the presence of bromide as a promoter in acetic acid at 175–225 °C under 15–30 atm of air, followed by hydrogenation of the crude TPA to remove 4-carboxybenzaldehyde (4-CBA) by a Pd catalyst [48–50]. This process was developed by Scientific Design and Amoco Ltd. (Amoco process). Currently, about 70% of TPA produced worldwide is based on the Amoco process, and almost all new plants adopt this method. However, there are several disadvantages to the Amoco process: (1) significant combustion of acetic acid, used as the solvent, to CO and CO_2; (2) use of the highly corrosive bromide ion, which calls for the use of vessels lined with expensive metals such as titanium; (3) contamination of 4-CBA in crude TPA, which necessitates elaborate hydrogenation and recrystallization procedures in manufacturing the purified TPA required for PET. Therefore, a new oxidation system for the production of TPA is required to overcome these disadvantages. Partenheimer recently published a review devoted to the aerobic oxidation of alkylbenzenes, especially PX, using the Co/Mn/Br system [50].

The aerobic oxidation of PX to TPA was examined by the NHPI catalyst to elaborate a halogen-free catalytic system [51]. The oxidation of PX with dioxygen (1 atm) in the presence of catalytic amounts of NHPI (20 mol%) and Co(OAc)$_2$ (0.5 mol%) in acetic acid at 100 °C for 14 h produced TPA in 67% yield and PTA (15%), together with small amounts of 4-CBA, 4-carboxybenzyl alcohol, 1,4-diacetoxymethylbenzene, and 4-acetoxymethylbenzoic acid, as well as several unidentified compounds in 1–2% yields, respectively, at over 99% conversion [Eq. (8)]. The yield of TPA is improved to 82%, when Mn(OAc)$_2$ (0.5 mol%) is added to the NHPI/Co(OAc)$_2$ system. The synergistic effect of Co and Mn salts in the aerobic oxidation of alkylbenzenes has been well documented [50–53]. From a practical point of view, it is important that the aerobic oxidation of PX under air (30 kg cm^{-2}) by the NHPI/Co/Mn system is completed within 3 h at 150 °C to form TPA in 84% yield [Eq. (9)]. Both *o*- and *m*-xylenes were also successfully converted into the corresponding dicarboxylic acids, isophthalic acid and phthalic acid, respectively, in high yields.

$$\text{PX} + \text{O}_2\ (1\ \text{atm}) \xrightarrow[\text{AcOH, 100 °C}]{\substack{\text{NHPI (20 mol\%)} \\ \text{Co(acac)}_2\ (0.5\ \text{mol\%}) \\ \text{Mn(OAc)}_2\ (0.5\ \text{mol\%})}} \text{p-toluic acid} + \text{TPA} \tag{8}$$

	p-toluic acid	TPA
NHPI/Co	15%	67%
NHPI/Co/Mn	4%	82%

$$\text{xylene} + \text{Air}\ (30\ \text{atm}) \xrightarrow[\text{AcOH, 150 °C}]{\substack{\text{NHPI (15 mol\%)} \\ \text{Co(acac)}_2\ (0.5\ \text{mol\%}) \\ \text{Mn(OAc)}_2\ (0.5\ \text{mol\%})}} \text{toluic acid} + \text{phthalic acid} \tag{9}$$

	toluic acid	diacid
para	4%	84%
meta	7%	86%
ortho	6%	73%

As shown in Eq. (8), ca. 20 mol% of NHPI must be used to obtain TPA in satisfactory yield (over 80%), because NHPI gradually decomposes to inert phthalimide and phthalic anhydride during the oxidation. If the NHPI used can be reduced by a simple modification, the present oxidation would be more desirable. Efforts to reduce the amount of the NHPI led to the discovery of an efficient catalyst, N-acetoxyphthalimide (NAPI), which can be easily prepared by the reaction of NHPI with acetic anhydride. Surprisingly, PX was oxidized to TPA in high yield (80%) even by the use of 5 mol% of NAPI, Co(OAc)$_2$ (0.5 mol%), and Mn(OAc)$_2$ (0.5 mol%) [Eq. (10)]. The effect of NAPI is considered to be resistant to the rapid decomposition to phthalimide or phthalic anhydride at the early stage of the reaction where violent chain reactions take place, since NAPI is gradually hydrolyzed to NHPI by water present in acetic acid as well as the water resulting during the oxidation.

$$\text{PX} + \text{O}_2\ (1\ \text{atm}) \xrightarrow[\text{AcOH, 100 °C}]{\substack{\text{NAPI (5 mol\%)} \\ \text{Co(OAc)}_2\ (0.5\ \text{mol\%}) \\ \text{Mn(OAc)}_2\ (0.5\ \text{mol\%})}} \underset{8\%}{\text{p-toluic acid}} + \underset{80\%}{\text{TPA}} \tag{10}$$

5.2.2.3 Oxidation of Methylpyridines and Methylquinolines

Pyridinecarboxylic acids are useful and important intermediates in pharmaceutical syntheses. Although the synthesis of these carboxylic acids by the aerobic oxidation of alkylpyridines is straightforward, the oxidation is usually difficult to carry out selectively owing to their low reactivities [50, 54]. Pyridinecarboxylic acids are readily prepared by the oxidation of alkylpyridines with nitric acid or by the hydrolysis of pyridinecarboxamides derived from pyridinecarbonitrile [55]. According to the recent literature, nicotinic acid is obtained in ca. 50% yield at 52% conversion by the oxidation of β-picoline in the presence of Co(OAc)$_2$ and Mn(OAc)$_2$ using LiCl as a promoter under air (16 atm) at 170 °C [56]. The aerobic oxidation of β-picoline to nicotinic acid catalyzed by NHPI has been examined [57]. Nicotinic acid is used as a precursor of vitamin B$_3$ and is manufactured commercially on a large scale by nitric acid oxidation of 5-ethyl-2-methylpyridine [56a]. The oxidation of β-picoline in the presence of NHPI (10 mol%) and Co(OAc)$_2$ (1.5 mol%) under dioxygen (1 atm) at 100 °C for 15 h in acetic acid affords nicotinic acid in 76% yield at 82% conversion [Eq. (11a)] [57]. This is the first successful oxidation of picolines with O$_2$ under mild conditions. In contrast to the oxidation of β-picoline by the NHPI/Co/Mn system, where nicotinic acid was formed in good yield, γ-picoline is oxidized with some difficulty under these conditions to form 4-pyridinecarboxylic acid in low yield (22%). After optimization of the reaction conditions, 4-pyridinecarboxylic acid was obtained by the use of NHPI (20 mol%), Co(OAc)$_2$ (1 mol%) and Mn(OAc)$_2$ (1 mol%) in 60% yield at 67% conversion [Eq. (11b)] [57].

$$\text{β-picoline} + O_2 \xrightarrow[\text{AcOH, 100 °C}]{\substack{\text{NHPI (10 mol\%)} \\ \text{Co(OAc)}_2 \text{ (1.5 mol\%)}}} \text{nicotinic acid} \quad (11a)$$

Conv. 82%, 92%

$$\text{γ-picoline} + \text{Air} \xrightarrow[\text{AcOH, 150 °C}]{\substack{\text{NAPI (20 mol\%)} \\ \text{Co(OAc)}_2 \text{ (0.5 mol\%)} \\ \text{Mn(OAc)}_2 \text{ (0.5 mol\%)}}} \text{4-pyridinecarboxylic acid} \quad (11b)$$

Conv. 67%, 90%

Quinolines and their derivatives are common in natural products, and have attractive applications as pharmaceuticals and agrochemicals [58]. For example, 3-quinolinecarboxylic acid derivatives are reported to be potent inhibitors of bacterial DNA gyrase. So far, there have been only limited methods for the preparation of quinolinecarboxylic acids, despite their potential importance [59]. The synthesis of quinolinecarboxylic acids from the corresponding methylquinolines by direct oxidation seems to be the simplest method, but the reaction has been difficult to carry out selectively because of the low reactivity of the methyl group bearing the quinoline ring. Classically, the oxidation was examined by the use of a stoichiometric amount of a metal oxidant, such as KMnO$_4$ [60], CrO$_3$ [60] and nickel peroxide [61], or by Pd-catalyzed oxidation with H$_2$O$_2$ [62].

Treatment of 3-methylquinoline (**10**) by the NHPI-Co-Mn system under the reaction conditions used for β-picoline, however, results in the recovery of the starting material **10**. This is believed to be because the activation of O$_2$ by the Co^{II} becomes difficult, probably due to the coordination of **10** to Co(OAc)$_2$. As described below, the nitration of alkanes with NO$_2$ is enhanced in the presence of NHPI catalyst, in which the generation of PINO from NHPI is easily achieved by NO$_2$ without any transition metal. Hence, the oxidation of **10**, through adding a small amount of NO$_2$, produced 3-quinolinecarboxylic acid. For instance, the oxidation of **10** with O$_2$ (1 atm) catalyzed by NHPI (20 mol%), Co(OAc)$_2$ (2 mol%) and Mn(OAc)$_2$ (0.1 mol%) in the presence of NO$_2$ (10 mol%) gave 3-quinolinecarboxylic acid in 75% yield at 90% conversion [Eq. (12)]. Other methylquinolines are also successfully oxidized under reaction conditions similar to those used for **10** [63].

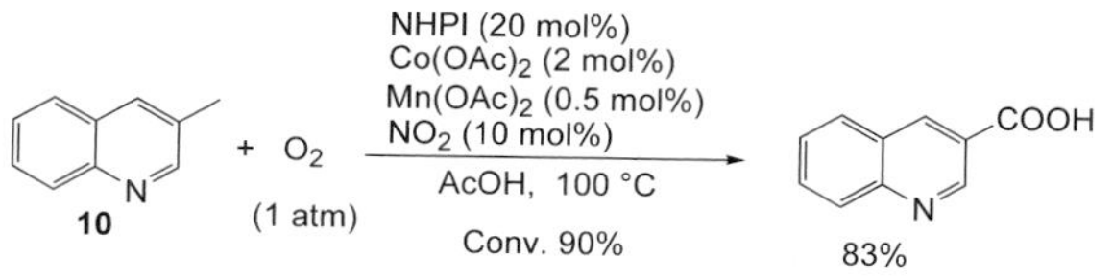

$$\text{10} + O_2 \text{ (1 atm)} \xrightarrow[\text{AcOH, 100 °C}]{\substack{\text{NHPI (20 mol\%)} \\ \text{Co(OAc)}_2 \text{ (2 mol\%)} \\ \text{Mn(OAc)}_2 \text{ (0.5 mol\%)} \\ \text{NO}_2 \text{ (10 mol\%)}}} \text{quinoline-COOH (83\%)} \qquad (12)$$

Conv. 90%

5.2.2.4 Oxidation of Hydroaromatic and Benzylic Compounds

Various hydroaromatic and benzylic compounds can be oxidized under a normal pressure of dioxygen, catalyzed by NHPI, even in the absence of a transition metal species, giving the corresponding oxygenated compounds in good yields. For example, treatment of fluorene with dioxygen in the presence of a catalytic amount of NHPI in benzonitrile at 100 °C for 20 h affords fluorenone in 80% yield. Similarly, xanthene produced xanthone in excellent yield [Eq. (13)] [8, 64]. After our findings on NHPI catalysis in the aerobic oxidation, Einhorn et al. reported the oxidation of these substrates with O$_2$ at room temperature in the presence of NHPI and acetaldehyde, and they concluded that the active species is the PINO formed by the reaction of NHPI with an acetylperoxy radical (Scheme 5.4) [65]. They prepared chiral N-hydroxyimides and used them as the catalyst for the asymmetric oxidation of indanes to give indanones in 8% ee (Figure 5.3) [66].

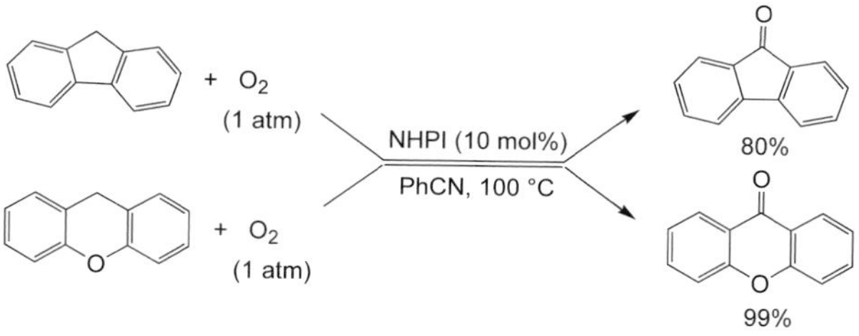

$$\text{fluorene} + O_2 \text{ (1 atm)} \xrightarrow[\text{PhCN, 100 °C}]{\text{NHPI (10 mol\%)}} \text{fluorenone (80\%)} \qquad (13)$$

$$\text{xanthene} + O_2 \text{ (1 atm)} \longrightarrow \text{xanthone (99\%)}$$

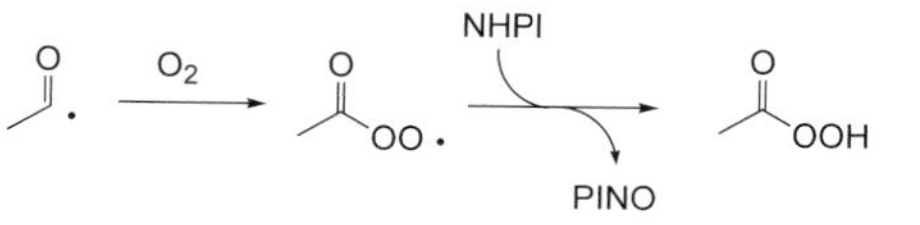

Scheme 5.4 Formation of PINO by reaction of NHPI with acetylperoxy radical

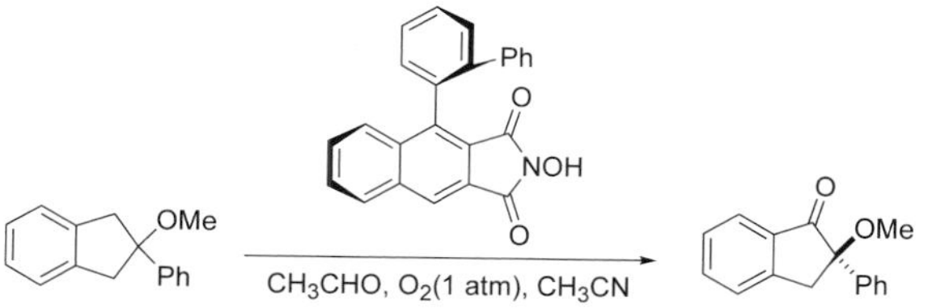

Fig. 5.3 Oxidation with chiral N-hydroxyimides prepared by Einhorn et al.

Hydroperoxides are used not only as oxidizing agents of olefins but also as important precursors for the synthesis of phenols. For instance, α-hydroperoxyethylbenzene obtained by aerobic oxidation of ethylbenzene is used as an active oxygen carrier in the epoxidation of propylene, which is known as the Halcon process [67]. The cumene–phenol process (Hock Process) based on the decomposition of cumene hydroperoxide with sulfuric acid to phenol and acetone is the current method for phenol synthesis that is used worldwide [68]. An efficient approach to phenols through the formation of hydroperoxides from alkylbenzenes is successfully achieved by aerobic oxidation using NHPI as a catalyst. The oxidation of several alkylbenzenes with dioxygen by NHPI followed by treatment with an acid affords phenols in good yields. For example, the aerobic oxidation of cumene in the presence of a catalytic amount of NHPI at 75 °C and subsequent treatment with H_2SO_4 leads to phenol in 81% selectivity at 90% conversion [Eq. (14)] [69]. Hydroquinone (61%) and 4-isopropylphenol (33%) are obtained from 1,4-diisopropylbenzene. Recently, Sheldon and coworkers have reported the highly selective oxidation of cyclohexylbenzene to cyclohexylbenzene-1-hydroperoxide (CHBH). The aerobic oxidation of cyclohexylbenzene in the presence of NHPI (0.5 mol%) and the CHBH (2 mol%) as an initiator without solvent affords the desired CHBH (98% selectivity) at 32% conversion [70]. They considered that this oxidation provides an overall co-product free route to phenol production. The acid-catalyzed decomposition of the CHBH would give a mixture of phenol and cyclohexanone, which is subsequently dehydrogenated with an appropriate catalyst to form phenol (Scheme 5.5).

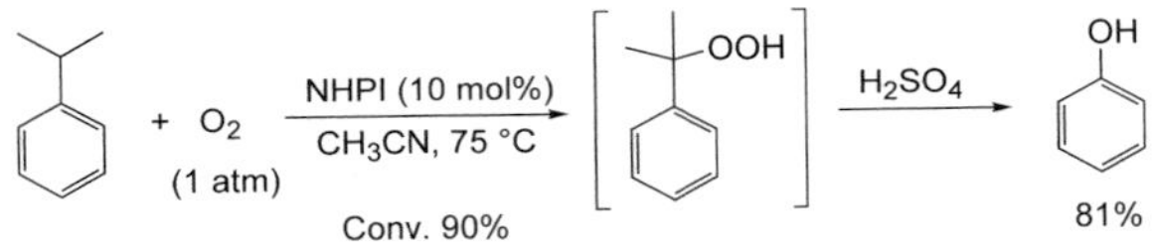

$$(14)$$

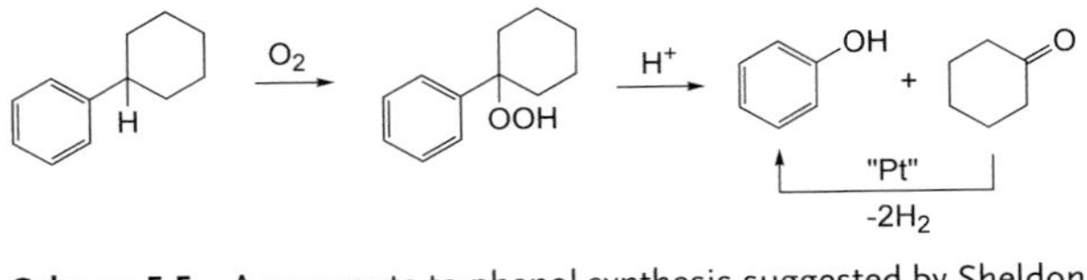

Scheme 5.5 A new route to phenol synthesis suggested by Sheldon et al.

5.2.3
Preparation of Acetylenic Ketones by Direct Oxidation of Alkynes

α,β-Acetylenic carbonyl compounds, ynones, are important intermediates in organic synthesis, since further elaboration of ynones can lead to highly valuable compounds such as heterocyclic compounds [71], α,β-unsaturated ketones [72], cyclopentenones [73], nucleosides [74], and chiral pheromones [75], etc. Several methods have been reported for the synthesis of conjugated acetylenic ketones: a coupling reaction of acetylenides with activated acylating reagents such as acid chloride or anhydrides [76].

On the other hand, the selective oxidation of alkynes to ynones is carried out by the use of CrO_3/TBHP [77], CrO_3(pyridine)$_2$ [78], Na_2CrO_4/acetic anhydride [78], and SeO_2/TBHP systems [79], but these oxidations are not completely successful [80]. An alternative approach to preparing ynones is the oxygenation of the propargylic C–H bonds of alkynes with dioxygen, since the bond dissociation energies of the propargyl C–H bonds (87 ± 2 kcal mol^{-1} for 2-pentyne) are approximately equal to those of the benzylic C–H bonds of alkylbenzenes (88 ± 1 kcal mol^{-1} for toluene) [81]. However, the conventional oxidation of alkynes with dioxygen at a higher temperature (around 150 °C) results in undesired over-oxidation products, such as carboxylic acids. Since the aerobic oxidation of alkylbenzenes by the NHPI catalyst could be effected even at room temperature, the NHPI-catalyzed oxidation of alkynes at a lower temperature is expected to suppress undesired side reactions.

Treatment of 4-octyne with dioxygen (1 atm) under the influence of NHPI (10 mol%) and Cu(acac)$_2$ (0.5 mol%) in acetonitrile at room temperature for 30 h produces 4-octyn-3-one (77%) and 4-octyn-3-ol (22%) at 70% conversion [Eq. (15)] [82]. The same reaction at 50 °C for 6 h gives the ynone in 84% selectivity based on 83% conversion. This oxidation would offer a facile catalytic method for the preparation of conjugated ynones from alkynes, since 1-decyne, upon treatment with TBHP in the presence of SeO_2, leads to the acetylenic alcohol, 1-decyn-3-ol, rather than the ynone, 1-decyn-3-one [79]. The NHPI/Cu^{II} system can also promote the oxidation of acetylenic alcohols to ketones. The reaction of 1-octyn-3-ol under dioxygen (1 atm) in the presence of NHPI and Cu(acac)$_2$ affords 1-octyn-3-one in 95% yield based on 57% conversion.

$$\text{4-octyne} + O_2 \;(1\text{ atm}) \xrightarrow[\text{CH}_3\text{CN}]{\text{NHPI (10 mol\%)},\; \text{Cu(acac)}_2\;(0.5\text{ mol\%})} \text{4-octyn-3-one} \quad (15)$$

25 °C, 30 h   77% (Conv. 70%)
50 °C, 6 h   84% (Conv. 83%)

5.2.4
Oxidation of Alcohols

The oxidation of alcohols to the corresponding carbonyl compounds is a frequently used transformation in organic synthesis [83]. There have been many catalytic methods for the aerobic oxidation of alcohols to the corresponding carbonyl compounds [84, 85]. However, some of these oxidations are carried out in the presence of a reducing agent, such as an aldehyde, which is eventually converted into carboxylic acid, or they are severely limited to particular reactive alcohols, such as benzylic and allylic alcohols. Recently, a few aerobic oxidations involving non-activated alcohols have appeared, although expensive metal catalysts such as Ru and Pd must be employed to effect the oxidation [86]. In 1996, Markó and coworkers developed an efficient aerobic oxidation system of aliphatic alcohols using an inexpensive $CuCl_2$/phenanthroline catalyst combined with azodicarboxylate [87]. Reusable heterogeneous catalysts consisting of Ru or Pd have been reported by the groups working with Kaneda [88] and Uemura [89], respectively. Sheldon and coworkers have succeeded in the aerobic oxidation of alcohols by a water-soluble Pd catalyst [90].

As described in the preceding sections, alkanes are oxidized by the NHPI/Co^{II} system with dioxygen under mild conditions. This catalytic system is expected to promote the aerobic oxidation of the hydroxy functions of alcohols to carbonyl functions [91, 92]. The oxidation of 2-octanol in ethyl acetate at 70 °C in the presence of NHPI (10 mol%) and Co(OAc)$_2$ (0.5 mol%) under dioxygen (1 atm) gives rise to 2-octanone in quantitative yield. Benzoic acids such as m-chlorobenzoic acid (MCBA) enhance the oxidation of alcohols to carbonyl compounds. 2-Octanol can be converted into 2-octanone with O$_2$ even at room temperature by adding a catalytic amount of MCBA to the NHPI/Co(OAc)$_2$ system [Eq. (16)]. The aerobic oxidation of aliphatic alcohols at room temperature has been reported only by Lenz and Ley who used [Bu$_4$N]$^+$[RuO$_4$]$^-$ assisted by 4 Å molecular sieves [86 b].

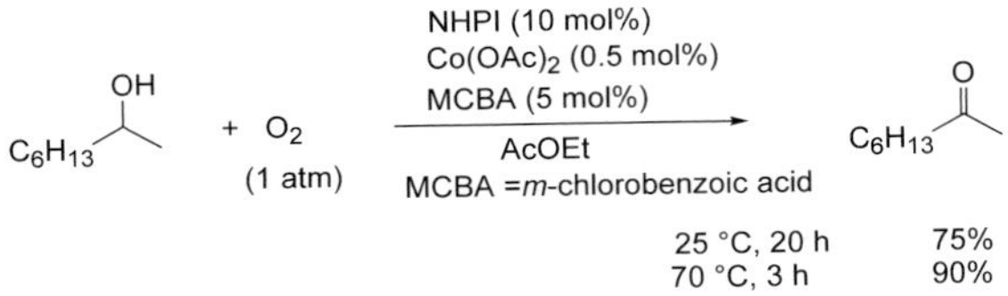

(16)

Figure 5.4 shows the oxidation of secondary and primary alcohols under ambient conditions by the NHPI/Co(OAc)$_2$/MCBA system. Aromatic and cyclic alcohols afford the corresponding ketones in good to quantitative yields. Primary alcohols are also oxidized to carboxylic acids in good yields, although MCPBA is added instead of MCBA. Lauryl alcohol led to lauric acid (66% yield), which is used as a surfactant source. In this oxidation, which proceeds through a free radical process, primary alcohols are rapidly converted into carboxylic acids without isolation of aldehydes, because the hydrogen atom abstraction from aldehydes to afford acyl radicals takes place more easily than that from alcohols to furnish α-hydroxyalkyl radicals [5 a]. The oxidation of allylic alcohols is acheived easily.

In contrast to oxidations of diols with stoichiometric oxidants such as NaIO$_4$, Pb(OAc)$_4$ [93], or hydrogen peroxide [94], which are often used in organic synthesis, little work has been done so far on the oxidation of diols with dioxygen [95]. Recently,

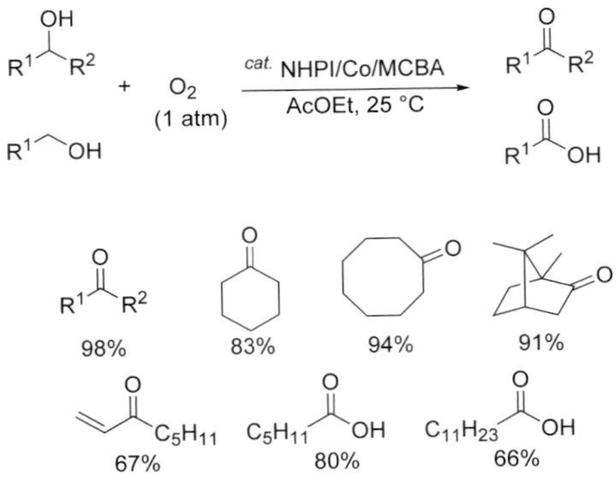

Fig. 5.4 Aerobic oxidation of alcohols by the NHPI/Co/MCBA system

5.2 NHPI-catalyzed Aerobic Oxidation

Uemura and coworkers have reported the Pd(OAc)$_2$-catalyzed lactonization of αω-primary diols with dioxygen in the presence of pyridine and 3 Å molecular sieves [96]. Oxidative cleavage of aliphatic and cyclic 1,2-diols with O$_2$ furnishes aldehydes and dialdehydes, respectively, using Ru(PPh$_3$)$_3$Cl$_2$ on active carbon [97]. The oxidation of 1,2-octanediol with dioxygen catalyzed by NHPI combined with Co(acac)$_3$ afforded heptanoic acid, in 88% selectivity at 80% conversion (Table 5.2) [92]. A precursor to heptanoic acid is an α-ketol, since 1-hydroxy-2-octanone is obtained as a principal product at the limited stage of the reaction (Scheme 5.6). An independent oxidation of the α-ketol leads to the carboxylic acid in good yield. Woodward and coworkers have applied the NHPI-catalyzed oxidation to the carbon–carbon bond cleavage of diols to carboxylic acids [98].

Unlike 1,2-diols, internal vic-diols, for example, 2,3-octanediol, are selectively oxidized to diketones such as 2,3-octanedione rather than cleaved to carboxylic acids.

Tab. 5.2 Oxidation of various diols with dioxygen[a]

Diol	Conv. (%)	Products (Yield/%)		
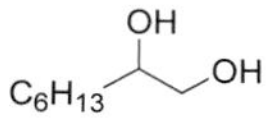 C$_6$H$_{13}$-CH(OH)-CH$_2$OH	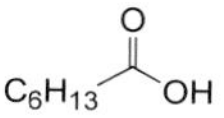 80	C$_6$H$_{13}$-COOH (70)		
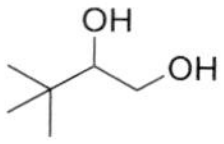 (CH$_3$)$_3$C-CH(OH)-CH$_2$OH	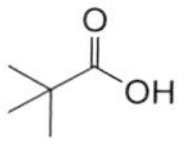 89	(CH$_3$)$_3$C-COOH (71)		
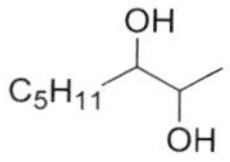 C$_5$H$_{11}$-CH(OH)-CH(OH)-CH$_3$	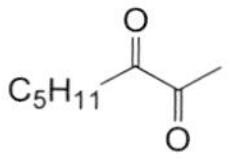 96	C$_5$H$_{11}$-CO-CO-CH$_3$ (86)		
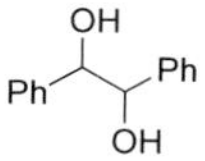 Ph-CH(OH)-CH(OH)-Ph	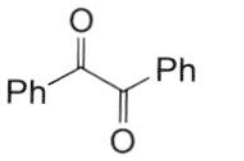 97	Ph-CO-CO-Ph (84)	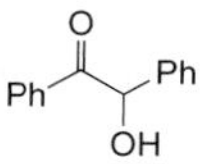 Ph-CO-CH(OH)-Ph (12)	
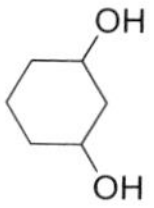 cyclohexane-1,2-diol	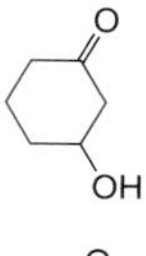 80	2-hydroxycyclohexanone (80)		
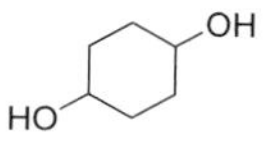 4-hydroxycyclohexanol	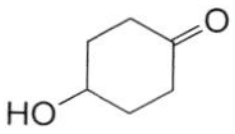 88	4-hydroxycyclohexanone (72)	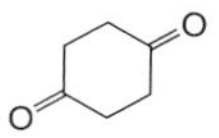 1,4-cyclohexanedione (16)	
 HO-(CH$_2$)$_5$-OH	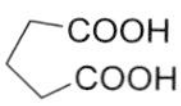 80	HOOC-(CH$_2$)$_3$-COOH (66)		

[a] Diols (3 mmol) were allowed to react with molecular oxygen (1 atm) in the presence of NHPI (10 mol%) and Co(acac)$_3$ (1 mol%) in CH$_3$CN (5 mL).

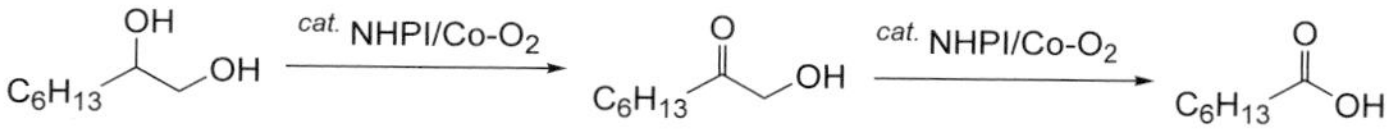

C$_6$H$_{13}$-CH(OH)-CH$_2$OH $\xrightarrow{\text{cat. NHPI/Co-O}_2}$ C$_6$H$_{13}$-CO-CH$_2$OH $\xrightarrow{\text{cat. NHPI/Co-O}_2}$ C$_6$H$_{13}$-COOH

Scheme 5.6 Oxidation of 1,2-octanediol to heptanoic acid

The conversion of *vic*-diols to diketones is usually performed by oxidation with metal oxidants such as $AgCO_3$ [99] and permanganate [100], and by a TEMPO/NaOCl system under electrochemical conditions [101] or by a catalytic method using heteropolyoxometalates and H_2O_2 [102]. Interestingly, 1,3- and 1,4-diols are selectively converted into the corresponding hydroxy ketones rather than the diketones. An α,ω-diol such as 1,5-pentanediol gives rise to the dicarboxylic acid in good yield. The present reaction provides an alternative and useful route to dicarboxylic acids from diols using dioxygen.

5.2.5
Epoxidation of Alkenes Using Dioxygen as Terminal Oxidant

The epoxidation of alkenes using dioxygen *via* a catalytic process is a challenging subject in the field of oxidation chemistry. Much effort has been devoted to the epoxidation of alkenes with dioxygen using transition metals as catalysts [5d, 5e, 6a, 103–110]. For instance, β-diketonate complexes of Ni, V, and Fe are reported to catalyze efficiently the epoxidation of alkenes with dioxygen in the presence of an aldehyde, alcohol, or acetal as a reducing agent under mild conditions [104j]. On the other hand, Ru–porphyrin complexes [108] and Ru-substituted polyoxometalates, $\{[WZnRu_2(OH)(H_2O)](ZnW_9O_{34})_2\}^{11-}$ [109], catalyze the epoxidation of alkenes without any reducing agents.

The hexafluoroacetone (HFA)-catalyzed epoxidation of alkenes utilizing H_2O_2 obtained *in situ* by the NHPI-catalyzed aerobic oxidation of alcohols has been examined (Scheme 5.7). A hydroperoxide derived from HFA and H_2O_2 has been reported to epoxidize various alkenes in fair to good yields [111, 112]. This epoxidation system seems to be an interesting industrial strategy, for it does not require the storage and transportation of explosive H_2O_2 [113]. In addition, the resulting ketones can be easily reduced to the original alcohols. 2-Octene was allowed to react under O_2 (1 atm) in the presence of 1-phenylethanol, influenced by catalytic amounts of NHPI (10 mol%) and HFA (10 mol%) in benzonitrile at 80 °C for 24 h, giving 2,3-epoxyoctane in 93% selectivity based on 93% conversion [Eq. (17)] [114]. This is the first successful epoxidation with H_2O_2 generated *in situ* from alcohols and O_2 without any metal catalysts. The important feature of this reaction is that the epoxida-

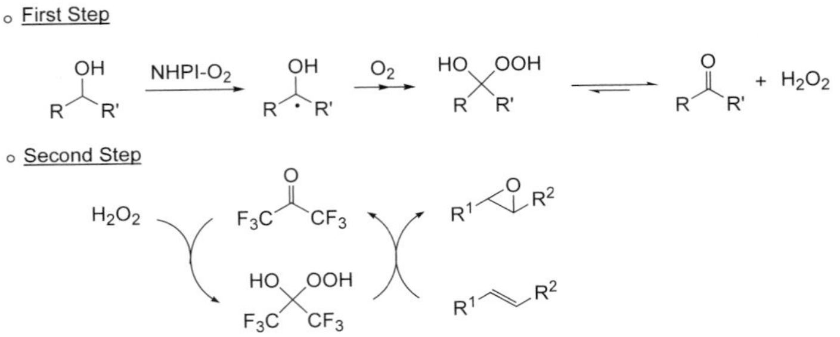

Scheme 5.7 A new strategy for the epoxidation of alkenes

tion of *cis*- and *trans*-2-octenes proceeded in a stereospecific manner to form *cis*- and *trans*-2,3-epoxyoctanes, respectively, in high yields, although O_2 is used as a terminal oxidant.

$$C_5H_{11}\text{-CH=CH-CH}_3 + O_2 \xrightarrow[\substack{\text{PhCN, 80 °C} \\ \text{1-phenylethanol (500 mol\%)}}]{\substack{\text{NHPI (10 mol\%)} \\ \text{HFA (10 mol\%)}}} C_5H_{11}\text{-CH(O)CH-CH}_3 \quad (17)$$

(1 atm)

trans isomer   Conv. 93%   93% (trans : cis = 99 : 1)
cis isomer   Conv. 94%   86% (cis : trans = 98 : 2)

5.2.6
Baeyer-Villiger Oxidation of KA-Oil

KA-oil, a mixture of cyclohexanone and cyclohexanol obtained by the aerobic oxidation of cyclohexane, is an important intermediate in petroleum industrial chemistry for the production of adipic acid and ε-caprolactam, which are key materials for manufacturing 6,6-nylon and 6-nylon, respectively [115]. Baeyer-Villiger oxidation is a frequently used synthetic tool for the conversion of cycloalkanones into lactones. Usually, this transformation is carried out by the use of peracids such as peracetic acid and *m*CPBA [116], hydrogen peroxide [117], and bis(trimethylsilyl)peroxide [118]. However, the catalytic Baeyer-Villiger oxidation using dioxygen is limited to the *in situ* generation of peracids using excess aldehydes and O_2 [119]. In industry, ε-caprolactone is manufactured by the reaction of cyclohexanone with peracetic acid generated by the aerobic oxidation of acetaldehyde [115]. From both synthetic and industrial points of view, it is very attractive that the KA-oil can be used as the starting material for the production of ε-caprolactone with molecular oxygen *via* a catalytic process.

A new strategy for ε-caprolactone synthesis is outlined in Scheme 5.8. The aerobic oxidation of cyclohexanol (**3**) catalyzed by NHPI gives a mixture of cyclohexanone (**2**) and hydrogen peroxide through the formation of 1-hydroxy-1-hydroperoxycyclohexane (**11**) (path 1). Treatment of the resulting reaction mixture with an appropriate catalyst would produce ε-caprolactone (**12**) (path 2). A KA-oil consisting of a 1 : 1 mixture of **3** and **2** was employed as a model starting material. If the aerobic oxidation of the KA-oil in the presence of NHPI is completed, 2 equiv. of **3** and 1 equiv. of H_2O_2 are expected to be formed. Treatment of a 1 : 1 mixture of **3** (6 mmol) and **2** (6 mmol) by catalytic amounts of NHPI (0.6 mmol) and 2,2′-azobisisobutyronitrile (AIBN) (0.3 mmol) under an O_2 atmosphere in CH_3CN at 75 °C for 15 h, followed by $InCl_3$ (0.45 mmol) at 25 °C for 6 h affords **12** in 57% selectivity based on the KA-oil reacted, and 77% of KA-oil was recovered [Eq. (18)]. Water-stable Lewis acids such as $Sc(OTf)_3$ and $Gd(OTf)_3$ afford ε-caprolactone in somewhat lower yields [120].

$$\mathbf{3} + \mathbf{2} + O_2 \xrightarrow[\substack{\text{CH}_3\text{CN, 75 °C} \\ \text{Conv. 23\%}}]{\substack{\text{NHPI (10 mol\%)} \quad \text{InCl}_3 \text{ (7.5 mol\%)} \\ \phantom{\text{CH}_3\text{CN, 75 °C}} \quad 75 \text{ °C}}} \mathbf{12} \quad (18)$$

(1 atm)   57%

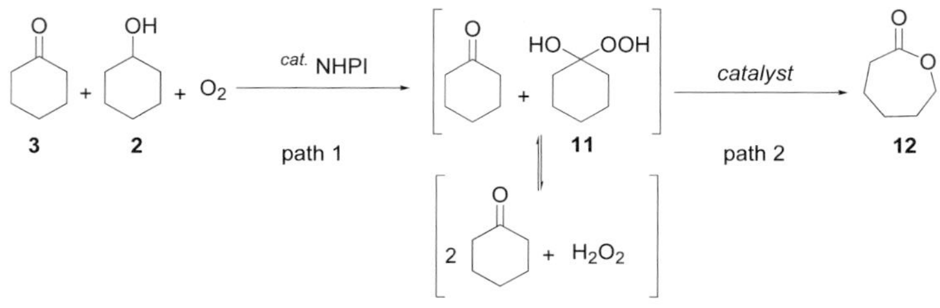

138 | *5 Aerobic Oxidations and Related Reactions Catalyzed by N-Hydroxyphthalimide*

Scheme 5.8 A new strategy for the Baeyer-Villiger oxidation of KA-oil

5.2.7
Preparation of ε-Caprolactam Precoursor from KA-Oil

ε-Caprolactam (CL) is a very important monomer for the production of nylon-6, and about 4.2 million tons of CL were manufactured worldwide in 1998 [121]. Most methods of current CL production involve the conversion of cyclohexanone with hydroxylamine sulfate into cyclohexanone oxime followed by Beckmann rearrangement by the action of oleum and then treatment with ammonia giving CL. A serious drawback of this process is the co-production of a large amount of ammonium sulfate waste [121, 122]. Thomas and coworkers reported a method for the one-step production of cyclohexanone oxime and CL by the reaction of cyclohexanone with ammonia under a pressure of air (34.5 atm) in the presence of a bifunctional molecular sieve catalyst [123]. Hydrogen peroxide oxidation of cyclohexanone in the presence of NH_3 catalyzed by titanium silicate is reported to produce CL [124]. In patent work, on the other hand, the transformation of 1,1′-peroxydicyclohexylamine (PDHA) to a 1:1 mixture of CL and cyclohexanone by LiBr has been reported [125].

It is interesting to develop a novel route to the CL precursor, PDHA, which has so far been prepared by hydrogen peroxide oxidation of cyclohexanone (**3**) followed by treatment with ammonia [121, 125]. Because of the easy transformation of PDHA to a 1:1 mixture of CL and **3** under the influence of an appropriate catalyst, such as lithium halides, the CL production *via* PDHA is considered to be a superior candidate for a next-generation waste-free process for CL. The NHPI-catalyzed aerobic oxidation of KA-oil was applied to the synthesis of PDHA without the formation of any ammonia sulfate waste. The strategy is outlined in Scheme 5.9. The NHPI-catalyzed oxidation of KA-oil, a mixture of **3** and **2**, with O_2 produces 1,1′-dihydroxydicyclohexyl peroxide, which appears to exist in equilibrium with cyclohexanone and H_2O_2 (path 1). Subsequent treatment of the resulting reaction mixture with NH_3 would afford PDHA (path 2). A 1:2 mixture of **3** and **2** was reacted under a dioxygen atmosphere (1 atm) in the presence of small amounts of NHPI and AIBN in ethyl acetate at 60 °C for 20 h, followed by the reaction with an atmosphere of ammonia at 70 °C for 2 h to give 84% of PDHA at 24% conversion of the KA-oil [Eq. (19)]. This route provides a more economical and environmentally friendly process than that of the current method using hydroxylamine sulfate.

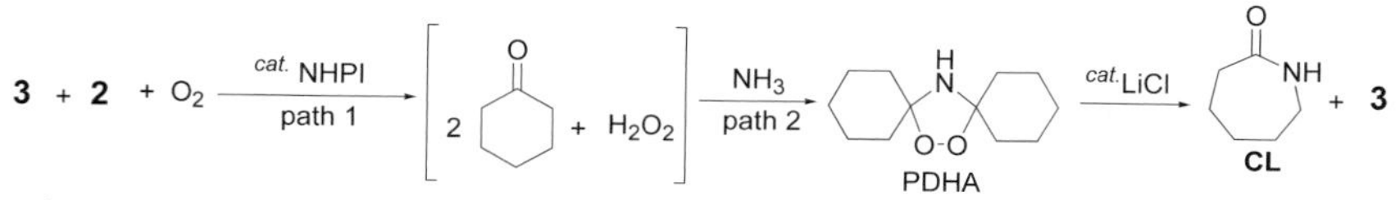

Scheme 5.9 A new strategy for the synthesis of the ε-caprolactam precursor, PDHA

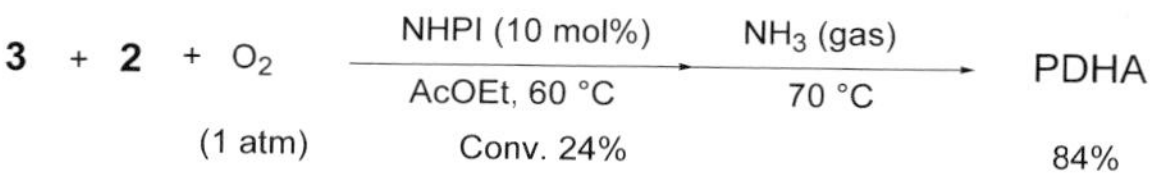

$$3 + 2\, O_2 \xrightarrow[\text{AcOEt, 60 °C}]{\text{NHPI (10 mol\%)}} \xrightarrow[\text{70 °C}]{\text{NH}_3\text{ (gas)}} \text{PDHA} \quad (19)$$

(1 atm)   Conv. 24%   84%

5.3
Functionalization of Alkanes Catalyzed by NHPI

5.3.1
Carboxylation of Alkanes with CO and O_2

Carbonylation as well as carboxylation of alkanes with carbon monoxide (CO) are challenging transformations in organic synthesis [126]. There have been several important discoveries including the Rh-catalyzed photocarbonylation of alkanes by Tanaka and coworkers [127], and Margl et al. [128], and the carboxylation of methane with CO/O_2 using Pd/Cu [129] or $RhCl_3$ [130] catalysts by the groups working with Fujiwara and Sen. The carbonylation of adamantanes under the influence of Lewis acid and superacids has also been reported [131, 132]. Following the first report on the free-radical mediated carbonylation by Coffmann and coworkers in 1952 [133], this type of reaction was then not investigated for a long period of time, probably because the reaction must be conducted under extremely high CO pressure (200–300 atm) [134]. In 1990, Ryu and coworkers performed a successful free-radical carbonylation of alkyl halides with CO mediated by tributyltin hydride [126 e, 135]. Sen and Lin disclosed a free-radical carboxylation of methane to acetic acid by the use of peroxydisulfate as a radical source [136]. Benzophenone- [137] and polyoxotungstate-photocatalyzed [141] as well as mercury-photosensitized [139] carbonylations of cyclohexane afford cyclohexanecarbaldehyde. The trapping of alkyl radicals generated from alkanes under the influence of NHPI catalyst by CO followed by O_2 leads to carboxylic acids (Scheme 5.10) [140]. The carboxylation of adamantane under CO/air (15/1 atm) in the presence of NHPI (10 mol%) in a mixed solvent of acetic acid and 1,2-dichloroethane at 95 °C for 4 h affords 1-adamantanecarboxylic acid, 1,3-adamantanedicarboxylic acid, 2-admantanecarboxylic acid, and several oxygenated products such as 1-adamantanol and 2-adamantanone [Eq. (20)].

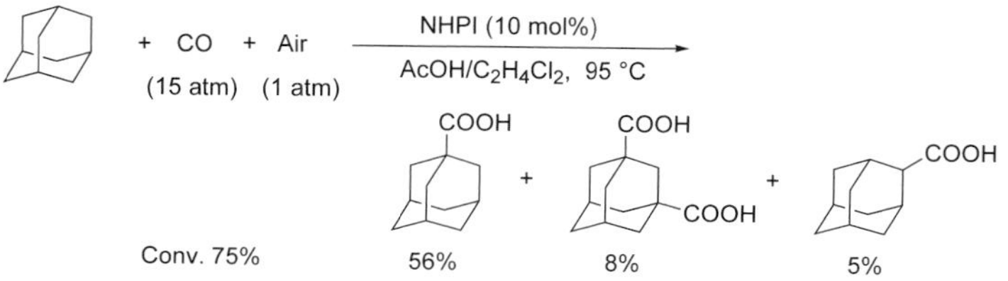

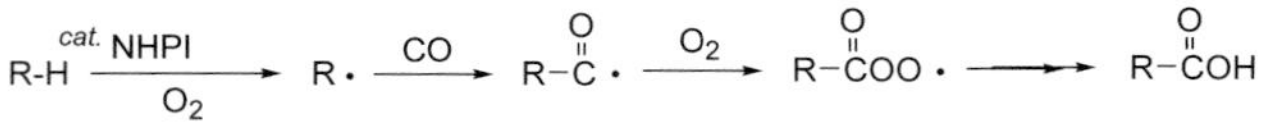

5 Aerobic Oxidations and Related Reactions Catalyzed by N-Hydroxyphthalimide

$$R-H \xrightarrow[O_2]{cat.\ NHPI} R\cdot \xrightarrow{CO} R-\overset{O}{\underset{\|}{C}}\cdot \xrightarrow{O_2} R-\overset{O}{\underset{\|}{C}}OO\cdot \longrightarrow R-\overset{O}{\underset{\|}{C}}OH$$

Scheme 5.10 Carbonylation of alkanes with CO and O_2 catalyzed by NHPI

The present strategy was successfully applied to the preparation of adamantanedicarboxylic acid, which is an interesting monomer in polymer chemistry, through a stepwise procedure [Eq. (21)], although the dicarboxylic acid is difficult to obtain by conventional methods. Similarly, 1,3-dimethyladamantane and *endo*-tricyclo[5.2.1.0]-decane were carboxylated to the respective mono- and dicarboxylic acids.

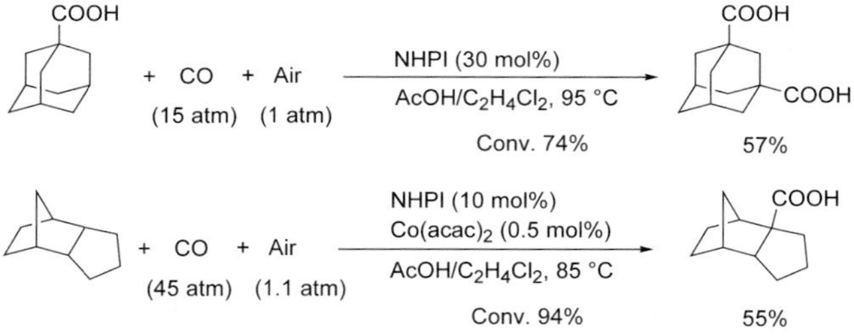

Adamantane + CO (15 atm) + Air (1 atm) → [NHPI (30 mol%), AcOH/C$_2$H$_4$Cl$_2$, 95 °C, Conv. 74%] → adamantane-1,3-dicarboxylic acid, 57%

endo-tricyclo[5.2.1.0]decane + CO (45 atm) + Air (1.1 atm) → [NHPI (10 mol%), Co(acac)$_2$ (0.5 mol%), AcOH/C$_2$H$_4$Cl$_2$, 85 °C, Conv. 94%] → carboxylic acid product, 55%

(21)

5.3.2
First Catalytic Nitration of Alkanes Using NO_2

Nitration of lower alkanes such as methane and ethane with nitric acid or nitrogen dioxide is practiced industrially to produce nitroalkanes [141, 142]. However, a major problem in current industrial nitration is that the reaction must be run at fairly high temperature (250–400 °C), because of the difficulty of obtaining C–H bond homolysis by NO_2 [141]. Under such high temperatures, higher alkanes undergo not only homolysis of the C–H bonds but also cleavage of the C–C skeleton. Hence, the nitration is limited to lower alkanes [142]. The nitration of propane results in all of the possible nitroalkanes, i.e., nitromethane, nitroethane, 1-nitropropane, and 2-nitropropane [141, 143]. Therefore, the catalytic nitration of alkanes under mild conditions should offer a promising and superior alternative. Since NO_2 is a paramagnetic molecule, the generation of PINO from NHPI by the action of NO_2 in analogy with O_2 is expected. Indeed, when NO_2 was added to NHPI in benzene at room temperature, an ESR signal attributable to PINO is instantly observed as a triplet. As a typical result, the nitration of cyclohexane with NO_2 by NHPI without any solvent under air (1 atm) proceeds smoothly even at 70 °C to give nitrocyclohexane (70% based on NO_2 used) and cyclohexyl nitrite (7%) along with a small amount of an oxygenated product, cyclohexanol (5%) [Eq. (22)] [144].

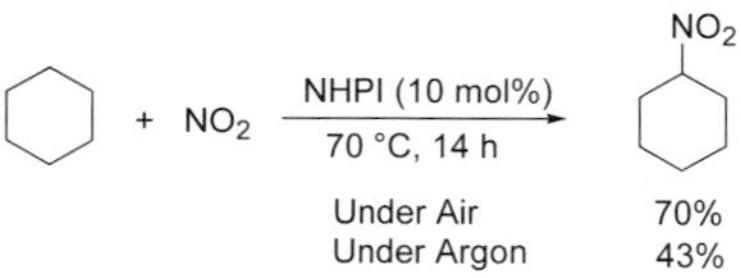

cyclohexane + NO_2 → [NHPI (10 mol%), 70 °C, 14 h] → nitrocyclohexane
Under Air: 70%
Under Argon: 43%

(22)

5.3 Functionalization of Alkanes Catalyzed by NHPI

It is important that the NHPI-catalyzed nitration is conducted under air, since NO generated in the course of the reaction can be readily reoxidized to NO_2 by O_2. In the absence of air, the yield of nitrocyclohexane decreases to 43%. After the nitration, the NHPI catalyst can be separated from the reaction mixture by simple filtration and reused repeatedly. Nitrocyclohexane is easily reduced to cyclohexanone oxime. Therefore, this nitration provides an alternative practical route to cyclohexanone oxime, which is a raw material for ε-caprolactam leading to nylon-6 [145–147].

A plausible pathway is shown in Scheme 5.11. The hydrogen atom abstraction from the hydroxyimide group of NHPI is induced by NO_2 to form PINO, a key radical species. The PINO abstracts the hydrogen atom from an alkane to give an alkyl radical, which is readily trapped by NO_2 to form a nitroalkane. The HNO_2 formed is converted into HNO_3, H_2O, and NO which is easily oxidized to NO_2 under air [148]. The most promising feature of the NHPI-catalyzed nitration of alkanes by NO_2 is that the nitration can be conducted under air at moderate temperature. Owing to the higher concentration of NO_2 than air, the alkyl radicals formed can react selectively with NO_2 rather than with O_2 to give nitroalkanes in preference to oxygenated products. The conventional nitration is difficult to carry out in air, because the nitration must be carried out at high temperature (250–400 °C). Under these temperatures, the resulting alkyl radicals react not only with NO_2 but also with O_2 to provide a complex mixture of products [142b]. Through the use of the NHPI catalyst, the highly selective nitration of higher alkanes with NO_2/air under mild conditions was realized for the first time.

NHPI + NO_2 ⟶ PINO + HNO_2
R–H + PINO ⟶ R• + NHPI
R• + NO_2 ⟶ R–NO_2
$3HNO_2$ ⟶ HNO_3 + H_2O + 2NO
2NO + O_2 ⟶ $2NO_2$

Scheme 5.11 A possible reaction path for alkane nitration catalyzed by NHPI

A wide variety of alkanes were successfully nitrated by the NHPI/NO_2 system (Figure 5.5). In addition, nitric acid instead of NO_2 was found to act as an efficient nitrating reagent. For example, the reaction of adamantane with concentrated HNO_3 in the presence of a catalytic amount of NHPI in $PhCF_3$ at 60 °C under Ar afforded nitroadamantane and 1,3-dinitroadamantane in 64% and 3% yields, respectively [Eq. (23)].

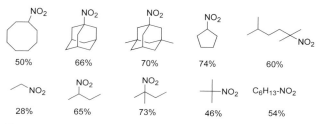

Fig. 5.5 Nitration of various alkanes by the NHPI/NO_2 system

adamantane + HNO₃ → (NHPI (10 mol%), PhCF₃, 60 °C, under Ar) → 1-nitroadamantane (64%) + 1,3-dinitroadamantane (3%)   (23)

5.3.3
Sulfoxidation of Alkanes Catalyzed by Vanadium

The sulfoxidation of aromatic hydrocarbons has been studied extensively, but work on the sulfoxidation of saturated hydrocarbons to alkanesulfonic acids remains at a less satisfactory level. The Strecker reaction using alkyl halides, preferably alkyl bromides, and alkali metal or ammonium sulfides, is commonly used for the synthesis of alkanesulfonic acids [149]. Another procedure, the oxidation of thiols with bromine in the presence of water or hydrogen peroxide and acetic acid, has been reported [150]. Attempts to realize the sulfoxidation of alkanes with SO_2 and O_2 have not been studied fully, in spite of their importance, because of the difficulty of selective cleavage of the C–H bond in alkanes. Only a few reactions have been reported for the sulfoxidation of alkanes such as cyclohexane *via* a radical process using a mixture of SO_2 and O_2 by means of the photo- and peroxide-initiated techniques [151]. However, the efficiency of the sulfoxidation by these methods is insufficient. Therefore, if alkanes can be sulfoxidated catalytically by SO_2/O_2 without irradiation with light or initiation by a peroxide, such a method has enormous synthetic potential and provides a very attractive route to alkanesulfonic acids. The direct sulfoxidation of alkanes using SO_2 and O_2 was efficiently catalyzed by a vanadium species in the presence or absence of NHPI [152]. The reaction of adamantane with a mixture of SO_2 and O_2 (0.5/0.5 atm) in the presence of NHPI (10 mol%) and $VO(acac)_2$ (0.5 mol%) in acetic acid at 40 °C for 2 h produced 1-adamantanesulfonic acid in 95% selectivity based on 65% conversion [Eq. (24)]. Smith and Williams obtained the same product in 15% yield by the photosulfoxidation of adamantane with SO_2/O_2 in the presence of H_2O_2 [153]. Surprisingly, 1-adamantanesulfonic acid was obtained with high selectivity and at moderate conversion even in the absence of the NHPI [Eq. (24)].

adamantane + SO_2 / O_2 (0.5/0.5 atm) → (NHPI (10 mol%), VO(acac)₂ (0.5 mol%), AcOH, 40 °C) → 1-adamantanesulfonic acid (SO₃H)   (24)

NHPI/VO(acac)₂ 95% (Conv. 65%)
VO(acac)₂   98% (Conv. 43%)

In order to assess the potential of various metal ions in this sulfoxidation, a series of first-row transition metal salts, $TiO(acac)_2$, $Cr(acac)_3$, $Mn(acac)_3$, $Fe(acac)_3$, $Co(acac)_2$, $Ni(OAc)_2$, and $Cu(OAc)_2$ were tested. It is interesting to note that no sulfoxidation was induced by these metal salts other than vanadium ions [154]. From a survey of vanadium compounds, $VO(acac)_2$ and $V(acac)_3$ were found to be efficient catalysts. $VO(acac)_2$ promotes the reaction even at room temperature affording the sulfonic acid in

81% selectivity at 64% conversion after 24 h. The addition of a small amount of hydroquinone stopped the reaction. This indicates that a radical chain process is involved in this catalytic sulfoxidation. A variety of alkanes were successfully sulfoxidized by a mixture of SO_2 and O_2 giving the corresponding alkanesulfonic acids in high selectivities (Figure 5.6). Adamantane having either an electron-withdrawing or electron-donating group was sulfoxidized in good selectivity in a range of ca. 60–70% conversion. The aliphatic hydrocarbon, octane, afforded a mixture of 2-, 3-, and 4-octanesulfonic acids. The sulfoxidation of alkanes seems to proceed via the following reaction steps (Scheme 5.12).

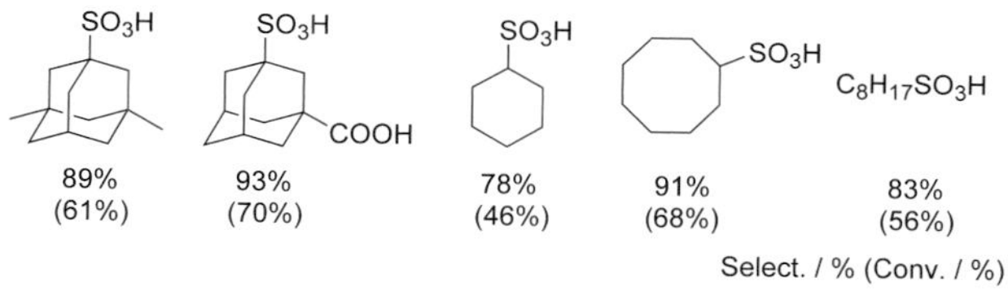

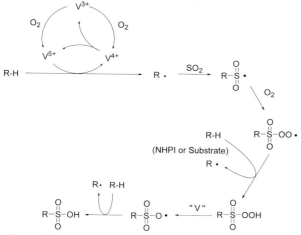

Fig. 5.6 Sulfoxidation of various alkanes catalyzed by $VO(acac)_2$

Scheme 5.12 A possible reaction mechanism for the sulfoxidation of alkanes

The sulfoxidation may be initiated by one-electron transfer from an alkane to a V^V species generated in situ from $VO(acac)_2$ and O_2 to form an alkyl cation radical which readily liberates a proton to form an alkyl radical. The V^{IV} species is reported to undergo disproportionation to V^V and V^{III} in the oxidative polymerization of diphenyl disulfide by the vanadium ion under a dioxygen atmosphere [155]. In addition, α-hydroxycarbonyl compounds are oxidized to α-dicarbonyl compounds by $VOCl_3$ and $VO(acac)_2$ under an oxygen atmosphere [156]. The resulting radical is trapped by SO_2 and then O_2 to generate an alkanesulfonylperoxy radical, which is finally converted into an alkanesulfonic acid through the well-known reaction pathway [157].

5.3.4
Reaction of NO with Organic Compounds

In recent years, much attention has been paid to nitric oxide (NO), a molecule having a free radical character, in the fields of biochemistry and medical science [158, 159]. However, its application to synthetic organic chemistry is quite limited because of the sparse information available on the chemical behavior of NO and the difficulty incurred in controlling its reactivity [160–164]. Recently, Kato and Mukaiyama and Yamada and coworkers have reported that NO can be used as a nitrogen source for the synthesis of nitrogen-containing compounds such as 2-nitrosocarboxamides [160] and nitroalkenes [161], respectively. A novel utilization of NO in organic synthesis with the use of NHPI has been developed [165, 166]. The reaction of adamantane with NO (1 atm) in the presence of NHPI (10 mol%) in a mixed solvent of benzonitrile and acetic acid at 100 °C for 20 h afforded 1-N-adamantylbenzamide in substantial yield along with a small amount of nitroadamantane [Eq. (25)] [165]. This reaction provides a novel and alternative modified Ritter-type reaction, although there are a few reports on the transformation of adamantane into the amide by means of the anodic oxidation [167] or nitronium tetrafluoroborate [168].

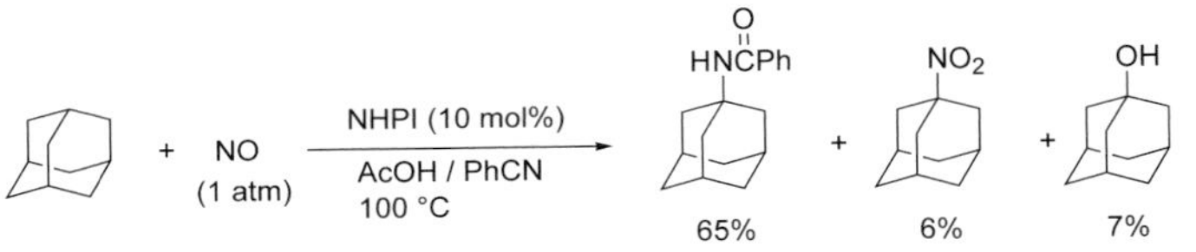

$$\text{adamantane} + \text{NO (1 atm)} \xrightarrow[\text{AcOH / PhCN} \atop 100\ °C]{\text{NHPI (10 mol\%)}} \text{HNCPh-adamantane (65\%)} + \text{NO}_2\text{-adamantane (6\%)} + \text{OH-adamantane (7\%)} \quad (25)$$

On the other hand, benzyl ethers react with NO in the presence of the NHPI catalyst to afford the corresponding aromatic aldehydes [Eq. (26)] [166]. The reaction of 4-methoxymethyltoluene catalyzed by NHPI (10 mol%) under NO (1 atm) for 5 h leads to p-tolualdehyde in 50% yield. tert-Butoxymethyltoluene and tert-butyl benzyl ethers are converted into the corresponding aldehydes in good yields.

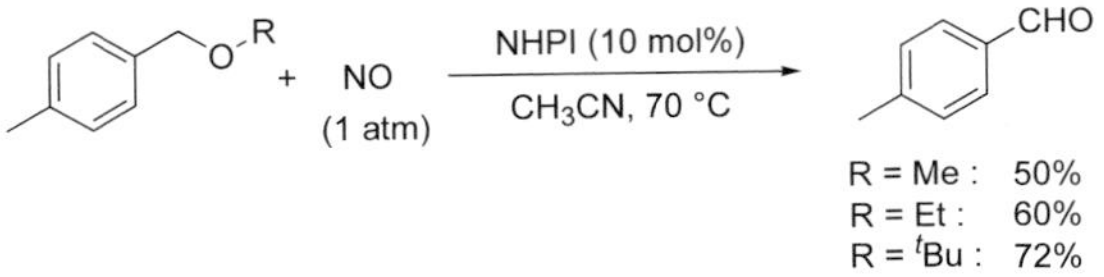

$$\text{Ar-CH}_2\text{-O-R} + \text{NO (1 atm)} \xrightarrow[\text{CH}_3\text{CN, 70 °C}]{\text{NHPI (10 mol\%)}} \text{Ar-CHO} \quad (26)$$

R = Me : 50%
R = Et : 60%
R = ^{t}Bu : 72%

The most important application of this procedure is the transformation of ethers into benzenedicarbaldehydes, which are attractive starting materials in pharmaceutical synthesis [169]. 1,3-Dihydro-2-benzofuran, 1,3-di-tert-butoxymethyl- and 1,4-dimethoxymethylbenzenes are converted into the respective dialdehydes in good yields [Eq. (27)] [166]. Of the various indirect procedures to obtain dialdehydes, hydrolysis of α,α,α′,α′-tetrabromoxylenes is usually used, although the preparation of bromides is troublesome [170]. Therefore, the present procedure provides a very convenient and direct route to benzenedicarbaldehydes.

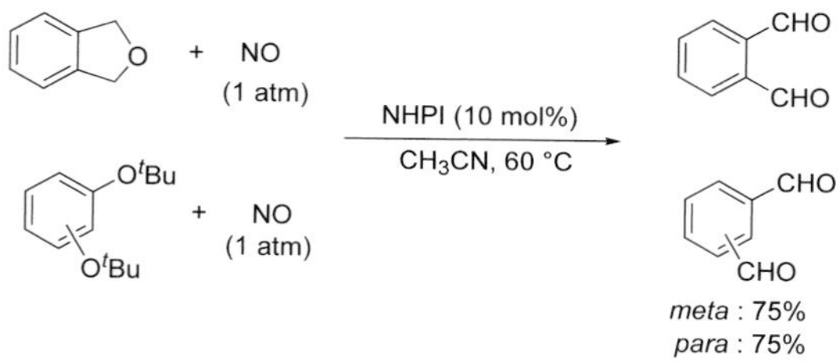

5.3 Functionalization of Alkanes Catalyzed by NHPI

$$\text{(27)}$$

(reaction scheme with NHPI (10 mol%), CH$_3$CN, 60 °C; products: dicarbaldehydes; meta : 75%, para : 75%)

Mechanistically, the reactions of adamantane and ethers with NO are rationally explained by considering the formation of carbocations as transient intermediates (Scheme 5.13). The generation of PINO from NHPI in the presence of NO is confirmed by ESR measurements [165], but the formation of PINO by this method may be due to traces of NO$_2$ contained in the reaction system. On the other hand, Suzuki has suggested the formation of a cationic species via a diazonium nitrate in the nitration of alkenes with NO [171]. The nucleophilic attack of the benzonitrile and water on the adamantyl and benzylic cations would result in the amide and aldehyde, respectively [166].

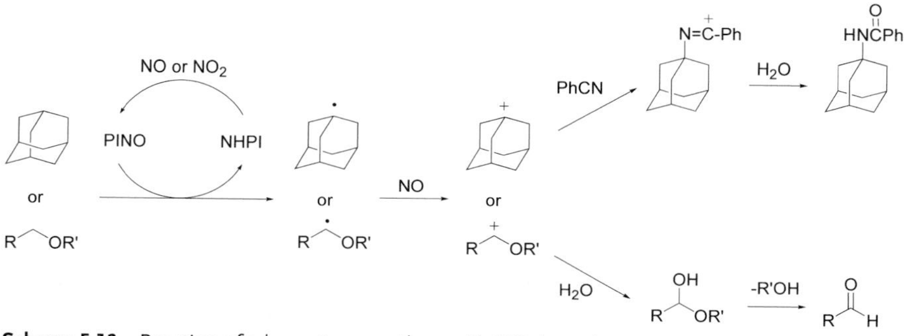

Scheme 5.13 Reaction of adamantane or ethers with NO through the formation of carbocations as transient intermediates

5.3.5
Ritter-type Reaction with Cerium Ammonium Nitrate (CAN)

Ritter-type reaction of an alkane with nitrile forming an amide has been accomplished by the use of Br$_2$/H$_2$SO$_4$ [172], NO$_2$BF$_4$ [173], AlCl$_3$/CH$_2$Cl$_2$ [174], electrolysis [175], Pb(OAc)$_4$ [176], and HNO$_3$/CCl$_4$ [177], these methods, however, are limited to the reaction of adamantane or its derivatives. Hill demonstrated that lower alkanes such as isobutane react with acetonitrile in the presence of a polyoxometalate under photo-assistance to form the corresponding acetoamide in high selectivity [178]. The reaction is postulated to proceed through the formation of an alkyl radical followed by one-electron oxidation by the W ion to a carbocation.

The Ritter-type reaction of adamantane is accomplished using the NHPI/NO system. In this section, we show that NHPI combined with cerium ammonium nitrate

(CAN) serves as an efficient system for the generation of both PINO from NHPI and carbocations from alkyl radicals. Thus, benzylic compounds first undergo the amidation with alkyl nitrile under mild conditions to form amides in good yields. The reaction of ethylbenzene in the presence of CAN and NHPI in EtCN under argon at 80 °C for 6 h produced N-(1-phenylethyl)propionamide in 84% selectivity at 61% conversion [Eq. (28)]. The NHPI/CAN system can apply to the Ritter-type reaction of various alkylbenzenes and adamanatanes.

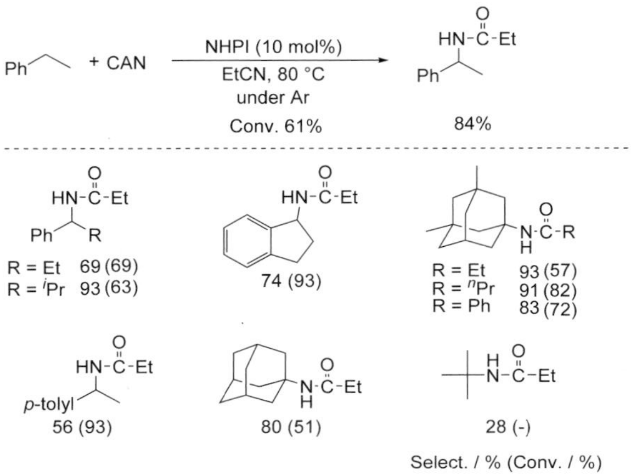

(28)

Select. / % (Conv. / %)

It is reasonable to assume that the present reaction is initiated by the reaction of NHPI with CAN to form PINO, which is thought to be a key species for the generation of alkyl radicals (Scheme 5.14). Indeed, PINO is generated upon treatment of NHPI with CAN in MeCN at 70 °C. The resulting PINO abstracts a hydrogen atom from these hydrocarbons to generate the corresponding alkyl radicals (**A**), which undergo the one-electron oxidation by Ce^{IV} to form carbocations (**B**). The carbocations **B** thus generated are trapped by nitriles, followed by H$_2$O to afford amide derivatives [179].

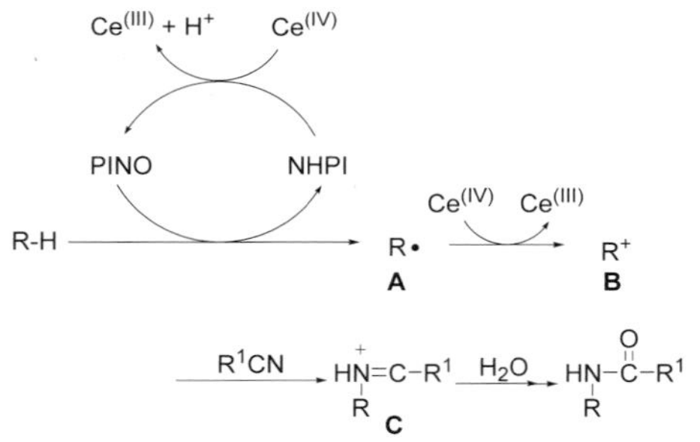

Scheme 5.14 A possible reaction path for the Ritter-type reaction by the NHPI/CAN system

5.4
Carbon–Carbon Bond Forming Reaction *via* Generation of Carbon Radicals Assisted by NHPI

Additions of carbon radicals to alkenes, which can lead to the formation of new carbon–carbon bonds, are of major synthetic interest in organic chemistry, because of many advantages of the reactions over ionic processes [180]. Nowadays, numerous methods for the generation of carbon radicals and their inter- or intramolecular additions to alkenes for the synthesis of fine chemicals and natural products have been developed [180, 181]. For instance, reactions of alkyl halides with tributyltin hydride or tris(trimethylsilyl)silane [182] and the thermal decomposition of Barton esters [183] are the most common methodologies for the generation of alkyl radicals. Although the peroxide- and photo-initiated reactions are often used as practical synthetic means, major problems of these methods are the lack of selectivity, generality and efficiency of the reaction [180]. Therefore, the carbon–carbon bond forming reaction through the carbon radical generation from alkanes is a worthwhile target in free radical chemistry.

5.4.1
Oxyalkylation of Alkenes with Alkanes and Dioxygen

The NHPI-catalyzed aerobic oxidation of alkanes proceeds through the formation of alkyl radicals as mentioned previously. If alkyl radicals generated from alkanes could add to alkenes smoothly, it would provide a powerful strategy for the construction of a C–C bond in which alkanes can be directly used as alkyl sources. Furthermore, since the generation of PINO from NHPI is performed under a dioxygen atmosphere, the concomitant introduction of both an alkyl group and an oxygen function to alkenes is possible. This new reaction type may be regarded as a catalytic oxyalkylation of alkenes. An approach to oxyalkylation is illustrated in Figure 5.7. The reaction involves an alkyl radical generation by the NHPI/Co/O_2 system and the addition of the resulting alkyl radical to an alkene to form an adduct radical, which is readily trapped by O_2. The reaction of methyl acrylate with 1,3-dimethyladamantane under a mixed gas of O_2 (0.5 atm) and N_2 (0.5 atm) catalyzed by NHPI (20 mol%) in the

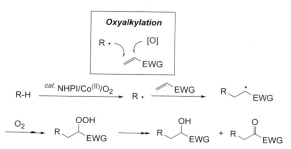

Fig. 5.7 Oxyalkylation of alkenes with alkanes and O_2 catalyzed by the NHPI/Co^{II} system

presence of Co(acac)$_3$ (1 mol%) in acetonitrile at 75 °C for 16 h gave a mixture that was about 7 : 3 in the oxyalkylated products, methyl 3-(3,3'-dimethyladamantyl)-2-hydroxypropionate and methyl 3-(3,3'-dimethyladamantyl)-2-oxopropionate, in 91% yield [Eq. (29)] [184]. This is the first simultaneous introduction of alkyl and oxygen functions to alkenes through a catalytic process [185]. Fukunishi and Tabushi reported the peroxide-initiated simple radical addition of 1,3-dimethyladamantane to maleate and fumaronitrile affording the adduct [186]. Additionally, the reaction of cyclohexane with methyl acrylate under dioxygen (1 atm) gives rise to the corresponding three-component coupling product in 75% selectivity at 80% conversion.

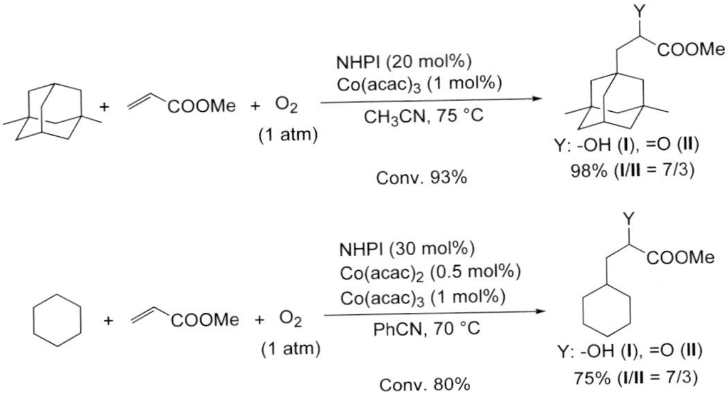

$$\text{(29)}$$

5.4.2
Synthesis of α-Hydroxy-γ-lactones by Addition of α-Hydroxy Carbon Radicals to Unsaturated Esters

α-Hydroxy-γ-lactones are valuable synthetic precursors to compounds such as α,β-butenolides, which have potent biological activities [187], efficient food intake-control substances [188], and monomers of biodegradable polymers as well as fine chemicals [189]. However, there are few practical methods for the synthesis of these lactones [190]. The strategy for the oxyalkylation of alkenes with alkanes and O$_2$ was extended to the reaction of alkenes with alcohols and O$_2$ leading to α-hydroxy-γ-lactones [191]. The reaction of 2-propanol with methyl acrylate under a dioxygen atmosphere was examined in the presence of NHPI (10 mol%) combined with Co(OAc)$_2$ (0.1 mol%) and Co(acac)$_3$ (1 mol%), and α-hydroxy-γ,γ-dimethyl-γ-butyrolactone was obtained in 78% yield (Scheme 5.15). Mori et al. prepared the same lactone *via* three steps from isobutene and trichloroacetaldehyde in 14% yield [190a]. This new method for the construction of α-hydroxy-γ-lactones is quite general for a variety of alcohols and α,β-unsaturated esters (Figure 5.8). The preparation of α-hydroxy-γ-spirolactones from cyclic alcohols is especially notable, because these spirolactones have been very difficult to synthesize up to now. The reaction can be explained by Scheme 5.15: (i) *in situ* generation of an α-hydroxy carbon radical from an alcohol assisted by NHPI/Co^{II}/O$_2$, (ii) the addition of the radical to methyl acrylate, (iii) trapping of the adduct radical by O$_2$, and (iv) the intramolecular cyclization to give α-hydroxy-γ-butyrolac-

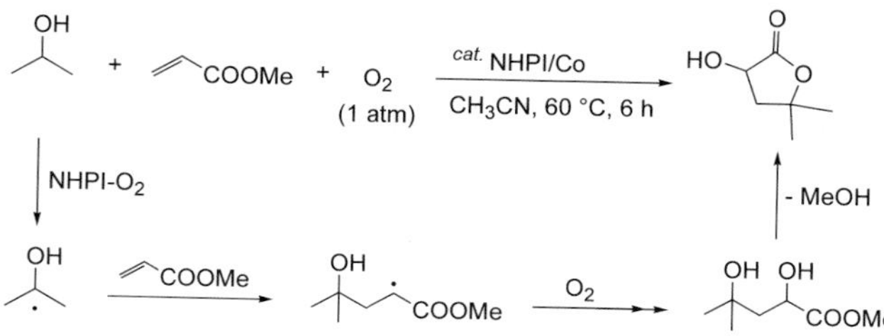

5.4 Carbon–Carbon Bond Forming Reaction via Generation of Carbon Radicals

Scheme 5.15 Radical addition of 2-propanol to methyl acrylate under O_2 catalyzed by the NHPI/Co(acac)$_2$ system

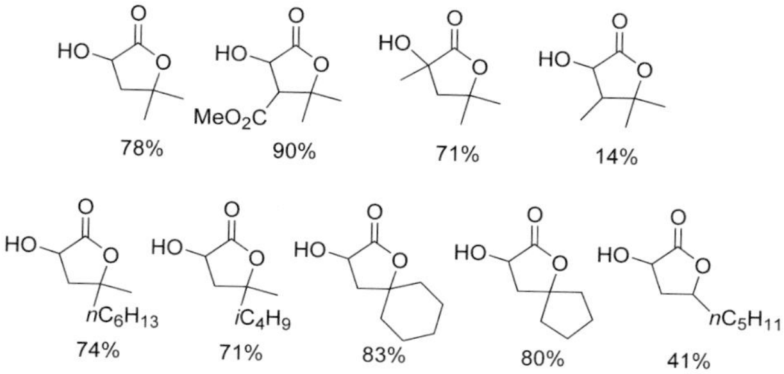

Fig. 5.8 Synthesis of α-hydroxy-γ-lactones

tone. Considering the low-cost material, reaction efficiency, and reaction simplicity, this method provides an innovative approach to α-hydroxy-γ-lactones which have considerable industrial potential.

5.4.3
Hydroxyacylation of Alkenes Using 1,3-Dioxolanes and Dioxygen

Addition of aldehydes through cleavage of the aldehydic carbon–hydrogen bond to terminal alkenes is known as hydroacylation (Figure 5.9) [192]. If the concomitant introduction of acyl and hydroxy moieties to alkenes, which is referred to as hydroxyacylation, can be achieved by a cascade reaction, it would provide a novel route to β-hydroxy carbonyl compounds (Figure 5.9). β-Oxycarbonyl arrays constitute important structural subunits in a variety of natural and non-natural materials and in key intermediates leading to pharmaceuticals [193]. There is one report on the hydroxyacylation of acrylates with acyl radicals derived from aldehydes using dioxygen as a hydroxy source assisted by a cobalt(II) Schiff-base complex, but the attempt was not fully successful due to the decarbonylation from the acyl radicals as well as the reaction of the acyl radicals with O_2 leading to carboxylic acids, which caused undesired side reactions [194]. To overcome these drawbacks, 1,3-dioxolanes, masked aldehydes, were used as the acyl source [195].

Since α,α-dioxaalkyl radicals corresponding to acyl radical equivalents are expected to be generated from 1,3-dioxolanes by the use of the NHPI/O_2 system, the apparent

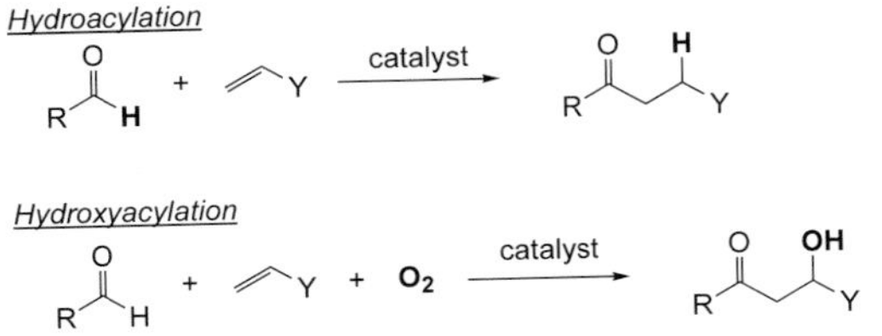

150 | *5 Aerobic Oxidations and Related Reactions Catalyzed by N-Hydroxyphthalimide*

Fig. 5.9 Hydroxyacylation of alkenes

Hydroacylation

Hydroxyacylation

hydroxyacylation of methyl acrylate using several 1,3-dioxolanes and O_2 was examined [196, 197]. Reaction of 2-methyl-1,3-dioxolane with methyl acrylate under O_2 (1 atm) in the presence of NHPI (5 mol%) and a small amount of $Co(OAc)_2$ (0.05 mol%) at room temperature for 3 h produced β-hydroxy ketal in 81% yield [Eq. (30)]. The dioxolane moiety can be easily deprotected under acidic conditions in quantitative yield [Eq. (31)]. This is a useful method for the introduction of formyl and hydroxy groups to alkenes. Although the direct use of formaldehyde in organic synthesis is restricted owing to its intractability, the reaction of 1,3-dioxolane, masked formaldehyde, with methyl acrylate, followed by deprotection of the dioxolane moiety, gave an adduct in good yield. α-Hydroxy esters are reported to be valuable precursors for the synthesis of attractive compounds possessing a variety of pharmacological properties, e. g., pyridazinones and the alkaloid epibatidine [Eq. (32)] [198].

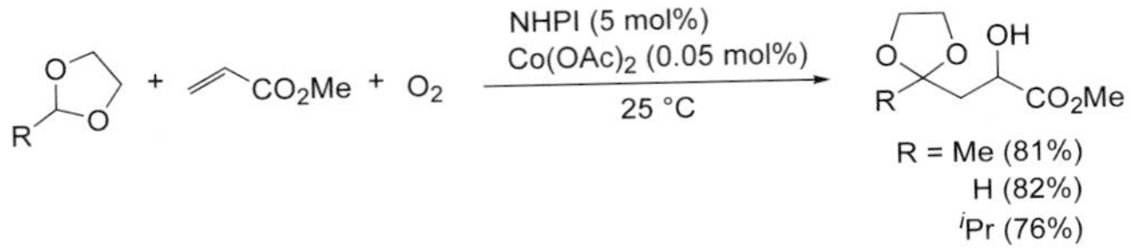

$$R = Me\ (81\%)$$
$$H\ (82\%)$$
$$^iPr\ (76\%)$$

(30)

$$R = Me\ (99\%)$$
$$H\ (83\%)$$

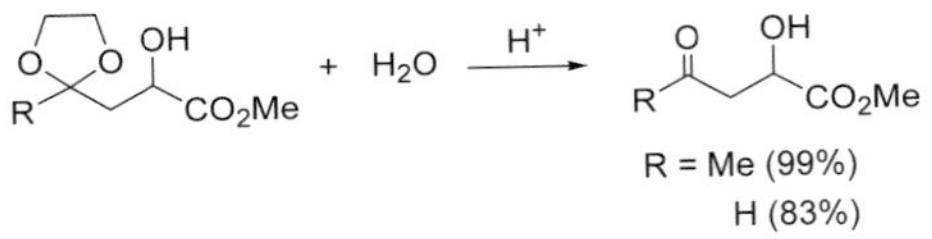

(31)

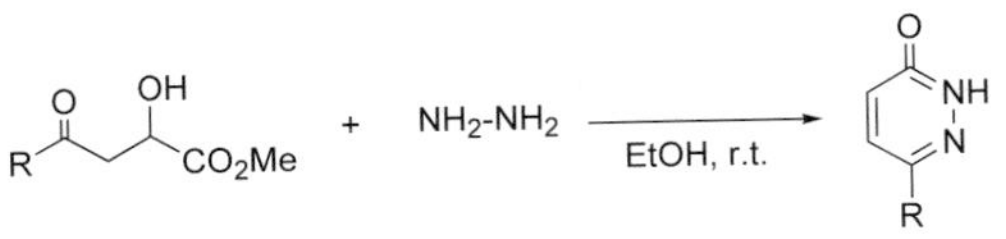

(32)

5.4.4
Hydroacylation of Alkenes Using NHPI as a Polarity-reversal Catalyst

Intermolecular radical-chain addition of aldehydes to alkenes is the simplest methodology for the synthesis of long-chain unsymmetrical ketones, but employment of this method is usually difficult for the synthesis of simple aliphatic ketones [199]. The hydroacylation between alkenes and aldehydes *via* a radical process involves the following reaction steps: (i) hydrogen abstraction from an aldehyde by a radical initiator to form an acyl radical (**A**), (ii) addition of the acyl radical to alkene leading to

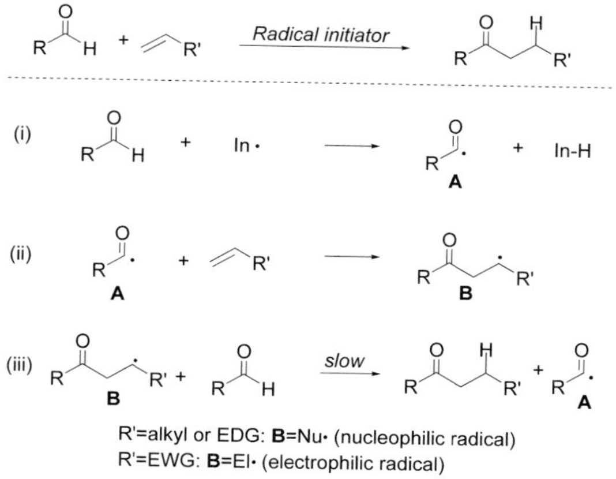

Scheme 5.16 Hydroacylation of alkenes via radical process

R'=alkyl or EDG: **B**=Nu• (nucleophilic radical)
R'=EWG: **B**=El• (electrophilic radical)

a β-oxocarbon radical (**B**), and (iii) abstraction of the aldehydic hydrogen atom from another aldehyde by **B**, generating a ketone and acyl radical **A** (Scheme 5.16).

If the R' is an alkyl or an electron-donating group (EDG) in this radical-chain reaction, step (iii) becomes a sluggish process, since the abstraction of the aldehydic hydrogen atom by nucleophilic radical **B** (Nu•) proceeds with difficulty. In contrast, if the R' is an electron-withdrawing group (EWG), this step proceeds smoothly because of the ease of aldehydic hydrogen abstraction by an electrophilic radical **B** (El•) [200]. Acyl radicals, which are nucleophilic in nature, are known to add more easily to electron-deficient alkenes than normal alkenes [201]. As a result, the hydroacylation of simple alkenes with aldehydes via a radical process proceeds with difficulty. Recently, the hydroacylation of alkenes, particularly electron-rich alkenes, with aldehydes was reported to be achieved by the use of methyl thioglycolate (HSCH$_2$CO$_2$Me) which acts as a polarity-reversal catalyst. For instance, the addition of butanal to isopropenyl acetate using di-tert-butyl hyponitrite (TBHN) as an initiator and methyl thioglycolate as the polarity-reversal catalyst at 60 °C produces 1-acetoxyhexan-3-on in good yield [202, 203].

Figure 5.10 shows the hydroacylation of alkenes with aldehydes using NHPI as a polarity-reversal catalyst. The hydrogen atom of the hydroxylimide moiety adjacent to two carbonyl groups may be easily abstracted by a nucleophilic radical rather than an electrophilic one, and the resulting PINO would behave as an electrophilic radical that can efficiently abstract the aldehydic hydrogen atom. The radical-chain hydroacylation of simple alkenes with aldehydes assisted by NHPI is carried out as follows: stirring a solution in dry toluene containing pentanal, oct-1-ene, NHPI and dibenzoyl peroxide (BPO) at 80 °C under argon for 12 h, followed by addition of BPO in toluene gives 5-tridecanone in 88% selectivity at 72% conversion (Figure 5.10). Several alkenes also react with pentanal under the influence of NHPI under an Ar atmosphere to form the corresponding ketones in moderate to good yields, respectively [204].

A possible reaction path for the NHPI-catalyzed hydroacylation of alkenes with aldehydes is shown in Scheme 5.17. The reaction may be initiated by the hydrogen atom abstraction from the aldehyde by the radical initiator (In•), giving an acyl radi-

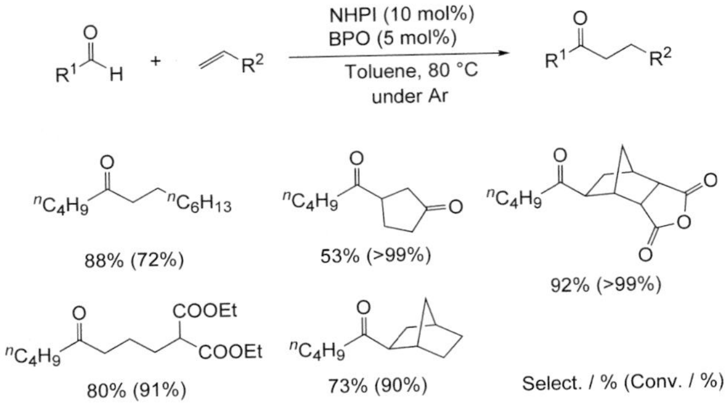

Fig. 5.10 Hydroacylation of alkenes using NHPI as a polarity-reversal catalyst

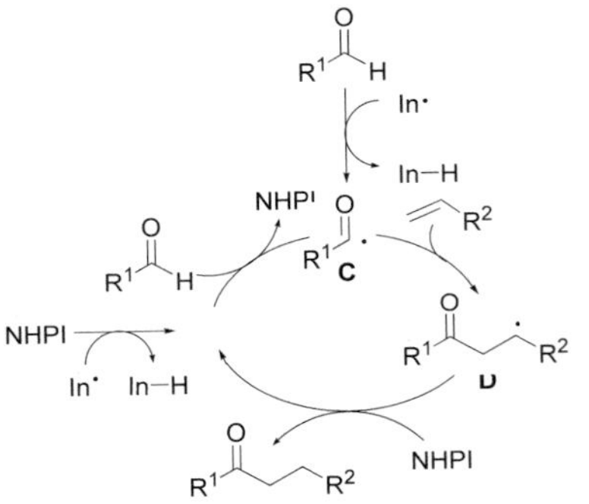

Scheme 5.17 A possible reaction path for hydroacylation using NHPI as a polarity-reversal catalys

cal **C** which then adds to an alkene to afford a β-oxocarbon radical **D**. The resulting radical **D** having a nucleophilic character abstracts the hydrogen atom from NHPI leading to a ketone and PINO. The abstraction of the hydrogen atom from an aldehyde by the PINO forms the acyl radical **C** and NHPI. An alternative formation of PINO from NHPI and the radical initiator (In•) may also be possible.

5.5
Conclusions

The achievement of highly efficient and selective transformations of hydrocarbons into useful chemical substances is an ambitious goal in synthetic chemistry. It is interesting to open up a new vista in organic synthesis and to confirm the catalytic method for the carbon radical generation from a C–H bond of a wide variety of compounds by the use of N-hydroxyphthalimide (NHPI) which serves as a carbon radical producing catalyst (CRPC). By employing NHPI as the catalyst, a novel aerobic oxidation of alkanes, which surpasses the conventional autoxidations in conversion and selectivity, has been achieved under mild conditions. This oxidation method provides access to a diverse array of significant oxygen-containing compounds. In particular, a success in the direct conversion of cyclohexane into adipic acid with dioxygen in high conversion and selectivity has greatly benefited the chemical industry as an environmentally benign process, because the current production of adipic acid *via* nitric acid oxidation causes the evolution of nitrogen oxides that are serious air-polluting materials. In addition, the finding that the NHPI catalyzes the aerobic oxidation of alkylbenzenes even at room temperature is notable. The epoxidation of alkenes by *in situ* generated hydroperoxides or hydrogen peroxide has been explored for the first time. This new methodology is applicable to the functionalization of alkanes to afford nitroalkanes, alkanesulfonic acids, and carboxylic acids by allowing them to react with NO_2, SO_2/O_2 and CO/O_2, respectively. Finally, a new type of reaction for the concomitant introduction of alkanes and O_2 into alkenes, which is referred to as catalytic oxyalkylation of alkenes, has been established. α-Hydroxy-γ-lactones, which are very difficult to synthesize by conventional methods, are easily prepared by the reaction of alcohols, alkenes and O_2 under the influence of the NHPI catalyst. The NHPI can be used as a polarity-reversal catalyst for hydroacylation of alkenes with aldehydes.

References

[1] C. N. Satterfield, *Heterogeneous Catalysis in Industrial Practice*, McGraw-Hill, New York, **1991**.

[2] (a) G. W. Parshall, S. D. Ittel, *Homogeneous Catalysis*, 2nd edn., Wiley, New York, **1992**; (b) G. W. Parshall, *J. Mol. Catal.* **1978**, *4*, 243.

[3] N. M. Emanuel, E. T. Denisov, Z. K. Marizus, *Liquid-Phase Oxidation of Hydrocarbons*, Plenum Press, New York, **1967**, pp. 309–346.

[4] H. A. Wittcoff, B. G. Reuben, *Industrial Organic Chemicals*, Wiley, New York, **1996**.

[5] (a) R. A. Sheldon, J. K. Kochi, *Metal-Catalyzed Oxidation of Organic Compounds*, Academic Press, New York, **1981**; (b) C. L. Hill, *Activation and Functionalization of Alkanes*, Academic Press, New York, **1989**; (c) R. A. Sheldon, *CHEMTECH* **1991**, *21*, 566; (d) L. I. Simándi, *Catalytic Activation of Dioxygen by Metal Complexes*, Kluwer Academic, Dordrecht, **1992**; (e) D. H. R. Barton, A. E. Martell, D. T. Sawyer, *The Activation of Dioxygen and Homogeneous Catalytic Oxidation*, Plenum, New York, **1993**.

[6] (a) S.-I. Murahashi, *Angew. Chem., Int. Ed. Engl.* **1995**, *34*, 2443; (b) A. E. Shilov, G. B. Shul'pin, *Chem. Rev.* **1997**, *97*, 2912; (c) A. E. Shilov, G. B. Shul'pin, *Activation and Catalytic Reactions of Saturated Hydrocarbons in the Presence of Metal Complexes*, Kluwer Academic, The Netherlands, **2000**.

[7] (a) E. Grochowski, T. Boleslawska, J. Jurczak, *Synthesis* **1977**, 718; (b) M. Masui, T. Ueshima, S. Ozaki, *Chem. Commun.* **1983**, 479; (c) S. Ozaki, T. Hamaguchi, K. Tsuchida, Y. Kimata, M. Masui, *J. Chem. Soc., Perkin Trans. 2* **1989**, 951; (d) J. Foricher, C. Fuerbringer, K. Pfoerther, EP198351, **1986**.

[8] Y. Ishii, K. Nakayama, M. Takeno, S. Sakaguchi, T. Iwahama, Y. Nishiyama, *J. Org. Chem.* **1995**, *60*, 3934.

[9] (a) R. A. Sheldon, *CHEMTECH* **1994**, *38*, 24; (b) I. W. C. E. Arends, R. A. Sheldon, M. Wallau, U. Schuchardt, *Angew. Chem., Int. Ed. Engl.* **1997**, *36*, 1144; (c) B. Mennier, *Chem. Rev.* **1992**, *92*, 1411; (d) D. H. Busch, N. W. Alcock, *Chem. Rev.* **1994**, *94*, 585; (e) *Dioxygen Activation and Homogeneous Catalytic Oxidation*, (Ed.: L. I.

Simándi), Elsevier, Amsterdam, **1991**, and references cited therein; (f) *New Developments in Selective Oxidation*, (Eds.: G. Centi, F. Trifiro), Elsevier, Amsterdam, **1990**.

[10] (a) I. TABUSHI, A. YAZAKI, *J. Am. Chem. Soc.* **1979**, *101*, 6456; (b) I. TABUSHI, A. YAZAKI, *J. Am. Chem. Soc.* **1981**, *103*, 7371; (c) I. TABUSHI, *Coord. Chem. Rev.* **1988**, *86*, 1.

[11] (a) D. H. R. BARTON, J. BOIVIN, M. GASTIGER, K. MORZYCKI, R. S. HAY-MOTHERWELL, W. B. MOTHERWELL, N. OZBALIK, K. M. SCHWARTZENTRUBER, *J. Chem. Soc., Perkin Trans. 1* **1986**, 947; (b) D. H. R. BARTON, D. DOLLER, *Acc. Chem. Res.* **1992**, *25*, 504; (c) D. H. R. BARTON, T. LI, J. MACKINNON, *Chem. Commun.* **1997**, 557, and references cited therein; (d) J. T. GROVES, R. NEUMANN, *J. Org. Chem.* **1988**, *53*, 3891; (e) P. BATTIONI, J. F. BARTOLI, P. LEDUC, M. FONTECAVE, D. MANSUY, *J. Chem. Soc., Chem. Commun.* **1987**, 791; (f) E. I. KARASEVICH, A. M. KHENKIN, A. E. SHILOV, *J. Chem. Soc., Chem. Commun.* **1987**, 731; (g) N. KITAJIMA, M. ITO, H. FUKUI, Y. MORO-OKA, *Chem. Commun.* **1991**, 102; (h) F. MINISCI, F. FONTANA, *Tetrahedron Lett.* **1994**, *35*, 1427; (i) Y. KURUSU, D. C. NECKERS, *J. Org. Chem.* **1991**, *56*, 1981.

[12] (a) S.-I. MURAHASHI, Y. ODA, T. NAOTA, *J. Am. Chem. Soc.* **1992**, *114*, 7913; (b) N. KOMIYA, T. NAOTA, S.-I. MURAHASHI, *Tetrahedron Lett.* **1996**, *37*, 1633; (c) N. KOMIYA, T. NAOTA, Y. ODA, S.-I. MURAHASHI, *J. Mol. Catal. A: Chemical* **1997**, *117*, 21; (d) T. PUNNIYAMURTHY, S. J. S. KARLA, J. IQBAL, *Tetrahedron Lett.* **1995**, *36*, 8497; (e) P. BATTIONI, R. IWANEJKO, D. MANSUY, T. MLODNICKA, J. POLTOWICZ, F. SANCHES, *J. Mol. Catal. A: Chemical* **1996**, *109*, 91; (f) A. BRAVO, F. FONTANA, F. MINISCI, A. SERRI, *Chem. Commun.* **1996**, 1843; (g) M. M. DELL'ANNA, P. MASTRORILLI, C. F. NOBILE, *J. Mol. Catal. A: Chemical* **1998**, *130*, 65.

[13] (a) M. LIN, A. SEN, *J. Am. Chem. Soc.* **1992**, *114*, 7307; (b) M. FONTECAVE, D. MANSUY, *Tetrahedron* **1984**, *40*, 4297; (c) W. Y. LU, J. F. BARTOLI, P. BATTIONI, D. MANSUY, *New J. Chem.* **1992**, *16*, 621; (d) J. F. BARTOLI, P. BATTIONI, W. R. DE FOOR, D. MANSUY, *Chem. Commun.* **1994**, 23; (e) A. S. GOLDSTEIN, R. H. BEER, R. S. DRAGO, *J. Am. Chem. Soc.* **1994**, *116*, 2424.

[14] (a) R. C. CHAMBERS, C. L. HILL, *Inorg. Chem.* **1989**, *28*, 2509; (b) C. L. HILL, C. M. PROSSER-MCCARTHA, *Coord. Chem. Rev.* **1995**, *143*, 407.

[15] D. ATTANASIO, L. SUBER, *Inorg. Chem.* **1989**, *28*, 3779.

[16] (a) G. B. SHUL'PIN, M. M. KATS, P. LEDERER, *J. Gen. Chem. USSR* **1989**, *59*, 2450; (b) A. MALDOTTI, C. BARTOCCI, R. AMADELLI, E. POLO, P. BATTIONI, D. MANSUY, *Chem. Commun.* **1991**, 1487; (c) G. B. SHUL'PIN, G. V. NIZOVA, YU. N. KOZLOV, *New. J. Chem.* **1996**, *20*, 1243, and references cited therein.

[17] (a) G. B. SHUL'PIN, M. C. GUERREIRO, U. SCHUCHARDT, *Tetrahedron* **1996**, *52*, 13051; (b) G. V. NIZOVA, G. SUSS-FINK, G. B. SHUL'PIN, *Chem. Commun.* **1997**, 397; (c) G. V. NIZOVA, G. SUSS-FINK, G. B. SHUL'PIN, *Tetrahedron* **1997**, *53*, 3603; (d) G. V. NIZOVA, G. SUSS-FINK, S. STANISLAS, G. B. SHUL'PIN, *Chem. Commun.* **1998**, 1885.

[18] (a) J. E. LYONS, P. E. ELLIS JR., *Metalloporphyrins in Catalytic Oxidations*, (Ed.: R. A. Sheldon), Dekker, New York, **1994**, 291, and references cited therein; (b) P. E. ELLIS JR., J. E. LYONS, *Chem. Commun.* **1989**, 1188; (c) P. E. ELLIS JR., J. E. LYONS, *Chem. Commun.* **1989**, 1190; (d) P. E. ELLIS JR., J. E. LYONS, *Chem. Commun.* **1989**, 1316; (e) P. E. ELLIS JR., J. E. LYONS, *Coord. Chem. Rev.* **1990**, *105*, 181; (f) H. L. CHEN, P. E. ELLIS JR., T. WIJESEKERA, T. E. HAGAN, S. E. GROH, J. E. LYONS, D. P. RIDGE, *J. Am. Chem. Soc.* **1994**, *116*, 1086; (g) M. W. GRINSTAFF, M. G. HILL, J. A. LABINGER, H. B. GRAY, *Science* **1994**, *264*, 1311.

[19] (a) N. MIZUNO, M. TATEISHI, T. HIROSE, M. IWAMOTO, *Chem. Lett.* **1993**, 2137; (b) N. MIZUNO, H. ISHIGE, Y. SEKI, M. MISONO, D.-J. SUH, W. HAN, T. KUDO, *Chem. Commun.* **1997**, 1295; (c) C. NOZAKI, M. MISONO, N. MIZUNO, *Chem. Lett.* **1998**, 1263; (d) N. MIZUNO, M. MISONO, *Chem. Rev.* **1998**, *98*, 199.

[20] (a) M. M. T. KHAN, R. S. SHUKLA, *J. Mol.*

Catal. **1988**, *44*, 85; (b) M. M. T. Khan, H. C. Bajaj, R. S. Shukla, S. A. Mirza, R. S. Shukla, *J. Mol. Catal.* **1988**, *44*, 51.

[21] R. Neumann, A. M. Khenkin, M. Dahan, *Angew. Chem., Int. Ed. Engl.* **1995**, *34*, 1587.

[22] A. S. Goldstein, R. S. Drago, *Inorg. Chem.* **1991**, *30*, 4506.

[23] (a) D. D. Davis, *Ullman's Encyclopedia of Industrial Chemistry*, 5th edn., Vol. A1 (Ed.: W. Gerhartz), Wiley, New York, **1985**, pp. 270–272; (b) D. D. Davis, D. R. Kemp, *Kirk-Othmer Encyclopedia of Chemical Technology*, 4th edn., Vol. 1 (Eds.: J. I. Kroschwitz, M. Howe-Grant), Wiley, New York, **1990**, pp. 471–480, and references cited therein.

[24] (a) J. W. M. Steeman, S. Kaarsemaker, P. Hoftyzer, *J. Chem. Eng. Sci.* **1961**, *14*, 139; (b) S. A. Miller, *Chem. Proc. Eng. London* **1969**, *50*, 63.

[25] K. Tanaka, *CHEMTECH* **1974**, 555, and references cited therein.

[26] K. Sato, M. Aoki, R. Noyori, *Science* **1998**, *281*, 1646.

[27] T. Iwahama, K. Syojyo, S. Sakaguchi, Y. Ishii, *Org. Proc. Res. Dev.* **1998**, *2*, 255.

[28] Y. Ishii, T. Iwahama, S. Sakaguchi, K. Nakayama, Y. Nishiyama, *J. Org. Chem.* **1996**, *61*, 4520.

[29] N. Sawatari, T. Yokota, S. Sakaguchi, Y. Ishii, *J. Org. Chem.* **2001**, *66*, 7889.

[30] (a) M. Lin, T. E. Hogan, A. Sen, *J. Am. Chem. Soc.* **1996**, *118*, 4574; (b) M. Lin, T. E. Hogan, A. Sen, *J. Am. Chem. Soc.* **1992**, *114*, 7307; (c) A. Sen, *Acc. Chem. Res.* **1998**, *31*, 550.

[31] (a) Y. Taniguchi, S. Horie, K. Takai, Y. Fujiwara, *J. Organomet. Chem.* **1995**, *504*, 137; (b) Y. Taniguchi, T. Kitamura, Y. Fujiwara, S. Horie, K. Takai, *Catal. Today* **1997**, *36*, 85; (c) C. Jia, T. Kitamura, Y. Fujiwara, *Acc. Chem. Res.* **2001**, *34*, 633.

[32] G. Süss-Fink, L. Gonzalez, G. B. Shul'pin, *App. Catal. A: General*, **2001**, *217*, 111.

[33] A. Shibamoto, S. Sakaguchi, Y. Ishii, *Tetrahedron Lett.* **2002**, *43*, 8859.

[34] *Kirk-Othmer Encyclopedia of Chemical Technology*, 4th edn., Vol. 1 (Eds.: J. I. Kroschwitz, M. Howe-Grant), Wiley, New York, **1990**. See the following sections: (a) E. Billig, Vol. 4, pp. 691–700; (b) H.-D. Hahn, G. Dambkes, N. Rupprich, Vol. A4, pp. 463–474; (c) C. E. Loeffler, L. Stautzenberber, J. D. Unrun, Vol. A4, pp. 358–405, and references cited therein.

[35] J. E. Lyons, P. E. Ellis Jr., H. K. Myers, *J. Catal.* **1995**, *155*, 59.

[36] S. Sakaguchi, S. Kato, T. Iwahama, Y. Ishii, *Bull. Chem. Soc. Jpn.* **1998**, *71*, 1240.

[37] (a) J. D. Bacha, J. K. Kochi, *J. Org. Chem.* **1965**, *30*, 3272; (b) F. D. Greene, M. L. Savitz, F. D. Osterholts, H. H. Fau, W. H. Smith, P. M. Zanet, *J. Org. Chem.* **1963**, *28*, 55; (c) J. K. Kochi, *J. Am. Chem. Soc.* **1962**, *84*, 1193.

[38] Y. Ishii, S. Kato, T. Iwahama, S. Sakaguchi, *Tetrahedron Lett.* **1996**, *37*, 4993.

[39] We have found that quaternary ammonium bromides accelerate the NHPI-catalyzed aerobic oxidation even in the absence of any metal catalyst: K. Matsunaka, T. Iwahama, S. Sakaguchi, Y. Ishii, *Tetrahedron Lett.* **1999**, *40*, 2165.

[40] H. Ohtake, T. Higuchi, M. Hirobe, *J. Am. Chem. Soc.* **1992**, *114*, 10660.

[41] R. A. Sheldon, *Dioxygen Activation and Homogeneous Catalytic Oxidation*, (Ed.: L. I. Simándi), Elsevier, Amsterdam, **1991**, pp. 573–594.

[42] W. T. Reichle, F. M. Konrad, J. R. Brooks, *Benzene and its Industrial Derivatives*, (Ed.: E. G. Hancock), Benn, London, **1975**.

[43] Y. Yoshino, Y. Hayashi, T. Iwahama, S. Sakaguchi, Y. Ishii, *J. Org. Chem.* **1997**, *62*, 6810.

[44] (a) E. S. Huyser, *Free-Radical Chain Reactions*, Wiley, New York, **1970**; (b) A. F. Parsons, *An Introduction to Free Radical Chemistry*, Blackwell Science, Oxford, **2000**.

[45] B. B. Wentzel, M. P. J. Donners, P. L. Alsters, M. C. Feiters, R. J. M. Nolte, *Tetrahedron* **2000**, *56*, 7797.

[46] A. Mackor, T. A. J. W. Wajer, T. de Boer, *Tetrahedron* **1968**, *24*, 1623.

[47] F. Minisci, C. Punta, F. Recupero, F. Fontana, G. F. Pedulli, *Chem. Commun.* **2002**, 688.

[48] (a) J. R. Sheehan, *Ullman's Encyclopedia of Industrial Organic Chemicals*, Vol. 8, VCH, Weinheim, **1999**, pp. 4573–4591; (b) C. Park, J. R. Sheehan, *Kirk-Othmer Encyclopedia of Chemical Technology*, 4th edn., Vol. 18 (Eds.: J. I. Kroschwitz, M. Howe-Grant), Wiley, New York, **1995**, pp. 991–1043.

[49] (a) W. F. Brill, *Ind. Eng. Chem.* **1960**, *52*, 837; (b) P. Raghavendrchar, S. Ramachandran, *Ind. Eng. Chem. Res.* **1992**, *31*, 453; (c) K. A. Roby, P. J. Kingsley, *CHEMTECH* **1996**, 39; (d) A. Cincotti, R. Orru, G. Cao, *Catal. Today* **1999**, *52*, 331.

[50] W. Partenheimer, *Catal. Today* **1995**, *23*, 69, and references cited therein.

[51] Y. Tashiro, T. Iwahama, S. Sakaguchi, Y. Ishii, *Adv. Synth. Catal.* **2001**, *343*, 220.

[52] (a) D. A. S. Ravens, *Trans. Faraday. Soc.* **1959**, *55*, 1768; (b) Y. Kamiya, T. Nakajima, K. Sakoda, *Bull. Chem. Soc. Jpn.* **1966**, *39*, 519.

[53] (a) S. A. Chavan, S. B. Halligudi, D. S. P. Ratnasamy, *J. Mol. Catal. A: Chemical* **2000**, *161*, 49, and references cited therein; (b) W. Partenheimer, V. V. Grushin, *Adv. Synth. Catal.* **2001**, *343*, 102, and references cited therein.

[54] (a) W. F. Hoelderich, *Appl. Catal. A: General* **2000**, 194–195, 487; (b) A. N. Roy, D. K. Guha, D. Bhattacharyya, *Indian Chem. Eng.* **1982**, *24*, 46, and references cited therein.

[55] (a) D. D. Davis, *Ullman's Encyclopedia of Industrial Chemistry*, 5th edn., Vol. A27 (Ed.: W. Gerhartz), VCH, Weinheim, **1985**, 584; (b) P. Paraskewas, *Synthesis* **1974**, 819; (c) S. Mukhopadhyay, S. B. Chandalia, *Org. Proc. Res. Dev.* **1999**, *3*, 455.

[56] S. Mukhopadhyay, S. B. Chandalia, *Org. Proc. Res. Dev.* **1999**, *3*, 227.

[57] A. Shibamoto, S. Sakaguchi, Y. Ishii, *Org. Proc. Res. Dev.* **2000**, *4*, 505.

[58] (a) J. P. Michael, *Nat. Prod. Rep.* **1997**, *14*, 605; (b) G. Jones, *Comprehensive Heterocyclic Chemistry II*, Vol. 5 (Eds.: A. R. Katritzky, C. W. Rees, E. F. V. Scriven), Pergamon Press, Oxford, **1996**, pp. 167–243.

[59] (a) J. N. Kim, H. J. Lee, K. Y. Lee, H. S. Kim, *Tetrahedron Lett.* **2001**, *42*, 3737; (b) K. Kobayashi, R. Nakahashi, A. Shimizu, T. Kitamura, O. Morikawa, H. Konishi, *J. Chem. Soc., Perkin Trans. 1*, **1999**, 1547, and references cited therein.

[60] C. K. Cain, J. N. Plampin, J. Sam, *J. Org. Chem.* **1955**, *20*, 466.

[61] D. W. Ladner, *Synth. Commun.* **1986**, *16*, 157.

[62] S. Paraskewas, *Synthesis* **1974**, 819.

[63] S. Sakaguchi, A. Shibamoto, Y. Ishii, *Chem. Commun.* **2002**, 180.

[64] Y. Ishii, T. Iwahama, S. Sakaguchi, K. Nakayama, Y. Nishiyama, *J. Org. Chem.* **1996**, *61*, 4520.

[65] C. Einhorn, J. Einhorn, C. Marcadal, J.-L. Pierre, *Chem. Commun.* **1997**, 447.

[66] C. Einhorn, J. Einhorn, C. Marcadal, J.-L. Pierre, *J. Org. Chem.* **1999**, *64*, 4542.

[67] R. Landau, G. A. Sullivan, D. Brown, *CHEMTECH*, **1979**, 602.

[68] (a) H. Hock, S. Lang, *Ber. Dtsch. Chem. Ges.* **1944**, *B77*, 257; (b) W. Jordan, H. van Barneveld, O. Gerlich, M. K. Boymann, J. Ullrich, *Ullmann's Encyclopedia Industrial Organic Chemicals*, Vol. A9, Wiley-VCH, Weinheim, **1985**, pp. 299–312.

[69] O. Fukuda, S. Sakaguchi, Y. Ishii, *Adv. Synth. Catal.* **2001**, *343*, 809.

[70] I. W. C. E. Arends, M. Sasidharan, A. Kühnle, M. Duda, C. Jost, R. A. Sheldon, *Tetrahedron*, **2002**, *58*, 9055.

[71] K. Utimoto, M. Miwa, H. Nozaki, *Tetrahedron. Lett.* **1981**, *22*, 4277.

[72] A. B. Smith III, P. A. Levenberg, J. Z. Suits, *Synthesis* **1986**, 184.

[73] M. Karpf, J. Huguet, A. S. Dreiding, *Helv. Chim. Acta* **1986**, *65*, 13.

[74] (a) S. T.-K. Tam, R. S. Klein, F. G. de las Heras, J. J. Fox, *J. Org. Chem.* **1979**, *44*, 4854, and references cited therein; (b) C. M. Gupta, G. H. Jones, J. G. Moffatt, *J. Org. Chem.* **1976**, *41*, 3000.

[75] (a) N. Sayo, K.-I. Azuma, K. Mikawa, T. Nakai, *Tetrahedron Lett.* **1984**, *25*, 565; (b) K. Midland, N. H. Nguyen, *J. Org. Chem.* **1981**, *46*, 4107.

[76] (a) H. C. Brown, U. S. Racherla, S. M. Singh, *Tetrahedron Lett.* **1984**, *25*,

2411; (b) J. F. NORMANT, M. BOURGAN, *Tetrahedron Lett.* **1970**, *11*, 2659; (c) A. G. DAVIES, R. J. PUDDEPHATT, *Tetrahedron Lett.* **1967**, *8*, 2265.

[77] J. E. SHAW, J. J. SHERRY, *Tetrahedron Lett.* **1971**, *12*, 4379.

[78] J. MUZART, O. PIVA, *Tetrahedron Lett.* **1988**, *29*, 2321.

[79] M. A. UMBREIT, K. B. SHARPLESS, *J. Am. Chem. Soc.* **1977**, *99*, 5527.

[80] (a) W. B. SHEATS, L. K. OLLI, R. STOUT, J. T. LUNDZEN, R. JUSTUS, W. G. NIGH, *J. Org. Chem.* **1979**, *44*, 4075; (b) A. MCKILLOP, O. H. OLDENZIEL, B. P. SWANN, E. C. TAYLOR, R. L. ROBEY, *J. Am. Chem. Soc.* **1973**, *95*, 1296; (c) M. SCHRODER, W. P. GRIFFITH, *J. Chem. Soc., Dalton Trans.* **1978**, 1599; (d) D. G. LEE, V. S. CHANG, *J. Org. Chem.* **1979**, *44*, 2726; (e) P. MULLER, A. J. GODOY, *Helv. Chim. Acta* **1981**, *64*, 2531.

[81] D. M. GOLDEN, *Annu. Rev. Phys. Chem.* **1982**, *33*, 493.

[82] S. SAKAGUCHI, T. TAKASE, T. IWAHAMA, Y. ISHII, *Chem. Commun.* **1998**, 2037.

[83] (a) R. C. LAROCK, *Comprehensive Organic Transformations*, VCH, New York, **1989**, p. 604; (b) S. LEY, A. MADIN, *Comprehensive Organic Synthesis*, Vol. 7 (Eds.: B. M. Trost, I. Flemings, S. V. Ley), Pergamon, Oxford, **1991**, p. 251.

[84] (a) L. L. SIMÁNDI, *Catalytic Activation of Dioxygen by Metal Complexes*, Kluwer Academic, Dordrecht, **1992**, pp. 297–317; (b) B. R. JAMES, *Dioxygen Activation and Homogeneous Catalytic Oxidation* (Ed.: L. L. Simándi), Elsevier, Amsterdam, **1991**, p. 195.

[85] (a) R. TANG, S. E. DIAMOND, N. NEARY, F. MARES, *J. Chem. Soc., Chem. Commun.* **1978**, 562; (b) M. MATSUMOTO, N. WATANABE, *J. Org. Chem.* **1984**, *49*, 3435; (c) C. BILGRIEN, S. DAVIS, R. S. DRAGO, *J. Am. Chem. Soc.* **1987**, *109*, 3786; (d) S.-I. MURAHASHI, T. NAOTA, J. HIRAI, *J. Org. Chem.* **1993**, *58*, 7318; (e) K. KANEDA, T. YAMASHITA, T. MATSUSHITA, K. EBITANI, *J. Org. Chem.* **1998**, *63*, 1750; (f) T. F. BLACKBURN, J. SCHWARTZ, *J. Chem. Soc., Chem. Commun.* **1977**, 158; (g) K. KANEDA, Y. FUJIE, K. EBITANI, *Tetrahedron Lett.* **1997**, *38*, 9023; (h) A. K. MANDAL, J. IQBAL, *Tetrahedron Lett.* **1997**, *53*, 7641; (i) K. S. COLEMAN, M. COPPE, C. THOMAS, J. A. OSBORN, *Tetrahedron Lett.* **1999**, *40*, 3723; (j) J.-E. BÄCKVALL, R. L. CHOWDHURY, U. KARLSSON, *Chem. Commun.* **1991**, 473; (k) G. Z. WANG, U. ANDREASSON, J.-E. BÄCKVALL, *Chem. Commun.* **1994**, 1037; (l) T. INOKUCHI, K. NAKAGAWA, S. TORII, *Tetrahedron Lett.* **1995**, *36*, 3223; (m) A. DIJKSMAN, A. MARINO-GONZALEZ, A. MARITA I. PAYERAS, I. W. C. E. ARENDS, R. SHELDON, *J. Am. Chem. Soc.* **2001**, *123*, 6826; (n) G. CSJERNYIK, A. H. ELL, L. FADINI, B. PUGIN, J.-E. BÄCKVALL, *J. Org. Chem.* **2002**, *67*, 1657.

[86] (a) RU: I. E. MARKÓ, P. R. GILES, M. TSUKAZAKI, M. I. CHELLÉ-REGNAUT, C. J. URCH, S. M. BROWN, *J. Am. Chem. Soc.* **1997**, *119*, 12661; (b) R. LENZ, S. V. LEY, *J. Chem. Soc., Perkin Trans. 1* **1997**, 3291; (c) A. HANYU, E. TAKEZAWA, S. SAKAGUCHI, Y. ISHII, *Tetrahedron Lett.* **1998**, *39*, 5557; (d) PD: A. DIJKSMAN, I. W. C. E. ARENDS, R. A. SHELDON, *Chem. Commun.* **1999**, 1591; (e) K. P. PETERSON, R. C. LAROCK, *J. Org. Chem.* **1998**, *63*, 3185; (f) T. NISHIMURA, T. ONOUE, K. OHE, S. UEMURA, *Tetrahedron Lett.* **1998**, *39*, 6011; (g) CO: T. YAMADA, T. MUKAIYAMA, *Chem. Lett.* **1989**, 519; (h) OTHER METALS: M. F. SEMMELHACK, C. R. SCHMID, D. A. CORTES, C. S. CHOU, *J. Am. Chem. Soc.* **1984**, *106*, 3374.

[87] (a) I. E. MARKÓ, P. R. GILES, M. TSUKAZAKI, C. J. URCH, *Science* **1996**, *274*, 2044; (b) I. E. MARKÓ, A. GAUTIER, I. C. REGNAUT, P. R. GILES, M. TSUKAZAKI, C. J. URCH, S. M. BROWN, *J. Org. Chem.* **1998**, *63*, 7576; (c) I. E. MARKÓ, P. R. GILES, M. TSUKAZAKI, I. C. REGNAUT, A. GAUTIER, M. BROWN, C. J. URCH, *J. Org. Chem.* **1999**, *64*, 2433.

[88] (a) K. YAMAGUCHI, K. MORI, T. MIZUGAKI, K. EBITANI, K. KANEDA, *J. Am. Chem. Soc.* **2000**, *122*, 7144; (b) K. YAMAGUCHI, N. Mizuno, *Angew. Chem., Int. Ed. Engl.* **2002**, *41*, 4538.

[89] T. NISHIMURA, N. KAKIUCHI, M. INOUE, S. UEMURA, *Chem. Commun.* **2000**, 1245.

[90] (a) G.-J. TEN BRINK, I. W. C. E. ARENDS, R. A. SHELDON, *Science* **2000**, *287*,

1636; (b) Y. Uozumi, R. Nakao, *Angew. Chem., Int. Ed. Engl.* **2003**, *42*, 194.

[91] T. Iwahama, S. Sakaguchi, Y. Nishiyama, Y. Ishii, *Tetrahedron Lett.* **1995**, *36*, 6923.

[92] T. Iwahama, Y. Yoshino, T. Keitoku, S. Sakaguchi, Y. Ishii, *J. Org. Chem.* **2000**, *65*, 6502.

[93] (a) T. V. Lee, *Comprehensive Organic Synthesis*, Vol. 7 (Eds.: B. M. Trost, I. Flemings, S. V. Ley), Pergamon, Oxford, **1991**, p. 291; (b) W. S. Trahanovsky, *Oxidation in Organic Chemistry* (Eds.: A. T. Blomquist, H. Wasserman), Academic Press, New York, **1978**; (c) M. Hudlicky, Oxidation, *Organic Chemistry*, ACS Monograph 186, American Chemical Society, Washington DC, **1986**, and references cited therein.

[94] (a) C. Venturello, M. Ricci, *J. Org. Chem.* **1986**, *51*, 1599; (b) Y. Sakata, Y. Ishii, *J. Org. Chem.* **1991**, *56*, 6233; (c) Y. Sakata, Y. Katayama, Y. Ishii, *Chem. Lett.* **1992**, 671, and references cited therein.

[95] (a) T. R. Felthouse, *J. Am. Chem. Soc.* **1987**, *109*, 7566; (b) T. Okamoto, K. Sakai, S. Oka, *J. Am. Chem. Soc.* **1988**, *110*, 1187; (c) G. D. Vries, S. Schors, *Tetrahedron Lett.* **1968**, *9*, 5689.

[96] T. Nishimura, T. Onoue, K. Ohe, S. Uemura, *J. Org. Chem.* **1999**, *64*, 6750.

[97] E. Takezawa, S. Sakaguchi, Y. Ishii, *Org. Lett.* **1999**, *1*, 713.

[98] M. A. Oakley, S. Woodward, K. Coupland, D. Parker, C. Temple-Heald, *J. Mol. Catal. A: Chemical* **1999**, *150*, 105.

[99] M. Fetizon, M. Golfier, J. M. Louis, *J. Chem. Soc., Chem. Commun.* **1969**, 1102.

[100] S. Baskaran, J. Das, S. Chandrasekaran, *J. Org. Chem.* **1989**, *54*, 5182.

[101] T. Inokuchi, S. Matsumoto, T. Nishiyama, S. Torii, *Synlett* **1990**, 57.

[102] T. Iwahama, S. Skaguchi, Y. Nishiyama, Y. Ishii, *Tetrahedron Lett.* **1995**, *36*, 1523.

[103] (a) S. Ito, K. Inoue, M. Matsumoto, *J. Am. Chem. Soc.* **1982**, *104*, 645; (b) F. P. Guengrich, T. L. Macdonald, *Acc. Chem. Res.* **1984**, *17*, 9; (c) I. Tabushi, M. Kodera, *J. Am. Chem. Soc.* **1986**, *108*, 1101; (d) D. Mansuy, M. Fontecve, J.-F. Bartoli, *J. Chem. Soc., Chem. Commun.* **1981**, 874; (e) J.-C. Marchon, R. Ramasseul, *Synthesis* **1989**, 389.

[104] Aldehydes as reducing agents:
(a) T. Mukaiyama, T. Takai, T. Yamada, O. Rhode, *Chem. Lett.* **1990**, 1661; (b) R. Irie, Y. Ito, T. Katsuki, *Tetrahedron Lett.* **1991**, *32*, 6891; (c) K. Kaneda, S. Haruna, T. Imanaka, M. Hamamoto, Y. Nishiyama, Y. Ishii, *Tetrahedron Lett.* **1991**, *33*, 6827; (d) S.-I. Murahashi, Y. Oda, T. Naota, N. Komiya, *Chem. Commun.* **1993**, 139; (e) M. Hamamoto, K. Nakayama, Y. Nishiyama, Y. Ishii, *J. Org. Chem.* **1993**, *58*, 6421; (f) N. Mizuno, T. Hirose, M. Tateishi, M. Iwamoto, *Chem. Lett.* **1993**, 1839; (g) E. Bouhlel, P. Laszlo, M. Levart, M.-T. Montaufier, G. P. Singh, *Tetrahedron Lett.* **1993**, *34*, 1133; (h) T. Punniyamurthy, B. Bhatia, J. Iqbal, *J. Org. Chem.* **1994**, *59*, 850; (i) P. Mastvorilli, C. F. Nobile, *J. Mol. Catal.* **1994**, *94*, 19; (j) T. Mukaiyama, T. Yamada, *Bull. Chem. Soc. Jpn.* **1995**, *68*, 17, and references cited therein.

[105] Ketones as reducing agents: (a) T. Takai, E. Hata, K. Yorozu, T. Mukaiyama, *Chem. Lett.* **1992**, 2077; (b) T. Punniyamurthy, B. Bhatia, J. Iqbal, *Tetrahedron Lett.* **1993**, *34*, 4657.

[106] Acetals as reducing agents: K. Yorozu, T. Takai, T. Yamada, T. Mukaiyama, *Bull. Chem. Soc. Jpn.* **1994**, *67*, 2195.

[107] Other reducing agents: (a) H. Sakurai, Y. Hataya, T. Goromaru, H. Matsuura, *J. Mol. Catal.* **1985**, *29*, 153; (b) M. Shimizu, H. Orita, T. Hayakawa, K. Takehira, *J. Mol. Catal.* **1988**, *45*, 85.

[108] J. T. Groves, R. Quinn, *J. Am. Chem. Soc.* **1985**, *107*, 5790.

[109] R. Neumann, M. Dahan, *Nature* **1997**, *388*, 24.

[110] R. Neumann, M. Dahan, *Chem. Commun.* **1995**, 171.

[111] 2-Hydroperoxyhexafluoro-2-propanol is reported to be easily derived from HFA (or HFA-hydrate) and H_2O_2; (a) R. D. Chambers, M. Clark, *Tetrahedron Lett.* **1970**, *11*, 2741; (b) B. Ganem, R. P. Heggs, A. J. Biloski,

D. R. Schwartz, *Tetrahedron Lett.* **1980**, *21*, 685; (c) B. Ganem, A. J. Biloski, R. P. Heggs, *Tetrahedron Lett.* **1980**, *21*, 689.

[112] (a) R. P. Heggs, B. Ganem, *J. Am. Chem. Soc.* **1979**, *101*, 2484; (b) A. J. Biloski, R. P. Heggs, B. Ganem, *Synthesis* **1980**, 810; (c) M. C. A. van Vliet, I. W. C. E. Arends, R. A. Sheldon, *Chem. Commun.* **1999**, 263.

[113] For storage and transportation of H_2O_2: G. Goor, W. Kunkel, *Ullmann's Encyclopedia of Industrial Chemistry*, 5th edn., Vol. A13 (Eds.: B. Elvers, S. Hawkins, M. Ravenscroft, G. Schulz), VCH, Weinheim, **1989**, pp. 461–463.

[114] (a) T. Iwahama S. Sakaguchi, Y. Ishii, *Chem. Commun.* **1999**, 727; (b) T. Iwahama S. Sakaguchi, Y. Ishii, *Heterocycles* **2000**, *52*, 693.

[115] (a) T. Michael, E. I. Musser, *Ullman's Encyclopedia Industrial Organic Chemicals*, Vol. 3, Wiley-VCH, Weinheim, **1999**, pp. 1807–1823; (b) W. B. Fisher, J. F. Van Peppen, *Kirk-Othmer Encyclopedia of Chemical Technology*, 4th edn., Vol. 7 (Eds.: J. L. Kroshwite, M. HomeGrant), John Wiley and Sons, New York, **1995**, pp. 851–859.

[116] (a) G. R. Krow, *Comprehensive Organic Synthesis*, Vol. 7 (Ed.: B. M. Trost), Pergamon Press, New York, **1991**, pp. 671–688; (b) *Organic Peroxides*, (Ed.; D. Swern), Wiley-Interscience, New York, **1971**.

[117] (a) J. Fischer, W. F. Hölderich, *Appl. Catal. A: General* **1999**, *180*, 435–443; (b) G. Strukul, A, Varagnolo, F. Pinna, *J. Mol. Catal. A: Chemical* **1997**, *117*, 413–423, (c) Z. B. Wang, T. Mizusaki, T. Sano, Y. Kawakami, *Bull. Chem. Soc. Jpn.* **1997**, *70*, 25967–2570; (d) M. Frisone, T. Del, F. Pinna, G. Strukul, *Organometallics* **1993**, *12*, 148–156.

[118] R. Göttlich, K. Yamakoshi, H. Sasai, M. Shibasaki, *Synlett* **1997**, 971–973.

[119] (a) A. Bolm, G. Schlingloff, K. Weickhardt, *Tetrahedron Lett.* **1993**, *34*, 3405–3408; (b) M. Hamamoto, K. Nakayama, Y. Nishiyama, Y. Ishii, *J. Org. Chem.* **1993**, *58*, 6421–6425; (c) S. Murahashi, Y. Oda, T. Naota, *Tetrahedron Lett.* **1992**, *33*, 7557–7560;

(d) T. Yamada, K. Takahashi, K. Kato, T. Takai, S. Inoki, T. Mukaiyama, *Chem. Lett.* **1991**, 641–644.

[120] O. Fukuda, S. Sakaguchi, Y. Ishii, *Tetrahedron Lett.* **2001**, *42*, 3479.

[121] (a) V. D. Luedeke, *Encyclopedia of Chemical Processing and Design*, (Eds.: J. J. Mcketta), Marcel Dekker, New York, **1978**, pp. 73–95; (b) W. N. Fisher, L. Crescentini, *Kirk-Othmer Encyclopedia of Chemical Technology*, Vol. 4 (Ed.: J. I. Kroschwitz), John Wiley & Sons, New York, **1992**, pp. 827–839; (c) J. Ritz, H. Fuchs, H. Kiecza, W. C. Moran, *Ullmann's Encyclopedia of Industrial Organic Chemicals*, Vol. 2, Wiley-VCH, Weinheim, **1999**, pp. 1013–1043.

[122] About 2.8 kg of ammonium sulfate are generated per kilogram of cyclohexanone oxime produced: G. Bellussi, C. Perego, *Cattech* **2000**, *4*, 4.

[123] R. Raja, G. Sankar, J. M. Thomas, *J. Am. Chem. Soc.* **2001**, *123*, 8153.

[124] (a) A. Cesana, M. A. Mantegazza, M. Pastori, *J. Mol. Catal. A: Chemical*, **1997**, *117*, 367; (b) M. A. Mantegazza, G. Petrini, A. Cesana, EP 00564040, **1993**.

[125] (a) N. Kakeya, T. Takai, JP 2001122851, **2001**; (b) E. G. E. Hawkins, US 3947406, **1976**; (c) C. W. Capp, B. W. Harris, E. G. E. Hawkins, G. E. Edwin, FR 1567559, **1969**; (d) J. S. Reddy, S. Sivasanker, P. Ratnasamy, *J. Mol. Catal.* **1991**, *69*, 383.

[126] For recent reviews: (a) D. H. R. Barton, D. Doller, *Acc. Chem. Res.* **1992**, *25*, 504; (b) J. Sommer, J. Bukala, *Acc. Chem. Res.* **1993**, *26*, 370; (c) B. A. Arndtsen, R. G. Bergman, T. A. Mobley, T. H. Peterson, *Acc. Chem. Res.* **1995**, *28*, 154; (d) C. L. Hill, *Synlett* **1995**, 127; (e) I. Ryu, N. Sonoda, *Angew. Chem. Int. Ed.* **1996**, *35*, 1051.

[127] T. Sakakura, T. Sodeyama, K. Sasaki, K. Wada, M. Tanaka, *J. Am. Chem. Soc.* **1990**, *112*, 7221.

[128] P. Margl, T. Ziegler, P. E. Blochl, *J. Am. Chem. Soc.* **1996**, *118*, 5412.

[129] (a) K.-I. Sato, J. Watanabe, K. Takaki, Y. Fujiwara, *Chem. Lett.* **1991**, 1433; (b) Y. Fujiwara, K. Takaki, Y. Tani-

guchi, *Synlett* **1996**, 591, and references therein.

[130] (a) M. Lin, A. Sen, *Nature* **1994**, *368*, 613; (b) M. Lin, T. E. Hogan, A. Sen, *J. Am. Chem. Soc.* **1996**, *118*, 4574.

[131] K.-I. Takeuchi, F. Akiyama, T. Miyazaki, I. Kitagawa, K. Okamoto, *Tetrahedron* **1987**, *43*, 701.

[132] (a) O. Farooq, M. Marcelli, G. K. S. Prakash, G. A. Olah, *J. Am. Chem. Soc.* **1988**, *110*, 864; (b) I. S. Akhrem, S. Z. Bernadyuk, M. E. Vol'pin, *Mendeleev Commun.* **1993**, 188.

[133] M. M. Brubaker, D. D. Coffman, H. H. Hoehn, *J. Am. Chem. Soc.* **1952**, *74*, 1509.

[134] (a) D. D. Coffman, R. Cramer, W. E. Mochel, *J. Am. Chem. Soc.* **1958**, *80*, 2882; (b) C. Walling, E. S. Savas, *J. Am. Chem. Soc.* **1960**, *82*, 1738; (c) W. A. Thaler, *J. Am. Chem. Soc.* **1967**, *89*, 1902; (d) T. Susuki, J. Tsuji, *J. Org. Chem.* **1970**, *35*, 2982.

[135] (a) I. Ryu, K. Kusano, A. Ogawa, N. Kambe, N. Sonoda, *J. Am. Chem. Soc.* **1990**, *112*, 1295; (b) I. Ryu, N. Sonoda, *Chem. Rev.* **1996**, *96*, 177; (c) C. Chatgilialoglu, D. Crich, M. Komatsu, I. Ryu, *Chem. Rev.* **1999**, *99*, 1991, and references cited therein.

[136] A. Sen, M. Lin, *Chem. Commun.* **1992**, 892.

[137] W. T. Boese, S. Goldman, *J. Am. Chem. Soc.* **1992**, *114*, 350.

[138] C. L. Hill, B. S. Jaynes, *J. Am. Chem. Soc.* **1995**, *117*, 4704.

[139] R. R. Ferguson, R. H. Crabtree, *J. Org. Chem.* **1991**, *56*, 5503.

[140] S. Kato, T. Iwahama, S. Sakaguchi, Y. Ishii, *J. Org. Chem.* **1998**, *63*, 222.

[141] (a) L. F. Albright, *Chem. Eng.* **1966**, *73*, 149, and references cited therein; (b) C. P. Spaeth, US Patent 2,883,432, **1959**.

[142] (a) S. B. Markofsky, *Ullmann's Encyclopedia Industrial Organic Chemicals*, Vol. 6 (Eds.: B. Elvers, S. Hawkins), Wiley-VCH, Weinheim, **1999**, pp. 3487–3501; (b) L. F. Albright, *Kirk-Othmer Encyclopedia of Chemical Technology*, Vol. 17 (Eds.: J. I. Kroschwite, M. Howe-Grant), Wiley, New York, **1995**, pp. 68–107.

[143] (a) G. G. Bachman, B. H. Hass, M. L. Addison, *J. Org. Chem.* **1952**, *17*, 914;
(b) G. G. Bachman, B. H. Hass, M. L. Addison, *J. Org. Chem.* **1952**, *17*, 928;
(c) G. G. Bachman, V. J. Hewett, A. Millikan, *J. Org. Chem.* **1952**, *17*, 935; (d) G. G. Bachman, L. Kohn, *J. Org. Chem.* **1952**, *17*, 942.

[144] S. Sakaguchi, Y. Nishiwaki, T. Kitamura, Y. Ishii, *Angew. Chem., Int. Ed. Engl.* **2001**, *40*, 222.

[145] (a) J. Rita, H. Fuchs, H. Kieczka, W. C. Moran, *Ullmann's Encyclopedia Industrial Organic Chemicals*, Vol. 2 (Eds.: B. Elvers, S. Hawkins), Wiley-VCH, Weinheim, **1999**, pp. 1013–1043; (b) W. B. Fisher, L. Crescen'tini, *Kirk-Othmer Encyclopedia of Chemical Technology*, Vol. 4 (Eds.: J. I. Kroschwite, M. Howe-Grant), Wiley, New York, **1995**, pp. 827–839.

[146] Although ε-caprolactam is currently produced by the reaction of cyclohexanone with hydroxylamine followed by a Beckmann rearrangement with sulfuric acid, the efficiency for the production of cyclohexanone by aerobic oxidation of cyclohexane is not high.

[147] The vapor-phase nitration of cyclohexane with NO_2 at 240 °C forms nitrocyclohexane in only 16 % yield together with large amounts of undesired nitrated products: R. Lee, F. L. Albright, *Ind. Eng. Chem. Process Res. Dev.* **1965**, *4*, 411.

[148] K. Jones, *Comprehensive Inorganic Chemistry*, (Ed.: A. F. Trotman-Dickenson), Pergamon, Oxford, **1973**, p. 147–388.

[149] F. M. Beringer, R. A. Falk, *J. Am. Chem. Soc.* **1959**, *81*, 2997.

[150] (a) H. A. Young, *J. Am. Chem. Soc.* **1937**, *59*, 811; (b) R. C. Murray, *J. Chem. Soc.* **1933**, 739.

[151] (a) B. Bjellqvist, *Acta. Chem. Scand.* **1973**, *27*, 3180; (b) R. R. Ferguson, R. H. Crabtree, *J. Org. Chem.* **1991**, *56*, 5503; (c) R. H. Crabtree, A. Habib, *Comprehensive Organic Synthesis*, Vol. 7 (Ed.: B. M. Trost), Pergamon, New York, **1991**, and references cited therein.

[152] Y. Ishii, K. Matsunaka, S. Sakaguchi, *J. Am. Chem. Soc.* **2000**, *122*, 7390.

[153] G. W. Smith, H. D. Williams, *J. Org. Chem.* **1961**, *26*, 2207.

[154] A review for the modern synthesis

using vanadium species as a catalyst has been reported: T. Hirao, *Chem. Rev.* **1997**, *97*, 2707.

[155] K. Yamamoto, E. Tsuchida, H. Nishide, M. Jikei, K. Oyaizu, *Macromolecules* **1993**, *26*, 3432.

[156] M. Kirihara, Y. Ochiai, S. Takizawa, H. Takahata, H. Nemoto, *Chem. Commun.* **1999**, 1387.

[157] R. Graf, *Justus Liebigs Ann. Chem.* **1952**, *578*, 50.

[158] (a) C. Walling, *Free Radicals in Solution*, Wiley, New York, **1957**;
(b) J. K. Kochi, *Free Radicals*, Vol. II, Wiley, New York, **1973**.

[159] (a) E. Culotta, D. E. Koshland Jr., *Science* **1992**, *258*, 1862; (b) M. N. Wrightham, A. J. Cann, H. F. Sewel, *Med. Hypotheses* **1992**, *38*, 236.

[160] K. Kato, T. Mukaiyama, *Chem. Lett.* **1990**, 1395.

[161] E. Hata, T. Yamada, T. Mukaiyama, *Bull. Chem. Soc. Jpn.* **1995**, *68*, 3629, and references cited therein.

[162] (a) T. Nagano, H. Takizawa, M. Hirobe, *Tetrahedron Lett.* **1995**, *36*, 8239; (b) T. Itoh, K. Nagata, Y. Matsuya, M. Miyazaki, A. Ohsawa, *J. Org. Chem.* **1997**, *62*, 3582, and references cited therein.

[163] E. Bosch, J. K. Kochi, *J. Chem. Soc., Perkin Trans. 1* **1995**, 1057.

[164] J. S. B. Park, J. C. Walton, *J. Chem. Soc., Perkin Trans. 2* **1997**, 2579.

[165] S. Sakaguchi, M. Eikawa, Y. Ishii, *Tetrahedron Lett.* **1997**, *38*, 7075.

[166] M. Eikawa, S. Sakaguchi, Y. Ishii, *J. Org. Chem.* **1999**, *64*, 4676.

[167] V. R. Koch, L. L. Miller, *J. Am. Chem. Soc.* **1973**, *95*, 8631.

[168] G. A. Olah, P. Ramaiah, C. B. Rao, G. Stanford, R. Golam, N. J. Trivedi, J. A. Olah, *J. Am. Chem. Soc.* **1993**, *115*, 7246.

[169] There have been few reports on practical methods for the synthesis of these dialdehydes so far: (a) J. S. Cha, K. D. Lee, O. O. Kwon, J. M. Kim, H. S. Lee, *Bull. Korean. Chem. Soc.* **1995**, *16*, 561; (b) R. J. P. Corriu, G. F. Lanneau, M. Perrot, *Tetrahedron Lett.* **1988**, *29*, 1271; (c) H. Firouzabadi, N. Iranpoor, F. Kiaeezadeh, J. Toofan, *Tetrahedron Lett.* **1986**, *42*, 719; (d) G. Green, A. P. Griffith, D. M. Hollinshead, S. V. Ley, M. Schrader, *J. Chem. Soc., Perkin Trans. 1* **1984**, 681.

[170] J. C. Bill, D. S. Tarbell, *Organic Syntheses, Collect*, Vol. IV, Wiley, New York, **1963**, 807.

[171] H. Suzuki, *Pharmacia* **1997**, *33*, 487.

[172] G. A. Olah, P. Ramaiah, C. B. Rao, G. Sandford, R. Golam, N. J. Trivedi, G. A. Olah, *J. Am. Chem. Soc.* **1993**, *115*, 7246.

[173] (a) R. D. Bach, J. W. Holubka, R. C. Badger, S. J. Rajan, *J. Am. Chem. Soc.* **1979**, *101*, 1979; (b) G. A. Olah, P. Ramaiah, C. B. Rao, G. Sandford, R. Golam, N. J. Trivedi, G. A. Olah, *J. Am. Chem. Soc.* **1993**, *115*, 7246.

[174] G. A. Olah, Q. Wang, *Synthesis* **1992**, 1090.

[175] V. R. Koch, L. L. Miller, *J. Am. Chem. Soc.* **1973**, *95*, 8631.

[176] S. R. Jones, J. M. Mellor, *J. Chem. Soc., Perkin Trans 1*, **1976**, 2576.

[177] J. M. Bakke, C. B. Storm, *Acta. Chem. Scand.* **1989**, *43*, 399.

[178] C. L. Hill, *Synlett* **1995**, *2*, 127.

[179] S. Sakaguchi, T. Hirabayashi, Y. Ishii, *Chem. Commun.* **2002**, 516.

[180] (a) B. Giese, Radicals, *Organic Synthesis: Formation of Carbon-Carbon Bonds*, Pergamon, Oxford, **1986**; (b) D. P. Curran, *Comprehensive Organic Synthesis*, Vol. 4 (Eds.: B. M. Trost, I. Fleming, M. F. Semmelhock), Pergamon, Oxford, **1991**, p. 715; (c) W. B. Motherwell, D. Crich, *Free-Radical Reactions in Organic Synthesis*, Academic Press, London, **1992**; (d) J. Fossey, D. Lefort, J. Sorba, *Free Radicals in Organic Synthesis*, Wiley, Chichester, **1995**; (e) C. Chatgilialoglu, P. Renaud, *General Aspects of the Chemistry of Radicals*, (Ed.: Z. B. Alfassi), Wiley, Chichester, **1999**, pp. 501–538.

[181] (a) D. J. Hart, *Science* **1984**, *223*, 883; (b) D. P. Curran, *Synthesis* **1988**, 489; (c) G. G. Melikyan, *Synthesis* **1993**, 833, and references therein; (d) C. Ollivier, P. Renaud, *Angew. Chem., Int. Ed. Engl.* **2000**, *39*, 925; (e) A. Studer, S. Amrein, *Angew. Chem., Int. Ed. Engl.* **2000**, *22*, 3080, and references cited therein.

[182] Reviews: (a) W. P. Neumann, *Synthesis* **1987**, 665; (b) D. P. Curran, *Synthesis*

1988, 417; (c) A. G. Davies, *Organotin Chemistry*, Wiley-VCH, Weinheim, **1997**; (d) B. Giese, *Angew. Chem., Int. Ed. Engl.* **1983**, *22*, 753; (e) C. Chatgilialoglu, *Acc. Chem. Res.* **1992**, *40*, 7019; (f) P. Dowd, W. Zhang, *Chem. Rev.* **1993**, *93*, 2091.

[183] (a) D. H. R. Barton, D. Crich, W. B. Motherwell, *Tetrahedron* **1985**, *41*, 3901; (b) D. Crich, *Aldrichim. Acta* **1986**, *20*, 35; (c) M. Ramaiah, *Tetrahedron* **1987**, *43*, 3541.

[184] T. Hara, T. Iwahama, S. Sakaguchi, Y. Ishii, *J. Org. Chem.* **2001**, *69*, 6425.

[185] Acrylates are known to be easily polymerized by radical initiation. Fischer et al. have determined the accurate rate constant for the addition of $CH_2CO_2Bu^t$ radical to methyl acrylate ($k = 6 \times 10^5$ M^{-1} s^{-1}) [180a]. Thus, such alkenes may be difficult to use as acceptors in the conventional radical additions of alkyl radicals [180b]. In contrast, the present oxyalkylation seems to provide the successful addition of 1,3-dimethyladamantane to acrylates, since O_2 existing *in situ* quickly quenches the radical intermediate to prevent the polymerization. (a) I. Beranek, H. Fischer, *Free Radicals in Synthesis and Biology*, (Ed.: F. Minisci), Kluwer, Dordrecht, **1989**, 303; (b) M. Itoh, T. Taguchi, V. V. Chung, M. Tokuda, A. Suzuki, *J. Org. Chem.* **1972**, *37*, 2357.

[186] K. Fukunishi, I. Tabushi, *Synthesis* **1988**, 826.

[187] (a) R. L. Hanson, H. A. Lardy, S. M. Kupchan, *Science* **1970**, *168*, 378; (b) Y. S. Rao, *Chem. Rev.* **1976**, *76*, 625; (c) H. M. R. Hoffman, J. Rabe, *Angew. Chem., Int. Ed. Engl.* **1985**, *24*, 94.

[188] K. Puthuraya, Y. Oomura, N. Shimizu, *Brain Res.* **1985**, *332*, 165.

[189] S. Li, M. Vert, *Degradable Polymers* (Eds.: G. Scott, D. Gilead), Chapman & Hall, London, **1995**, pp. 43–52.

[190] (a) K. Mori, T. Takigawa, T. Matsuo, *Tetrahedron* **1979**, *35*, 933; (b) G. A. Garcia, H. Mufioz, J. Tamariz, *Synth. Commun.* **1983**, *13*, 569; (c) B. B. Jarvis, K. M. Wells, T. Kaufmann, *Synthesis* **1990**, 1079; (d) A. H. Munoz, J. Tamariz, R. Jiminez, G. G. Mora, *J. Chem. Res. (S)* **1993**, 68; (e) H. Laurent-Robert, C. Le Roux, J. Dubac, *Synlett* **1998**, 1138.

[191] T. Iwahama, S. Sakaguchi, Y. Ishii, *Chem. Commun.* **2000**, 613.

[192] (a) C.-H. Jun, J.-B. Hong, D.-Y. Lee, *Synlett* **1999**, 1, and references cited therein; (b) J. Schwartz, J. B. Cannon, *J. Am. Chem. Soc.* **1974**, *96*, 4721; (c) C. F. Lochow, R. G. Miller, *J. Am. Chem. Soc.* **1976**, *98*, 1281; (d) M. Tracy, J. Patrick, *J. Org. Chem.* **1952**, *17*, 1009.

[193] (a) M. Braun, *Angew. Chem., Int. Ed. Engl.* **1987**, *26*, 24; (b) C. H. Heathcock, *Comprehensive Carbanion Chemistry*, Part B (Eds.: E. Buncel, T. Durst), Elsevier, Amsterdam, **1984**, Chap. 4; (c) D. A. Evans, J. M. Takacs, L. R. McGee, M. D. Ennis, D. J. Mathre, J. Bartroli, *Pure Appl. Chem.* **1981**, *53*, 1109.

[194] T. Punniyamurthy, B. Bhatia, J. Iqbal, *J. Org. Chem.* **1994**, *59*, 850.

[195] The use of dioxolanes as masked aldehydes has been extensively studied: (a) H.-S. Dang, B. P. Roberts, *Tetrahedron Lett.* **1999**, *40*, 8929; (b) A. Gross, L. Fensterbank, S. Bogen, R. Thouvenot, M. Malacria, *Tetrahedron* **1997**, *53*, 13797.

[196] K. Hirano, T. Iwahama, S. Sakaguchi, Y. Ishii, *Chem. Commun.* **2000**, 2457.

[197] As another approach to the introduction of a carbonyl function to alkenes, we have reported the catalytic radical addition of ketones to terminal alkenes using an $Mn/Co/O_2$ redox system: T. Iwahama, S. Sakaguchi, Y. Ishii, *Chem. Commun.* **2000**, 2317.

[198] (a) W. J. Coates, A. McKillop, *Synthesis* **1993**, 334; (b) P. G. Baraldi, A. Bigoni, B. Cacciari, C. Caldari, S. Manfredini, G. Spalluto, *Synthesis* **1994**, 1158; (c) A. Hernandez, M. Marcos, H. Rapoport, *J. Org. Chem.* **1995**, *60*, 2683.

[199] M. S. Kharasch, W. H. Urry, B. M. Kuderna, *J. Org. Chem.* **1949**, *14*, 248.

[200] (a) M. Tracy, J. Patrick, *J. Org. Chem.* **1952**, *17*, 1009; (b) P. Gottschalk, D. C. Neckers, *J. Org. Chem.* **1985**, *50*, 3498.

[201] (a) H.-S. Dang, B. P. Roberts, *Chem. Commun.* **1996**, 2201; (b) H.-S. Dang,

B. P. ROBERTS, *J. Chem. Soc., Perkin Trans. 1* **1998**, 67.

[202] (a) V. Paul, B. P. Roberts, C. R. Willis, *J. Chem. Soc., Perkin Trans. 2* **1989**, 1953; (b) R. P. ALLEN, B. P. ROBERTS, C. R. WILLIS, *Chem. Commun.* **1989**, 1387; (c) B. P. ROBERTS, *Chem. Soc. Rev.* **1999**, *28*, 25.

[203] Another approach to understand controlling factors of the reactivity for the radical hydrogen abstraction has been made by A. A. Zavitsas: A. A. ZAVITSAS, *J. Chem. Soc., Perkin Trans. 2* **1998**, 499.

[204] S. TSUJIMOTO, T. IWAHAMA, S. SAKAGUCHI, Y. ISHII, *Chem. Commun.* **2001**, 2352.

6
Ruthenium-catalyzed Oxidation of Alkenes, Alcohols, Amines, Amides, β-Lactams, Phenols, and Hydrocarbons

Shun-Ichi Murahashi and Naruyoshi Komiya

6.1
Introduction

Ruthenium complexes have great potential for catalytic oxidation reactions of various compounds [1–7]. The reactivity of ruthenium complexes can be controlled by its oxidation state and ligand. The highest valent ruthenium complex is ruthenium(VIII) tetroxide (RuO_4), which is known as a strong oxidant and is useful for the cleaving of carbon–carbon double bonds. On the other hand, middle-valent oxo-ruthenium (Ru=O) species can be generated upon treatment of low-valent ruthenium complexes with a variety of oxidants. An important feature of these active species is their high capability to oxidize various substrates such as alkenes, alcohols, amines, amides, β-lactams, phenols, and unactivated hydrocarbons under mild conditions. Ruthenium-catalyzed oxidations and their applications to organic synthesis will be presented in this chapter.

6.2
RuO$_4$-promoted Oxidation

RuO_4 has been widely used as a powerful oxidant for oxidative transformation of a variety of organic compounds [8]. RuO_4 can be generated on treatment of $RuCl_3$ or RuO_2 with an oxidant. The oxidation reaction can be carried out conveniently in a biphasic system (Scheme 6.1) using a catalytic amount of $RuCl_3$ or RuO_2 with the combined use of an oxidant such as $NaIO_4$, HIO_4, $NaOCl$, or $NaBrO_3$, or under electrochemical conditions.

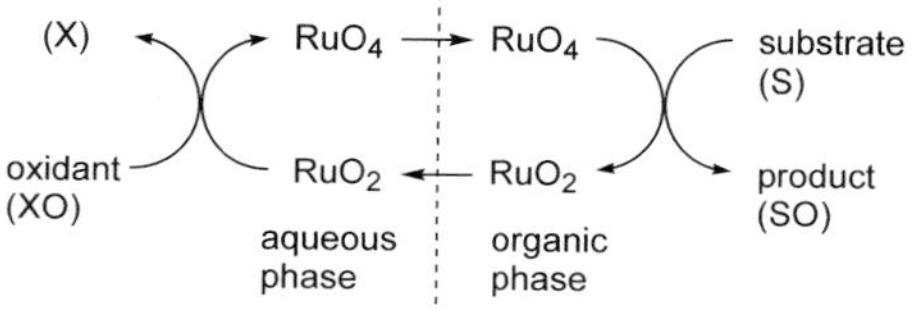

Scheme 6.1

Modern Oxidation Methods. Edited by Jan-Erling Bäckvall
Copyright © 2004 WILEY-VCH Verlag GmbH & Co. KGaA, Weinheim
ISBN: 3-527-30642-0

Problems such as very slow and incomplete reaction have been often encountered in the oxidations with RuO_4. These sluggish reactions are due to inactivation of ruthenium catalysts because of the formation of low-valent ruthenium carboxylate complexes. The inactivation can be prevented by addition of CH_3CN. Thus, various oxidations with RuO_4 are improved considerably by employing a solvent system consisting of CCl_4-H_2O-CH_3CN [8c]. Typically, oxidative cleavage of (*E*)-5-decene (**1**) with an $RuCl_3$/$NaIO_4$ system in a CCl_4-H_2O-CH_3CN system gave pentanoic acid in 88% yield, while the same reaction in a conventional CCl_4-H_2O system gave pentanal (17%) along with 80% of recovered **1** [Eq. (1)].

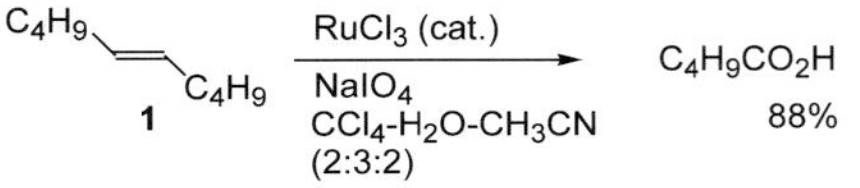

$$C_4H_9\text{−CH=CH−}C_4H_9 \xrightarrow[\substack{NaIO_4 \\ CCl_4\text{-}H_2O\text{-}CH_3CN \\ (2:3:2)}]{RuCl_3\text{ (cat.)}} C_4H_9CO_2H \quad 88\% \tag{1}$$

Primary and secondary alcohols are oxidized to the corresponding carboxylic acids and ketones, respectively, [Eqs. (2) and (3)] [9]. Olefins undergo oxidative cleavage to afford the carbonyl compounds [Eqs. (4) and (5)] [10], while *cis*-dihydroxylation occurs selectively when the reaction is carried out in a short period of time (0.5 min) at 0 °C in EtOAc-CH_3CN-H_2O [Eq. (6)] [10g].

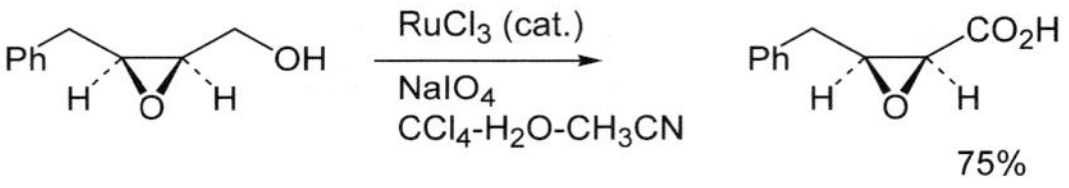

$$\text{Ph-CH(O)-CH(OH)} \xrightarrow[\substack{NaIO_4 \\ CCl_4\text{-}H_2O\text{-}CH_3CN}]{RuCl_3\text{ (cat.)}} \text{Ph-CH(O)-CO}_2\text{H} \quad 75\% \tag{2}$$

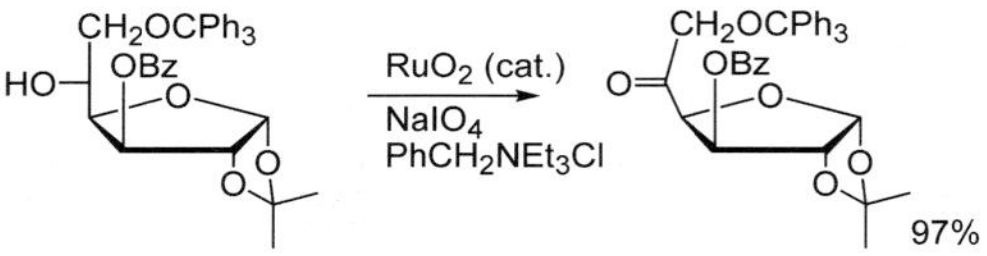

(3) Sugar derivative oxidation with RuO_2 (cat.), $NaIO_4$, $PhCH_2NEt_3Cl$ — 97%

(4) Glycal oxidation with RuO_2 (cat.), $NaIO_4$, CCl_4-H_2O — 97%

$$\text{cyclohexene} \xrightarrow[CCl_4\text{-}H_2O]{\substack{RuCl_3\text{ (cat.)} \\ NaOCl}} \text{HO}_2\text{C(CH}_2\text{)}_4\text{CO}_2\text{H} \quad 86\% \tag{5}$$

$$\text{cyclohexene} \xrightarrow[\substack{NaIO_4 \\ EtOAc\text{-}H_2O\text{-}CH_3CN \\ 0.5\text{ min, 0 °C}}]{RuCl_3\text{ (cat.)}} \textit{cis}\text{-cyclohexane-1,2-diol} \quad 58\% \tag{6}$$

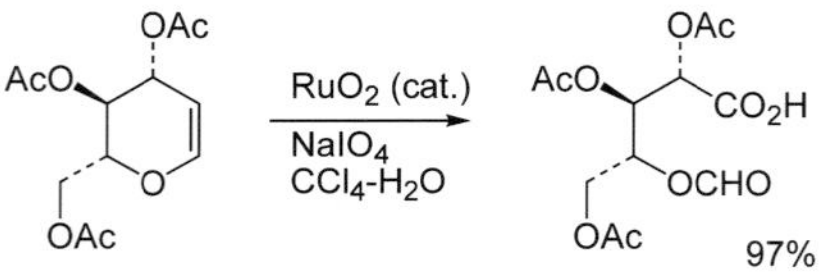

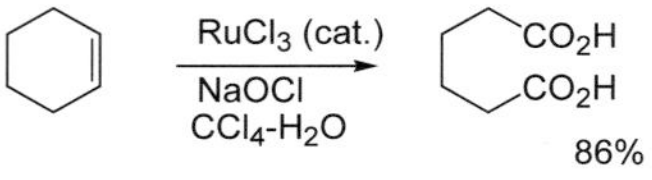

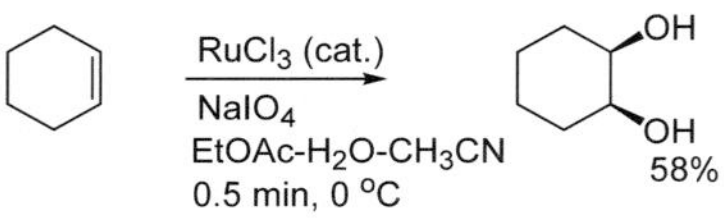

Octavalent RuO_4 generated from a $RuCl_3$/hypochlorite or periodate system is usually too reactive, and the C=C bond cleavage is often a major reaction; however, the addition of a bipyridine ligand facilitates the epoxidation of alkenes, because it works as an electron-donating ligand to enhance the electron density on the metal and to modulate the reactivity of RuO_4 [11–14]. $RuCl_3$ associated with bipyridyl [11] and phenanthrolines [12], catalyzes the epoxidation of alkenes with sodium periodate [Eq. (7)]. The reactions are stereospecific for both *cis*- and *trans*-alkenes. The dioxoruthenium(IV) complex $\{RuO_2(bpy)[IO_3(OH)_3]\} \cdot 1.5H_2O$ (2) was isolated by the reaction of RuO_4 with bipyridyl in the presence of $NaIO_4$, and the complex acts as an efficient epoxidation catalyst under similar conditions [Eq. (7)] [13].

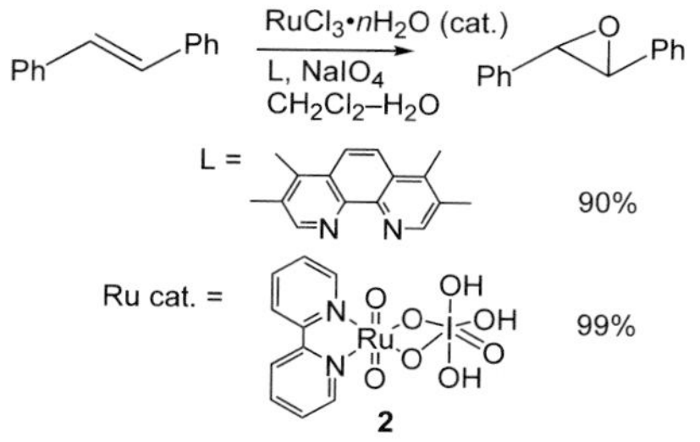

$$Ph\text{-CH=CH-}Ph \xrightarrow[CH_2Cl_2-H_2O]{RuCl_3 \cdot nH_2O \text{ (cat.)}, L, NaIO_4} Ph\text{-epoxide-}Ph \qquad (7)$$

L = 4,4′-dimethyl-2,2′-bipyridyl, 90%

Ru cat. = complex **2**, 99%

1,2-Dihaloalkenes are oxidized to α-diketones in a variety of norbornyl derivatives, which can serve as highly potent and inextricable templates for strained polycyclic unnatural compounds [Eq. (8)] [15].

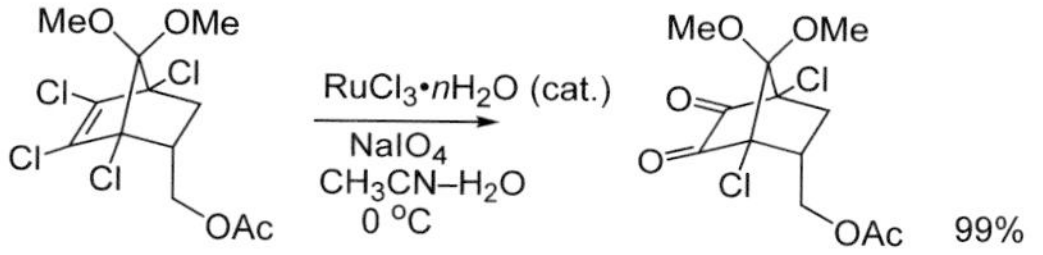

$$\text{(dichloro-norbornyl-OAc with MeO, OMe)} \xrightarrow[CH_3CN-H_2O, 0°C]{RuCl_3 \cdot nH_2O \text{ (cat.)}, NaIO_4} \text{(diketone product, OAc)} \qquad 99\% \qquad (8)$$

Aromatic rings are converted smoothly into carboxylic acids [Eqs. (9) and (10)] [16]. Terminal alkynes undergo a similar oxidative cleavage to afford carboxylic acids, while internal alkynes are converted into diketones [Eq. (11)] [17].

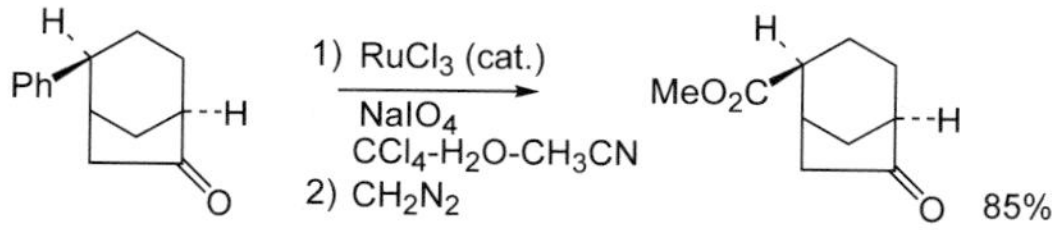

$$Ph\text{-norbornanone} \xrightarrow[\substack{1) RuCl_3 \text{ (cat.)}, NaIO_4, CCl_4-H_2O-CH_3CN \\ 2) CH_2N_2}]{} MeO_2C\text{-norbornanone} \qquad 85\% \qquad (9)$$

$$Ph\text{-CH(OCOPh)-CH}_2\text{-} \xrightarrow[CCl_4-H_2O-CH_3CN]{RuCl_3 \text{ (cat.)}, HIO_4} HO_2C\text{-CH(OCOPh)-CH}_2\text{-} \qquad 80\% \qquad (10)$$

$$C_{13}H_{27}\text{-}\!\!\equiv\!\!\text{-TBDMS} \xrightarrow[CCl_4-H_2O-CH_3CN]{RuO_2 \text{ (cat.)}, NaIO_4} C_{13}H_{27}\text{-CO-CO-TBDMS} \qquad 95\% \qquad (11)$$

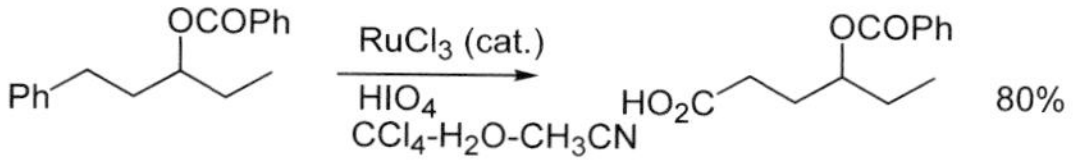

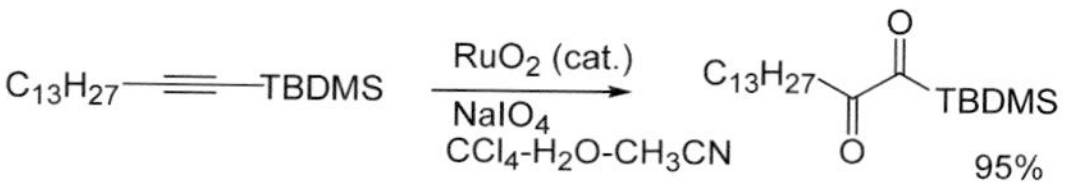

The oxidation of allenes gives α,α-dihydroxy ketones [Eq. (12)] [18]. Various heteroatom-containing compounds undergo oxidation of methylene groups at the α-position. Ethers are converted into esters and lactones [19]. The efficiency of the α-oxidation of ethers can be improved by pH control using hypochlorite in biphasic media [Eq. (13)] [19a]. Tertiary amines [20] and amides [21] undergo similar oxygenation reactions at the α-position of nitrogen to afford the corresponding amides and imides, respectively [Eq. (14)]. Carbon–carbon side-chain fragmentation occurs when N,C-protected serine and threonine are subjected to oxidation. The method has been successfully applied to the N–C bond scission of peptides at serine or threonine residue [Eq. (15)] [22].

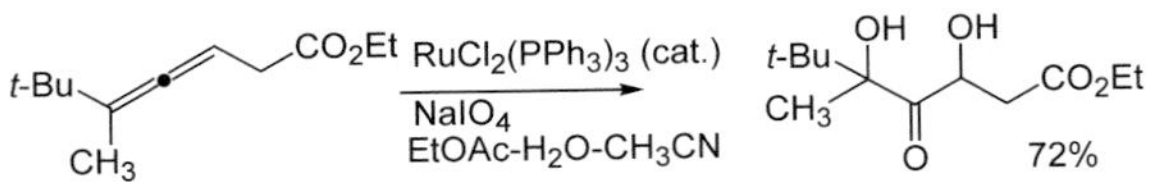

$$t\text{-Bu}\diagup\!\!\!\diagdown\!\!\!\diagup CO_2Et \xrightarrow[\text{EtOAc-H}_2\text{O-CH}_3\text{CN}]{\text{RuCl}_2(\text{PPh}_3)_3 \text{ (cat.)} \atop \text{NaIO}_4} t\text{-Bu}\diagup\!\!\!\diagup_{CH_3}^{OH\;OH}\!\!\!\diagdown\!\!\!\diagup CO_2Et \quad 72\% \tag{12}$$

$$C_3H_7\diagup\!\!\!\diagdown OC_4H_9 \xrightarrow[\text{CH}_2\text{Cl}_2\text{-H}_2\text{O (pH 9.5)}]{\text{RuCl}_2(\text{dppp})_2 \text{ (cat.)} \atop \text{NaOCl}} C_3H_7\diagup\!\!\!\overset{O}{\diagdown}OC_4H_9 \quad 93\% \tag{13}$$

$$\underset{\text{Boc}}{\text{pyrrolidine}}\text{-OSiMe}_2\text{-}t\text{-Bu} \xrightarrow[\text{CCl}_4\text{-H}_2\text{O}]{\text{RuO}_2 \text{ (cat.)} \atop \text{NaIO}_4} \underset{\text{Boc}}{\text{2-oxopyrrolidine}}\text{-OSiMe}_2\text{-}t\text{-Bu} \quad 90\% \tag{14}$$

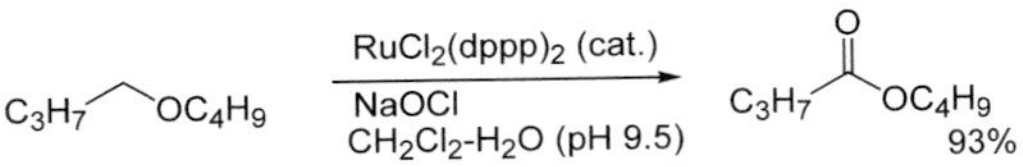

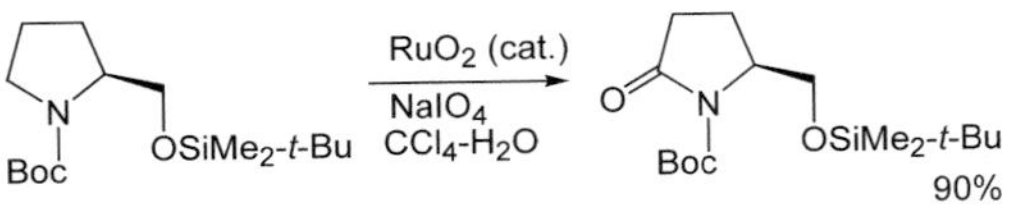

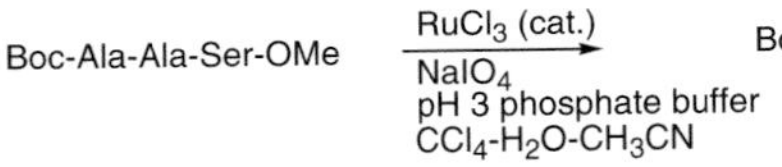

$$\text{Boc-Ala-Ala-Ser-OMe} \xrightarrow[\text{CCl}_4\text{-H}_2\text{O-CH}_3\text{CN}]{\text{RuCl}_3 \text{ (cat.)} \atop \text{NaIO}_4 \atop \text{pH 3 phosphate buffer}} \text{Boc-Ala-Ala-NH}_2 \quad 78\% \tag{15}$$

Unactivated alkanes can also be oxidized with the RuCl$_3$/NaIO$_4$ system. Tertiary carbon–hydrogen bonds undergo chemoselective hydroxylation to afford the corresponding tertiary alcohols [Eq. (16)] [23].

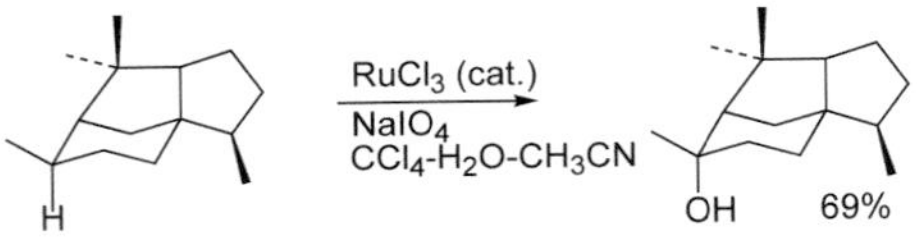

$$\text{tricyclic alkane-H} \xrightarrow[\text{CCl}_4\text{-H}_2\text{O-CH}_3\text{CN}]{\text{RuCl}_3 \text{ (cat.)} \atop \text{NaIO}_4} \text{tricyclic alkane-OH} \quad 69\% \tag{16}$$

Bridgehead carbons of adamantane [24], pinane [25], and fused norbornanes [26] undergo selective hydroxylation under similar reaction conditions. Methylene groups of cycloalkanes undergo hydroxylation and then subsequent oxidation to afford the corresponding ketones [27]. In general, methyl groups of alkanes undergo no reaction with RuO$_4$, while the methyl group of toluene can be converted into the corresponding carboxylic acids [Eq. (17)] [28].

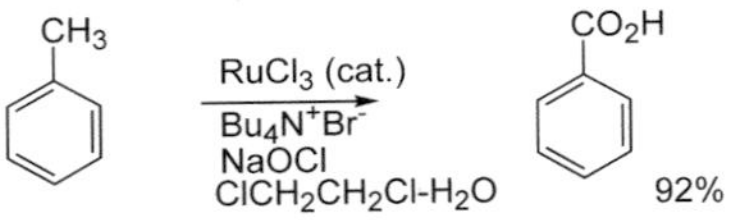

$$\text{PhCH}_3 \xrightarrow[\text{ClCH}_2\text{CH}_2\text{Cl-H}_2\text{O}]{\text{RuCl}_3 \text{ (cat.)} \atop \text{Bu}_4\text{N}^+\text{Br}^- \atop \text{NaOCl}} \text{PhCO}_2\text{H} \quad 92\% \tag{17}$$

6.3
Oxidation with Low-valent Ruthenium Catalysts and Oxidants

6.3.1
Oxidation of Alkenes

The treatment of a low-valent ruthenium catalyst with an oxidant generates middle-valent Ru=O species, which often show different reactivity to that of the RuO_4 oxidation. The epoxidation of alkenes with metalloporphyrins have been studied as model reactions of cytochrome P-450 [29]. Ruthenium porphyrins such as $Ru(OEP)(PPh_3)Br$ (OEP = octaethylporphyrinato) have been examined for the catalytic oxidation of styrene with PhIO [30]. Hirobe and coworkers [31] and Groves and coworkers [32] reported that the ruthenium porphyrin-catalyzed oxidation of alkenes with 2,6-dichloropyridine N-oxide gave the corresponding epoxides in high yields [Eqs. (18) and 19)]. The substituents at the 2- and 6-positions on pyridine N-oxide are necessary for high efficiency, because simple pyridine coordinates to the ruthenium more strongly and retards the catalytic activity. Nitrous oxide (N_2O) can be also used as an oxidant for the epoxidation of trisubstituted olefins in the presence of a ruthenium porphyrin catalyst [33].

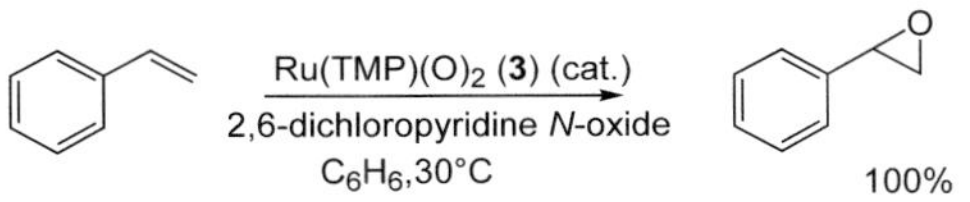

$$\text{styrene} \xrightarrow[\substack{\text{2,6-dichloropyridine N-oxide} \\ C_6H_6, 30°C}]{Ru(TMP)(O)_2 \text{ (3) (cat.)}} \text{styrene oxide} \quad 100\% \tag{18}$$

$$\text{1-octene} \xrightarrow[\substack{\text{2,6-dichloropyridine N-oxide} \\ CH_2Cl_2, 65°C}]{Ru(TPFPP)(CO) \text{ (4) (cat.)}} \text{1,2-epoxyoctane} \quad 96\% \tag{19}$$

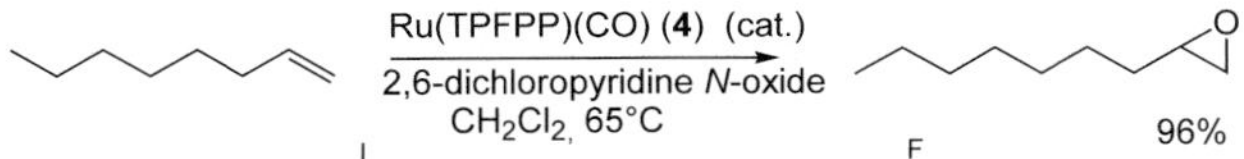

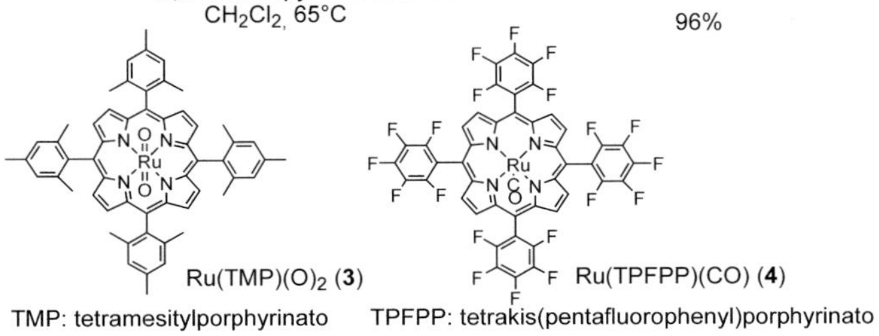

Ru(TMP)(O)₂ (3)
TMP: tetramesitylporphyrinato

Ru(TPFPP)(CO) (4)
TPFPP: tetrakis(pentafluorophenyl)porphyrinato

Non-porphyrin ruthenium complexes such as $[RuCl(DPPP)_2]$ (DPPP = 1,3-bis(diphenylphosphino)propane) and $[Ru(6,6-Cl_2bpy)_2(H_2O)_2]$ catalyze oxidations of alkenes with PhIO [34] or t-BuOOH [35] to give the corresponding epoxides in moderate yields.

The ruthenium-catalyzed aerobic oxidation of alkenes has been explored by several groups. Groves and coworkers reported that $Ru(TMP)(O)_2$ (3) catalyzed aerobic epoxidation of alkenes proceeds under 1 atm of molecular oxygen without any reducing agent [32b]. A Ru-containing polyoxometalate, $\{[WZnRu_2(OH)(H_2O)](ZnW_9O_{34})_2\}^{11-}$ [36] and a sterically hindered ruthenium complex, $[Ru(dmp)_2(CH_3CN)_2](PF_6)$ (dmp = 2,9-dimethyl-1,10-phenanthroline) [37] are also effective for epoxidation with molecular oxygen. Knochel and coworkers reported that the ruthenium catalyst bearing perfluorinated 1,3-diketone ligands catalyzes the aerobic epoxidation of alkenes in a perfluorinated solvent in the presence of *i*-PrCHO [38]. Asymmetric epoxidations have been reported using ruthenium complexes and oxidants such as PhIO, PhI(OAc)$_2$, 2,6-dichloropyridine N-oxide, and molecular oxygen [39–44].

It was postulated that one of the possible intermediates for metalloporphyrin promoted epoxidation is the intermediate **5** (Scheme 6.2) [45].

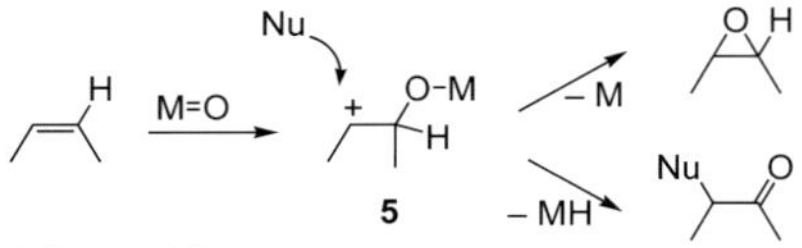

Scheme 6.2

If one could trap the intermediate **5** with an external nucleophile, such as water, a new type of catalytic oxidation of alkenes could be performed. Indeed, the transformation of alkenes into α-ketols was discovered to proceed highly efficiently. Thus, the low-valent ruthenium-catalyzed oxidation of alkenes with peracetic acid in an aqueous solution under mild conditions gives the corresponding α-ketols, which are important key structures of various biologically active compounds [Eq. (20)] [46].

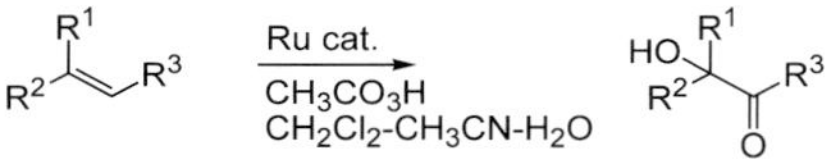

$$\text{(20)}$$

Typically, the RuCl$_3$-catalyzed oxidation of 3-acetoxy-1-cyclohexene (**6**) with peracetic acid in H$_2$O-CH$_3$CN-CH$_2$Cl$_2$ (1:1:1) gave (2R^*,3S^*)-3-acetoxy-2-hydroxycyclohexanone (**7**) chemo- and stereoselectively in 78% yield [Eq. (21)].

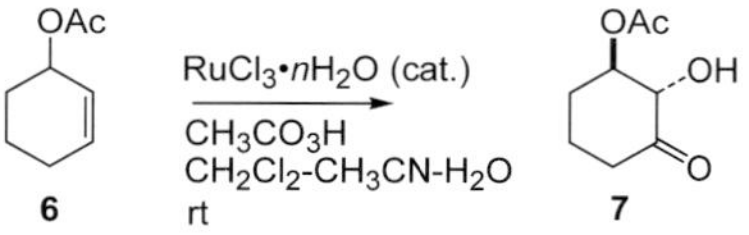

$$\text{(21)}$$

The oxidation is highly efficient and quite different from that promoted by RuO$_4$. Indeed, the oxidation of 1-methylcyclohexane **8** under the same conditions gives 2-hydroxy-2-methylcyclohexanone (**9**) (67%), while the oxidation of the same substrate **8** under conditions in which the RuO$_4$ is generated catalytically gives 6-oxoheptanoic acid (**10**) (91%) [Eq. (22)].

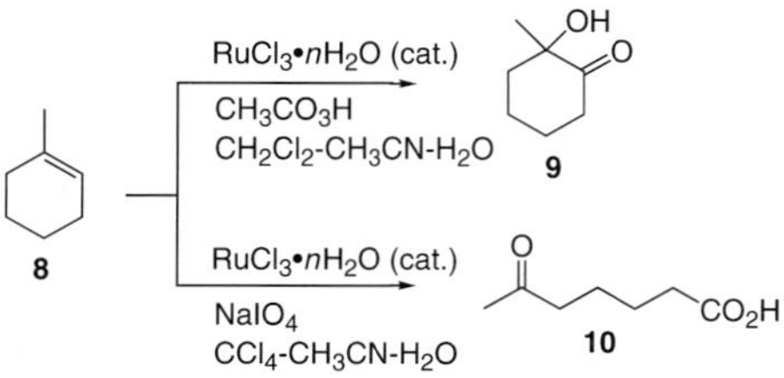

$$\text{(22)}$$

This particular oxidation can be applied to the oxidation of substituted alkenes having functional groups such as acetoxy, methoxycarbonyl, and azide groups to give the corresponding α-ketols in good to excellent yields. The oxidation of 3-azide-1-cyclohexene (11) gave (2S^*,3R^*)-3-azide-2-hydroxycyclohexanone (12) chemo- and stereoselectively (65%) [Eq. (23)].

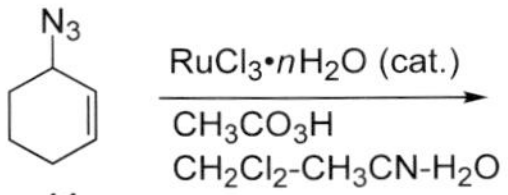

$$\text{11} \xrightarrow[\text{CH}_2\text{Cl}_2\text{-CH}_3\text{CN-H}_2\text{O}]{\text{RuCl}_3 \cdot n\text{H}_2\text{O (cat.)}, \text{CH}_3\text{CO}_3\text{H}} \text{12} \qquad (23)$$

The efficiency of the present reaction has been demonstrated by the synthesis of cortisone acetate 15 [47], which is a valuable anti-inflammatory agent. The oxidation of 3β,21-diacetoxy-5α-pregn-17-ene (13) proceeds stereoselectively to give 20-oxo-5α-pregnane-3β,17α,21-triol 3,21-diacetate (14) (57%) [Eq. (24)]. Conventional treatment of 14 followed by microbial oxidation with *Rhizopus nigricaus* gave 15 [48].

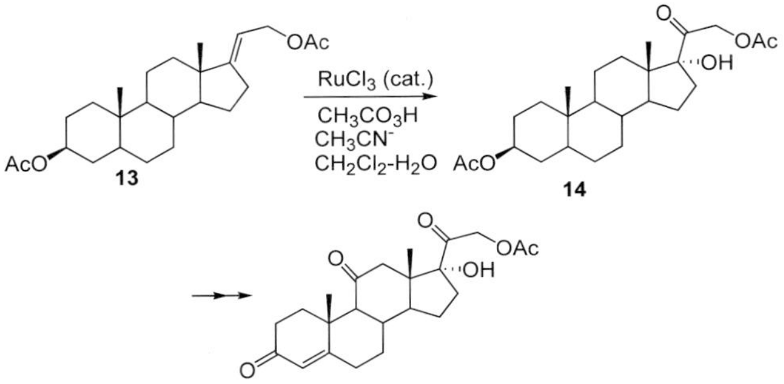

$$13 \xrightarrow[\text{CH}_2\text{Cl}_2\text{-H}_2\text{O}]{\text{RuCl}_3 \text{ (cat.)}, \text{CH}_3\text{CO}_3\text{H}, \text{CH}_3\text{CN}} 14 \longrightarrow \text{Cortisone acetate (15)} \qquad (24)$$

Furthermore, the method can be applied to the synthesis of 4-demethoxyadriamycinone, which is the side-chain of the cancer drugs adriamycins, such as idarubicin and annamycin (16). The ruthenium-catalyzed oxidation of allyl acetate 17 gives the corresponding α-hydroxyketone 18 in 60% yield [Eq. (25)] [49]. The reaction was also applied to the oxidation of α,β-unsaturated carbonyl compounds 19, and this provides a new method for the synthesis of α-oxo-ene-diols 20 [Eq. (26)] [50].

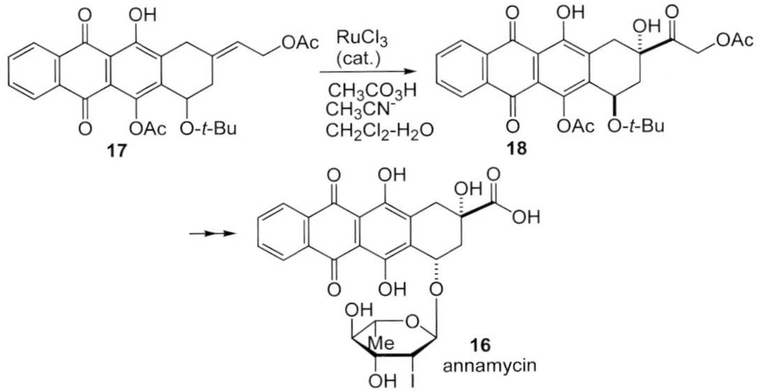

$$17 \xrightarrow[\text{CH}_2\text{Cl}_2\text{-H}_2\text{O}]{\text{RuCl}_3 \text{ (cat.)}, \text{CH}_3\text{CO}_3\text{H}, \text{CH}_3\text{CN}} 18 \longrightarrow \text{16 annamycin} \qquad (25)$$

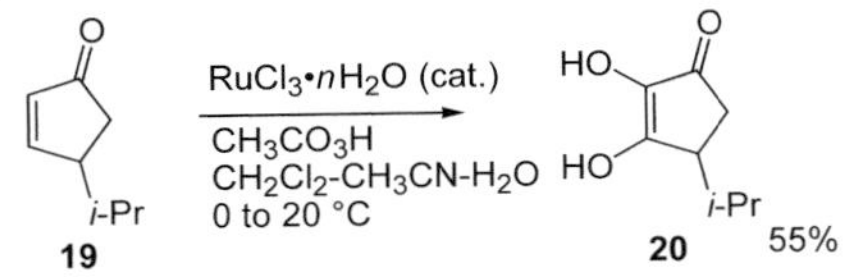

$$\text{(26)}$$

6.3.2
Oxidation of Alcohols

The ruthenium-catalyzed oxidation of alcohols has been reported using various catalytic systems which include the $RuCl_2(PPh_3)_3$ catalyst with oxidants such as N-methylmorpholine N-oxide (NMO) [51], iodosylbenzene [52], TMSOOTMS [53], $RuCl_3$ with hydrogen peroxide [54], K_2RuO_4 with peroxodisulfate [55], and the Ru(-pybox)(Pydic) complex with diacetoxyiodosylbenzene [56]. The salt of the perruthenate ion with a quaternary ammonium salt, $(n\text{-}Pr_4N)(RuO_4)$ (TPAP), which is soluble in a variety of organic solvents, shows far milder oxidizing properties than RuO_4 [57]. One of the key features of the TPAP system is its ability to tolerate other potentially reactive groups. For example, double bonds, polyenes, enones, halides, cyclopropanes, epoxides, and acetals all remain intact during TPAP oxidation [57]. The oxidation of primary alcohols with TPAP gives the corresponding aldehydes [Eqs. (27) and (28)], whereas RuO_4 oxidation results in the formation of carboxylic acid. NaOCl can be also used as an oxidant for the TPAP-catalyzed oxidation of secondary alcohols [58].

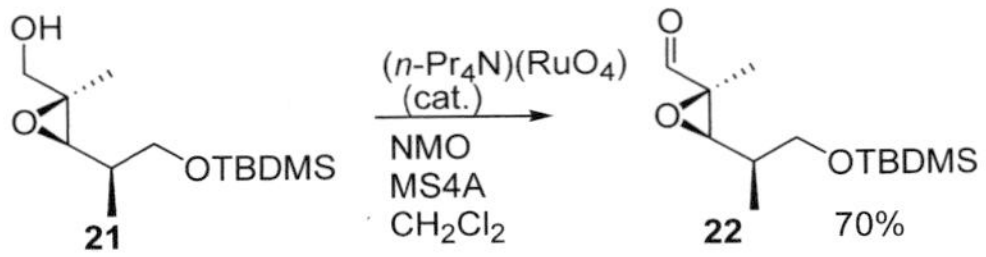

$$\text{(27)}$$

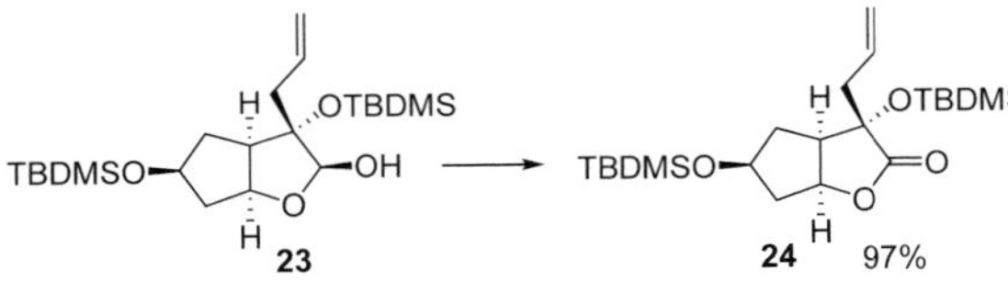

$$\text{(28)}$$

The $RuCl_2(PPh_3)_3$-catalyzed reaction of secondary alcohols with t-BuOOH gives ketones under mild conditions [59, 60]. This oxidation can be applied to the transformation of cyanohydrins into acyl cyanides [59], which are excellent acylating reagents. Typically, the oxidation of cyanohydrin **25** with 2 equiv. of t-BuOOH in dry benzene at room temperature gives benzoyl cyanide (**26**) in 92% yield [Eq. (29)]. It is worth noting that the acyl cyanides thus obtained are excellent reagents for the chemoselective acylation reaction. The reaction of amino alcohols with acyl cyanides selectively gives N-acylated amino alcohols. Furthermore, primary amines are selectively acylated in the presence of secondary amines [61]. Utility of the reaction has been illustrated by the short-step synthesis of maytenine (**27**) [Eq. (29)]. A ruthenium complex [Cn*Ru(CF$_3$CO$_2$)$_3$(H$_2$O)] (Cn* = N,N',N''-trimethyl-1,4,7-triazacyclono-nane) catalyst can be used for the oxidation of alcohols with t-BuOOH [62].

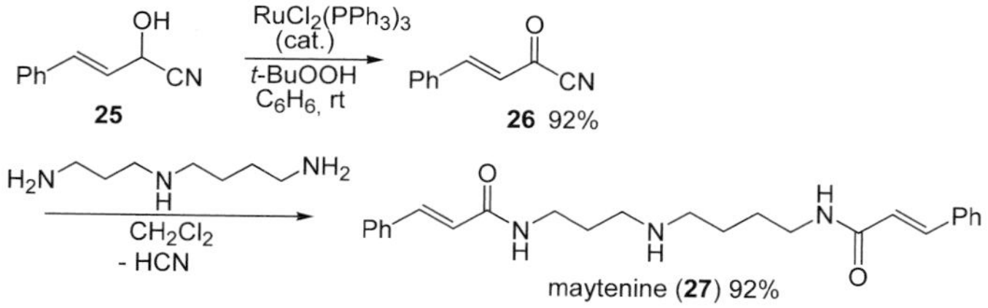

Various aliphatic and aromatic secondary alcohols can be oxidized with peracetic acid in the presence of RuCl$_3$ catalyst to give the corresponding ketones with high efficiency [63].

The generation of peracetic acid *in situ* provides an efficient method for the aerobic oxidation of alcohols. The oxidation of various aliphatic and aromatic alcohols can be carried out at room temperature with molecular oxygen (1 atm) in the presence of acetaldehyde and a RuCl$_3$–Co(OAc)$_2$ bimetallic catalyst [Eq. (30)] [64]. This method is highly convenient, because the products can be readily isolated simply by removal of both acetic acid and the catalyst by washing with a small amount of water. Under the same reaction conditions, primary alcohols are oxidized smoothly to the corresponding carboxylic acids. The present aerobic oxidation can be rationalized by assuming the following two sequential pathways: (1) formation of peracid by a cobalt-catalyzed radical chain reaction of aldehyde with molecular oxygen and (2) ruthenium-catalyzed oxidation of alcohol with the peracetic acid thus formed.

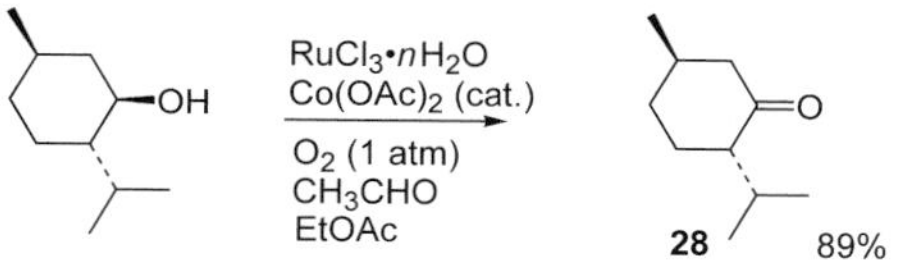

An alternative method for the oxidation of alcohols is dehydrogenative oxidation *via* a hydrogen transfer reaction. Alcohols undergo dehydrogenation in the presence of a ruthenium catalyst and hydrogen acceptors such as acetone [Eq. (31)] [65]. By regenerating the hydrogen acceptor in the presence of the co-catalyst and oxygen, the hydrogen transfer reaction can be extended to the catalytic aerobic oxidation. Thus, the ruthenium hydride formed during the hydrogen transfer can be converted into ruthenium by a multi-step electron-transfer process including hydroquinone, the metal complex, and molecular oxygen (Scheme 6.3). On the basis of this process, aerobic oxidation of alcohols to aldehydes and ketones can be performed at ambient pressure of O$_2$ in the presence of a ruthenium–cobalt bimetallic catalyst and hydroquinone [66]. Typically, cycloheptanol is oxidized to cycloheptanone under an O$_2$ atmosphere with a catalytic system consisting of ruthenium complex **29**, cobalt complex **30**, and a 1,4-benzoquinone [Eq. (32)] [66c]. Using trifluoromethyltoluene as a solvent, the aerobic oxidation of a primary alcohol can be performed by a RuCl$_2$(PPh$_3$)$_3$/hydroquinone system [67].

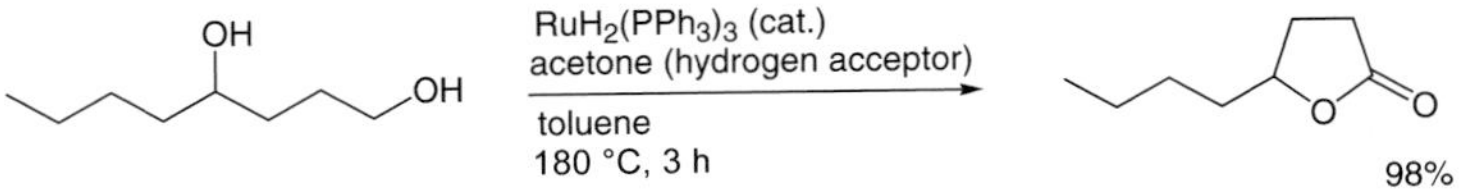

Scheme 6.3

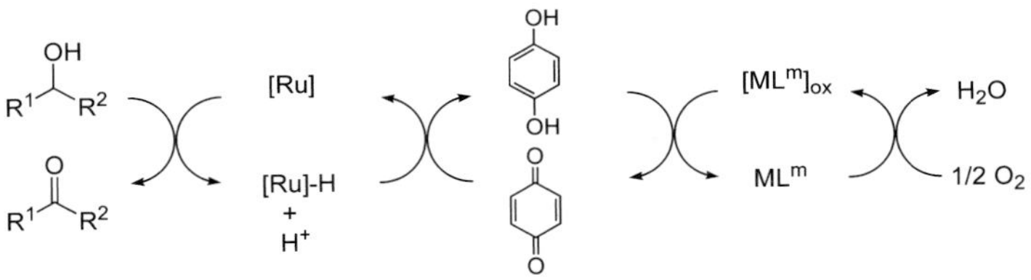

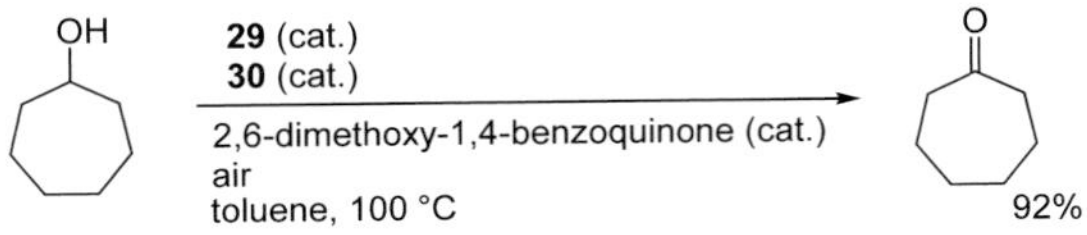

(31) 98%

(32) 92%

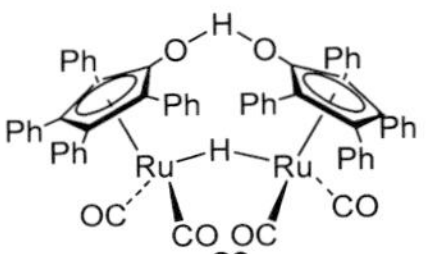

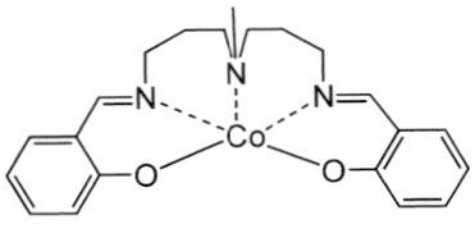

29, 30 structures shown.

The oxidation of secondary alcohols by a RuCl$_2$(PPh$_3$)$_3$–BzOTEMPO–O$_2$ system gives the corresponding ketones [Eq. (33)] [68]. The combination of RuCl$_2$(PPh$_3$)$_3$–TEMPO affords an efficient catalytic system for the aerobic oxidation of a broad range of primary and secondary alcohols at 100 °C, giving the corresponding aldehydes and ketones, respectively, in >99% selectivity in all cases [69]. The reoxidation of the ruthenium hydride species with TEMPO was proposed in the latter system [69c].

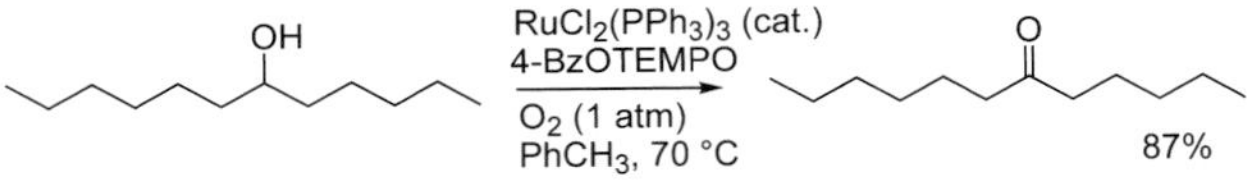

(33) 87%

Allylic alcohols can be converted into α,β-unsaturated aldehydes with 1 atm of molecular oxygen in the presence of RuO$_2$ catalyst [70].

TPAP can be used as an effective catalyst for the aerobic oxidation of alcohols to give the corresponding carbonyl compounds [Eq. (34)] [71]. A polymer supported perruthenate (PSP) and a perruthenate immobilized within MCM-41 can be used for heterogeneous oxidation of alcohols [72].

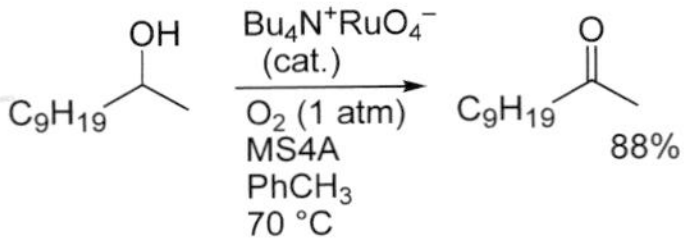

(34) 88%

Heterogeneous catalysts such as Ru–Al–Mg–hydrotalcites, Ru–Co–Al–hydrotalcites, Ru-hydroxyapatite (RuHAP) [73], and Ru–Al$_2$O$_3$ [74] are highly efficient catalysts for aerobic oxidation of alcohols [Eq. (35)]. In these oxidation reactions, the key step is postulated as the reaction of Ru–H with O$_2$ to form Ru–OOH, analogous to Pd–OOH, which has been shown to operate in the palladium-catalyzed Wacker-type asymmetric oxidation reaction [75].

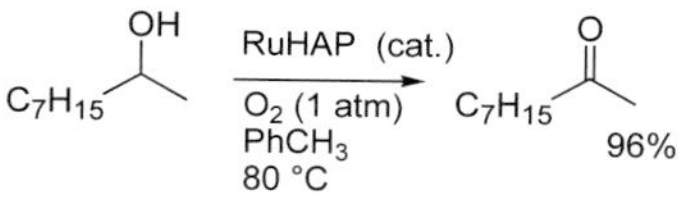

$$C_7H_{15}\text{-CH(OH)-} \xrightarrow[\substack{\text{PhCH}_3 \\ 80\,°C}]{\text{RuHAP (cat.)} \\ O_2\ (1\ \text{atm})} C_7H_{15}\text{-C(=O)-} \quad 96\% \tag{35}$$

RuHAP is also effective for the oxidation of organosilanes to the corresponding silanols [73e]. Catalytic oxidative cleavage of vicinal-diols to aldehydes with dioxygen was reported with RuCl$_2$(PPh$_3$)$_3$ on active carbon [76]. Ionic liquids such as tetramethyl ammonium hydroxide and Aliquate® 336 can be used as the solvent for the RuCl$_2$(PPh$_3$)$_3$-catalyzed aerobic oxidation of alcohols [77].

Kinetic resolution of secondary alcohols was reported after asymmetric oxidation using chiral (nitrosyl)Ru(salen) chloride (**31**) [Eq. (36)] [78].

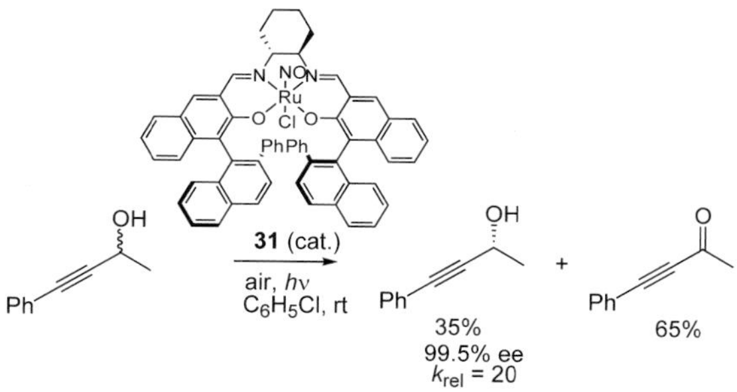

$$\text{Ph-C≡C-CH(OH)-Me} \xrightarrow[\substack{\text{C}_6\text{H}_5\text{Cl, rt}}]{\text{31 (cat.)} \\ \text{air, }h\nu} \text{Ph-C≡C-CH(OH)-Me} + \text{Ph-C≡C-C(=O)-Me} \tag{36}$$

35%, 99.5% ee, $k_{rel} = 20$; 65%

6.3.3
Oxidation of Amines

Selective oxidative demethylation of tertiary methyl amines is one of the specific and important functions of cytochrome P-450. Novel cytochrome P-450 type oxidation behavior with tertiary amines has been found in the catalytic systems of low-valent ruthenium complexes with peroxides. These systems exhibit specific reactivity toward oxidations of nitrogen compounds such as amines and amides, differing from that with RuO$_4$. Low-valent ruthenium complex-catalyzed oxidation of tertiary methylamines with *t*-BuOOH gives the corresponding α-(*tert*-butyldioxy)alkylamines efficiently [Eq. (37)] [79]. The hemiaminal type product has a similar structure to the α-hydroxymethylamine intermediate derived from the oxidation with cytochrome P-450.

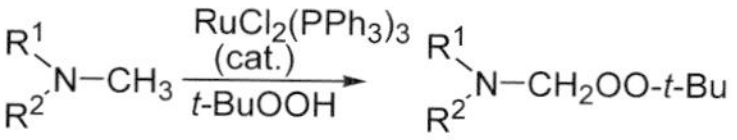

$$R^1R^2N\text{-CH}_3 \xrightarrow[\substack{\text{(cat.)} \\ t\text{-BuOOH}}]{\text{RuCl}_2(\text{PPh}_3)_3} R^1R^2N\text{-CH}_2\text{OO-}t\text{-Bu} \tag{37}$$

As shown in Scheme 6.4, the catalytic oxidation reactions can be rationalized by assuming the formation of oxo-ruthenium species by the reaction of low-valent ruthenium complexes with peroxides. α-Hydrogen abstraction from amines and the subsequent electron transfer gives the iminium ion ruthenium complex **32**. Trapping **32** with *t*-BuOOH would afford the corresponding α-*tert*-butylhydroxyamines, water, and low-valent ruthenium complex to complete the catalytic cycle.

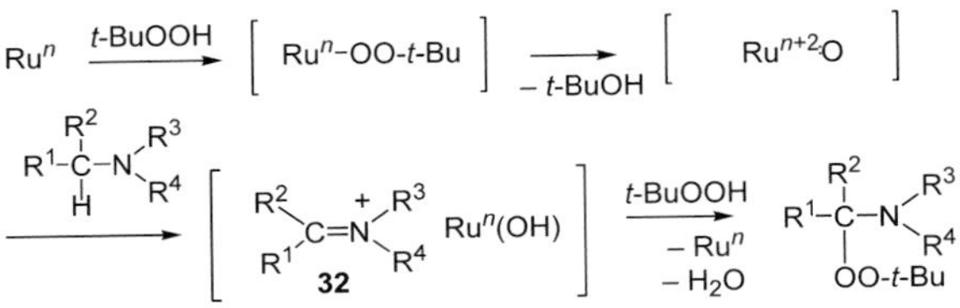

Scheme 6.4

The oxidation of N-methylamines provides various useful methods for organic synthesis. Selective demethylation of tertiary methylamines can be performed by the ruthenium-catalyzed oxidation and subsequent hydrolysis [Eq. (38)] [79]. This is the first practical synthetic method for the N-demethylation of tertiary amines. The methyl group is removed chemoselectively in the presence of various alkyl groups.

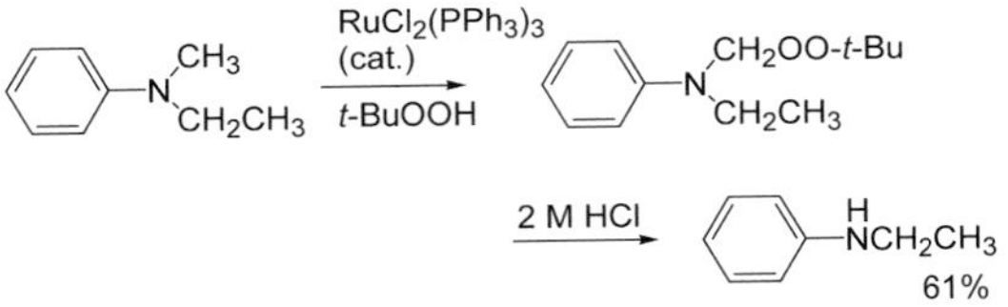

(38)

Biomimetic construction of piperidine skeletons from N-methylhomoallylamines is performed by means of the ruthenium-catalyzed oxidation and a subsequent olefin-iminium ion cyclization reaction. *trans*-1-Phenyl-3-propyl-4-chloropiperidine **34** was obtained from N-methyl-N-(3-heptenyl)aniline stereoselectively *via* **33** (55%) upon treatment with a 2 M HCl solution [Eq. (39)].

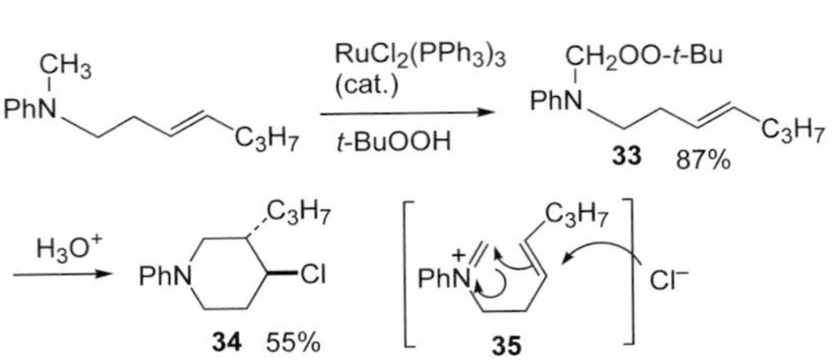

(39)

This cyclization can be rationalized by assuming the formation of iminium ion **35** by protonation of the oxidation product **33**, subsequent elimination of *t*-BuOOH, nucleophilic attack of an alkene, giving a carbonium ion, which is trapped by the Cl⁻

6.3 Oxidation with Low-valent Ruthenium Catalysts and Oxidants

nucleophile from the less hindered side. Similar treatment using CF_3CO_2H in place of HCl gave the corresponding hydroxy derivative [Eq. (40)].

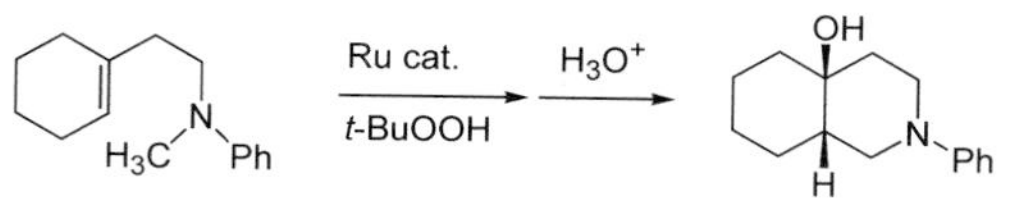

$$\text{cyclohexene-CH}_2\text{-N(CH}_3\text{)Ph} \xrightarrow[\text{t-BuOOH}]{\text{Ru cat.}} \xrightarrow{H_3O^+} \text{decahydroisoquinoline-OH-N-Ph} \quad (40)$$

α-Methoxylation of tertiary amines can be carried out by treatment with hydrogen peroxide in the presence of $RuCl_3$ catalyst in MeOH [80]. Thus, the oxidation of tertiary amine **36** gave the corresponding α-methoxyamine **37** in 60% yield [Eq. (41)].

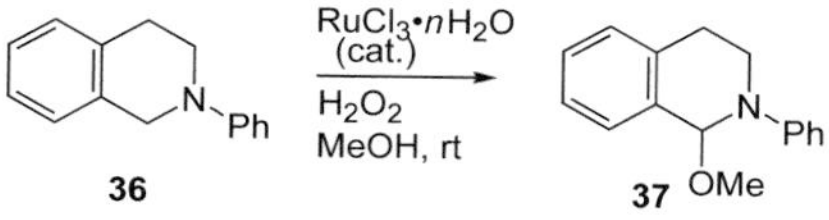

$$\mathbf{36} \xrightarrow[\text{MeOH, rt}]{\substack{RuCl_3 \cdot nH_2O \\ (cat.) \\ H_2O_2}} \mathbf{37} \quad (41)$$

Tertiary amine N-oxides can be prepared from the corresponding tertiary amines by $RuCl_3$-catalyzed oxidation with molecular oxygen [81].

Secondary amines can be converted into the corresponding imines in a single step highly efficiently through treatment with 2 equiv. of t-BuOOH in benzene in the presence of $RuCl_2(PPh_3)_3$ catalyst at room temperature [Eq. (42)] [82]. This is the first catalytic oxidative transformation of secondary amines to imines. In some cases a 4 Å molecular sieve is required to prevent the hydrolysis of the imines produced. The oxidations of tetrahydroisoquinoline **40** and allylamine **42** gave the corresponding cyclic imine **41** and azadiene **43** in 93% and 69% yields, respectively [Eqs. (43) and (44)].

$$\underset{\mathbf{38}}{R^1R^2\text{CHNHR}^3} \longrightarrow \underset{\mathbf{39}}{R^1R^2\text{C=N-R}^3} \quad (42)$$

$$\mathbf{40} \xrightarrow[\substack{\text{t-BuOOH} \\ C_6H_6}]{RuCl_2(PPh_3)_3 \text{ (cat.)}} \mathbf{41} \quad 93\% \quad (43)$$

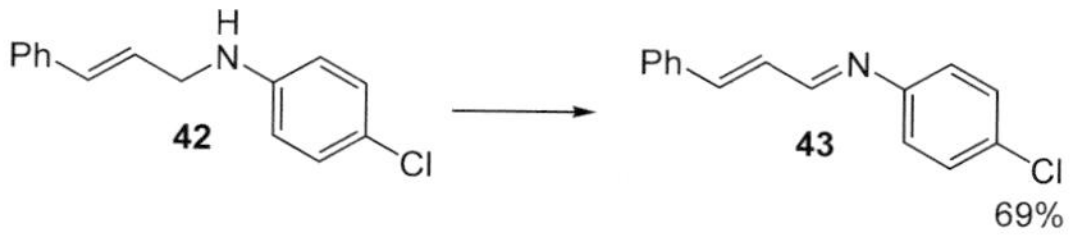

$$\mathbf{42} \longrightarrow \mathbf{43} \quad 69\% \quad (44)$$

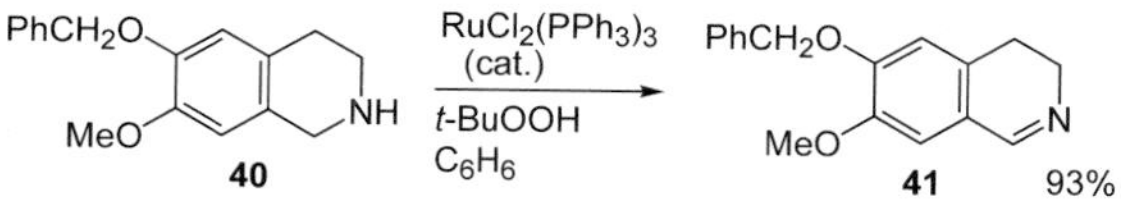

Aromatization takes place when an excess amount of t-BuOOH is used. For example, tetrahydroquinoline **44** can be converted into quinoline (**45**) (73%) [Eq. (45)].

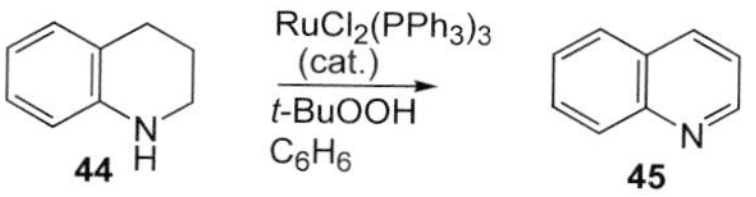

$$\mathbf{44} \xrightarrow[\substack{\text{t-BuOOH} \\ C_6H_6}]{RuCl_2(PPh_3)_3 \text{ (cat.)}} \mathbf{45} \quad (45)$$

It is worth noting that tungstate-catalyzed oxidation of the secondary amine **38** with hydrogen peroxide gives nitrone **46** [Eq. (46)] [83]. These two catalytic transformations of secondary amines [Eqs. (42) and (46)] are particularly useful for the introduction of a substituent at the α-position of the amines, because either imines or nitrones undergo diastereo and enantioselective reactions with nucleophiles to give chiral α-substituted amines highly efficiently [84].

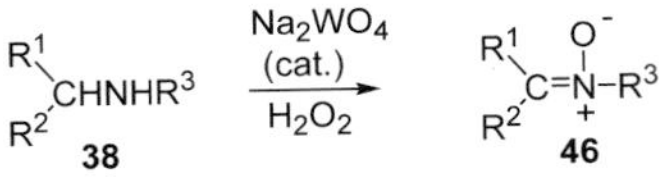

$$\underset{\underset{38}{R^2}}{\overset{R^1}{>}}\!CHNHR^3 \xrightarrow[H_2O_2]{Na_2WO_4 \text{ (cat.)}} \underset{\underset{46}{R^2}}{\overset{R^1}{>}}\!C=\overset{+}{N}\!-\!R^3 \quad \overset{O^-}{} \tag{46}$$

The catalytic system consisting of $(n\text{-}Pr_4N)(RuO_4)$ and N-methylmorpholine N-oxide (NMO) can also be used for the oxidative transformation of secondary amines into imines [Eq. (47)] [85 a, b].

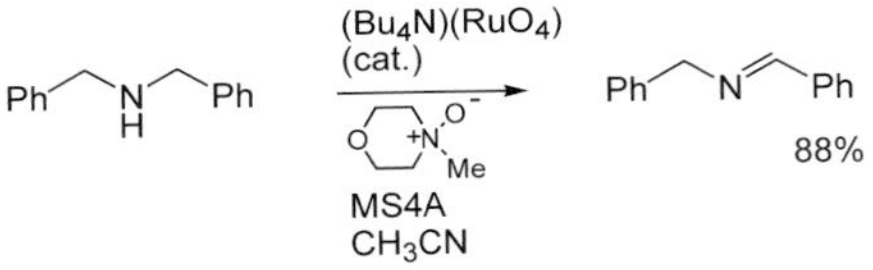

$$Ph\!\smallsetminus\!\underset{H}{N}\!\smallsetminus\!Ph \xrightarrow[\underset{CH_3CN}{MS4A}]{(Bu_4N)(RuO_4)\text{ (cat.)}} Ph\!\smallsetminus\!N\!=\!Ph \quad 88\% \tag{47}$$

The oxidation of secondary amines to imines can be performed by a hydrogen transfer reaction under mild conditions using a catalytic amount of 2,6-dimethoxy benzoquinone/MnO_2 [Eq. (48)] [85 c].

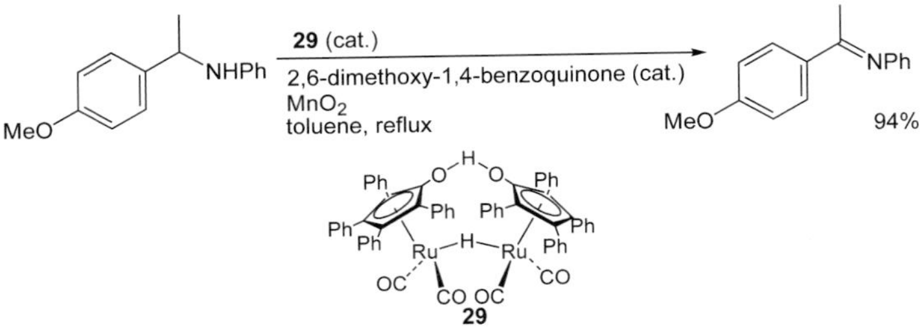

$$\text{MeO-C}_6H_4\text{-CH(Me)-NHPh} \xrightarrow[\underset{\text{toluene, reflux}}{\underset{MnO_2}{\text{2,6-dimethoxy-1,4-benzoquinone (cat.)}}}]{\textbf{29} \text{ (cat.)}} \text{MeO-C}_6H_4\text{-C(Me)=NPh} \quad 94\% \tag{48}$$

29 = $[Ph_4(\eta^4\text{-C}_4)Ru(CO)_2]_2(\mu\text{-H})(\mu\text{-OH})$ type diruthenium complex

The ruthenium-catalyzed oxidation of diphenylmethylamine with t-BuOOH gave benzophenone (88%), which was formed by hydrolysis of the imine intermediate [Eq. (49)] [82].

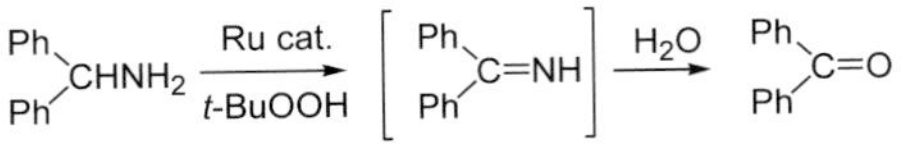

$$\underset{Ph}{\overset{Ph}{>}}\!CHNH_2 \xrightarrow[t\text{-BuOOH}]{Ru \text{ cat.}} \left[\underset{Ph}{\overset{Ph}{>}}\!C=NH\right] \xrightarrow{H_2O} \underset{Ph}{\overset{Ph}{>}}\!C=O \tag{49}$$

Potassium ruthenate (K_2RuO_4) was used as a catalyst for the oxidation of benzylamine with $K_2S_2O_8$ [86]. James and coworkers reported that aerobic oxidation of primary amines can be performed in the presence of a ruthenium porphyrin complex $Ru(TMP)(O)_2$ to give nitriles (100%) [Eq. (50)] [87].

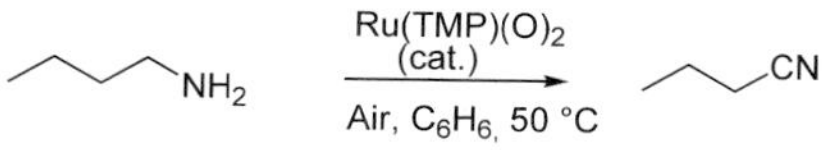

$$\text{CH}_3\text{CH}_2\text{CH}_2\text{CH}_2\text{NH}_2 \xrightarrow[\text{Air, C}_6\text{H}_6,\ 50\ °\text{C}]{\text{Ru(TMP)(O)}_2\ \text{(cat.)}} \text{CH}_3\text{CH}_2\text{CH}_2\text{CN} \qquad (50)$$

Heterogeneous catalysts such as hydroxyapatite-bound Ru complex [88a] and Ru/Al$_2$O$_3$ [88b, c] can also be used for the aerobic oxidation of primary amines to nitriles [Eqs. (51) and (52)].

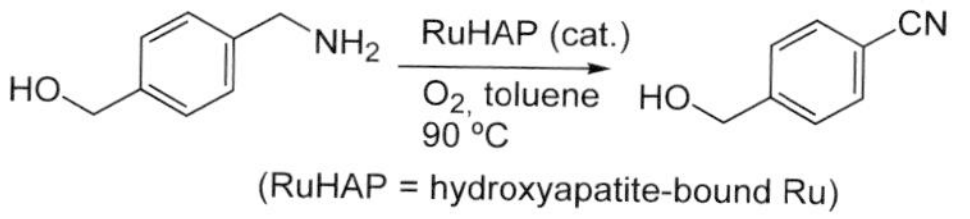

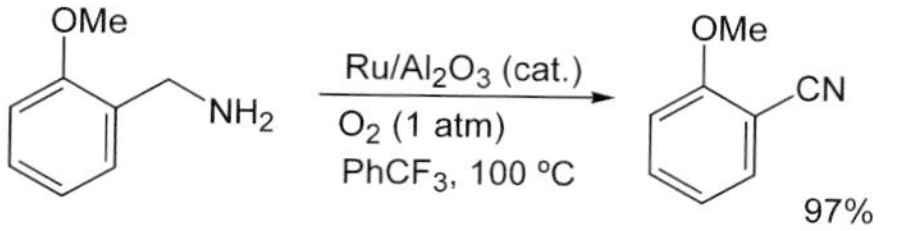

$$\text{HO-C}_6\text{H}_4\text{-CH}_2\text{NH}_2 \xrightarrow[\text{O}_2,\ \text{toluene}]{\text{RuHAP (cat.)}} \text{HO-C}_6\text{H}_4\text{-CN} \qquad (51)$$

(RuHAP = hydroxyapatite-bound Ru)

$$\text{2-MeO-C}_6\text{H}_4\text{-CH}_2\text{NH}_2 \xrightarrow[\text{PhCF}_3,\ 100\ °\text{C}]{\text{Ru/Al}_2\text{O}_3\ \text{(cat.)}} \text{2-MeO-C}_6\text{H}_4\text{-CN} \qquad (52)$$

97%

6.3.4
Oxidation of Amides and β-Lactams

The oxidation of the α-C–H bond of amides is an attractive strategy for the synthesis of biologically active nitrogen compounds. Selective oxidation of amides is difficult because of low reactivity. The RuCl$_2$(PPh$_3$)$_3$-catalyzed oxidation of amides with *t*-BuOOH proceeds under mild conditions to give the corresponding α-(*tert*-butyl-dioxy)amides **47** with high efficiency [Eq. (53)] [89].

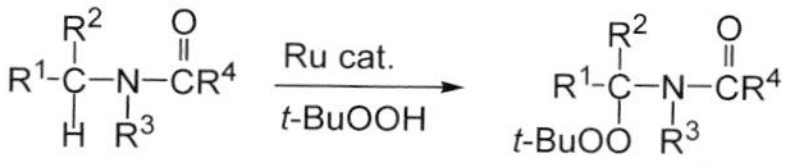

$$\underset{\underset{\text{H}}{|}\ \ \underset{\text{R}^3}{|}}{\text{R}^1\text{-C-N-CR}^4} \xrightarrow[\text{\textit{t}-BuOOH}]{\text{Ru cat.}} \underset{\underset{\textit{t}\text{-BuOO}}{|}\ \ \underset{\text{R}^3}{|}}{\text{R}^1\text{-C-N-CR}^4} \qquad (53)$$

47

The ruthenium-catalyzed oxidation of 1-(methoxycarbonyl)pyrrolidine with *t*-BuOOH gives 2-(*t*-butyldioxy)-1-(methoxycarbonyl)pyrrolidine (**48**) in 60% yield [Eq. (54)].

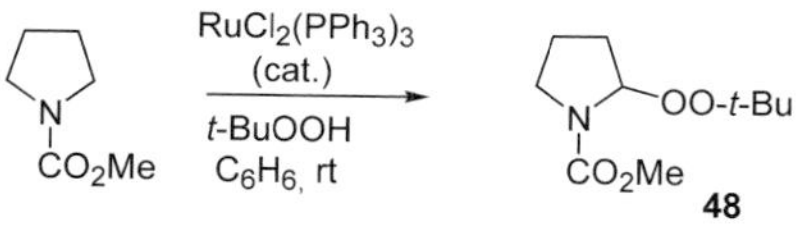

$$\underset{\text{CO}_2\text{Me}}{\text{pyrrolidine-N}} \xrightarrow[\text{\textit{t}-BuOOH, C}_6\text{H}_6,\ \text{rt}]{\text{RuCl}_2(\text{PPh}_3)_3\ \text{(cat.)}} \underset{\text{CO}_2\text{Me}}{\text{2-OO-\textit{t}-Bu-pyrrolidine-N}} \qquad (54)$$

48

The *tert*-butyldioxy amides of the isoquinoline **49** and indole **50**, which are important synthetic intermediates of natural products, were obtained in excellent yields [Eqs. (55) and (56)]. Since the Lewis acid-promoted reactions of these oxidized products with nucleophiles give the corresponding *N*-acyl-α-substituted amines efficiently, the present reactions provide versatile methods for selective carbon–carbon bond formation at the α-position of amides [90].

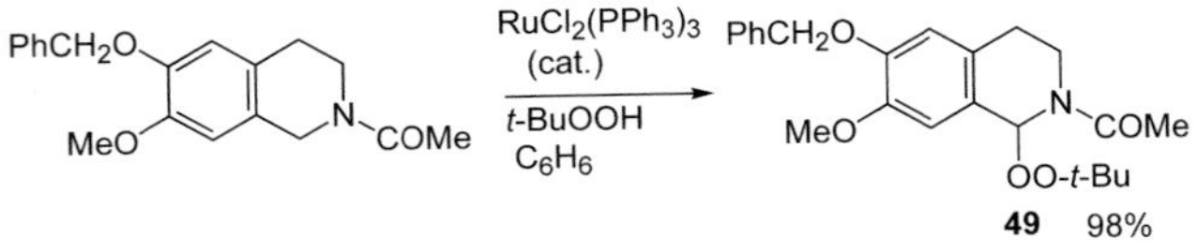

180 | *6 Ruthenium-catalyzed Oxidation of Alkenes, Alcohols, Amines, Amides, ...*

(55)

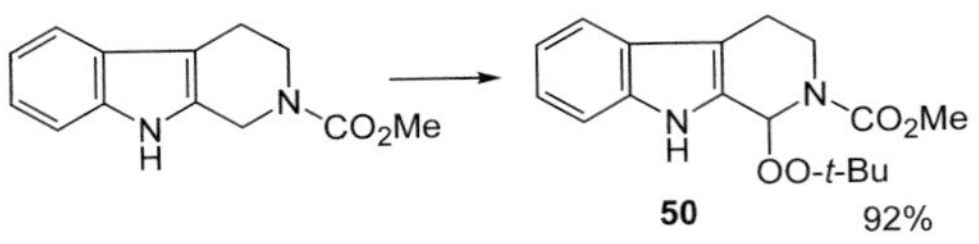

(56)

Typically, a TiCl$_4$-promoted reaction of α-*t*-butyldioxypyrrolidine **48** with a silyl enol ether gave the keto amide **51** (81%), while a similar reaction with less reactive 1,3-diene gave the α-substituted amide **52** [Eq. (57)].

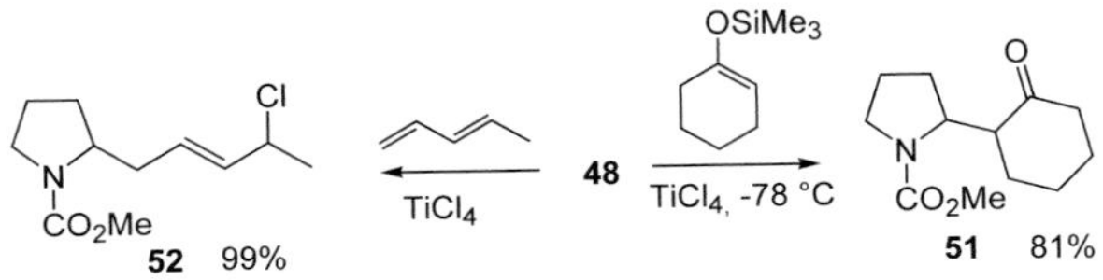

(57)

Oxidative modification of peptides has been performed by the ruthenium-catalyzed oxidation with peracetic acid. For example, the reaction of N,C-protected peptides containing glycine residues with peracetic acid in the presence of RuCl$_3$ catalyst gives α-ketoamides derived from the selective oxidation at the C-α position of the glycine residue (81%, conversion 70%) [Eq. (58)] [91].

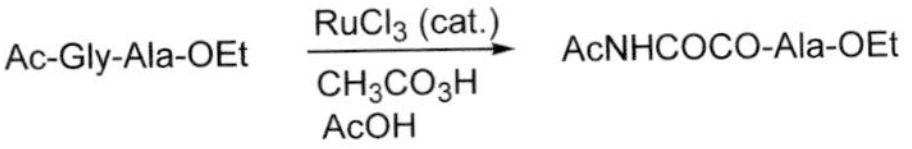

Ac-Gly-Ala-OEt $\xrightarrow[\text{AcOH}]{\text{RuCl}_3 \text{ (cat.)} \atop \text{CH}_3\text{CO}_3\text{H}}$ AcNHCOCO-Ala-OEt (58)

One of the most challenging topics amongst the oxidations of amides is the catalytic oxidation of β-lactams. Oxidation of β-lactams requires specific reaction conditions because of the high strain of the four-membered rings. Direct oxidative acyloxylation of β-lactams was successfully carried out by the ruthenium-catalyzed oxidation with peracetic acid in acetic acid under mild conditions. The products obtained are highly versatile and key intermediates in the synthesis of antibiotics. Thus, the ruthenium-catalyzed oxidation of 2-azetidinones with peracetic acid in acetic acid in the presence of sodium acetate at room temperature gives the corresponding 4-acetoxy-2-azetidinones **53** in 94% yield [Eq. (59)] [89]. One can use RuCl$_2$(PPh$_3$)$_3$ or RuCl$_3$, but for practical synthesis ruthenium on charcoal can be used conveniently. Although peracetic acid is the best oxidant, other oxidants such as *m*-chloroperbenzoic acid, methyl ethyl ketone peroxide, and iodosylbenzene can be used for the acyloxylation of β-lactams.

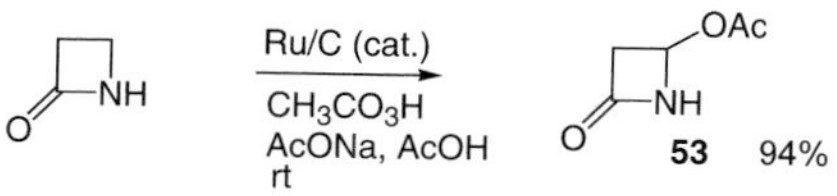

[Scheme, Eq. (59)]: Azetidinone-NH + Ru/C (cat.), CH₃CO₃H, AcONa, AcOH, rt → 4-OAc azetidinone **53**, 94%

Importantly, (1′R,3S)-3-[1′-(*tert*-butyldimethylsilyloxy)ethyl]azetidin-2-one (**54**) can be converted into the corresponding 4-acetoxyazetidinone **55** with extremely high diastereoselectivity (94%, >99% de) [Eq. (60)]. The product **55** is a versatile and key intermediate for the synthesis of carbapenems of antibiotics.

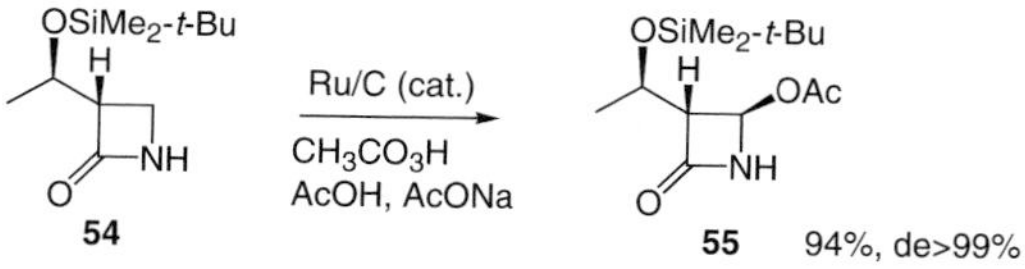

[Scheme, Eq. (60)]: **54** (OSiMe₂-*t*-Bu azetidinone) $\xrightarrow{\text{Ru/C (cat.), CH}_3\text{CO}_3\text{H, AcOH, AcONa}}$ **55** (4-OAc), 94%, de>99%

This method was applied to the stereoselective synthesis of 3-amino-4-acetoxyazetidinones **56** in 85% yield [Eq. (61)] [92].

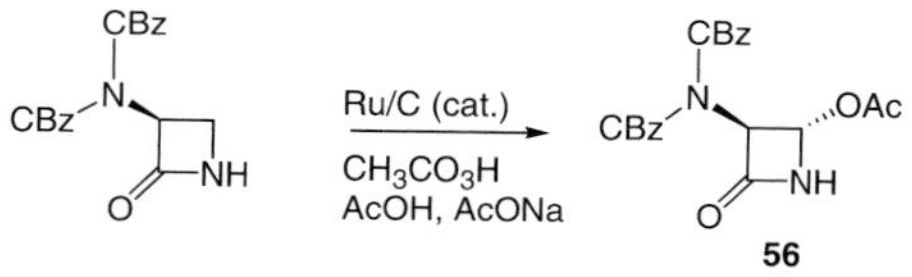

[Scheme, Eq. (61)]: CBz-N(CBz)-azetidinone-NH $\xrightarrow{\text{Ru/C (cat.), CH}_3\text{CO}_3\text{H, AcOH, AcONa}}$ **56** (4-OAc)

Aerobic oxidation of β-lactams can be performed highly efficiently in the presence of acetaldehyde, an acid, and sodium acetate [93]. Typically, the RuCl₃-catalyzed oxidation of β-lactam **54** with molecular oxygen (1 atm) in the presence of acetaldehyde and sodium carboxylate gave the corresponding 4-acyloxy β-lactam **55** in 91% yields (de >99%) [Eq. (62)]. This aerobic oxidation shows similar reactivity to the ruthenium-catalyzed oxidation with peracetic acid.

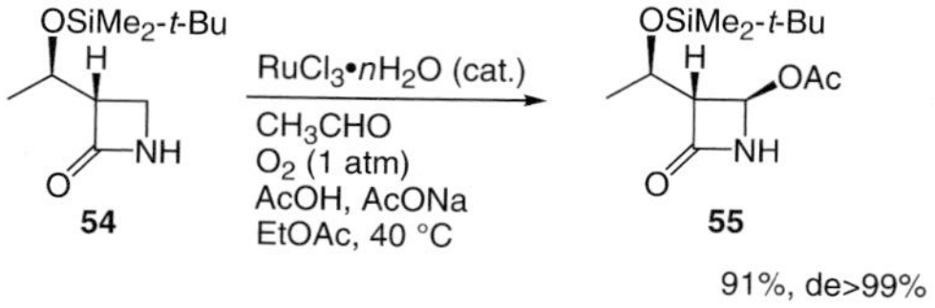

[Scheme, Eq. (62)]: **54** $\xrightarrow{\text{RuCl}_3 \cdot n\text{H}_2\text{O (cat.), CH}_3\text{CHO, O}_2 \text{(1 atm), AcOH, AcONa, EtOAc, 40 °C}}$ **55**, 91%, de>99%

6.3.5
Oxidation of Phenols

The oxidative transformation of phenols is of importance with respect to the biological and synthetic aspects. However, the oxidation of phenols generally lacks selectivity because of coupling reactions caused by phenoxyl radicals [94], and selective oxidation of phenols is limited to phenols bearing bulky substituents at the 2- and 6-positions [95]. Using ruthenium catalysts, a biomimetic and selective oxidation of phenols can be performed. Thus, the oxidation of *p*-substituted phenols bearing no substituent at the 2- and 6-positions with *t*-BuOOH in the presence of $RuCl_2(PPh_3)_3$ catalyst gives 4-(*tert*-butyldioxy)-4-alkylcyclohexadienones selectively [Eq. (63)] [96].

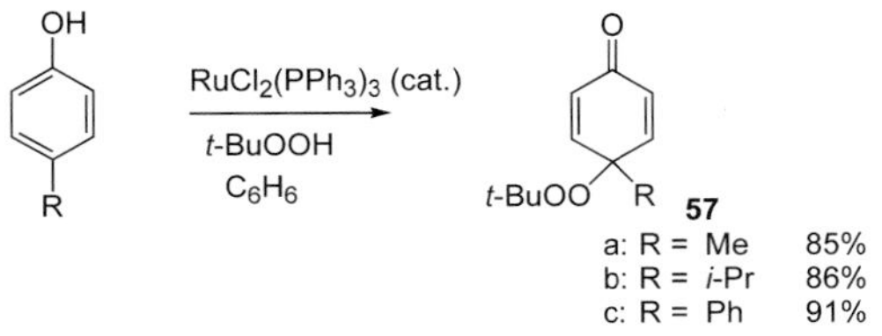

(63)

a: R = Me   85%
b: R = *i*-Pr   86%
c: R = Ph   91%

The reaction can be rationalized by assuming that the mechanism is that which involves an oxo-ruthenium complex (Scheme 6.5). Hydrogen abstraction with oxo-ruthenium species gives the phenoxyl radical **58**, which undergoes fast electron transfer to the ruthenium to give a cationic intermediate **59**. Nucleophilic reaction with the second molecule of *t*-BuOOH gives the product **57**.

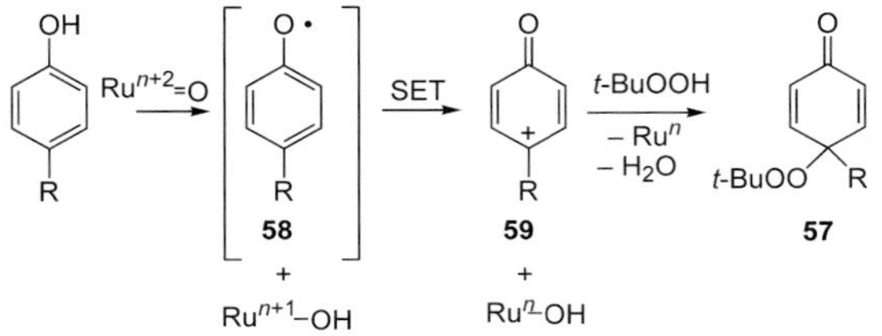

Scheme 6.5

The 4-(*tert*-butyldioxy)-4-alkylcyclohexadienones **57** thus obtained are versatile synthetic intermediates. The $TiCl_4$-promoted transformation of **60**, obtained from the oxidation of 3-methyl-4-isopropylphenol gives 2,6-disubstituted quinone **61**, which is derived from rearrangement of the *i*-Pr group (93%) [Eq. (64)].

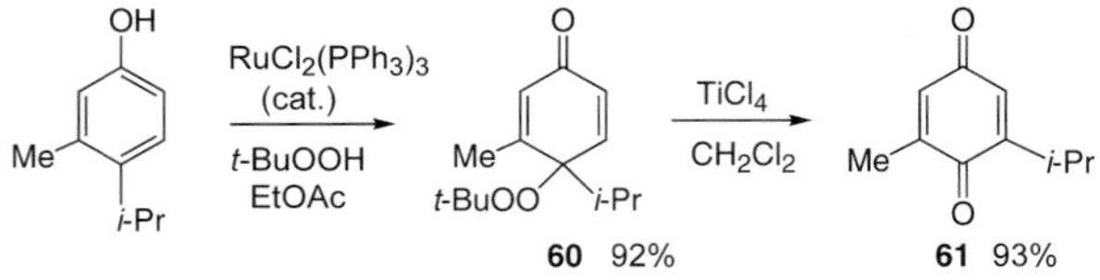

(64)

60 92%   61 93%

Interestingly, sequential migration Diels–Alder reactions of *tert*-butyldioxy dienone **63** in the presence of *cis*-1,3-pentadiene gave *cis*-fused octahydroanthraquinone **64** stereoselectively (73%) [Eq. (65)].

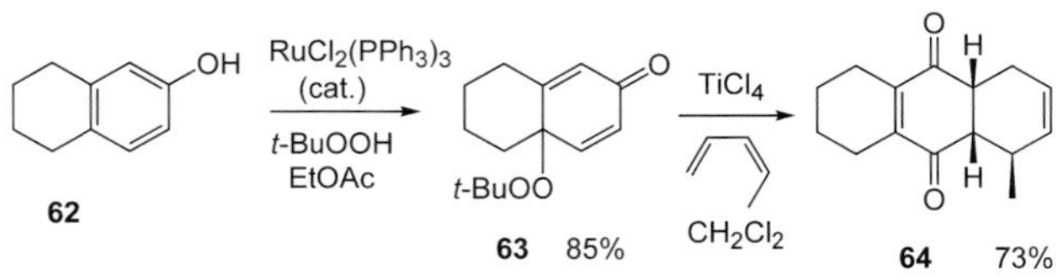

(65)

62   63 85%   64 73%

Ruthenium-catalyzed oxygenation of catechols gives muconic acid anhydride (**65**) and 2H-pyran-2-one (**66**) [Eq. (66)] [97].

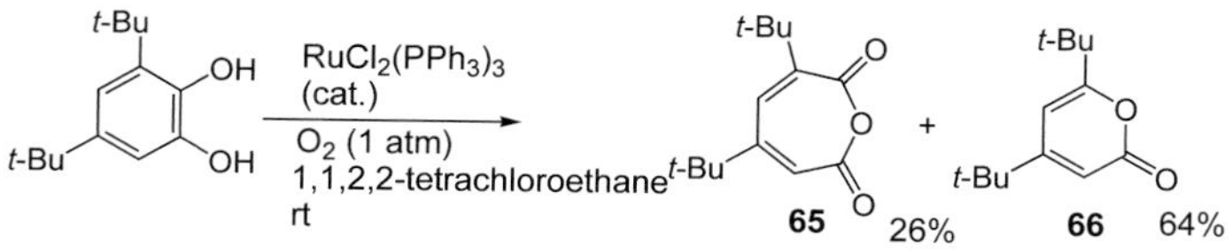

(66)

Oxidation of aromatic ring bearing methoxy groups was performed using a ruthenium porphyrin catalyst. The Ru(TPP)(O)$_2$-catalyzed (TPP = 5,10,15,20-tetraphenylporphyrinato) oxidation of polymethoxybenzene with 2,6-dichloropyridine N-oxide gives the corresponding p-quinone derivatives [Eq. (67)] [98]. The ^{18}O labeling experiments showed that the reaction proceeds via selective hydroxylation of the aromatic ring by oxo-ruthenium porphyrins to afford phenol derivatives, which undergo the subsequent oxidation to give the corresponding quinones.

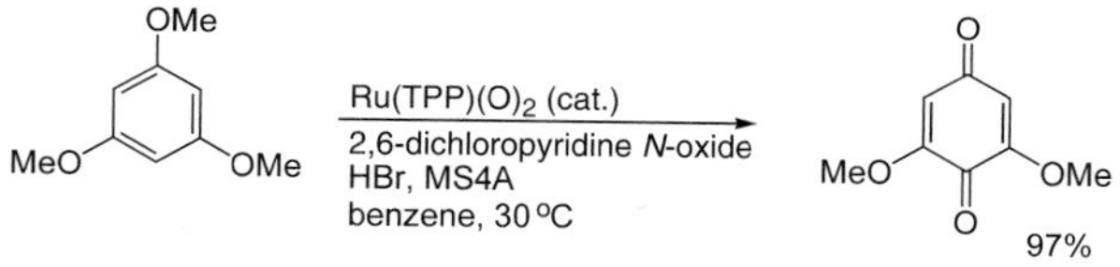

(67)

6.3.6
Oxidation of Hydrocarbons

The catalytic oxidation of unactivated hydrocarbons remains a challenging topic. Ruthenium porphyrins, such as Ru(OEP)(PPh$_3$)$_3$, show the catalytic activity for the oxidation of alkanes with PhIO [30]. The oxidation of alkanes with 2,6-dichloropyridine N-oxide in the presence of Ru(TMP)(O)$_2$ (TMP, tetramesitylporphyrinato) (**3**) and HBr [99] and Ru(TPFPP)(CO) [TPFPP, tetrakis(pentafluorophenyl)porphyrinato] (**4**) [32a] gives the corresponding oxidized compounds. Hydroxylation of adamantane was achieved with high selectivity and high efficiency (12 300 turnovers) [Eq. (68)]. Zeolite-encapsulated perfluorinated ruthenium phthalocyanines catalyze the oxidation of cyclohexane with t-BuOOH [100]. The addition of Lewis acids such as ZnCl$_2$ greatly accelerated the reaction rates in the stoichiometiric oxidation of alkanes by BaRu(O)$_2$(OH)$_3$ [101]. A dioxoruthenium complex with a D$_4$-chiral porphyrin ligand has been used for the enantioselective hydroxylation of ethylbenzene to give α-phenylethyl alcohol with 72% ee [102].

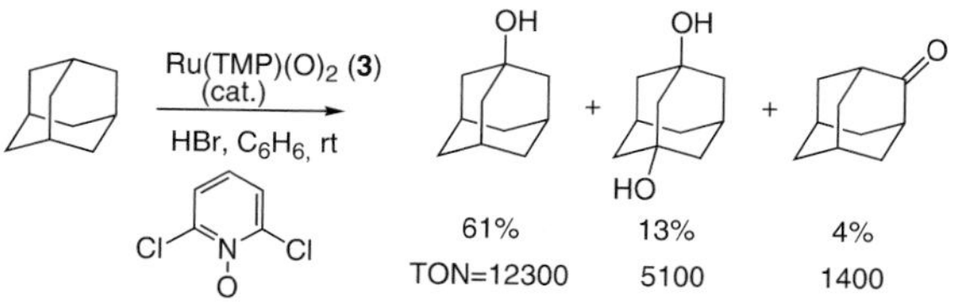

(68)

Non-porphyrin ruthenium complexes can be used for the catalytic oxidation of alkanes with peroxides. BaRuO$_3$(OH)$_2$-catalyzed oxidation of cyclohexane with PhIO gives oxidation products [101a]. [RuCl(dpp)$_2$]$^+$ can be used for the oxidation of alkanes with PhIO or LiClO [34]. The combinations of cis-[Ru(dmp)$_2$(MeCN)$_2$]$^{2+}$/H$_2$O$_2$ [37a] [Eq. (69)], cis-[Ru(Me$_3$tacn)(O)$_2$(CF$_3$CO$_2$)]$^+$/t-BuOOH (Me$_3$tacn, N,N′,N″-1,4,7-trimethyl-1,4,7-triazacyclononane) [35c], and cis-[Ru(6,6-Cl$_2$bpy)$_2$(OH$_2$)$_2$]$^{2+}$/t-BuOOH [35b] are efficient for the oxidation of cyclohexane. Ruthenium(III) complexes such as [RuCl$_2$(TPA)]$^+$ and [RuCl(Me$_2$SO)(TPA)]$^+$ bearing the tripodal ligand TPA [TPA = tris(2-pyridylmethyl)amine] were synthesized, and catalytic oxidation of adamantane with m-chloroperbenzoic acid was reported [103, 104]. Polyoxometalate [SiRu (H$_2$O)W$_{11}$O$_{39}$]$^{5-}$ also works as an oxidation catalyst using KHSO$_5$ [105a] and H$_2$O$_2$ [105b].

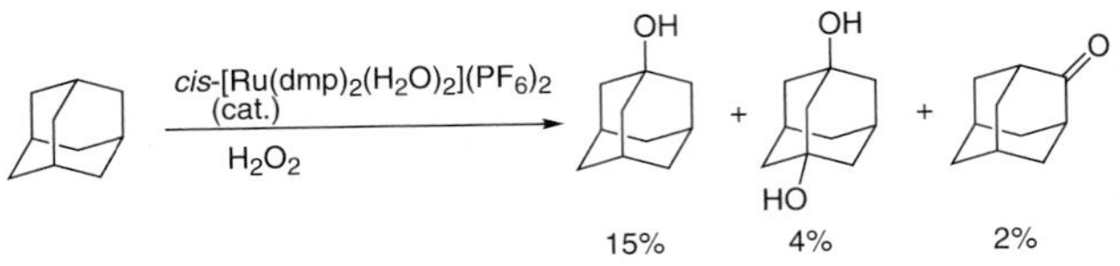

(69)

The oxidation of hydrocarbons with ruthenium catalysts bearing a simple ligand is highly effective. The oxidations of hydrocarbons with peroxides such as t-BuOOH and peracetic acid in the presence of ruthenium catalysts such as RuCl$_2$(PPh$_3$)$_3$ [106a, b] or Ru/C [106a, c] actually gave the corresponding ketones and alcohols. For example, RuCl$_2$(PPh$_3$)$_3$-catalyzed oxidation of fluorene with t-BuOOH gives fluorenone in 87% yield [Eq. (70)]. The Ru/C-catalyzed oxidation of cyclohexane with peracetic acid in ethyl acetate gives cychohexanone and cyclohexanol with 67% conversion [106c].

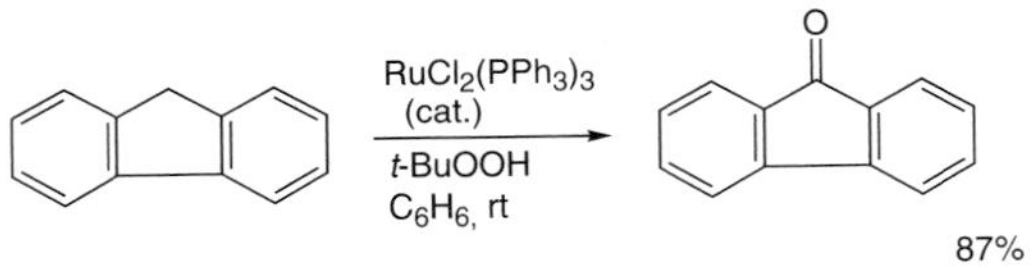

(70)

It is expected that more reactive species will be generated in the presence of a strong acid. Indeed, the RuCl$_3$·nH$_2$O-catalyzed oxidation of cyclohexane in trifluoroacetic acid and dichloromethane (5:1) with peracetic acid gives cyclohexyl trifluoroacetate in 77% [Eq. (71)] [106a].

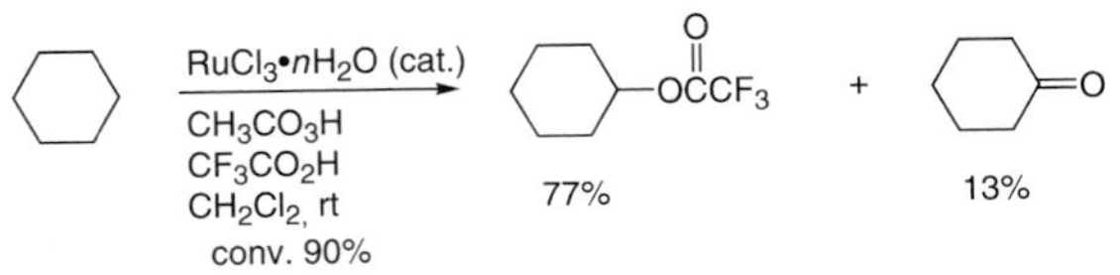

(71)

The ruthenium-catalyzed oxidation of nitriles takes place at the α-position. For example, the RuCl$_3$·nH$_2$O-catalyzed oxidation of benzylcyanide with t-BuOOH gives benzoylcyanide in 94% yield [107].

The allylic position of steroidal alkene can be oxidized with t-BuOOH in the presence of a RuCl$_3$ catalyst [Eq. (72)] [108].

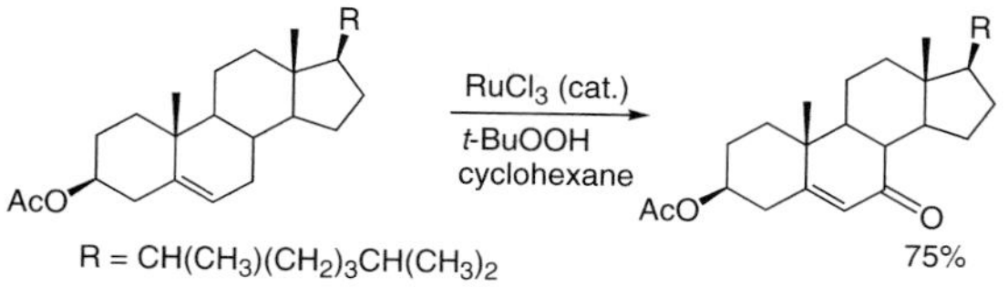

$$\text{AcO-steroid (R)} \xrightarrow[\text{cyclohexane}]{\text{RuCl}_3 \text{ (cat.)}, \text{ t-BuOOH}} \text{AcO-enone (R)} \quad 75\% \tag{72}$$

R = CH(CH$_3$)(CH$_2$)$_3$CH(CH$_3$)$_2$

The catalytic oxidation of alkanes with molecular oxygen under mild conditions is an especially rewarding goal, since direct functionalization of the unactivated C–H bonds of saturated hydrocarbons usually requires drastic conditions such as high temperature.

Oxo-metal species can be generated by the reaction of a low-valent ruthenium complex with molecular oxygen in the presence of an aldehyde [93]. Thus, the ruthenium-catalyzed oxidation of alkanes with molecular oxygen in the presence of acetaldehyde gives alcohols and ketones efficiently [109a]. Typically, RuCl$_3 \cdot n$H$_2$O-catalyzed oxidation of cyclooctane with molecular oxygen in the presence of an aldehyde gives the corresponding alcohol and ketone selectively [Eq. (73)].

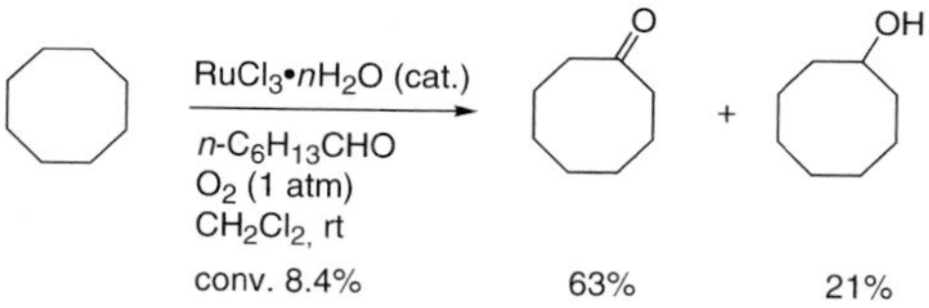

$$\text{cyclooctane} \xrightarrow[\substack{n\text{-C}_6\text{H}_{13}\text{CHO} \\ \text{O}_2 \text{ (1 atm)} \\ \text{CH}_2\text{Cl}_2, \text{ rt}}]{\text{RuCl}_3 \cdot n\text{H}_2\text{O (cat.)}} \text{cyclooctanone} + \text{cyclooctanol} \tag{73}$$

conv. 8.4%   63%   21%

These aerobic oxidations can be rationalized by assuming the sequence shown in Scheme 6.6. The metal-catalyzed radical chain reaction of an aldehyde with molecular oxygen affords the corresponding peracid. The reaction of a metal catalyst with the peracid thus formed would give an oxo-metal intermediate, followed by oxygen atom transfer to afford the corresponding alcohols. The alcohol is further oxidized to the corresponding ketone under these conditions.

RCHO + O$_2$ $\xrightarrow{\text{M cat.}}$ RCO$_3$H

M^n + RCO$_3$H $\longrightarrow$ M^{n+2}=O + RCO$_2$H

M^{n+2}=O + RH $\longrightarrow$ M^n + ROH

Scheme 6.6

A Ru(TPFPP)(CO) (**4**) complex has been prepared, and it was found that **4** is an efficient catalyst for the aerobic oxidation of alkanes using acetaldehyde [110]. Thus, the **4**-catalyzed oxidation of cyclohexane with molecular oxygen in the presence of acetaldehyde gave cyclohexanone and cyclohexanol in 62% yields, based on acetaldehyde with high turnover numbers of 14 100 [Eq. (74)]. It is worth noting that iron

[109] and copper [111] catalysts are also efficient for the oxidation of non-activated hydrocarbons at room temperature under 1 atm of molecular oxygen.

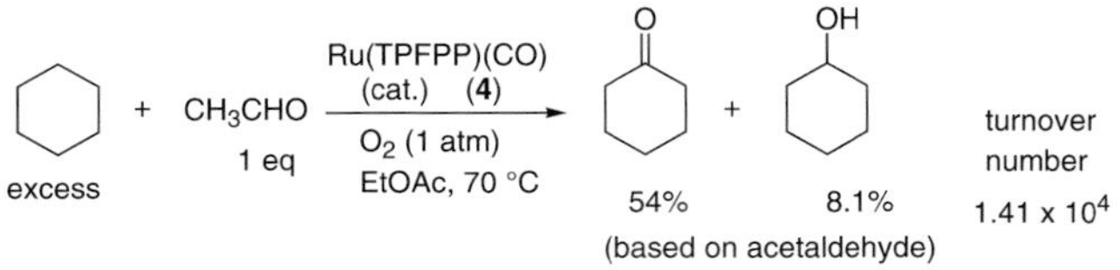

$$\text{cyclohexane (excess)} + CH_3CHO \text{ (1 eq)} \xrightarrow[\text{O}_2 \text{ (1 atm)}]{\text{Ru(TPFPP)(CO) (cat.) (4)}} \text{cyclohexanone (54\%)} + \text{cyclohexanol (8.1\%)} \quad \text{turnover number } 1.41 \times 10^4 \quad (74)$$

(based on acetaldehyde)

These oxidation reactions provide a powerful strategy for the synthesis of cyclohexanone by combination of the Wacker oxidation of ethylene with the present metal-catalyzed oxidation of cyclohexane (Scheme 6.7).

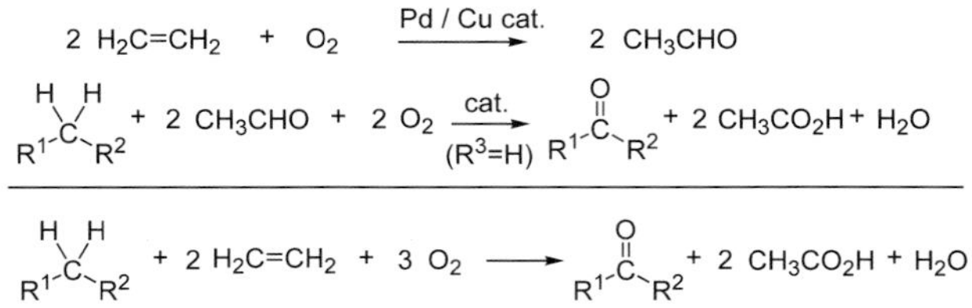

Scheme 6.7

Very few methods for the direct aerobic oxidation of alkanes have been reported using a perfluorinated ruthenium catalyst [Ru$_3$O(OCOCF$_2$CF$_2$CF$_3$)$_6$(Et$_2$O)$_3$]$^+$ [37c] and a ruthenium substituted polyoxometalate [WZnRu$_2$(OH)(H$_2$O)(ZnW$_9$O$_{34}$)$_2$]$^{11-}$ [Eq. (75)] [112, 113].

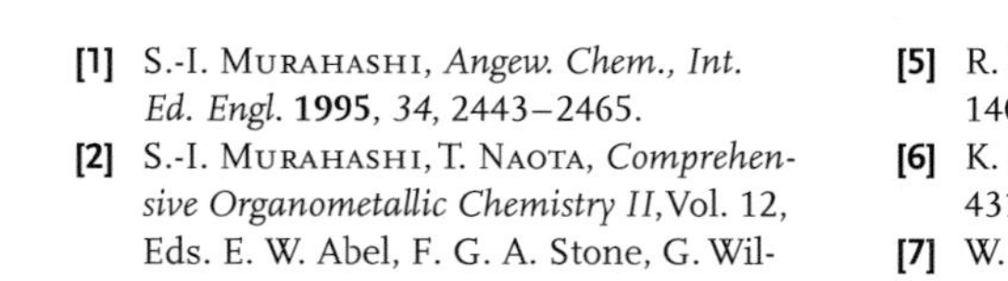

$$\text{adamantane} \xrightarrow[\text{O}_2 \text{ (1 atm)}]{\text{Na}_{11}[\text{WZnRu}_2(\text{OH})(\text{H}_2\text{O})(\text{ZnW}_9\text{O}_{34})_2] \text{ (cat.)}} \text{1-adamantanol} \quad (75)$$

1,2-dichloroethane, 80 °C, 72 h

57% TON = 568

References

[1] S.-I. MURAHASHI, *Angew. Chem., Int. Ed. Engl.* **1995**, *34*, 2443–2465.

[2] S.-I. MURAHASHI, T. NAOTA, *Comprehensive Organometallic Chemistry II*, Vol. 12, Eds. E. W. Abel, F. G. A. Stone, G. Wilkinson, Pergamon, Oxford, **1995**, pp. 1177–1192.

[3] T. NAOTA, H. TAKAYA, S.-I. MURAHASHI, *Chem. Rev.* **1998**, *98*, 2599–2660.

[4] E. A. SEDDON, K. R. SEDDON, *The Chemistry of Ruthenium*, Elsevier, Amsterdam, **1984**.

[5] R. H. HOLM, *Chem. Rev.* **1987**, *87*, 1401–1449.

[6] K. A. JØRGENSEN, *Chem. Rev.* **1989**, *89*, 431–458.

[7] W. P. GRIFFITH, *Chem. Soc. Rev.* **1992**, *21*, 179–185.

[8] (a) D. G. LEE, M. VAN DEN ENGH, *Oxidation in Organic Chemistry*, Ed. W. S. Trahanovski, Academic Press, New York, **1973**, part B, Chapter 4; (b) J. T. COURTNEY, *Organic Synthesis by Oxidation with Metal Compounds*, Eds. W. J. Mijs,

C. R. H. I. de Jonge, Plenum Press, New York, **1984**, Chapter 8; (c) P. H. J. CARLSEN, T. KATSUKI, V. S. MARTIN, K. B. SHARPLESS, *J. Org. Chem.* **1981**, *46*, 3936–3938.

[9] (a) P. E. MORRIS, JR., D. E. KIELY, *J. Org. Chem.* **1987**, *52*, 1149–1152; (b) Y, YAMAMOTO, H. SUZUKI, Y. MORO-OKA, *Tetrahedron Lett.* **1985**, *26*, 2107–2108; (c) S. KANEMOTO, H. TOMIOKA, K. OSHIMA, H. NOZAKI, *Bull. Chem. Soc. Jpn.* **1986**, *59*, 105–108; (d) S. GIDDINGS, A. MILLS, *J. Org. Chem.* **1988**, *53*, 1103–1107; (e) S. TORII, T. INOKUCHI, T. SUGIURA, *J. Org. Chem.* **1986**, *51*, 155–161.

[10] (a) D. G. LEE, T. CHEN, *Comprehensive Organic Synthesis*, Ed. B. M. Trost, I. Fleming, Pergamon Press, Oxford, **1991**, Vol. 7, Chapter 3.8, pp. 541–591; (b) S. TORII, T. INOKUCHI, K. KONDO, *J. Org. Chem.* **1985**, *50*, 4980–4982; (c) F. X. WEBSTER, J. RIVAS-ENTERRIOS, R. M. SILVERSTEIN, *J. Org. Chem.* **1987**, *52*, 689–691; (d) L. ALBARELLA, F. GIORDANO, M. LASALVIA, V. PICCIALLI, D. SICA, *Tetrahedron Lett.* **1995**, *36*, 5267–5270; (e) M. V. DENISENKO, N. D. POKHILO, L. E. ODINOKOVA, V. A. DENISENKO, N. I. UVAROVA, *Tetrahedron Lett.* **1996**, *37*, 5187–5190; (f) H. ORITA, T. HAYAKAWA, K. TAKEHIRA, *Bull. Chem. Soc. Jpn.* **1986**, *59*, 2637–2638; (g) T. K. M. SHING, V. W.-F. TAI, E. K. W. TAM, *Angew. Chem., Int. Ed. Engl.* **1994**, *33*, 2312–2313.

[11] G. BALAVOINE, C. ESKÉNAZI, F. MEUNIER, H. RIVIÈRE, *Tetrahedron Lett.* **1984**, *25*, 3187–3190.

[12] C. ESKÉNAZI, G. BALAVOINE, F. MEUNIER, H. RIVIÈRE, *J. Chem. Soc., Chem. Commun.* **1985**, 1111–1113.

[13] (a) A. J. BAILEY, W. P. GRIFFITH, A. J. P. WHITE, D. J. WILLIAMS, *J. Chem. Soc., Chem. Commun.* **1994**, 1833–1834; (b) A. J. BAILEY, W. P. GRIFFITH, P. D. SAVAGE, *J. Chem. Soc., Dalton Trans.* **1995**, 3537–3542.

[14] C. AUGIER, L. MALARA, V. LAZZERI, B. WAEGELL, *Tetrahedron Lett.* **1995**, *36*, 8775–8778.

[15] (a) F. A. KHAN, J. DASH, N. SAHU, C. SUDHEER, *J. Org. Chem.* **2002**, *67*, 3783–3787; (b) F. A. KHAN, J. DASH, *J. Am. Chem. Soc.* **2002**, *124*, 2424–2425; (c) F. A. KHAN, B. PRABHUDAS, N. SAHU, J. DASH, *J. Am. Chem. Soc.* **2000**, *122*, 9558–9559.

[16] (a) J. A. CAPUTO, R. FUCHS, *Tetrahedron Lett.* **1967**, 4729–4731; (b) D. M. PIATAK, G. HERBST, J. WICHA, E. CASPI, *J. Org. Chem.* **1969**, *34*, 116–120; (c) A. K. CHAKRABORTI, U. R. GHATAK, *Synthesis*, **1983**, 746–748; (d) M. KASAI, H. ZIFFER, *J. Org. Chem.* **1983**, *48*, 2346–2349; (e) U. A. SPITZER, D. G. LEE, *J. Org. Chem.* **1974**, *39*, 2468–2469; (f) M. T. NUÑEZ, V. S. MARTÍN, *J. Org. Chem.* **1990**, *55*, 1928–1932; (g) S. WOLFE, S. K. HASAN, J. R. CAMPBELL, *J. Chem. Soc., Chem. Commun.* **1970**, 1420–1421.

[17] (a) S. TORII, T. INOKUCHI, Y. HIRATA, *Synthesis*, **1987**, 377–379; (b) R. ZIBUCK, D. SEEBACH, *Helv. Chim. Acta* **1988**, *71*, 237–240.

[18] M. LAUX, N. KRAUSE, *Synlett* **1997**, 765–766.

[19] (a) L. GONSALVE, I. W. C. E. ARENDS, R. A. SHELDON, *Chem. Commun.* **2002**, 202–203; (b) P. F. SCHUDA, M. B. CICHOWICZ, M. R. HEIMANN, *Tetrahedron Lett.* **1983**, *24*, 3829–3830.

[20] (a) R. PERRONE, G. BETTONI, V. TORTORELLA, *Synthesis*, **1976**, 598–600; (b) G. BETTONI, G. CARBONARA, C. FRANCHINI, V. TORTORELLA, *Tetrahedron* **1981**, *37*, 4159–4164.

[21] (a) J. C. SHEEHAN, R. W. TULIS, *J. Org. Chem.* **1974**, *39*, 2264–2267; (b) S. YOSHIFUJI, K. TANAKA, Y. NITTA, *Chem. Pharm. Bull.* **1985**, *33*, 1749–1751.

[22] D. RANGANATHAN, N. K. VAISH, K. SHAH, *J. Am. Chem. Soc.* **1994**, *116*, 6545–6557.

[23] (a) A. TENAGLIA, E. TERRANOVA, B. WAEGELL, *Tetrahedron Lett.* **1989**, *30*, 5271–5274; (b) A. TENAGLIA, E. TERRANOVA, B. WAEGELL, *J. Org. Chem.* **1992**, *57*, 5523–5528.

[24] J. M. BAKKE, M. LUNDQUIST, *Acta Chem. Scand. B* **1986**, *40*, 430–433.

[25] (a) J.-L. COUDRET, B. WAEGELL, *Inorg. Chim. Acta* **1994**, *222*, 115–122; (b) J.-L. COUDRET, S, ZÖLLNER, B. J. RAVOO, L. MALARA, C. HANISCH, K. DÖRVE, A. DE MEIJERE, B. WAEGELL, *Tetrahedron*

Lett. **1996**, *37*, 2425–2428; (c) T. Hasegawa, H. Niwa, K. Yamada, *Chem. Lett.* **1985**, 1385–1386.

[26] A. Tenaglia, E. Terranova, B. Waegell, *J. Chem. Soc., Chem. Commun.* **1990**, 1344–1345.

[27] U. A. Spitzer, D. G. Lee, *J. Org. Chem.* **1975**, *40*, 2539–2540.

[28] Y. Sasson, G. D. Zappi, R. Neumann, *J. Org. Chem.* **1986**, *51*, 2880–2883.

[29] (a) *Metalloporphyrins Catalyzed Oxidations*, Eds. F. Montanari, L. Casella, Kluwer Academic Publishers, Dordrecht, **1994**; (b) *Metalloporphyrins in Catalytic Oxidations*, Ed. R. A. Sheldon, Marcel Dekker, New York, **1994**.

[30] (a) T. Leung, B. R. James, D. Dolphin, *Inorg. Chim. Acta* **1983**, *79*, 180–181; (b) D. Dolphin, B. R. James, T. Leung, *Inorg. Chim. Acta* **1983**, *79*, 25–27.

[31] (a) T. Higuchi, H. Ohtake, M. Hirobe, *Tetrahedron Lett.* **1989**, *30*, 6545–6548; (b) H. Ohtake, T. Higuchi, M. Hirobe, *Tetrahedron Lett.* **1992**, *33*, 2521–2524.

[32] (a) J. T. Groves, M. Bonchio, T. Carofiglio, K. Shalyaev, *J. Am. Chem. Soc.* **1996**, *118*, 8961–8962; (b) J. T. Groves, R. Quinn, *J. Am. Chem. Soc.* **1985**, *107*, 5790–5792.

[33] T. Yamada, K. Hashimoto, Y. Kitaichi, K. Suzuki, T. Ikeno, *Chem. Lett.* **2001**, 268–269.

[34] (a) M. Bressan, A. Morvillo, *J. Chem. Soc., Chem. Commun.* **1988**, 650–651; (b) M. Bressan, A. Morvillo, *J. Chem. Soc., Chem. Commun.* **1989**, 421–423; (c) K. Jitsukawa, H. Shiozaki, H. Masuda, *Tetrahedron Lett.* **2002**, *43*, 1491–1494.

[35] (a) W.-C. Cheng, W.-Y. Yu, K.-K. Cheung, C. M. Che, *J. Chem. Soc., Chem. Commun.* **1994**, 1063–1064; (b) T. C. Lau, C.-M. Che, W. O. Lee, C. K. Poon, *J. Chem. Soc., Chem. Commun.* **1988**, 1406–1407; (c) P. K. K. Ho, K.-K. Cheung, C.-M. Che, *Chem. Commun.* **1996**, 1197–1198.

[36] R. Neumann, M. Dahan, *Nature* **1997**, *388*, 353–355.

[37] (a) A. S. Goldstein, R. H. Beer, R. S. Drago, *J. Am. Chem. Soc.* **1994**, *116*, 2424–2429; (b) A. S. Goldstein, R. S. Drago, *J. Chem. Soc., Chem. Commun.* **1991**, 21–22; (c) S. Davis, R. S. Drago, *J. Chem. Soc., Chem. Commun.* **1990**, 250–251; (d) M. H. Robbins, R. S. Drago, *J. Chem. Soc., Dalton Trans.* **1996**, 105–110.

[38] I. Klement, H. Lütjens, P. Knochel, *Angew. Chem., Int. Ed. Engl.* **1997**, *36*, 1454–1456.

[39] Z. Gross, S. Ini, M. Kapon, S. Cohen, *Tetrahedron Lett.* **1996**, *37*, 7325–7328.

[40] A. Berkessel, M. Frauenkron, *J. Chem. Soc., Perkin Trans. 1* **1997**, 2265–2266.

[41] (a) T.-S. Lai, R. Zhang, K.-K. Cheung, H.-L. Kwong, C.-M. Che, *Chem. Commun.* **1998**, 1583–1584; (b) R. Zhang, W.-Y. Yu, H.-Z. Sun, W.-S. Liu, C.-M. Che, *Chem. Eur. J.* **2002**, *8*, 2495–2507.

[42] H. Nishiyama, T. Shimada, H. Itoh, H. Sugiyama, Y. Motoyama, *Chem. Commun.* **1997**, 1863–1864.

[43] (a) R. I. Kureshy, N. H. Khan, S. H. R. Abdi, K. N. Bhatt, *Tetrahedron: Asymmetry* **1993**, *4*, 1693–1701; (b) R. I. Kureshy, N. H. Khan, S. H. R. Abdi, A. K. Bhatt, *J. Mol. Catal. A: Chemical* **1996**, *110*, 33–40.

[44] T. Takeda, R. Irie, Y. Shinoda, T. Katsuki, *Synlett* **1999**, 1157–1159.

[45] (a) D. Ostovic, T. C. Bruice, *Acc. Chem. Res.* **1992**, *25*, 314–320; (b) A. J. Castellino, T. C. Bruice, *J. Am. Chem. Soc.* **1988**, *110*, 158–162.

[46] S.-I. Murahashi, T. Saito, H. Hanaoka, Y. Murakami, T. Naota, H. Kumobayashi, S. Akutagawa, *J. Org. Chem.* **1993**, *58*, 2929–2930.

[47] Y. Horiguchi, E. Nakamura, I. Kuwajima, *J. Am. Chem. Soc.* **1989**, *111*, 6257–6265.

[48] D. H. Peterson, S. H. Eppstein, P. D. Meister, B. J. Magerlein, H. C. Murray, H. M. Leigh, A. Weintraub, L. M. Reineke, *J. Am. Chem. Soc.* **1953**, *75*, 412–415.

[49] T. Hotopp, H.-J. Gutke, S.-I. Murahashi, *Tetrahedron Lett.* **2001**, *42*, 3343–3346.

[50] U. Beifuss, A. Herde, *Tetrahedron Lett.* **1998**, *39*, 7691–7692.

[51] K. B. Sharpless, K. Akashi, K. Oshima, *Tetrahedron Lett.* **1976**, 2503–2506.

[52] P. Müller, J. Godoy, *Tetrahedron Lett.* **1981**, *22*, 2361–2364.

[53] S. Kanemoto, K. Oshima, S. Matsubara, K. Takai, H. Nozaki, *Tetrahedron Lett.* **1983**, *24*, 2185–2088.
[54] (a) G. Barak, J. Dakka, Y. Sasson, *J. Org. Chem.* **1988**, *53*, 3553–3555; (b) G. Rothenberg, G. Barak, Y. Sasson, *Tetrahedron* **1999**, *55*, 6301–6310.
[55] K. S. Kim, S. J. Kim, Y. H. Song, C. S. Hahn, *Synthesis* **1987**, 1017–1018.
[56] S. Iwasa, K. Morita, K. Tajima, A. Fakhruddin, H. Nishiyama, *Chem. Lett.* **2002**, 284–285.
[57] (a) W. P. Griffith, S. V. Ley, G. P. Whitcombe, A. D. White, *J. Chem. Soc., Chem. Commun.* **1987**, 1625–1627; (b) An excellent review, S. V. Ley, J. Norman, W. P. Griffith, S. P. Marsden, *Synthesis* **1994**, 639–666, and references cited therein.
[58] L. Gonsalvi, I. W. C. E. Arends, R. A. Sheldon, *Org. Lett.* **2002**, *4*, 1659–1661.
[59] (a) S.-I. Murahashi, T. Naota, T. Nakajima, *Tetrahedron Lett.* **1985**, *26*, 925–928; (b) S.-I. Murahashi, T. Naota, *Synthesis* **1993**, 433–440; (c) S.-I. Murahashi, T. Naota, *Zh. Org. Khim.* **1996**, *32*, 223–232; (d) S.-I. Murahashi, T. Naota, *Reviews on Heteroatom Chemistry*, Vol. 1, MYU Research, Tokyo, **1988**, pp. 257–276; (e) S.-I. Murahashi, T. Naota, *Advances in Metal-Organic Chemistry*, Vol. 3, Ed. L. S. Liebeskind, JAI Press, London, **1994**, pp. 225–254.
[60] (a) Y. Tsuji, T. Ohta, T. Ido, H. Minbu, Y. Watanabe, *J. Organomet. Chem.* **1984**, *270*, 333–341; (b) M. Tanaka, T. Kobayashi, T. Sakakura, *Angew. Chem., Int. Ed. Engl.* **1984**, *23*, 518–518.
[61] S.-I. Murahashi, T. Naota, N. Nakajima, *Chem. Lett.* **1987**, 879–882.
[62] W.-H. Fung, W.-Y. Yu, C.-M. Che, *J. Org. Chem.* **1998**, *63*, 2873–2877.
[63] S.-I. Murahashi, T. Naota, Y. Oda, N. Hirai, *Synlett* **1995**, 733–734.
[64] S.-I. Murahashi, T. Naota, N. Hirai, *J. Org. Chem.* **1993**, *58*, 7318–7319.
[65] S.-I. Murahashi, T. Naota, K. Ito, Y. Maeda, H. Taki, *J. Org. Chem.* **1987**, *52*, 4319–4327.
[66] (a) J.-E. Bäckvall, R. L. Chowdhury, U. Karlsson, *J. Chem. Soc., Chem. Commun.* **1991**, 473–475; (b) G.-Z. Wang, U. Andreasson, J.-E. Bäckvall, *J. Chem. Soc., Chem. Commun.* **1994**, 1037–1038; (c) G. Csjernyik, A. H. Éll, L. Fadini, B. Pugin, J.-E. Bäckvall, *J. Org. Chem.* **2002**, *67*, 1657–1662.
[67] A. Hanyu, E. Takezawa, S. Sakaguchi, Y. Ishii, *Tetrahedron Lett.* **1998**, *39*, 5557–5560.
[68] T. Inokuchi, K. Nakagawa, S. Torii, *Tetrahedron Lett.* **1995**, *36*, 3223–3226.
[69] (a) A. Dijksman, I. W. C. E. Arends, R. A. Sheldon, *Chem. Commun.* **1999**, 1591–1592; (b) A. Dijksman, A. Marino-González, A. M. i Payeras, I. W. C. E. Arends, R. A. Sheldon, *J. Am. Chem. Soc.* **2001**, *123*, 6826–6833; (c) R. A. Sheldon, I. W. C. E. Arends, G. J. ten Brink, A. Dijksman, *Acc. Chem. Res.*, **2002**, *35*, 774–781.
[70] M. Matsumoto, N. Watanabe, *J. Org. Chem.* **1984**, *49*, 3435–3436.
[71] I. E. Markó, P. R. Giles, M. Tsukazaki, I. Chellé-Regnaut, C. J. Urch, S. M. Brown, *J. Am. Chem. Soc.* **1997**, *119*, 12661–12662.
[72] (a) B. Hinzen, R. Lenz, S. V. Ley, *Synthesis* **1998**, 977–979; (b) A. Bleloch, B. F. G. Johnson, S. V. Ley, A. J. Price, D. S. Shephard, A. W. Thomas, *Chem. Commun.* **1999**, 1907–1908.
[73] (a) K. Kaneda, T. Yamashita, T. Matsushita, K. Ebitani, *J. Org. Chem.* **1998**, *63*, 1750–1751; (b) T. Matsushita, K. Ebitani, K. Kaneda, *Chem. Commun.* **1999**, 265–266; (c) K. Yamaguchi, K. Mori, T. Mizugaki, K. Ebitani, K. Kaneda, *J. Am. Chem. Soc.* **2000**, *122*, 7144–7145; (d) H. Ji, T. Mizugaki, K. Ebitani, K. Kaneda, *Tetrahedron Lett.* **2002**, *43*, 7179–7183; (e) K. Mori, M. Tano, T. Mizugaki, K. Ebitani, K. Kaneda, *New. J. Chem.* **2002**, *26*, 1536–1538.
[74] K. Yamaguchi, N. Mizuno, *Angew. Chem., Int. Ed. Engl.* **2002**, *41*, 4538–4541.
[75] (a) T. Hosokawa, S.-I. Murahashi, *Acc. Chem. Res.* **1990**, *23*, 49–54; (b) T. Hosokawa, T. Uno, S. Inui, S.-I. Murahashi, *J. Am. Chem. Soc.* **1981**, *103*, 2318–2323.
[76] E. Takezawa, S. Sakaguchi, Y. Ishii, *Org. Lett.* **1999**, *1*, 713–715.
[77] A. Wolfson, S. Wuyts, D. E. De Vos,

I. F. J. Vankelecom, P. A. Jacobs, *Tetrahedron Lett.* **2002**, *43*, 8107–8110.

[78] (a) K. Masutani, T. Uchida, R. Irie, T. Katsuki, *Tetrahedron Lett.* **2000**, *41*, 5119–5123; (b) A. Miyata, M. Murakami, R. Irie, T. Katsuki, *Tetrahedron Lett.* **2001**, *42*, 7067–7070.

[79] S.-I. Murahashi, T. Naota, K. Yonemura, *J. Am. Chem. Soc.* **1988**, *110*, 8256–8258.

[80] S.-I. Murahashi, T. Naota, N. Miyaguchi, T. Nakato, *Tetrahedron Lett.* **1992**, *33*, 6991–6994.

[81] (a) D. P. Riley, *J. Chem. Soc., Chem. Commun.* **1983**, 1530–1532; (b) S. L. Jain, B. Sain, *Chem. Commun.* **2002**, 1040–1041.

[82] S.-I. Murahashi, T. Naota, H. Taki, *J. Chem. Soc., Chem. Commun.* **1985**, 613–614.

[83] (a) H. Mitsui, S. Zenki, T. Shiota, S.-I. Murahashi, *J. Chem. Soc., Chem. Commun.* **1984**, 874–875; (b) S.-I. Murahashi, H. Mitsui, T. Shiota, T. Tsuda, S. Watanabe, *J. Org. Chem.* **1990**, *55*, 1736–1744; (c) S.-I. Murahashi, T. Shiota, Y. Imada, *Org. Synth.* **1991**, *70*, 265–271; (d) S.-I. Murahashi, Y. Imada, H. Ohtake, *J. Org. Chem.* **1994**, *59*, 6170–6172; (e) S.-I. Murahashi, T. Shiota, *Tetrahedron Lett.* **1987**, *28*, 2383–2386.

[84] (a) T. Kawakami, H. Ohtake, H. Arakawa, T. Okachi, Y. Imada, S.-I. Murahashi, *Bull. Chem. Soc. Jpn.*, **2000**, *73*, 2423–2444; (b) S.-I. Murahashi, Y. Imada, T. Kawakami, K. Harada, Y. Yonemushi, T. Tomita, *J. Am. Chem. Soc.* **2002**, *124*, 2888–2889; (c) S.-I. Murahashi, T. Tsuji, S. Ito, *Chem. Commun.*, **2000**, 409–410.

[85] (a) A. Goti, M. Romani, *Tetrahedron Lett.* **1994**, *35*, 6567–6570; (b) A. Goti, F. De Sarlo, M. Romani, *Tetrahedron Lett.* **1994**, *35*, 6751–6574; (c) A. H. Éll, J. S. M. Samec, C. Brasse, J. E. Bäckvall. *Chem. Comm.* **2002**, 1144–1145.

[86] M. Schröder, W. P. Griffith, *J. Chem. Soc., Chem. Commun.* **1979**, 58–59.

[87] (a) A. J. Bailey, B. R. James, *Chem. Commun.* **1996**, 2343–2344; (b) S. Y. S. Cheng, N. Rajapakse, S. J. Rettig, B. R. James, *J. Chem. Soc., Chem. Commun.* **1994**, 2669–2670.

[88] (a) K. Mori, K. Yamaguchi, T. Mizugaki, K. Ebitani, K. Kaneda, *Chem. Commun.* **2001**, 461–462; (b) K. Yamaguch, N. Mizuno, *Angew. Chem., Int. Ed. Engl.* **2003**, *42*, 1480–1483; (c) K. Yamaguch, N. Mizuno, *Chem. Eur. J.* **2003**, *9*, 4353–4361.

[89] S.-I. Murahashi, T. Naota, T. Kuwabara, T. Saito, H. Kumobayashi, S. Akutagawa, *J. Am. Chem. Soc.* **1990**, *112*, 7820–7822.

[90] T. Naota, T. Nakato, S.-I. Murahashi, *Tetrahedron Lett.* **1990**, *31*, 7475–7478.

[91] S.-I. Murahashi, A. Mitani, K. Kitao, *Tetrahedron Lett.* **2000**, *41*, 10245–10249.

[92] G. Cainelli, M. DaCol, P. Galletti, D. Giacomini, *Synlett* **1997**, 923–924.

[93] S.-I. Murahashi, T. Saito, T. Naota, H. Kumobayashi, S. Akutagawa, *Tetrahedron Lett.* **1991**, *32*, 5991–5994.

[94] (a) K. U. Kingold, *Free Radicals*, Vol. 1, Ed. Kochi J. K., John Wiley & Sons, New York, **1973**, pp. 37–112; (b) A. Brovo, F. Fontana, F. Minisci, *Chem. Lett.* **1996**, 401–402.

[95] M. Shimizu, H. Orita, T. Hayakawa, Y. Watanabe, K. Takehira, *Bull. Chem. Soc. Jpn.* **1991**, *64*, 2583–2584.

[96] S.-I. Murahashi, T. Naota, N. Miyaguchi, S. Noda, *J. Am. Chem. Soc.* **1996**, *118*, 2509–2510.

[97] M. Matsumoto, K. Kuroda, *J. Am. Chem. Soc.* **1982**, *104*, 1433–1434.

[98] T. Higuchi, C. Satake, M. Hirobe, *J. Am. Chem. Soc.* **1995**, *117*, 8879–8880.

[99] (a) H. Ohtake, T. Higuchi, M. Hirobe, *J. Am. Chem. Soc.* **1992**, *114*, 10660–10662. (b) T. Shingaki, K. Miura, T. Higuchi, M. Hirobe, T. Nagano, *Chem. Commun.* **1997**, 861–862.

[100] K. J. Balkus, Jr., M. Eissa, R. Levado, *J. Am. Chem. Soc.* **1995**, *117*, 10753–10754.

[101] (a) T.-C. Lau, C.-K. Mak, *J. Chem. Soc., Chem. Commun.* **1995**, 943–944; (b) T.-C. Lau, C.-K. Mak, *J. Chem. Soc., Chem. Commun.* **1993**, 766–767.

[102] R. Zhang, W.-Y. Yu, T.-S. Lai, C.-M. Che, *Chem. Commun.* **1999**, 1791–1792.

[103] (a) T. Kojima, *Chem. Lett.* **1996**, 121–

122; (b) T. Kojima, Y. Matsuda, *Chem. Lett.* **1999**, 81–82.

[104] (a) M. Yamaguchi, H. Kousaka, T. Yamagishi, *Chem. Lett.* **1997**, 769–770; (b) M. Yamaguchi, Y. Ichii, S. Kosaka, D. Masui, T. Yamagishi, *Chem. Lett.* **2002**, 434–435.

[105] (a) R. Neumann. C. Abu-Gnim, *J. Chem. Soc., Chem. Commun.* **1989**, 1324–1325; (b) Y. Matsumoto, M. Asami, M. Hashimoto, M. Misono, *J. Mol. Catal. A: Chemical* **1996**, *114*, 161–168.

[106] (a) S.-I. Murahashi, N. Komiya, Y. Oda, T. Kuwabara, T. Naota, *J. Org. Chem.* **2000**, *65*, 9186–9193; (b) S.-I. Murahashi, Y. Oda, T. Naota, T. Kuwabara, *Tetrahedron Lett.* **1993**, *34*, 1299–1302; (c) S.-I. Murahashi, Y. Oda, N. Komiya, T. Naota, *Tetrahedron Lett.* **1994**, *35*, 7953–7956.

[107] S.-I. Murahashi, T. Naota, T. Kuwabara, *Synlett* **1989**, 62–63.

[108] R. A. Miller, W. Li, G. R. Humphrey, *Tetrahedron Lett.* **1996**, *37*, 3429–3432.

[109] (a) S.-I. Murahashi, Y. Oda, T. Naota, *J. Am. Chem. Soc.* **1992**, *114*, 7913–7914; (b) S.-I. Murahashi, X.-G. Zhou, N. Komiya, *Synlett* **2003**, 321–324.

[110] S.-I. Murahashi, T. Naota, N. Komiya, *Tetrahedron Lett.* **1995**, *36*, 8059–8062.

[111] (a) S.-I. Murahashi, Y. Oda, T. Naota, N. Komiya, *J. Chem. Soc., Chem. Commun.* **1993**, 139–140; (b) N. Komiya, T. Naota, S.-I. Murahashi, *Tetrahedron Lett.* **1996**, *37*, 1633–1636; (c) N. Komiya, T. Naota, Y. Oda, S.-I. Murahashi, *J. Mol. Catal. A: Chemical* **1997**, *117*, 21–37.

[112] R. Neumann, A. M. Khenkin, M. Dahan, *Angew. Chem., Int. Ed. Engl.* **1995**, *34*, 1587–1589.

[113] R. Neumann, M. Dahan, *J. Am. Chem. Soc.* **1998**, *120*, 11969–11976.

7
Selective Oxidation of Amines and Sulfides
Jan-E. Bäckvall

7.1
Introduction

Heteroatom oxidation is of great importance in organic synthesis, and among such reactions oxidations of amines and sulfides are the most common. Amines and sulfides can be oxidized to a number of different products and various reagents have been developed for these transformations. This chapter will deal with selective oxidations of sulfides (thioethers) to sulfoxides and of tertiary amines to N-oxides.

7.2
Oxidation of Sulfides to Sulfoxides

The oxidation of sulfides has been reviewed previously [1–3]. Organosulfur compounds, such as sulfoxides and sulfones, are useful synthetic reagents in organic chemistry. In particular, sulfoxides are important intermediates in the synthesis of natural products and biologically significant molecules [4] and they have also been employed as ligands in asymmetric catalysis [5] and as oxotransfer reagents [6]. The synthesis and utilization of chiral sulfoxides was reviewed recently [5a]. A large number of methods are available for the oxidation of sulfides to sulfoxides and an important issue is to obtain high selectivity for sulfoxide over sulfone. Usually there is a reasonably good selectivity for the oxidation to sulfoxide, since the sulfide is much more nucleophilic than the sulfoxide and hence reacts faster with the electrophilic reagent/catalyst. However, there are large variations between the different oxidation systems.

In this chapter, oxidations of sulfides to sulfoxides have been divided into three sub-sections: (1) stoichiometric reactions, (2) chemocatalytic reactions, and (3) biocatalytic reactions.

Modern Oxidation Methods. Edited by Jan-Erling Bäckvall
Copyright © 2004 WILEY-VCH Verlag GmbH & Co. KGaA, Weinheim
ISBN: 3-527-30642-0

7.2.1
Stoichiometric Reactions

A large number of methods are available in the literature for the oxidation of sulfides (thioethers) to sulfoxides by electrophilic reagents. These include the use of peracids [7], $NaIO_4$, MnO_2, CrO_3, SeO_2 and iodosobenzene. The use of these reagents up to 1989 has been reviewed [3]. Also, hydrogen peroxide can be used as a direct oxidant, although this reaction is slow in the absence of a catalyst. The use of this oxidant in catalytic reactions will be discussed below in Section 7.2.2.

7.2.1.1 Peracids
Peracids are commonly used oxidants in the oxidation of sulfides to sulfoxides [7]. The reaction proceeds with a good rate at room temperature and gives the sulfoxide together with some sulfone. The selectivity for sulfoxide over sulfone is usually sufficient for synthetic purposes since the oxidation of sulfoxide is considerably slower than the sulfide oxidation.

7.2.1.2 Dioxiranes
Dioxiranes have been successfully used as oxidants for the selective oxidation of sulfides to sulfoxides [Eq. (1)] [8].

$$R^1\text{-}S\text{-}R^2 + \underset{\text{O-O}}{\bigtriangledown} \longrightarrow R^1\text{-}S(\text{=O})\text{-}R^2 + \text{acetone} \tag{1}$$

These reactions are rapid and the sulfide is efficiently oxidized to sulfoxide as the only product with no over-oxidation to sulfone. The oxirane reaction [Eq. (1)] is thought to proceed *via* a direct oxygen transfer from the oxirane to the sulfide. In a recent mechanistic study [9] it was found that a hypervalent sulfurane is an intermediate in the oxidation of sulfides by dioxiranes. This intermediate is in equilibrium with the electrophilic zwitterionic intermediate formed as a result of electrophilic attack by the peroxide on the sulfide.

The oxirane oxidation of sulfides has found applications in organic synthesis [10, 11]. For example, sulfoxidation of disulfide **1** with dioxirane afforded disulfoxide **2** in 98% yield [Eq. (2)] [10].

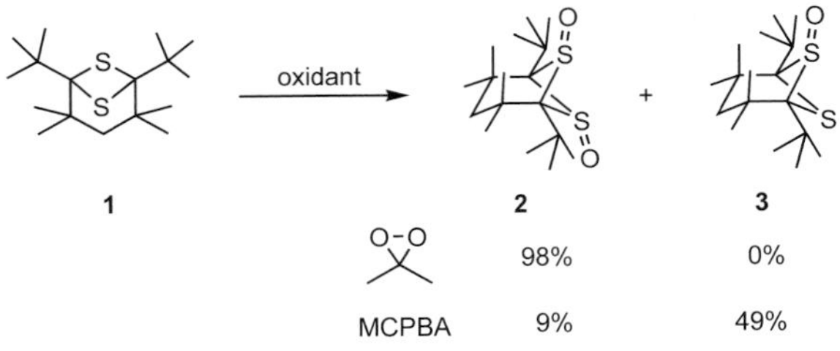

$$\mathbf{1} \xrightarrow{\text{oxidant}} \mathbf{2} + \mathbf{3} \tag{2}$$

	2	3
O-O dioxirane	98%	0%
MCPBA	9%	49%

The corresponding sulfoxidation of **1** with MCPBA was unsuccessful and gave only 9% of disulfoxide **2** together with substantial amounts of degraded dicarbonyl compound (29%) and monosulfoxide **3** (49%). Other applications of sulfoxidation with the use of dioxirane are given in ref. [11].

7.2.1.3 Oxone and Derivatives

Oxone, which is commercially available as a 2:1:1 mixture of $KHSO_5$, $KHSO_4$ and K_2SO_4, has been used for the oxidation of sulfides to sulfoxides [12]. It shows a tendency for over-oxidation and was originally used for the oxidation of sulfides to sulfones [13]. More recently, some improvements were obtained with surface-mediated Oxone oxidations when the Oxone is bound to a silica gel surface [Eq. (3)]. However, some sulfone was still formed.

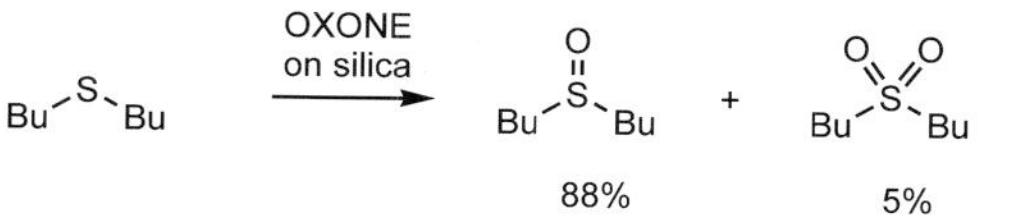

$$Bu\text{-}S\text{-}Bu \xrightarrow{\text{OXONE on silica}} Bu\text{-}S(=O)\text{-}Bu \; (88\%) + Bu\text{-}S(=O)_2\text{-}Bu \; (5\%) \quad (3)$$

In a recent modification to the Oxone type oxidation, the quaternary salt benzyltriphenyl phosphonium peroxymonosulfate $PhCH_2Ph_3P^+HSO_5^-$ was employed [Eq. (4)]. No over-oxidation to sulfone was detected, according to the authors.

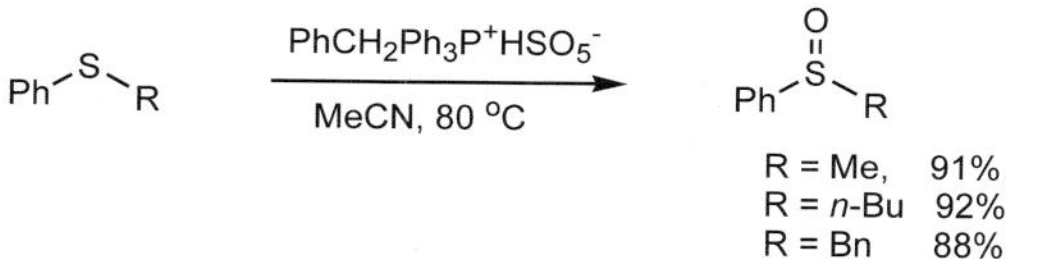

$$Ph\text{-}S\text{-}R \xrightarrow[\text{MeCN, 80 °C}]{PhCH_2Ph_3P^+HSO_5^-} Ph\text{-}S(=O)\text{-}R \quad (4)$$

R = Me, 91%
R = n-Bu 92%
R = Bn 88%

7.2.1.4 H_2O_2 in "Fluorous Phase"

Oxidation of sulfides to sulfoxides by H_2O_2 in hexafluoro-2-propanol has been reported to occur with an exceptionally high rate and selectivity [14]. Reaction of ethyl phenyl sulfide with 2 equiv. of 30% aqueous H_2O_2 in hexafluoro-2-propanol at 25 °C was complete within 5 min and gave the corresponding sulfoxide in 97% yield. No sulfone was formed and in a control experiment the sulfoxide was stirred with 2 equiv. of 30% H_2O_2 for 8 h without any sulfone being formed. The normally slow-reacting sulfides diphenyl sulfide (**4**), **5**, and **6**, reacted fast and the procedure tolerates double bonds (Table 7.1). This seems to be an excellent method for the efficient and highly selective oxidation of sulfides (thioethers) to the corresponding sulfoxides. Sulfides having double bonds also underwent a selective sulfoxidation (e.g., **7** and **8**, entries 4 and 5).

7 Selective Oxidation of Amines and Sulfides

Tab. 7.1 Hydrogen peroxide oxidation of sulfides to sulfoxides in a "fluorous phase"

Entry	Substrate	rxn time (min)	Yield of sulfoxide (%)
1	Ph-S-Ph (**4**)	5	99
2	4-Me-C$_6$H$_4$-S-CH$_2$-(2-pyridyl) (**5**)	10	93
3	tBu-S-tBu (**6**)	20	97
4	Ph-S-CH$_2$CH=CH$_2$ (**7**)	5	99
5	Ph-S-CH$_2$C≡CH (**8**)	15	94

7.2.2
Chemocatalytic Reactions

In almost all catalytic reactions reported for the oxidation of sulfides to sulfoxides, a peroxide compound (usually H_2O_2) or molecular oxygen is employed. The most common oxidant is H_2O_2, usually as a 30% aqueous solution.

7.2.2.1 H_2O_2 as Terminal Oxidant

A large number of catalysts have been reported in the literature for the H_2O_2 oxidation of sulfides to sulfoxides. These include various metal complexes (of transition metals and lanthanides), flavins and benzthiazoles.

Transition metals as catalysts

Oxidation of various sulfides to sulfones by H_2O_2 mediated by $TiCl_3$ was reported by Oae and coworkers [15]. Quantitative yields of sulfoxide were obtained within 5–15 min with no over-oxidation to sulfone. However, a 7-fold excess of hydrogen peroxide and 2 equiv. of $TiCl_3$ per mole of substrate were employed.

Tungsten-based catalytic systems for H_2O_2 oxidations of sulfides have attracted considerable interest and some early reports include the use of H_2WO_4 [16]. More recently, various tungsten-catalyzed methods have been used [17–21].

The Venturello-type peroxo complex $Q_3\{(PO_4)[W(O)(O_2)_2]_4\}$, with $Q = N\text{-}(n\text{-}C_{16}H_{33})$ pyridinium, was employed as the catalyst for the oxidation of sulfides to sulfoxides and sulfones by hydrogen peroxide [17]. The selectivity for sulfoxides was low with this catalyst, which gave only sulfone. The corresponding molybdenum complex $Q_3\{(PO_4)[M(O)(O_2)_2]_4\}$ and $Q_3PMo_{12}O_{40}$ as catalyst in the H_2O_2 oxidations gave mixtures of sulfoxide and sulfone that ranged from 3:1 to 1:3 depending on the substrate. Finally, $Q_3PW_{12}O_{40}$ as the catalyst in the H_2O_2 oxidations gave good selectivity [17].

Related peroxo-tungstates and -molybdates $(Ph_2PO_2)[MO(O_2)_2]_2^-$ (M = W or Mo) were studied as catalysts for H_2O_2 oxidation of sulfoxides. The selectivity for sulfoxide was low for these catalysts [18].

Polyoxymetalates $WZnMn_2(ZnW_9O_{34})_2^{2-}$ were employed to oxidize sulfides to sulfoxides in moderate selectivity (85–90% selectivity with 10–15% of sulfone) [19]. The catalytic effect was strong for the oxidation of aromatic sulfides but weak for the oxidation of aliphatic sulfides.

Noyori and coworkers have recently reported on a tungsten-based halogen-free system for the oxidation to sulfoxides with 30% aqueous H_2O_2, which gives sulfoxides with good to moderate selectivity [20]. The reactions were run without organic solvent and the catalyst employed was Na_2WO_4 together with $PhPO_3H_2$ and $CH_3(n\text{-}C_8H_{17})_3NHSO_4$. For example, thioanisole gave sulfoxide and sulfone in a ratio of 94:6 after 9 h at 0 °C with the use of a substrate to catalyst ratio of 1000:1.

Choudary et al. [21] subsequently extended the tungsten system to the use of layered double hydroxides (LDHs) with water as the solvent and 30% H_2O_2 as the oxidant. Under these conditions the new catalysts $LDH\text{-}WO_4^{2-}$ gave good turnover rates; however the selectivity of the oxidation of thioanisole was moderate with a sulfoxide:sulfone ratio of 88:12 [Eq. (5)]. Other sulfide oxidations also gave a moderate selectivity [Eq. (5)].

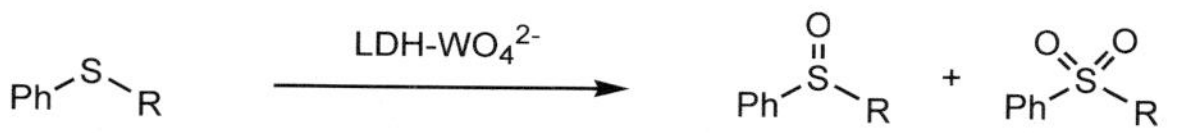

R = Me	94 : 6	ref 20
R = Me	88 : 12	ref 21
R = vinyl	71 : 29	ref 21

(5)

The advantage with the latter system, in spite of the moderate selectivity for sulfoxide over sulfone, is that the immobilized catalyst can be recovered and reused. The catalyst showed consistent activity and selectivity for six recyclings.

The catalytic cycle with the WO_4^{2-} catalysts is thought to involve tungsten peroxy complexes (Scheme 7.1). The tungsten peroxy complexes generated react fast with the sulfide with transfer of an oxygen to give the sulfoxide.

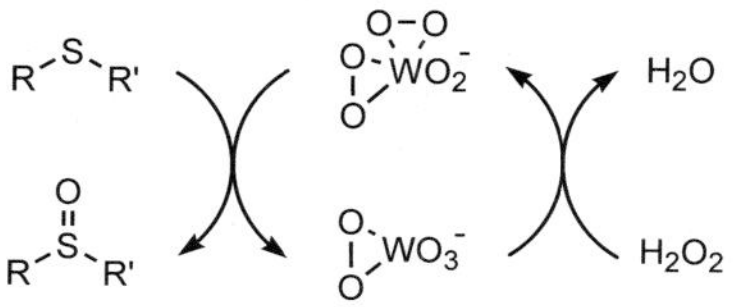

Scheme 7.1

Feringa investigated various nitrogen ligands for the selective Mn-catalyzed oxidation of sulfides to sulfoxides with 30% aqueous H_2O_2 [22]. The use of $Mn(OAc)_3 \cdot 2\,H_2O$ with bipyridine ligand **9** in the oxidation of thioanisole gave sulfoxide free from sulfone in 55% yield. With ligand **10** (Scheme 7.2) the same yield was obtained and now with the sulfoxide in 18% ee.

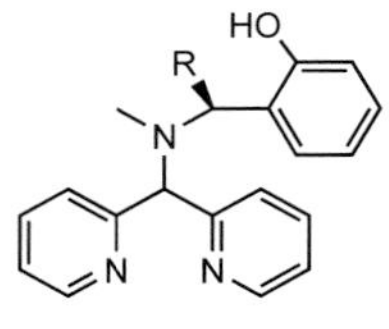

9 R = H
10 R = Me

Scheme 7.2

Titanium-catalyzed oxidations with 35% aqueous H_2O_2 using Schiff-base (salen)titanium oxo complexes as catalysts showed a very high activity [23]. The oxidation of methyl phenyl sulfide required only 0.1 mol% catalyst. The use of chiral salen complexes gave a low enantioselectivity (<20% ee).

Vanadium-catalyzed H_2O_2 oxidations of sulfides to sulfoxides have been reported by several groups. These reactions have been shown to work well for asymmetric sulfoxidation. In 1995 Bolm and Bienewald reported on the use of vanadium–chiral Schiff base complexes as catalysts for asymmetric sulfoxidation by 30% H_2O_2 [24]. Oxidation of thioanisole using ligand **11** afforded the corresponding sulfoxide in 70% ee in high yield without any significant over-oxidation to sulfone [Eq. (6)].

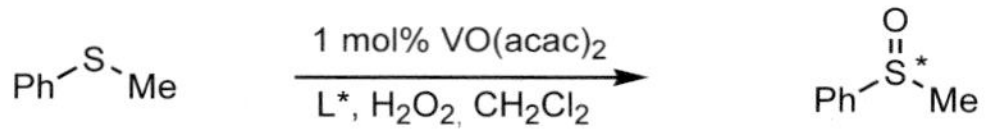

$$Ph\text{-}S\text{-}Me \xrightarrow[\text{L*, }H_2O_2,\text{ }CH_2Cl_2]{1 \text{ mol\% VO(acac)}_2} Ph\text{-}S(\text{=O})\text{-}Me \quad (6)$$

L* = **11** 94% (70% ee) [24]
L* = **12** 92% (78% ee) [25]
L* = **13** 83% (83% ee) [26]

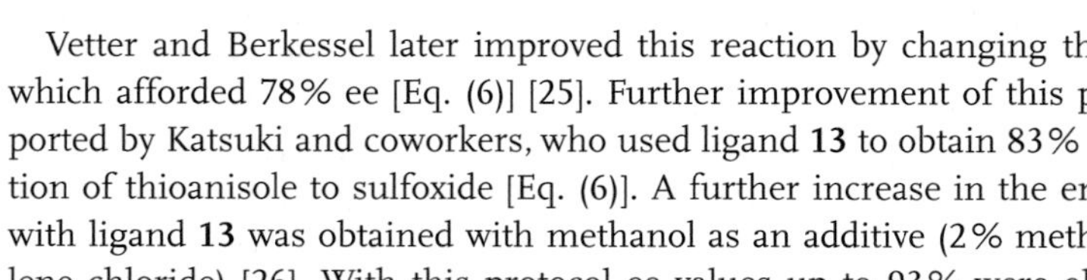

11 **12** **13**

Vetter and Berkessel later improved this reaction by changing the ligand to **12**, which afforded 78% ee [Eq. (6)] [25]. Further improvement of this protocol was reported by Katsuki and coworkers, who used ligand **13** to obtain 83% ee in the oxidation of thioanisole to sulfoxide [Eq. (6)]. A further increase in the enantioselectivity with ligand **13** was obtained with methanol as an additive (2% methanol in methylene chloride) [26]. With this protocol ee values up to 93% were obtained for aryl methyl sulfides [Eq. (7)]. In all of the reactions, except for Ar = p-$NO_2C_6H_4$, only traces of sulfone were formed. The latter substrate gave 10% sulfone.

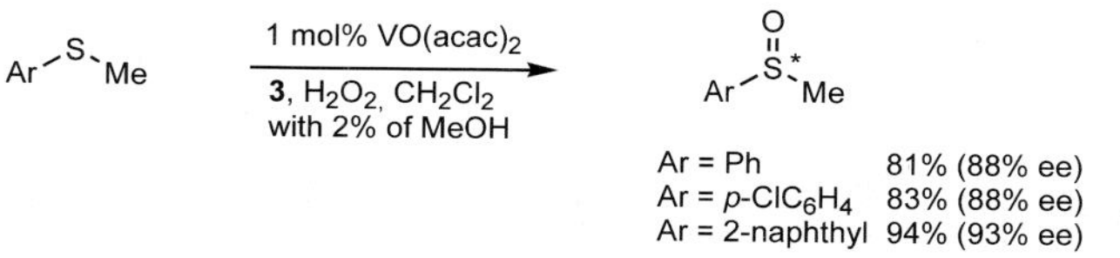

7.2 Oxidation of Sulfides to Sulfoxides

$$\text{Ar}-\text{S}-\text{Me} \xrightarrow[\text{3, H}_2\text{O}_2\text{, CH}_2\text{Cl}_2 \text{ with 2\% of MeOH}]{1 \text{ mol\% VO(acac)}_2} \text{Ar}-\overset{\overset{O}{\|}}{\underset{*}{S}}-\text{Me} \quad (7)$$

Ar = Ph   81% (88% ee)
Ar = p-ClC$_6$H$_4$ 83% (88% ee)
Ar = 2-naphthyl 94% (93% ee)

Jackson and coworkers [27] developed immobilized Schiff-base ligands inspired by those used by Bolm. A peptide Schiff-base library with ligands **14** bound to a solid support was investigated, where two amino acids (AA1 and AA2) and the salicylaldehyde were varied. A library of 72 ligands was prepared using six different salicylaldehydes, six different amino acids as amino acid 1 (AA1) and two different amino acids as amino acid 2 (AA2). Screening of these ligands in the VO(acac)$_2$-catalyzed H$_2$O$_2$ oxidation of sulfides in CH$_2$Cl$_2$ gave only a moderate enantioselectivity of 11% for thioanisole with the best ligand (R^1 = Ph, R^2 = H). Screening the ligands with Ti(O^{i}Pr)$_4$ as catalyst gave a better result and thioanisole afforded the sulfoxide in quantitative yield in 64% ee with ligand **15**. The best result with this ligand was obtained with 2-naphthyl methyl sulfide, which gave 72% ee in a high yield [Eq. (8)].

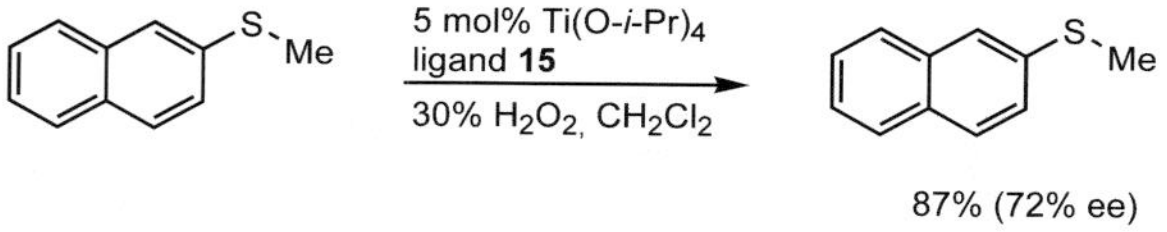

$$\text{2-Naphthyl-S-Me} \xrightarrow[\text{30\% H}_2\text{O}_2\text{, CH}_2\text{Cl}_2]{5 \text{ mol\% Ti(O-}i\text{-Pr)}_4 \text{, ligand 15}} \text{2-Naphthyl-S-Me} \quad (8)$$

87% (72% ee)

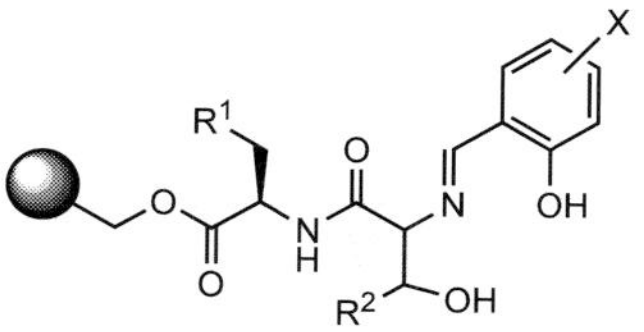

14 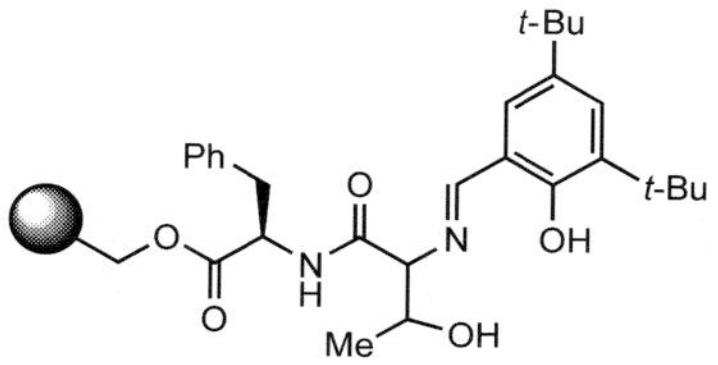   **15**

Chiral salen(Mn^{III}) complexes have been used as catalysts in the oxidation of sulfides to sulfoxides by 30% aqueous H$_2$O$_2$ in acetonitrile [28]. The use of 2–3 mol% of catalyst led to an efficient reaction with enantioselectivities of up to 68% ee.

Oxidation of sulfides in the presence of electron-rich double bonds is problematic with many of the traditional oxidants such as MCPBA, NaIO$_4$ and Oxone, due to interference with double bond oxidation (e.g., epoxidation). Koo and coworkers [29] addressed this problem and studied the selective oxidation of allylic sulfides having electron-rich double bonds. They tested various stoichiometric oxidants and a number of catalytic reactions with 30% aqueous H$_2$O$_2$ as the oxidant. Of all the oxidation systems tested for the sulfoxidation, they found that the use of LiNbMoO$_6$ as the catalyst with H$_2$O$_2$ as the oxidant gave the best result. With this system no epoxidation took place and a reasonably good selectivity for sulfoxide over sulfone was obtained (Table 7.2).

Tab. 7.2 Selective oxidation of allylic sulfides with electron-rich double bonds

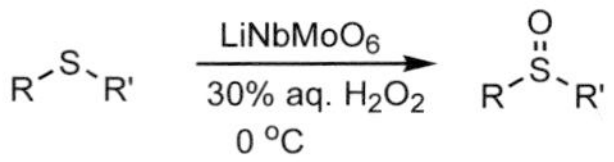

$$R\text{-}S\text{-}R' \xrightarrow[0\,°C]{\text{LiNbMoO}_6,\ 30\%\ \text{aq.}\ H_2O_2} R\text{-}S(O)\text{-}R'$$

Entry	Sulfide	Yield (%) Sulfoxide	Sulfone
1	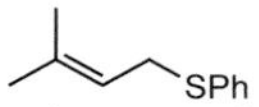 ⟋⟍⟋SPh	77	14
2	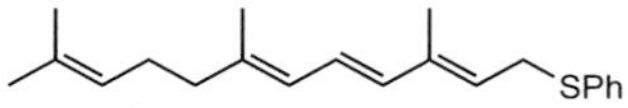 ⟋⟍⟋⟍⟋⟍SPh	80	4
3	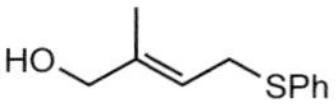 HO⟋⟍⟋SPh	75	12
4	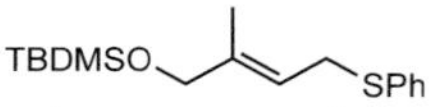 TBDMSO⟋⟍⟋SPh	82	12
5	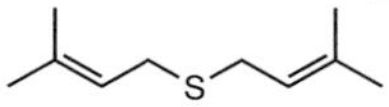 ⟋⟍S⟋⟍	54	8

Lanthanides as catalysts

Catalytic amounts of scandium triflate [Sc(OTf)$_3$] were found to greatly increase the rate of oxidation of sulfides by 60% H$_2$O$_2$ [30] The reaction is run at room temperature in methylene chloride containing 10% ethanol. The reaction shows quite a high selectivity for sulfoxide with sulfones being formed in only 2–4%.

Tab. 7.3 Oxidation of sulfides in the presence of catalytic amounts of scandium triflate

$$R\text{-}S\text{-}R' \xrightarrow[\text{room temp}]{\text{cat. Sc(OTf)}_3,\ 60\%\ \text{aq.}\ H_2O_2,\ CH_2Cl_2/10\%EtOH} R\text{-}S(O)\text{-}R'$$

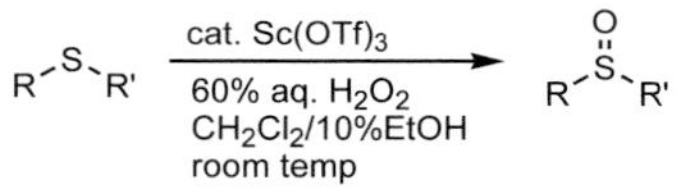

Entry	Substrate	rxn time (h)	Yield (%) Sulfoxide	Sulfone
1	Ph-S-Me	3	94	4
2	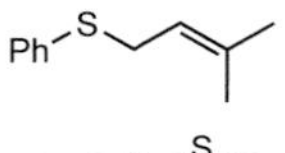 Ph-S-CH$_2$CH=CMe$_2$	2	98	2
3	p-Br-C$_6$H$_4$-S-Me	3	98	2
4	p-MeO-C$_6$H$_4$-S-Me	1.3	98	2

The reaction was applied to the oxidation of various cystein derivatives to their corresponding sulfoxides [Eq. (9)].

$$\text{Boc-Phe-Cys(Me)Ala-OAll} \xrightarrow[\substack{60\%\ \text{aq.}\ H_2O_2 \\ 15\ \text{min}}]{20\ \text{mol\%\ Sc(OTf)}_3} \text{Boc-Phe-Cys(O)(Me)Ala-OAll} \quad 100\% \qquad (9)$$

7.2 Oxidation of Sulfides to Sulfoxides

Flavins as catalysts

Flavins are organic molecules that are part of the FADH$_2$ cofactor of flavoenzymes (Scheme 7.3). In the FAD-containing monooxygenases (FADMO) molecular oxygen is activated to generate a flavin hydroperoxide.

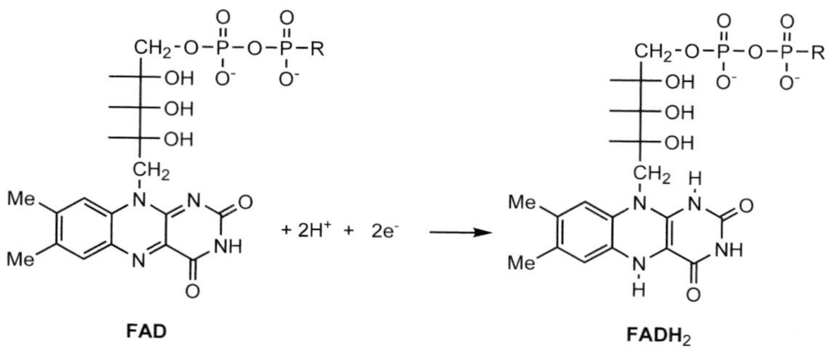

Scheme 7.3 The FAD/FAD$_2$ redox system is a cofactor in flavoenzymes (R = adenosine)

Model compounds of the natural flavins were studied by Bruice in the end of the 1970s [31]. In these studies N,N,N,-3,5,10-trialkylated flavins **16** and **17** were used. It was demonstrated that reactive flavin hydroperoxides can be generated from the reduced form **16** and molecular oxygen or from the oxidized form **17** and H$_2$O$_2$ (Scheme 7.4).

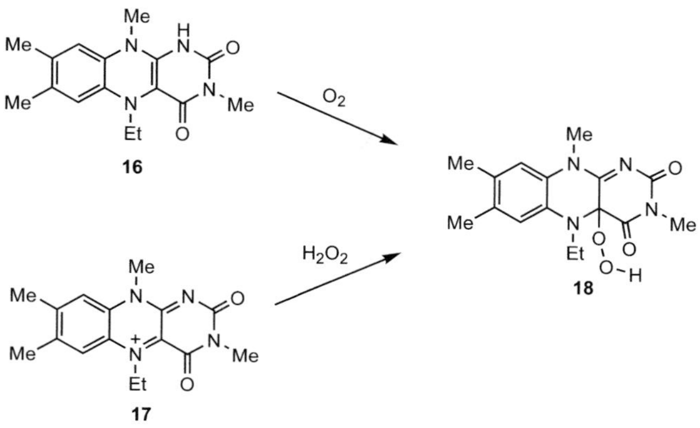

Scheme 7.4

Stoichiometric oxidation reactions with hydroperoxide **18** were studied and it was found that **18** oxidizes sulfides to sulfoxides in a highly selective manner [31, 32]. It was later demonstrated that these flavins can participate as catalysts in the H$_2$O$_2$ oxidation of sulfides to sulfoxides [33, 34].

More recently a modification of the structure of these flavins gave more efficient and robust organocatalysts for the H_2O_2-based sulfoxidations [35]. These new flavin catalysts **19** (Scheme 7.5) are superior compared with the previous "natural-based" flavin catalysts and have the advantage that they also give excellent results for the oxidation of tertiary amines to amine oxides [36] (see later, Section 7.3.2).

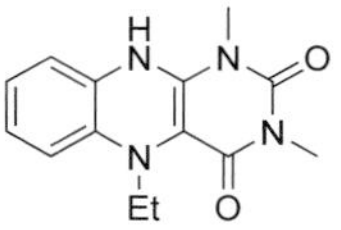

Scheme 7.5

Various thioethers were oxidized when using flavin **19** as the catalyst (Table 7.4) [35]. Only sulfoxide was formed and no over-oxidation to sulfone could be detected.

Tab. 7.4 Oxidation of thioethers using flavin **19** as the catalyst

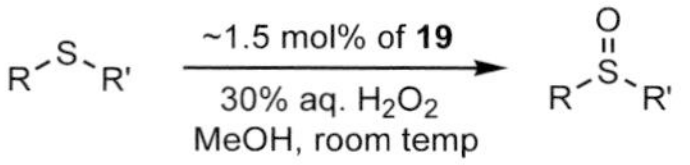

Entry	Substrate	Mol%	rxn time	Yield of sulfoxide (%)
1	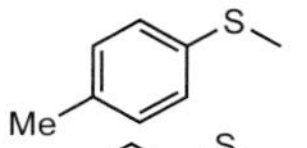 Me-C6H4-S-R	1.8	1 h	100
2	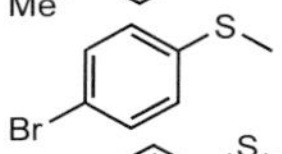 Br-C6H4-S-R	1.6	2 h 40 min	96
3	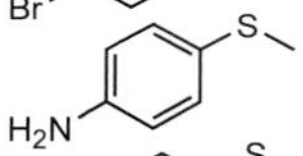 H2N-C6H4-S-R	1.3	23 min	99
4	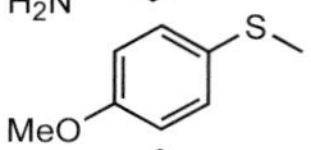 MeO-C6H4-S-R	1.6	45 min	92
5	 dithiane	1.7	20 min	99
6	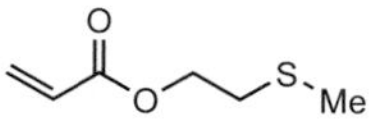 acrylate-S-Me	1.1	30 min	99

The structure of the flavin was studied and the new structures (N,N,N-1,3,5-trialkyl) were compared with those used previously (N,N,N-3,5,10-trialkyl). It was found that the new flavins were between one and two orders of magnitude faster than the previously used flavins.

The flavin-catalyzed sulfoxidation was recently extended to allylic and vinylic sulfides. It was found that the oxidation of allylic sulfides having electron-rich double bonds proceed with an exceptional selectivity for sulfoxidation (Table 7.5) [37]. Sulfone formation was depressed below the level of detection (<0.5%) and only in one single case could sulfone be observed (1.5% relative yield, for entry 6). No epoxide could be detected.

Tab. 7.5 Flavin catalyzed sulfoxidation of sulfides

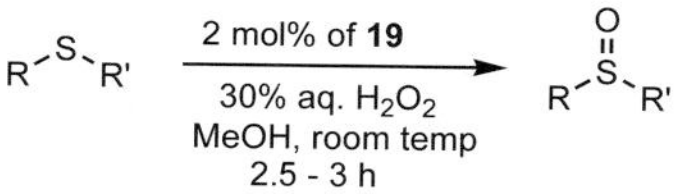

$$R\text{-}S\text{-}R' \xrightarrow[\substack{30\% \text{ aq. } H_2O_2 \\ \text{MeOH, room temp} \\ 2.5\text{-}3\text{ h}}]{2 \text{ mol\% of } \mathbf{19}} R\text{-}S(=O)\text{-}R'$$

Entry	Sulfide	Yield (%)	
		Sulfoxide	Sulfone[a]
1	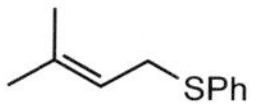 (prenyl-SPh)	92	n.d.
2	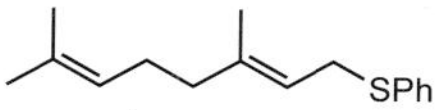 (geranyl-SPh)	76	n.d.
3	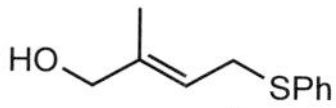 HO-...-SPh	77	n.d.
4	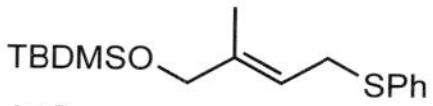 TBDMSO-...-SPh	87	n.d.
5	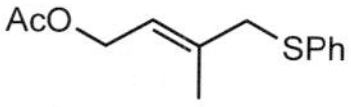 AcO-...-SPh	96	n.d.
6	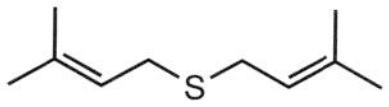 (diprenyl sulfide)	85	1.3

[a] n.d. = not detected.

The mechanism of the flavin-catalyzed oxidation of sulfides by hydrogen peroxide is shown in Scheme 7.6.

The reaction is initiated by reaction of catalyst **19** with molecular oxygen to give flavin hydroperoxide **20**. Once in the cycle this hydroperoxide can be regenerated by H_2O_2. The hydroperoxide **20** transfers an oxygen to the sulfide via the hydrogen bonded transition state **21**, to give sulfoxide and hydroxyflavin intermediate **22**. Elimination of OH⁻ from **22** produces the aromatic 1,4-diazine **23**. The latter species, which becomes the catalytic intermediate, reacts with hydrogen peroxide to regenerate the catalytic flavin hydroperoxide. The advantage of the N,N,N-1,3,5-trialkyl flavin system over the N,N,N-3,5,10-trialkylated analogues is that the elimination of OH⁻ (**22** → **23**) is fast due to aromatization. In the N,N,N-3,5,10-trialkylated system this step is slow and has been found to be the rate-determining step [33].

Chiral flavins have been used to obtain an asymmetric sulfoxidation with H_2O_2 as the oxidant [34a, 38]. Flavin **24** with planar chirality was used to oxidize various aryl methyl sulfides with 35% H_2O_2. The hydrogen peroxide was added slowly over 5 days at –20 °C to the substrate and the catalyst. The best result was obtained with the p-tolyl derivative, which gave 65% ee [Eq. (10)].

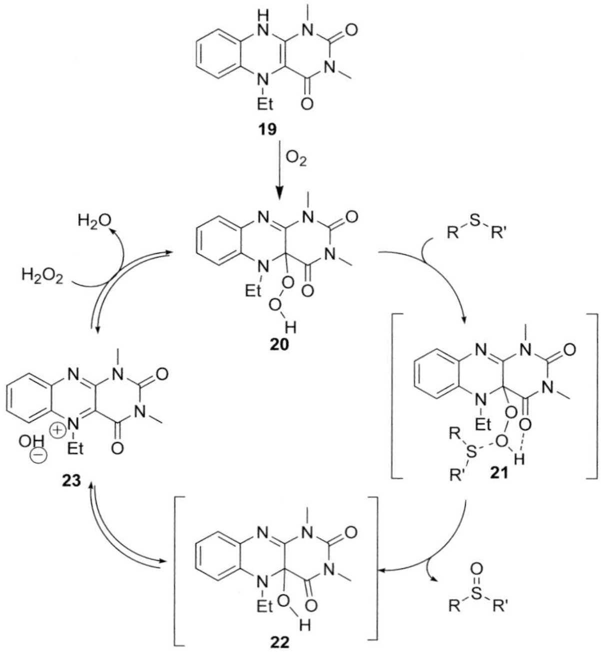

Scheme 7.6

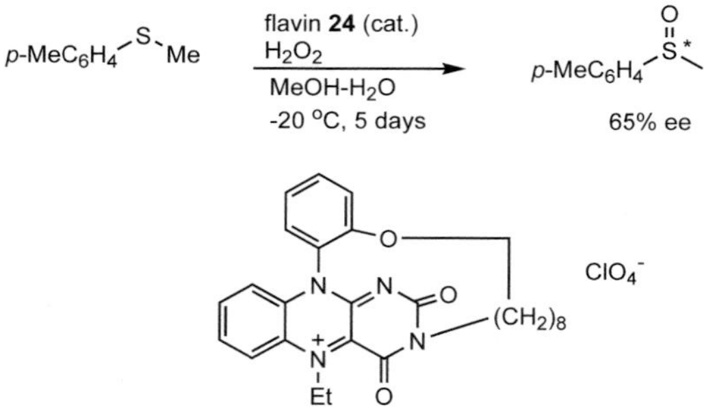

$$p\text{-MeC}_6\text{H}_4\text{-S-Me} \xrightarrow[\text{MeOH-H}_2\text{O} \\ -20\ ^\circ\text{C, 5 days}]{\text{flavin } \mathbf{24} \text{ (cat.)} \\ \text{H}_2\text{O}_2} p\text{-MeC}_6\text{H}_4\text{-S(=O)}^* \quad \text{65\% ee} \tag{10}$$

More recently, Murahashi has used flavin **25** to oxidize naphthyl methyl sulfide to its sulfoxide in 72% ee [Eq. (11)] [38].

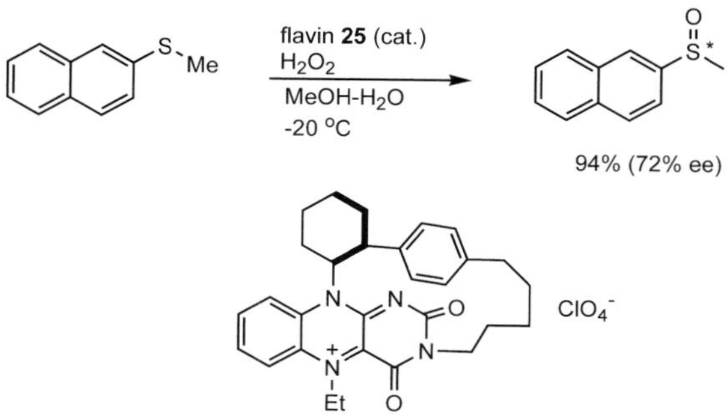

$$\text{naphthyl-S-Me} \xrightarrow[\substack{\text{MeOH-H}_2\text{O} \\ -20\,°\text{C}}]{\substack{\text{flavin 25 (cat.)} \\ \text{H}_2\text{O}_2}} \text{naphthyl-S*(=O)-Me} \quad (11)$$

94% (72% ee)

25 (flavinium catalyst structure with ClO$_4^-$ counterion)

7.2.2.2 Molecular Oxygen as Terminal Oxidant

Aerobic oxidation of sulfides to sulfoxides by molecular oxygen is of importance from an environmental point of view. This transformation can be achieved by non-catalyzed direct reaction with molecular oxygen, however, only at high oxygen pressure and elevated temperatures [39]. More recently, catalytic procedures that work at atmospheric pressure of molecular oxygen have been reported [40–44]. Various alkyl and aryl thioethers were selectively oxidized to sulfoxides by molecular oxygen in the presence of catalytic amounts of nitrogen dioxide (NO$_2$) [40]. The catalytic amount of NO$_2$ employed was between 4 and 36 mol%. Some examples are given in Eq. (12). The reaction, which is run at room temperature, is highly selective for sulfoxide over sulfone and no sulfone could be detected.

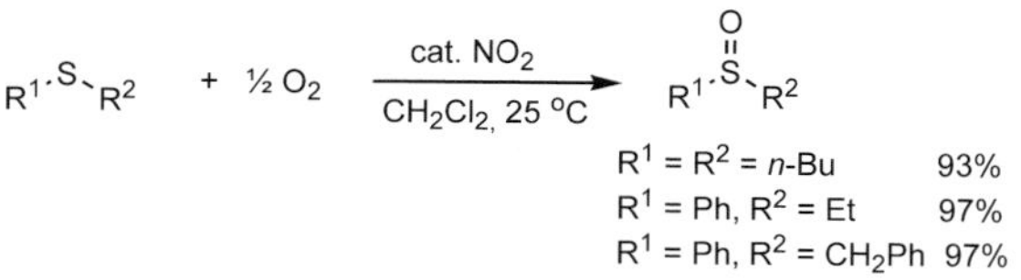

$$R^1\text{-S-}R^2 + \tfrac{1}{2}O_2 \xrightarrow[\text{CH}_2\text{Cl}_2,\ 25\,°\text{C}]{\text{cat. NO}_2} R^1\text{-S(=O)-}R^2 \quad (12)$$

$R^1 = R^2 = n\text{-Bu}$ 93%
$R^1 = \text{Ph},\ R^2 = \text{Et}$ 97%
$R^1 = \text{Ph},\ R^2 = \text{CH}_2\text{Ph}$ 97%

Ishi reported on the aerobic oxidation of sulfides in the presence of N-hydroxyphthalimide (NHPI) and alcohols [41]. The reaction works at the atmospheric pressure of oxygen; however, it requires 80–90 °C and the selectivity for sulfoxide over sulfone is moderate (~85–90%).

The binary system Fe(NO$_3$)$_3$-FeBr$_3$ was used as an efficient catalytic system for the selective aerobic oxidation of sulfides to sulfoxides [42]. The reaction works with air at room temperature at ambient pressure and employs 10 mol% of Fe(NO$_3$)$_3$ as the nonahydrate and 5 mol% of FeBr$_3$ [Eq. (13)].

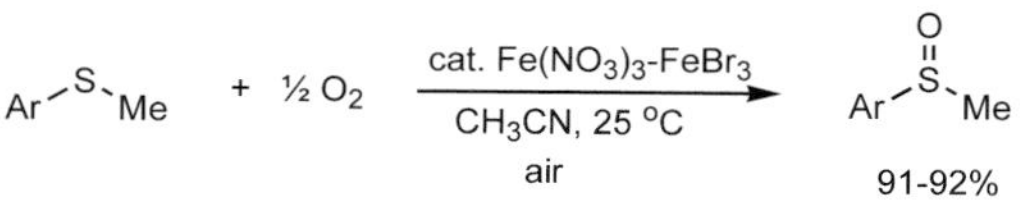

$$\text{Ar-S-Me} + \tfrac{1}{2}O_2 \xrightarrow[\substack{\text{CH}_3\text{CN},\ 25\,°\text{C} \\ \text{air}}]{\text{cat. Fe(NO}_3)_3\text{-FeBr}_3} \text{Ar-S(=O)-Me} \quad (13)$$

91-92%

Ar = p-X-C$_6$H$_4$ (X = H, OMe, Br, CN, NO$_2$)

7 Selective Oxidation of Amines and Sulfides

The mechanism of this aerobic oxidation involves the oxidation of bromide to bromine. The procedure may therefore be limited to sulfides that lack olefinic functionality.

A method for the mild and efficient aerobic oxidation of sulfides catalyzed by $HAuCl_4/AgNO_3$ was reported by Hill and coworkers [43]. The active catalyst is thought to be $Au^{III}Cl_2NO_3$(thioether). A very high selectivity for sulfoxide was observed under these oxidations and no sulfone was detected. Isotope labeling studies with $H_2^{18}O$ shows that water and not O_2 is the source of oxygen in the sulfoxide product.

Recently, Murahashi and coworkers have reported on an interesting flavin-catalyzed aerobic oxidation of sulfides to sulfoxides (Scheme 7.7) [44]. Flavin hydroperoxides can be generated from reaction of the lowest reduced form of the flavin (16) and molecular oxygen. These hydroperoxides (18) have been studied in the stoichiometric oxidation of sulfides to sulfoxides by Bruice [31].

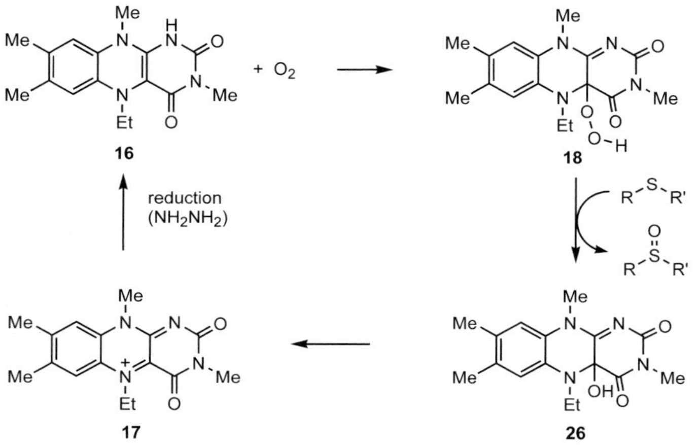

Scheme 7.7

They react rapidly with sulfides with the transfer of an oxygen to give sulfoxides. The 4a-hydroxyflavin 26 generated in this process can lose a hydroxide, to give 17. The latter flavin, which is the 2-electron-oxidized flavin, is unreactive towards molecular oxygen compared with 16. On the other hand it can react with a hydrogen peroxide to give 18, but in an aerobic process it is inert. In nature the corresponding molecule in the flavoenzyme (FAD-containing monooxygenase) is reduced by NADPH. In the process developed by Murahashi, hydrazine (NH_2NH_2) is employed for the reduction of the flavin 17 back to the reduced form 16. In this way a flavin-catalyzed aerobic oxidation of sulfides to sulfoxides was obtained [Eq. (14)].

$$R^{-S_{-R'}} + O_2 + \tfrac{1}{2}\,NH_2NH_2 \xrightarrow[\substack{35\,^\circ C \\ 1\text{ atm }O_2}]{1\text{ mol\% }\mathbf{17}} \underset{96\text{-}99\%}{R^{-\overset{\overset{O}{\|}}{S}_{-R'}}} + \tfrac{1}{2}\,N_2 + H_2O \quad (14)$$

This catalytic system mimics the flavoenzymatic aerobic oxidation and the reaction is highly selective for sulfoxide without over-oxidation to sulfone.

7.2.2.3 Alkyl Hydroperoxides as Terminal Oxidant

Alkyl hydroperoxides are known to oxidize sulfides slowly in a non-catalyzed reaction [3, 12b, 45]. If silica gel is present there is a significant rate acceleration of the reaction showing that there is a catalytic effect by the silica [12b].

Most applications of sulfide oxidations by alkyl hydroperoxides have involved titanium catalysis together with chiral ligands for enantioselective transformations. The groups of Kagan in Orsay [46] and Modena in Padova [47] reported independently on the use of chiral titanium complexes for the asymmetric sulfoxidation when using ^{t}BuOOH as the oxidant. A modification of the Sharpless reagent with the use of Ti(O^{i}Pr)$_4$ and (R,R)-diethyl tartrate [(R,R)-DET] afforded chiral sulfoxides up to 90% ee [Eq. (15)].

$$p\text{-tolyl}^{-S_{-Me}} \xrightarrow[^t\text{BuOOH}]{\text{Ti(O}^i\text{Pr)}_4,\ (R,R)\text{-DET}} \underset{89\%\text{ ee}}{p\text{-tolyl}^{-\overset{\overset{O}{\|}}{S}^{*}_{-Me}}} \quad (15)$$

The outcome of the reaction was later improved by replacing ^{t}BuOOH with cumene hydroperoxide [48].

An improved catalytic reaction with the use of 10 mol% of titanium using a ratio Ti(O^{i}Pr)$_4$/(R,R)-DET/^{i}PrOH = 1:4:4 in the presence of molecular sieves gave an efficient sulfoxidation with ee values up to 95% with various aryl methyl sulfoxides [49]. The asymmetric Ti-catalyzed sulfoxidations with alkyl hydroperoxides have been reviewed by Kagan [50].

The asymmetric titanium-catalyzed sulfoxidations with ^{t}BuOOH also works with chiral diols as ligands [51–53]. Various 1,2-diaryl-1,2-ethanediols were employed as ligands, and the use of 15 mol% of Ti(O^{i}Pr)$_4$ with 1,2-diphenyl-1,2-ethanediol gave ee values up to 90% [51b]. Also the use of BINOL and derivatives gave asymmetric sulfoxidations [52].

The use of (S,S)-1,2-bis-*tert*-butyl-1,2-ethanediol [(S,S)-**27**] in the titanium-catalyzed oxidation of various aryl methyl sulfides by cumene hydroperoxide afforded sulfoxides in ee values up to 95% [53]. Interestingly, the authors observed that the ee of the sulfoxide increased with the reaction time indicating a kinetic resolution of the sulfoxide product. A control experiment with racemic p-tolyl sulfoxide showed that the (R)-enantiomer is oxidized to sulfone three times faster than the (S)-enantiomer by the catalytic system employed. For this reason the yields of the chiral sulfoxides are moderate and in the range of 40–50% (Scheme 7.8).

A similar observation of kinetic resolution of the sulfoxide by over-oxidation to sulfone leading to an amplification of the ee has been previously reported by Uemura [52].

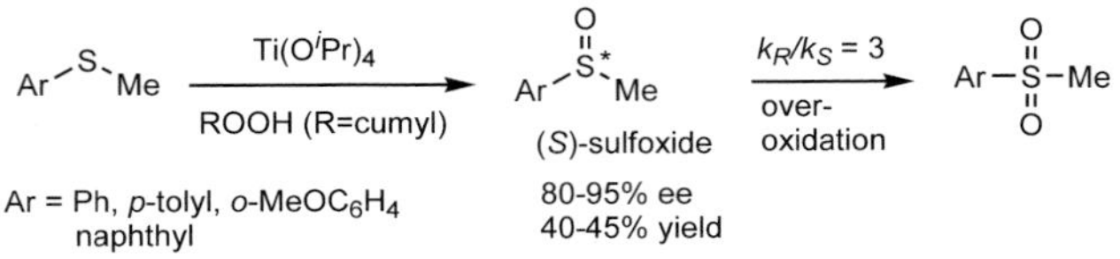

Scheme 7.8

Ar–S–Me →[Ti(O^iPr)_4, ROOH (R=cumyl)] Ar–S(=O)–Me (S)-sulfoxide →[$k_R/k_S = 3$, over-oxidation] Ar–S(=O)(=O)–Me

Ar = Ph, p-tolyl, o-MeOC$_6$H$_4$, naphthyl

80–95% ee
40–45% yield

An application of the Kagan-Modena procedure for the synthesis of the enantiomerically pure S-enantiomer of omeprazol was reported by Cotton et al. [54]. This enantiomer is called esomeprazol (Scheme 7.9) and is the active component of Nexium®. It is a highly potent gastric acid secretion inhibitor.

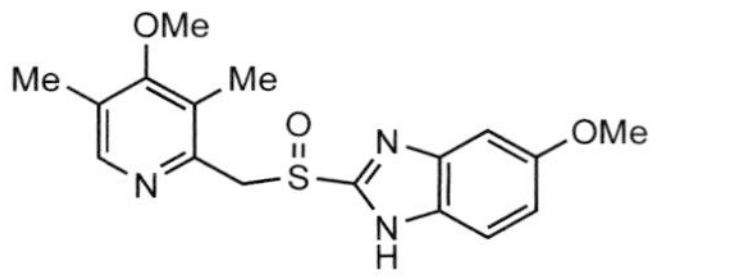

Scheme 7.9

Omeprazol (racemix sulfoxide)
Esomeprazol (S-form of sulfoxide)

It was found that a modification of the original procedure by addition of N,N-diisopropyl ethylamine had a dramatic effect on the enantioselectivity of the reaction. The role of the added amine is unclear. A large-scale oxidation of sulfide **28** (6.2 kg) using 30 mol% of the titanium catalyst in the presence of (S,S)-DET and N,N-diisopropyl ethylamine gave 92% of a crude product, which was >94% ee (Scheme 7.10). The ratio of sulfoxide to sulfone was 76:1. Recrystallization gave 3.83 kg of a product that was >99.5% ee. It is possible to run the reaction with less catalyst but this gives a slightly lower ee. Thus, the use of 4 mol% Ti(O^{i}Pr)$_4$ with Ti/(S,S)-DET/EtN(^{i}Pr)$_2$ = 1:2:1 gave esomeprazol in 96% yield that was 91% ee. The ratio of sulfoxide to sulfone was 35:1.

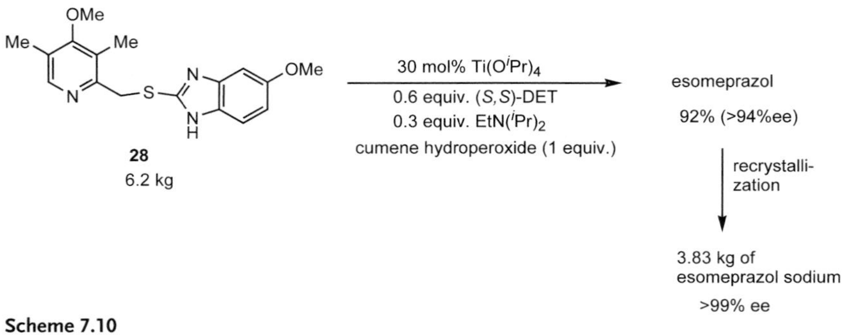

28
6.2 kg

→ 30 mol% Ti(O^{i}Pr)$_4$
0.6 equiv. (S,S)-DET
0.3 equiv. EtN(^{i}Pr)$_2$
cumene hydroperoxide (1 equiv.)

esomeprazol
92% (>94%ee)

→ recrystallization

3.83 kg of esomeprazol sodium
>99% ee

Scheme 7.10

7.2.2.4 Other Oxidants in Catalytic Reactions

Several chemocatalytic systems for sulfoxidation that employ oxidants other than hydrogen peroxide, molecular oxygen or alkyl hydroperoxide have been reported. Manganese-catalyzed oxidation of sulfides with iodosobenzene (PhIO) using chiral Mn(salen) complexes was used to obtain chiral sulfoxides in up to 90% ee [55]. PhIO was also employed as the oxidant in sulfoxidations catalyzed by quaternary ammonium salts [56]. The use of cetyltrimethylammonium bromide (n-$C_{16}H_{33}Me_3N^+Br^-$) gave the best result and with 5–10 mol% of this catalyst, high yields (90–100%) of sulfoxide were obtained from various sulfides.

A mild and chemoselective oxidation of sulfides to sulfoxides by o-iodooxybenzoic acid (IBX) catalyzed by tetraethylammonium bromide (TEAB) was recently reported [57]. The reaction is highly selective and no over-oxidation to sulfone was observed. Simple aryl alkyl sulfides are oxidized in 93–98% yield in 0.3–2 h at room temperature with the use of 5 mol% of TEAB. Diphenyl sulfide and phenyl benzyl sulfide took 30 and 36 h, respectively, to go to completion under these conditions.

7.2.3 Biocatalytic Reactions

Various peroxidases and monooxygenases have been used as biocatalysts for the oxidation of sulfides to sulfoxides [58, 59]. Haloperoxidases have been studied in the oxidations of sulfides and these reactions work with hydrogen peroxide as the oxidant. Baeyer-Villiger monooxygenases, the natural role of which is to oxidize ketones to esters, are NAD(P)H-dependent flavoproteins that have been used for sulfoxidations. Until recently only cyclohexanone monooxygenase (CHMO) had been cloned and overexpressed, but new developments have made a number of other Bayer-Villiger monooxygenases available.

7.2.3.1 Haloperoxidases

Oxidaton of sulfides catalyzed by haloperoxidases has recently been reviewed [58]. The natural biological role of haloperoxidases is to catalyze the oxidation of chloride, bromide, or iodide by hydrogen peroxide. There are three classes of haloperoxidases that have been identified: (1) those without a prosthetic group, found in bacteria, (2) heme-containing peroxidases, such as chloroperoxidase (CPO) and (3) vanadium containing peroxidases.

Asymmetric H_2O_2 oxidations of aryl methyl sulfides catalyzed by CPO occur in excellent enantioselectivity [Eq. (16)] [60, 61]. Electronic and in particular steric factors dramatically affect the yield of the reaction. Thus, small aromatic groups gave high yields in 99% ee, whereas a slight increase in size led to a dramatic drop in the yield, however, still in a high ee (99% ee).

$$\text{Ar}-\text{S}-\text{Me} + \text{H}_2\text{O}_2 \xrightarrow{\text{CPO}} \text{Ar}-\overset{*}{\underset{\|}{\text{S}}}(\text{O})-\text{Me} + \text{H}_2\text{O} \quad (16)$$

99% ee (R-form)

Ar = Ph, benzothiophene, 100% yield
Ar = o-Me-C$_6$H$_4$ benzothiazole 3% yield
Ar = p-Br-C$_6$H$_4$ 15% yield

The analogous oxidations of cyclic sulfides with the same biocatalyst (CPO) were studied by Allenmark and coworkers [62]. Only 1-thiaindane gave a synthetically useful outcome with high yield (99.5%) and high enantioselectivity (99% ee).

Allenmark and coworkers also studied the asymmetric sulfoxidation catalyzed by vanadium-containing bromoperoxidase (VBrPO) from *Corallina officinalis* [63, 64]. The practical use of this reaction is limited since the enzyme only accepts very few substrates such as 2,3-dihydrobenzo[b]thiophene, 2-(carboxy)phenyl methyl sulfide and 2-(carboxy)vinyl methyl sulfides [58, 63].

7.2.3.2 Ketone Monooxygenases

A number of ketone monooxygenases are available for synthetic transformations today [65]. Cyclohexanone monooxygenase (CHMO) had already been cloned and overexpressed in 1988 and has until recently been the only ketone monooxygenase studied extensively. In new developments a number of other monooxygenases such as cyclopentanone monooxygenase (CPMO), cyclododecanone monooxygenase (CDMO), steroid monooxygenase (SMO), and 4-hydroxyacetophenone monooxygenase (HAPMO) have been cloned [65]. Of the ketone monooxygenases known today CHMO, CPMO, and HAPMO have been used for sulfoxidation.

Oxidation of sulfides by molecular oxygen catalyzed by cyclohexanone monooxygenase (CHMO) has been studied by an Italian team [59, 66, 67]. CHMO is a flavin-dependent enzyme of about 60 kDa and is active as a monomer. It has found application in Baeyer-Villiger oxidation [65, 68a] and in the oxidation of sulfides to sulfoxides [69]. The aerobic oxidation with these monooxygenases requires NADPH to reduce the oxidized flavin back to the reduced form so that it can react again with molecular oxygen. CHMO-catalyzed oxidation of various sulfides by molecular oxygen in the presence of NADPH afforded sulfoxides in high yields and in most cases good to high enantioselectivity. The NADPH was employed in catalytic amounts by recycling of NADP by glucose-6-phosphate or L-malate. Results from aerobic oxidations of some methyl substituted sulfides are given in Eq. (17).

$$\text{R}-\text{S}-\text{Me} + \tfrac{1}{2}\text{O}_2 \xrightarrow[\text{NADPH}]{\text{CHMO}} \text{R}-\overset{*}{\underset{\|}{\text{S}}}(\text{O})-\text{Me} \quad (17)$$

R = Ph   88% (99% ee)
R = o-MeC$_6$H$_4$ 90% (87% ee)
R = 2-pyridyl   86% (87% ee)
R = p-FC$_6$H$_4$ 96% (92% ee)
R = ^{t}Bu   98% (99% ee)
R = p-MeC$_6$H$_4$ 94% (37% ee)

Also the corresponding ethyl derivatives p-FC$_6$H$_4$SEt and p-MeC$_6$H$_4$SEt gave high yields in 93 and 89% ee, respectively. On the other hand the p-MeC$_6$H$_4$SMe behaved differently and gave the corresponding sulfoxide in only 37% ee. The CHMO system with NADPH and glucose-6-phosphate was subsequently applied to the oxidation of various dialkyl sulfides. Thus, methyl sulfides RSMe with R = cyclopentyl, cyclohexyl and allyl gave the corresponding sulfoxides in 82–86% yield and in >98% ee [66].

Recently, Jansen and coworkers [70] used 4-hydroxyacetophenone monooxygenase (HAPMO) for aerobic oxidation of sulfides. Interestingly, both PhSMe and p-MeC$_6$H$_4$SMe gave the corresponding sulfoxides in >99% ee, which should be compared with the 99 and 37% ee, respectively, obtained with CHMO [66]. The flavoenzyme HAPMO, which was recently cloned [70, 71], is a promising biocatalyst for enantioselective oxidation of sulfides to sulfoxides.

Recombinants of baker's yeast expressing CHMO from *Actinobactersp.* NCIP 9871 have been used as whole cell biocatalysts for the oxidation of sulfides to their corresponding sulfoxides [Eq. (18)] [72].

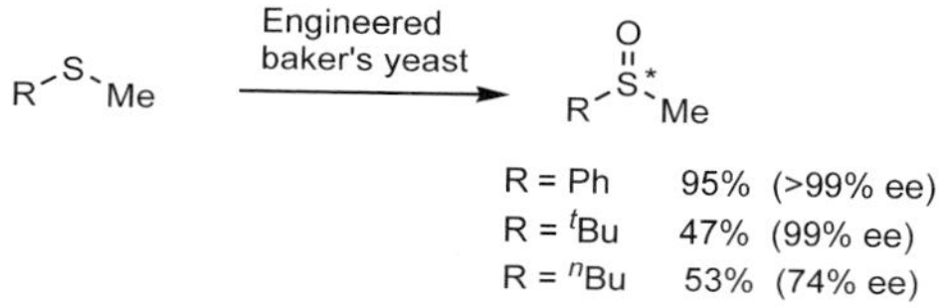

$$R-S-Me \xrightarrow{\text{Engineered baker's yeast}} R-S(=O)^*-Me \quad (18)$$

R = Ph   95% (>99% ee)
R = ^{t}Bu   47% (99% ee)
R = ^{n}Bu   53% (74% ee)

Some peroxidases are sensitive to an excess of hydrogen peroxide, which may complicate synthetic procedures with these enzymes. For example, *Coprinus cinerem* peroxidase (Cip) has been used for the enantioselective oxidation of sulfides to sulfoxides either by continuous slow addition of hydrogen peroxide [73] or by the use of an alkyl hydroperoxide [74]. Sulfoxidation with Cip as the catalyst was recently developed into an aerobic procedure by combining the peroxidase (Cip) with a glucose oxidase [75]. The glucose oxidase and molecular oxygen gives a slow production of hydrogen peroxide, which is slow enough to avoid degradation of the enzyme. This is a convenient procedure and aryl methyl sulfides, where the aryl group is phenyl, p-MeC$_6$H$_4$ or naphthyl, gave good yields (85–91%) in 79, 88, and 90% ee, respectively.

7.3
Oxidation of Tertiary Amines to N-Oxides

Previous reviews have dealt with metal-catalyzed [76] and stoichiometric [77] oxidation of amines in a broad sense. This section will be limited to the selective oxidation of tertiary amines to N-oxides. Amine N-oxides are synthetically useful compounds [78, 79] and are frequently used stoichiometric oxidants in osmium- [81–82] manganese- [83] and ruthenium-catalyzed [84, 85] oxidations, as well as in other organic transformations [86–88]. Aliphatic *tert*-amine N-oxides are useful surfactants [79] and are essential components in hair conditioners, shampoos, toothpaste, cosmetics, etc. [89].

Because of their importance, various methods have been reported for the oxidation of tertiary amines to N-oxides. The oxidations of amines will be divided into the following sections: (1) stoichiometric reactions, (2) chemocatalytic reactions, and (3) biocatalytic reactions. Finally we provide some examples where N-oxides are generated *in situ* as catalytic oxotransfer species in catalytic transformations.

7.3.1
Stoichiometric Reactions

Amine N-oxides can be prepared from amines with 30% aqueous hydrogen peroxide in a non-catalytic slow reaction [80, 90]. At elevated temperatures this oxidation proceeds at a reasonable rate and has been used in industrial applications. Various other oxidants have also been employed for N-oxidation of tertiary amines such as peracids [91], 2-sulfonyloxaziridines [92] and α-azohydroperoxides [93].

Messeguer and coworkers reported the use of dioxiranes for the oxidation of amines to N-oxides [94]. Oxidation of various tertiary aromatic amines with dimethyldioxirane (DMD) afforded amine N-oxides in quantitative yields. A few examples are given in Eq. (19).

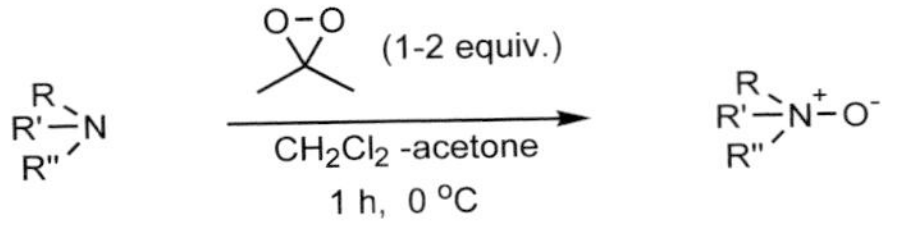

$$\begin{array}{c}R\\R'-N\\R''\end{array} \xrightarrow[\text{1 h, 0 °C}]{\text{O-O} \;\; (1\text{-}2 \text{ equiv.})} \begin{array}{c}R\\R'-\overset{+}{N}-O^-\\R''\end{array} \qquad (19)$$

R = R' = R" = Bu
R = R' = Me; R" = PhCH$_2$
R = PhCH$_2$; R', R" = -(CH$_2$)$_5$-
R = Me; R', R" = -(CH$_2$)$_2$O(CH$_2$)$_2$-

Interestingly, the reaction is chemoselective and oxidation of alkene-amines gave the N-oxides selectively without any epoxide formation. One example is given in Eq. (20). A number of substituted pyridines were also oxidized to the pyridine N-oxides by DMD in quantitative yields [94].

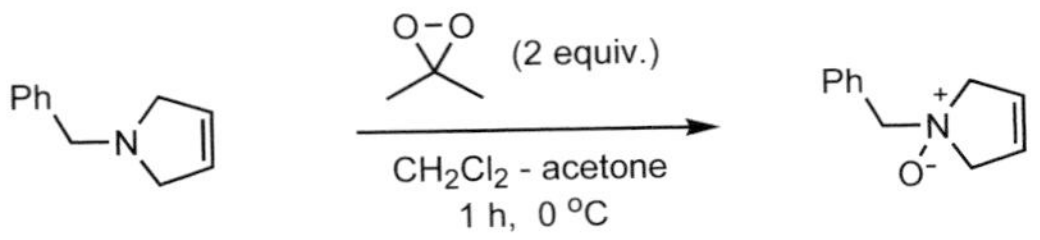

$$\text{Ph-N-(pyrrolidine with alkene)} \xrightarrow[\text{1 h, 0 °C}]{\text{O-O} \;\; (2 \text{ equiv.})\atop\text{CH}_2\text{Cl}_2 \text{ - acetone}} \text{Ph-N}^+(\text{O}^-)\text{-(pyrrolidine with alkene)} \qquad (20)$$

Selective oxidation of tertiary amines to N-oxides by HOF·CH$_3$CN was reported by Rozen and coworkers [95]. The reaction is rapid and amine N-oxides were isolated in high yields [Eq. (21)]. The HOF·CH$_3$CN complex was also employed to oxidize a number of substituted pyridines to their corresponding N-oxides [Eq. (22)].

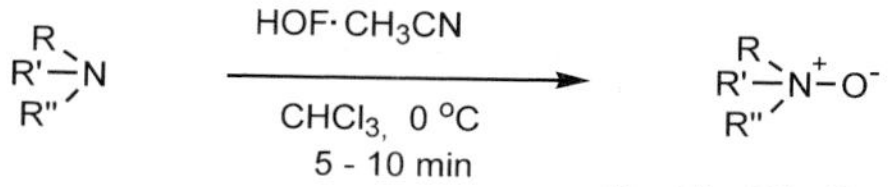

$$\underset{R''}{\overset{R}{\underset{|}{R'-N}}} \xrightarrow[\text{CHCl}_3, \, 0\,°\text{C}]{\text{HOF}\cdot\text{CH}_3\text{CN}} \underset{R''}{\overset{R}{\underset{|}{R'-\overset{+}{N}-O^-}}} \tag{21}$$

R = R' = R" = Bu   82%
R = R' = C$_8$H$_{17}$; R" = Me   95%
R = R' = cyclohexyl; R" = Me   95%
R = Ph; R' = Bn, R" = Et   85%
R = Ph R' = R" = -(CH$_2$)$_5$-   85%

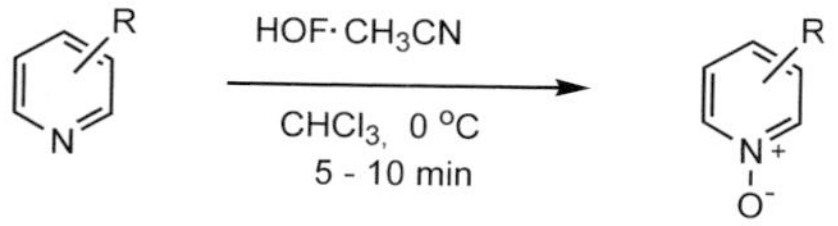

$$\xrightarrow[\text{CHCl}_3, \, 0\,°\text{C}]{\text{HOF}\cdot\text{CH}_3\text{CN}} \tag{22}$$

7.3.2
Chemocatalytic Oxidations

Some early work described the vanadium-catalyzed oxidations of tertiary amines to N-oxides by *tert*-butylhydroperoxide, but these reactions require elevated temperature [96]. In 1998a mild flavin-catalyzed oxidation of tertiary amines to N-oxides by 30% aqueous H$_2$O$_2$ was reported [36]. The flavin **19** was employed as the catalyst and the reaction occurs at room temperature. With the use of 2.5 mol% of the flavin the reaction takes 25–60 min to go to high conversion. The flavin hydroperoxide generated *in situ* (cf. Scheme 7.4) show a very high reactivity towards amines and it was estimated that the flavin hydroperoxide reacts >8000 times faster than H$_2$O$_2$ with the amine in entry 4 of Table 7.6. The success with N,N,N-1,3,5-trialkylated flavin **19** is that the flavin–OH (**22**, Scheme 7.6) formed after oxo transfer to the amine from flavin–OOH can easily lose the OH group and give the aromatized flavin **23**, which is the active catalyst.

Tab. 7.6 Oxidation of tertiary amines according to Eq. (23)

Entry	Amine	rxn time for >85% conversion	Product	Rate enhancement cat.: non-cat.[a]
1	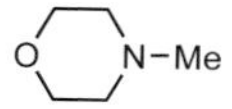 O(CH$_2$CH$_2$)$_2$N-Me (morpholine)	1 h	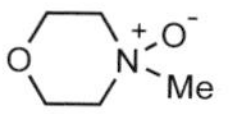 morpholine N-oxide	61:1 (6344:1)[b]
2	n-C$_{12}$H$_{25}$-N(Me)$_2$	27 min	n-C$_{12}$H$_{25}$-N$^+$(Me)$_2$O$^-$	49:1 (5096:1)[b]
3	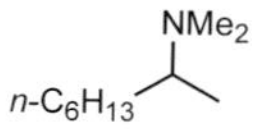 n-C$_6$H$_{13}$CH(NMe$_2$)–	25 min	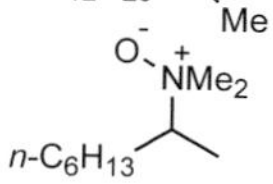 n-C$_6$H$_{13}$CH(N$^+$Me$_2$O$^-$)–	51:1 (5304:1)[b]
4	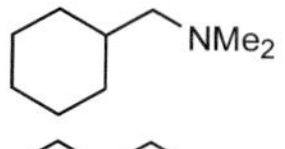 cyclohexyl-NMe$_2$	50 min	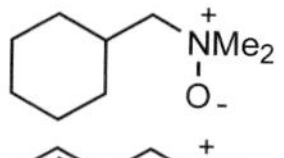 cyclohexyl-N$^+$Me$_2$O$^-$	83:1 (8632:1)[b]
5	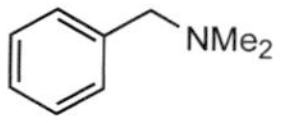 PhCH$_2$-NMe$_2$	31 min	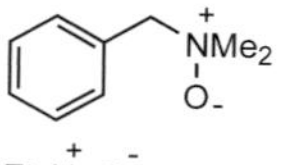 PhCH$_2$-N$^+$Me$_2$O$^-$	67:1 (6968:1)[b]
6	Et$_3$N	54 min	Et$_3$N$^+$-O$^-$	61:1 (2507:1)[b]

[a] Calculated by division of the initial rates of catalyzed and non-catalyzed reactions. [b] Estimated ratio of the reactivities of catalytic flavin hydroperoxide and H$_2$O$_2$.

$$\underset{R''}{\overset{R}{\underset{|}{R'-N}}} \xrightarrow[\text{MeOH, air} \atop \text{room temp, 0.5-1 h}]{\text{2.5 mol\% of 19} \atop \text{30\% aqueous H}_2\text{O}_2} \underset{R''}{\overset{R}{\underset{|}{R'-N^+-O^-}}} \qquad (23)$$

$$>85\% \text{ yield}$$

Structure **19** (flavin catalyst)

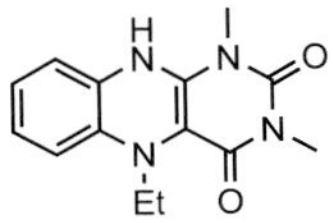

In preparative experiments the N,N,-dimethylamino derivatives **29a** and **29b** [Eq. (24)] were oxidized by H_2O_2 at room temperature for 2 h to the corresponding N-oxides **30a** and **30b** in 85 and 82% yield, respectively, using 1 mol% of flavin **19** [36].

$$R-NMe_2 \xrightarrow[\text{MeOH, air} \atop \text{room temp, 2 h}]{\text{1 mol\% of 19} \atop \text{30\% aqueous H}_2\text{O}_2} R-\overset{+}{N}Me_2 \qquad (24)$$
$$ \qquad\qquad\qquad\qquad\qquad\qquad \overset{|}{O^-}$$

29a R = n-C₁₂H₂₅   **30a** 85%
b R = PhCH₂   **b** 82%

A tungstate-exchanged Mg–Al layered double hydroxide (LDH) was employed to catalyze oxidation of aliphatic tertiary amines to N-oxides by 30% aqueous H_2O_2 [97]. The reaction takes 1–3 h at room temperature. The use of dodecylbenzenesulfonic acid sodium salt as an additive increased the rate of the oxidation by a factor of 2–3, except for N-methyl morpholine. An advantage of the LDH-WO_4^{2-} catalyst is that it can be reused (Table 7.7).

Tab. 7.7 Oxidation of *tert*-amines catalyzed by LDH-WO_4^{2-}

Entry	Amine	Procedure[a]	Product	rxn time (h)	Yield (%)
1	O(CH₂CH₂)₂N–Me (morpholine)	I	morpholine N-oxide	1.0	96
		II		1.0	96
2	n-C₁₀H₂₁–NMe₂	I	n-C₁₀H₂₁–N⁺(Me)₂–O⁻	1.0	97
3	PhCH₂–NMe₂	I	PhCH₂–N⁺Me₂–O⁻	1.5	96
		II		1.0	96
4	Et₃N	I	Et₃N⁺–O⁻	3.0	96
		II		1.5	96
5	N-phenylpiperidine	I	N-phenylpiperidine N-oxide	3.0	97
		II		1.0	97

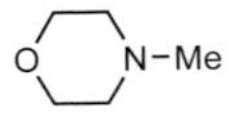

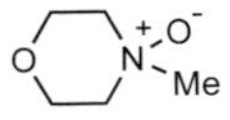

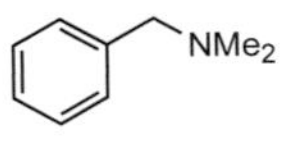

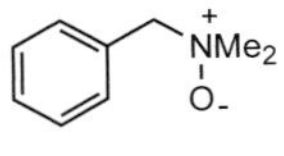

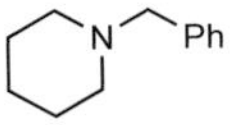

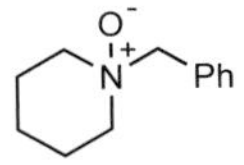

[a] I: 2 mmol of substrate were oxidized by H_2O_2 in water with LDH-WO_4^{2-} as catalyst; II: as in I but 6 mg of dodecylbenzenesulfonic acid sodium salt were added.

7.3 Oxidation of Tertiary Amines to N-Oxides

The aerobic flavin system with NH_2NH_2 as a reducing agent that was employed for the sulfoxidation (Section 7.2.2.2) can also be used for the N-oxidation of tertiary amines [44]. However, the reaction requires elevated temperature (60 °C), most likely because elimination of OH from the flavin-OH (cf. **26** → **17**, Scheme 7.7) is difficult for the N,N,N-3,5,10-trialkylated flavin at the higher pH caused by the amine.

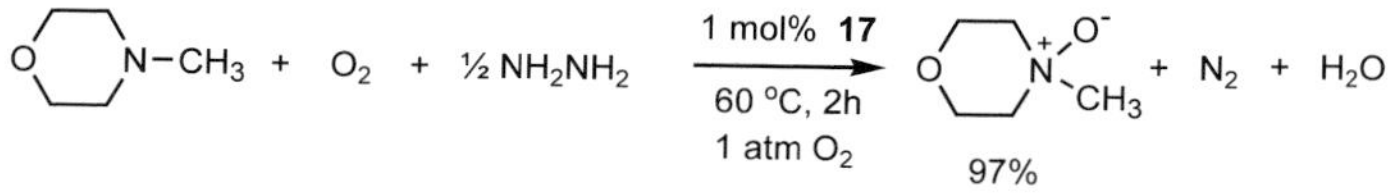

$$\text{O} \diagdown \text{N-CH}_3 + O_2 + \tfrac{1}{2} NH_2NH_2 \xrightarrow[\substack{60 \,°C,\, 2h \\ 1 \text{ atm } O_2}]{1 \text{ mol\% } \mathbf{17}} \text{O} \diagdown \text{N}^+(\text{O}^-)\text{CH}_3 + N_2 + H_2O \quad (25)$$

97%

17: N,N,N-3,5,10-trialkylated flavin (Me, Me, Et substituents)

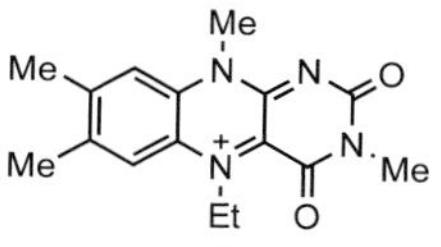

An aerobic oxidation of tertiary amines to N-oxides catalyzed by Cobalt Schiff-base complexes was recently reported [98]. The reaction was run at room temperature with 0.5 mol% of the cobalt catalyst [Eq. (26)]. The presence of molecular sieves (5 Å) enhanced the rate of the reaction. With this procedure various pyridines were oxidized to their corresponding N-oxides in yields ranging from 50 to 85%. Electron-deficient pyridines such as 4-cyanopyridine gave a slow reaction with only 50% yield.

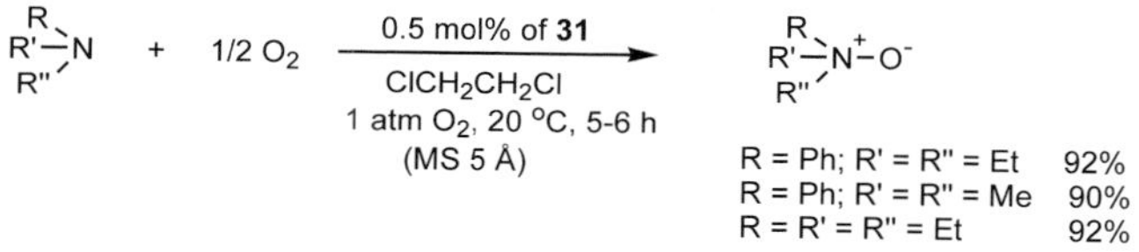

$$R'R''R\text{N} + \tfrac{1}{2} O_2 \xrightarrow[\substack{ClCH_2CH_2Cl \\ 1 \text{ atm } O_2,\, 20\,°C,\, 5\text{-}6\text{ h} \\ (MS\ 5\ \text{Å})}]{0.5 \text{ mol\% of } \mathbf{31}} R'R''R\text{N}^+\text{-O}^- \quad (26)$$

R = Ph; R' = R" = Et   92%
R = Ph; R' = R" = Me   90%
R = R' = R" = Et   92%

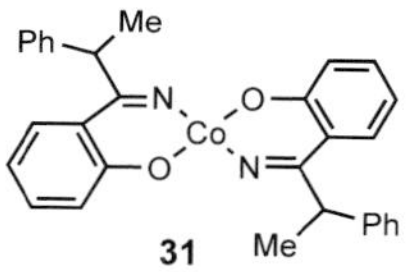

31: Co Schiff-base complex

The same authors also reported on an aerobic oxidation of tertiary amines and pyridines to their corresponding N-oxides catalyzed by ruthenium trichloride [99].

The oxidation of pyridines to their corresponding N-oxides catalyzed by methyltrioxorhenium (MTO) was reported to occur with various substituted pyridines [100, 101]. The oxidant employed is either H_2O_2 [100] or $Me_3SiOOSiMe_3$ [101]. The reaction gives high yields with both electron-rich and electron-deficient pyridines [Eq. (27)].

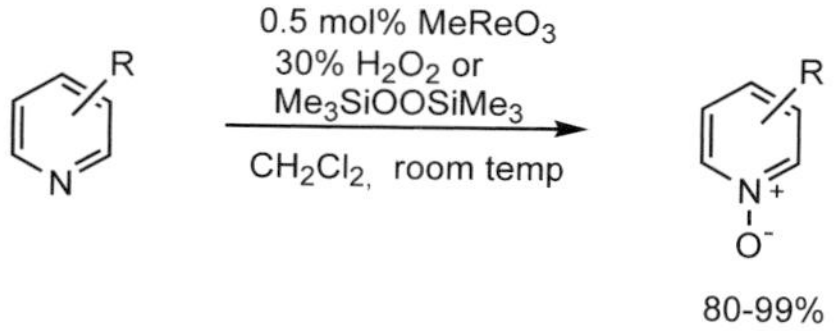

$$\text{R-pyridine} \xrightarrow[\substack{30\% \text{ H}_2O_2 \text{ or} \\ Me_3SiOOSiMe_3 \\ CH_2Cl_2,\text{ room temp}}]{0.5 \text{ mol\% } MeReO_3} \text{R-pyridine N-oxide} \quad (27)$$

80–99%

7.3.3
Biocatalytic Oxidation

There have only been limited examples reported on the biocatalytic oxidation of tertiary amines. Colonna et al. used bovine serum albumin (BSA) as a biocatalyst for asymmetric oxidation of tertiary amines to N-oxides [102]. Oxone, NaIO$_4$, H$_2$O$_2$, and MCPBA were tested as oxidants and the best results were obtained with NaIO$_4$ and H$_2$O$_2$ [Eq. (28)]. Thus, BSA-catalyzed N-oxidation of **32** with these oxidants afforded N-oxide **33** in high yield in 64–67% ee. The reaction is formally a dynamic kinetic resolution since the enantiomers of the starting material are in rapid equilibrium.

$$n\text{-}C_5H_{12}\text{-}\underset{Bn}{\overset{Me}{N}}\quad\xrightarrow[\text{72 h, room temp}]{\text{BSA, oxidant}}\quad n\text{-}C_5H_{12}\text{-}\underset{Bn}{\overset{O^-}{\underset{|}{\overset{|}{\overset{+}{N}}}}}\text{Me} \tag{28}$$

32 → **33**

H$_2$O$_2$ 100% (67% ee)
NaIO$_4$ 87% (64% ee)

Recently, Colonna and coworkers reported that cyclohexanone monooxygenase (CHMO) from *Actinobacter calcoaceticus* NCIMB9871 catalyzed the aerobic oxidation of amines [103].

7.3.4
Applications of Amine N-oxidation in Coupled Catalytic Processes

In 1976 N-methyl morpholine N-oxide (NMO) was introduced by VanRheenen et al. as a stoichiometric oxidant in osmium-catalyzed dihydroxylation of alkenes (the Upjohn procedure) [104]. This made it possible to use osmium tetroxide in only catalytic amounts. However, more environmentally friendly oxidants were sought and one idea was to recycle the amine to N-oxide *in situ* either by hydrogen peroxide or molecular oxygen. In this way it would be possible to use the NMO or even better the N-methyl morpholine (NMM) in only catalytic amounts. In 1999 a biomimetic dihydroxylation of alkenes based on this principle was reported in which the tertiary amine (NMM) in catalytic amounts is continuously reoxidized to N-oxide (NMO) by a cat. flavin/H$_2$O$_2$ system [Eq. (29), Scheme 7.11] [105]. The coupled electron transfer resembles oxidation processes occurring in biological systems. OsO$_4$, NMM, and flavin **19** are used in catalytic amounts and this leads to an efficient H$_2$O$_2$-based dihydroxylation of alkenes in high yields at room temperature.

$$\text{R}\diagup\hspace{-0.3em}=\hspace{-0.3em}\diagdown\text{R'} \;+\; H_2O_2 \;\xrightarrow[\text{room temp}]{\text{cat. OsO}_4,\text{ cat. NMM, cat. flavin }\mathbf{19}}\; \underset{R\quad R'}{HO\diagdown\hspace{-0.3em}\diagup OH} \tag{29}$$

In the presence of chiral ligand (DHQD)$_2$PHAL, high enantioselectivity was obtained in the dihydroxylation with hydrogen peroxide (Scheme 7.12) [106].

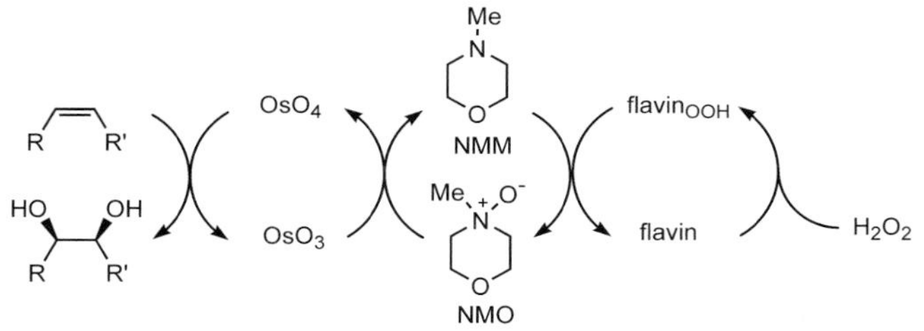

7.3 Oxidation of Tertiary Amines to N-Oxides

Scheme 7.11

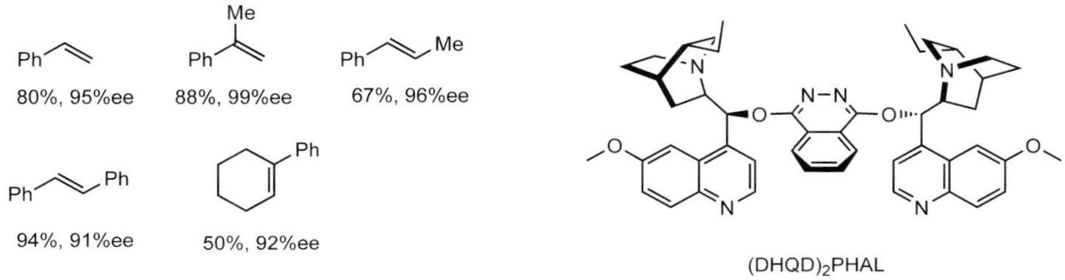

Ph⁀ 80%, 95%ee

Ph–C(Me)=CH₂ 88%, 99%ee

Ph–CH=CH–Me 67%, 96%ee

Ph–CH=CH–Ph 94%, 91%ee

1-Ph-cyclohexene 50%, 92%ee

(DHQD)₂PHAL

Scheme 7.12 Biomimetic dihydroxylation of alkenes to diols using the biomimetic system of Scheme 7.11 with chiral ligand (DHQD)₂PHAL

In a model study it was demonstrated that the catalytic system also works with MCPBA as the oxidant in place of cat. flavine/H_2O_2 [107].

It was later found that the chiral amine can be used as the tertiary amine generating the N-oxide required for reoxidation of Os^{VI}. Thus, a simplified procedure for osmium-catalyzed asymmetric dihydroxylation of alkenes by H_2O_2 was developed in which the tertiary amine NMM is omitted [108]. A robust version of this reaction, where the flavin **19** has been replaced by MeReO₃ (MTO), was recently reported [109]. The chiral ligand has a dual role in these reactions: it acts as a chiral inductor as well as an oxotransfer mediator (Scheme 7.13). The amine oxide of the chiral amine ligand was isolated and characterized by high resolution mass spectrometry [109].

An efficient osmium-catalyzed dihydroxylation of alkenes by H_2O_2 with NMM and MTO as electron transfer mediators under acidic conditions was recently reported [110]. Under these conditions alkenes, which are normally difficult to dihydroxylate, gave high yields of diol [Eq. (30)]. The N-oxide NMO is generated from NMM by cat. MTO/H_2O_2 in analogy with Schemes 7.11 and 7.13.

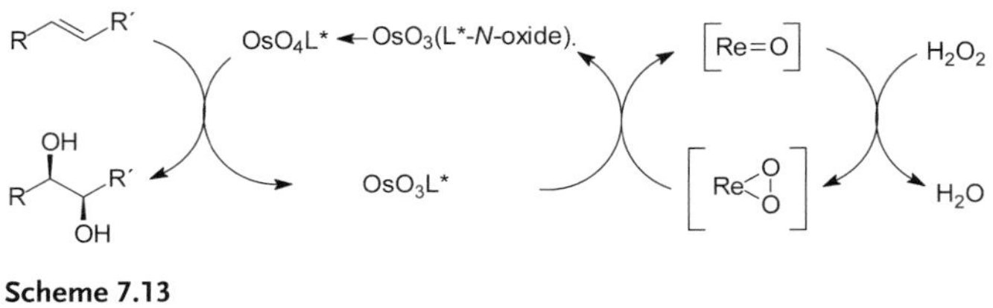

Scheme 7.13

$$R\diagdown\!\!\diagup R' + H_2O_2 \xrightarrow[\substack{\text{citric acid (5 mol\%)} \\ {}^t\text{BuOH:H}_2\text{O}}]{\substack{0.5 \text{ mol\% OsO}_4 \\ 1 \text{ mol\% MTO} \\ 20 \text{ mol\% NMM}}} R\diagdown\!\!\underset{\text{OH}}{\overset{\text{OH}}{\diagup}}\!\!R' \qquad (28)$$

R = H, R' = CH$_2$SO$_2$Ph   93%
R = Me, R' = CO$_2$Et   90%
R = CO$_2$Et, R' = CO$_2$Et   85%
R = Ph, R' = CO$_2$Et   84%

The coupled catalytic system of Scheme 7.11 was recently immobilized in an ionic liquid [bmim]PF$_6$ [111]. After completion of the reaction the product diol is extracted from the ionic liquid and the osmium, NMO and flavin stays in the ionic liquid. The immobilized catalytic system was reused five times without any loss of activity.

Choudary et al. [112] have also used the principle of *in situ* generation of *N*-oxide in catalytic amounts in osmium-catalyzed dihydroxylation. They used LDH-WO$_4^{2-}$ as the catalyst for H$_2$O$_2$ reoxidation of the amine to *N*-oxide.

7.4
Concluding Remarks

The selective oxidation of sulfides and amines to sulfoxide and amine oxides can be obtained with a variety of oxidants. In particular, catalytic oxidations employing environmentally benign oxidants such as hydrogen peroxide and molecular oxygen have attracted considerable interest recently. A number of catalytic methods that give highly selective and mildly selective oxidations with the latter oxidants are known today. Organocatalysts (e.g., flavins), biocatalysts and metal catalysts have been used for these transformations.

References

[1] S. PATAI, Z. RAPPOPORT, *Synthesis of Sulfones, Sulfoxides and Cyclic Sulfides*, Wiley, Chichester, **1994**.

[2] M. MADESCLAIR, *Tetrahedron* **1986**, *42*, 5459–5495.

[3] M. HUDLICK'Y, *Oxidation in Organic Chemistry*, ACS, Washington DC, **1990**.

[4] (a) E. N. PRILEZHAEVA, *Russ. Chem. Rev.* **2001**, *70*, 897–920; (b) M. C. CARREÑO, *Chem. Rev.* **1995**, *95*, 1717–1760; (c) S. BURRAGE, T. RAYNHAM, G. WILLIAMS, J. W. ESSEZ, C. ALLEN, M. CARDNO, V. SWALI, M. BRADLEY, *Chem. Eur. J.* **2000**, *6*, 1455–1466; (d) A. PADWA, M. D. DANCA, *Org. Lett.* **2002**, *4*, 715–717.

[5] (a) I. FERNANDÉZ, N. KHIAR, *Chem. Rev.* **2003**, *103*, 3651; (b) A. THORARENSEN, A. PALMGREN, K. ITAMI, J. E. BÄCKVALL, *Tetrahedron Lett.*, **1997**, *38*, 8541; (c) R. TOKUNOH, M. SODEOKA, K. AOE, M SHHIBASAKI, *Tetrahedron Lett.* **1995**, *36*, 8035–8038; (d) D. G. I. PETRA, P. C. J. KAMER, A. L. SPEK, H. E. SCHOEMAKER, P. W. N. M. VAN LEEUWEN, *J. Org. Chem.* **2000**, *65*, 3010–3017; (e) J. G. ROWLANDS, *Synlett* **2003**, 236–240; (f) R. DORTA, L. J. W. Shimon, D. Milstein, *Chem. Eur. J.* **2003**, *9*, 5237–5249.

[6] A. M. KHENKIN, R. NEUMANN, *J. Am. Chem. Soc.* **2002**, *124*, 4198–4199.

[7] (a) K. C. NICOLAU, R. L. MAGOLDA, W. J. SIPIO, W. E. BARNETTE, Z. LYZENKO,

M. M. Joullie, *J. Am. Chem. Soc.* **1980**, *102*, 3784–3793; (b) L. A. Paquette, R. V. C. Carr, *Org. Synth.* **1985**, *64*, 157.

[8] (a) R. W. Murray, R. Jeyaraman, *J. Org. Chem.* **1985**, *50*, 2487–2853; (b) R. W. Murray, *Chem. Rev.* **1989**, *89*, 1187; (c) S. Colonna, N. Gaggero, *Tetrahedron Lett.* **1989**, *30*, 6233.

[9] M. E. González-Núñez, R. Mello, J. Royo, J. V. Rios, G. Asensio, *J. Am. Chem. Soc.* **2002**, *124*, 9154–9163.

[10] A. Ishii, C. Tsuchiya, T. Shimada, K. Furusawa, T. Omata, J. Nakayama, *J. Org. Chem.* **2000**, *65*, 1799–1806.

[11] (a) J. Gildersleeve, A. Smith, K. Sakurai, S. Raghavan, D. Kahne, *J. Am. Chem. Soc.* **1999**, *121*, 6176–6182; (b) Y.-N. J. Jin, A. Ishii, Y. Sugihara, J. Nakayama, *Tetrahedron Lett.* **1998**, *39*, 3525–3528; (c) R. L. Beddoes, J. E. Painter, P. Quayle, P. Patel, *Tetrahedron* **1997**, *53*, 17297–17306; (d) W. A. Schenk, J. Frisch, M. Durr, N. Burzlaff, D. Stalke, R. Fleischer, W. Adam, F. Prechtl, A. K. Smerz, *Inorg. Chem.* **1997**, *36*, 2372–2378; (e) W. A. Schenk, M. Durr, *Chem. Eur. J.* **1997**, *3*, 713.

[12] C. J. Foti, J. D. Fields, *Org. Lett.* **1999**, *1*, 903–904; P. J. Kropp, G. W. Breton, J. D. Fields, J. C. Tung, B. R. Loomis, *J. Am. Chem. Soc.* **2000**, *122*, 4280–4285.

[13] (a) B. M. Trost, D. P. Curran, *Tetrahedron Lett.* **1981**, *22*, 1287–1290; (b) B. M. Trost, R. Braslau, *J. Org. Chem.* **1988**, *53*, 532.

[14] (a) K. S. Ravikumar, J.-P. Bégué, D. Bonnet-Delpon *Tetrahedron Lett.* **1998**, *39*, 3141–3144; (b) K. S. Ravikumar, Y. M. Zhang, J.-P. Bégué, D. Bonnet-Delpon *Eur. J. Org. Chem.* **1998**, 2937–2940.

[15] Y. Watanabe, T. Numata. S. Oae, *Synthesis*, **1981**, 204.

[16] H. S. Schultz, H. B. Freyermuth, S. R. Bue, *J. Org. Chem.* **1963**, *28*, 1140.

[17] Y. Ishi, H. Tanaka, Y. Nisiyama, *Chem. Lett.* **1994**, 1.

[18] N. M. Gresley, W. P. Griffith, A. C. Laemmel, H. I. S. Nogueira, B. C. Parkin, *J. Mol.Catal.* **1997**, *117*, 185–198.

[19] R. Neumann, D. Juviler, *Tetrahedron*, **1996**, *52*, 8781–8788.

[20] K. Sato, M. Hyodo, M. Aoki, X. Zheng, R. Noyori, *Tetrahedron*, **2001**, 57, 2469–2476.

[21] B. M. Choudary, B. Bharathi, Ch. Venkat Reddy, M. Lakshmi Kantam, *J. Chem. Soc., Perkin Trans. 1* **2002**, 2069–2074.

[22] J. Brinksma, R. La Crois, B. L. Feringa, M. I. Donnoli, C. Rosini, *Tetrahedron Lett.* **2001**, *42*, 4049–4052.

[23] A. Colombo, G. Marturano, A. Pasini, *Gazz. Chim. Ital.* **1986**, *116*, 35–40.

[24] (a) C. Bolm, F. Bienewald, *Angew. Chem., Int. Ed. Engl.* **1995**, *34*, 2640–2642; (b) C. Bolm, F. Bienewald, *SynLett* **1998**, 1327–1328.

[25] A. H. Vetter, A. Berkessel, *Tetrahedron Lett.* **1998**, *39*, 1741–1744.

[26] C. Ohta, H. Shimizu, A. Kondo, T. Katsuki, *SynLett* **2002**, 161–163.

[27] S. D. Green, C. Monti, R. F. W. Jackson, M. S. Anson, S. J. F. Macdonald, *Chem. Commun.* **2001**, 2594–2595.

[28] M. Palucki, P. Hanson, E. N. Jacobsen, *Tetrahedron Lett.* **1992**, *33*, 7177–7114.

[29] S. Choi, J.-D. Yang, M. Ji, M. Choi, M. Kee, K.-H. Ahn, S.-H. Byeon, W. Baik, S. Koo, *J. Org. Chem.* **2001**, *66*, 8192–8198.

[30] M. Matteucci, G. Bhalay, M. Bradley, *Org. Lett.* **2003**, *5*, 235–237.

[31] (a) S. Ball, T. C. Bruice, *J. Am. Chem. Soc.* **1979**, *101*, 4017–4019; (b) T. C. Bruice, *Acc. Chem. Res.* **1980**, *13*, 256–262.

[32] (a) S. Ball, T. C. Bruice, *J. Am. Chem. Soc.* **1980**, *102*, 6498–6503; (b) A. E. Miller, J. J. Bischoff, C. Bizup, P. Luninoso, S. Smiley, *J. Am. Chem. Soc.* **1986**, *108*, 7773–7778; (c) S. Oae, K. Asada, T. Yoshimura, *Tetrahedron Lett.* **1983**, *24*, 1265–1268.

[33] S.-I. Murahashi, T. Oda, Y. Masui, *J. Am. Chem. Soc.* **1989**, *111*, 5002–5003.

[34] (a) S. Shinkai, T. Yamaguchi, O. Manabe, F. Toda, *J. Chem. Soc., Chem. Commun.* **1988**, 1399–1401; (b) S. Shinkai, T. Yamaguchi, A. Kawase, A. Kitamura, O. Manabe,

J. Chem. Soc., Chem. Commun. **1987**, 1506–1508.

[35] A. B. E. Minidis, J. E. Bäckvall, Chem. Eur. J. **2001**, 7, 297–302.

[36] K. Bergstad, J. E. Bäckvall, J. Org. Chem. **1998**, 63, 6650–6655.

[37] A. A. Lindén, L. Krüger, J. E. Bäckvall, J. Org. Chem. **2003**, 68, 5890–5896.

[38] S.-I Murahashi, Angew. Chem., Int. Ed. Engl. **1995**, 34, 2443–2465.

[39] P. E. Correa, D. P. Riley, J. Org. Chem. **1985**, 50, 1787–1788.

[40] E. Bosch, J. K. Kochi, J. Org. Chem. **1995**, 60, 3172–3183.

[41] T. Iwahama, S. Sakaguchi, Y. P. Ishi, Tetrahedron Lett. **1998**, 39, 9059–9062.

[42] S. E. Martín, L. I. Rossi, Tetrahedron Lett. **2001**, 42, 7147–7151.

[43] E. Boring, Y. V. Geletii, G. L. Hill, J. Am. Chem. Soc. **2001**, 123, 1625–1635.

[44] Y. Imada, H. Ida, S. Ono, S.-I. Murahashi, J. Am. Chem. Soc. **2003**, 125, 2868–2869.

[45] C. R. Jonsons, D. McCants, Jr., J. Am. Chem. Soc. **1965**, 87, 1109–1114.

[46] P. Pitchen, E. Dunach, M. N. Deshmukh, H. B. Kagan, J. Am. Chem. Soc. **1984**, 106, 8188–8193.

[47] F. Di Furia, G. Modena, R. Seraglia, Synthesis **1984**, 325–326.

[48] S. Zhao, O. Samuel, H. B. Kagan, Org. Synth. **1989**, 68, 49–56.

[49] (a) J. M. Brunel, H. B. Kagan, SynLett **1996**, 404–406; (b) J. M. Brunel, H. B. Kagan, Bull. Soc. Chim. Fr. **1996**, 133, 1109–1115.

[50] (a) H. Kagan, T. Lukas, In Transition Metals for Organic Synthesis, Eds. M. Beller, C. Bolm, VCH-Wiley, Weinheim, **1998**, pp. 361–373; (b) H. B. Kagan, In Catalytic Asymmetric Synthesis, Ed. I. Ojima, Wiley-VCH, Weinheim, **2000**, pp. 327–356.

[51] (a) K. Yamamoto, H. Ands, T. Shuetas, H. Chikamatsu, Chem. Comm. **1989**, 754–755; (b) S. Superchi, C. Rosini, Tetrahedron: Asymmetry **1997**, 8, 349–352.

[52] (a) Umemura, J. Org. Chem. **1993**, 58, 4529–4533; (b) Reetz, Tetrahedron Lett. **1997**, 38, 5273–5276.

[53] Y. Yamanoi, T. Imamoto, J. Org. Chem. **1997**, 62, 8560–8564.

[54] H. Cotton, T. Elebring, M. Larsson, L. Li, H. Sörensen, S. von Unge, Tetrahedron: Asymmetry **2000**, 11, 3819–3825.

[55] (a) K. Noda, N. Hosoya, Yanai, R. Irie, T, Katsuki, Tetrahedron Lett. **1994**, 35, 1887–1890; (b) K. Noda, N. Hosoya, Yanai, R. Irie, Y. Yamashita, T, Katsuki, Tetrahedron **1994**, 50, 9609–9618.

[56] H. Tohma, S. Takizawa, H. Watanabe, Y. Kita, Tetrahedron Lett. **1998**, 39, 4547–4550.

[57] V. G. Shukla, D. Slagaonkar, K. G. Akamanchi, J. Org. Chem. **2003**, 68, 5422–5425.

[58] V. M. Dembitsky Tetrahedron, **2003**, 59, 4701–4720

[59] S. Collonna, N. Gaggero, P. Pasta, G. Ottolina, Chem. Commun. **1996**, 2303–2307.

[60] S. Collonna, N. Gaggero, C. Richelmi, P. Pasta, Trends Biotechnol. **1999**, 17, 163–168.

[61] M. P. J. Van Deurzen, I. J. Remkes, F. Van Rantwijk, R. A. Sheldon, J. Mol. Catal. A: Chem. **1997**, 117, 329–337

[62] S. G. Allenmark, M.A. Andersson, Tetrahedron: Asymmetry **1996**, 7, 1089–1094.

[63] M. A. Andersson, A. Willets, S. G. Allenmark, J. Org. Chem. **1997**, 62, 8455–8458; (b) M. A. Andersson, S. G. Allenmark, Tetrahedron **1998**, 54, 15293–15304.

[64] M. A. Andersson, S. G. Allenmark, Biocatal. Biotransform. **2000**, 18, 79–86.

[65] N. M. Kamerbeek, D. B. Janssen, W. J. H. van Berkel, M. W. Fraaije, Adv. Synth Catal. **2003**, 345, 667–678.

[66] S. Colonna, N. Gaggero, G. Carrea, P. Pasta, Chem. Commun. **1997**, 439–440.

[67] F. Secundo, G. Carrea, S. Dallavalle, G. Franzosi, Tetrahedron: Asymmetry **1993**, 4, 1981–1982.

[68] C. T. Walsh, Y. C. J. Chen, Angew. Chem., Int. Ed. Engl. **1988**, 27, 333.

[69] G. Carrea, B. Redigolo, S. Riva, S. Collonna, N. Gaggero, E. Battistel, D. Bianchini, Tetrahedron: Asymmetry **1992**, 3, 1063–1068.

[70] N. M. Kamerbeek, A. J. J., M. W.

Fraaije, D. B. Janssen, *Appl. Environ. Microbiol.* **2003**, *69*, 419–426

[71] N. M. Kamerbeek, M. J. Moonen, J. G. Van Der Ven, W. J. H. Van Berkel, M. W. Fraaije, D. B. Janssen, *Eur. J. Biochem.* **2001**, *268*, 2547–2557.

[72] G. Chen, M. M. Kayser, M. D. Mihovilovic, M. E. Mrstik, C. A. Martinez, J. D. Stewart, *New J. Chem.* **1999**, *23*, 827–832.

[73] A. Tuynman, M. K. S. Vink, H. L. Dekker, H. E. Schoemaker, R. Wever, *Eur. J. Biochem.* **1998**, *258*, 906–913.

[74] W. Adam, C. Mock-Knoblauch, C. R. Saha-Möller, *J. Org.Chem.* **1999**, *64*, 4834–4839.

[75] K. Okrasa, E. Guibé-Jampel, M. Therisod, *J. Chem. Soc., Perkin Trans. 1*, **2000**, 1077–1079.

[76] (a) S.-I. Murahashi, Y. Imada, In *Transition Metals for Organic Synthesis*, Eds. M. Beller, C. Bolm, VCH-Wiley, Weinheim, **1998**, Vol. 2, pp. 373–383; (b) Chapter 5 of this book, Section 5.3.3.

[77] T. L. Gilchrist In *Comprehensive Organic Synthesis*, Eds. B. M. Trost, I. Fleming, Pergamon Press, London, **1991**, Vol. 7, pp. 735–756.

[78] A. Albini, *Synthesis* **1993**, 263–267.

[79] J. D. Sauer, *Surfactant Sci. Ser.* **1990**, *34*, 275.

[80] V. VanRheenen, D. Y. Cha, W. M. Hartley, *Org. Syn. Coll. Vol.* **1988**, *VI*, 342.

[81] M. Schröder, *Chem. Rev.* **1980**, *80*, 187.

[82] L. Ahrgren, L. Sutin, *Org. Proc. Res. Dev.* **1997**, *1*, 425.

[83] S. Cicchi, F. Cardona, A. Brandi, M. Corsi, A. Goti, *Tetrahedron Lett.* **1999**, *40*, 1989.

[84] W. P. Griffith, S. V. Ley, G. P. Whitcombe, A. D. White, *J. Chem. Soc., Chem. Commun.* **1987**, 1625.

[85] S. V. Ley, J. Norman, W. P. Griffith, S. P. Marsden, *Synthesis* **1994**, 639.

[86] V. Franzen, S. Otto, *Chem. Ber.* **1961**, *94*, 1360.

[87] S. Suzuki, T. Onishi, Y. Fujita, H. Misawa, J. Otera, *Bull. Chem. Soc. Jpn.* **1986**, *59*, 3287.

[88] A. G. Godfrey, B. Ganem, *Tetrahedron Lett.* **1990**, *31*, 4825.

[89] (a) *Kirk-Othmer Encyclopedia of Chemical Technology*, John Wiley and Sons, Wiley-Interscience, New York, 4th edn., **1997**, Vol. 23, p. 524; (b) T. A. Isbell, T. P. Abbott, J. A. Dvorak, US Pat. 6,051,214, **2000**.

[90] A. C. Cope, E. Ciganek, *Org. Syn. Coll. Vol.* **1963**, *IV*, 612.

[91] (a) H. S. Mosher, L. Turner, A. Carlsmith, *Org. Syn. Coll. Vol.* **1963**, *IV*, 828; (b) P. Brougham, M. S. Cooper, D. A. Cummerson, H. Heaney, N. Thompson, *Synthesis* **1987**, 1015; (c) J. C. Craig, K. K. Purushothaman, *J. Org. Chem.* **1970**, *35*, 1721.

[92] W. W. Zajac, T. R. Walters, M. G. Darcy, *J. Org. Chem.* **1988**, *53*, 5856.

[93] A. L. Baumstark, M. Dotrong, P. C. Vasquez, *Tetrahedron Lett.* **1987**, *28*, 1963.

[94] M. Ferrer, F. Sanchez-Baeza, A. Messeguer, *Tetrahedron*, **1997**, *53*, 15877–15888.

[95] S. Dayan, M. Kol, S. Rozen, *Synthesis* **1999**, (Spec. Issue.), 1427–1430.

[96] (a) L. Kuhnen, *Chem. Ber.* **1966**, *99*, 3384; (b) M. N. Sheng, J. G. Zajacek, *Org. Synth. Collective Volume* **1988**, *6*, 501.

[97] B. M. Coudary, B. Bharathi, V. Reddy, M. L. Kantam, K. V. Raghavan, *Chem. Commun.* **2001**, 1736–1738.

[98] S. L. Jain, B. Sain, *Angew. Chem., Int. Ed. Engl.* **2003**, *42*, 1265–1267.

[99] S. L. Jain, B. Sain, *Chem Commun.* **2002**, 1040–1041.

[100] C. Coperet, H. Adolfsson, T.-A. V. Khuong, K. A. Yudin, K. B. Sharpless, *J. Org. Chem.* **1998**, *63*, 1740–1741.

[101] C. Coperet, H. Adolfsson, J. P. Chiang, A. K. Yudin, K. B. Sharpless, *Tetrahedron Lett.* **1998**, *39*, 761–764.

[102] S. Colonna, N. Gaggero, J. Drabowicz, P. Lyzwa, M. Mikolajczyk, *Chem. Commun.* **1999**, 1787–1788.

[103] (a) S. Colonna, V. Pironti, P. Pasta, F. Zambianchi, *Tetrahedron Lett.* **2003**, *44*, 869–871; (b) S. Colonna, V. Pironti, G. Carrea, P. Pasta, F. Zambianchi, *Tetrahedron* **2004**, *60*, 569–575.

[104] V. VanRheenen, R. C. Kelly, D. Y. Cha, *Tetrahedron Lett.* **1976**, 1973–1976.

[105] K. Bergstad, S. Y. Jonsson, J. E. Bäck-

vall, *J. Am. Chem. Soc.* **1999**, *121*, 10424–10425.

[106] S. Y. Jonsson, K. Färnegårdh, J. E. Bäckvall, *J. Am. Chem. Soc.* **2001**, *123*, 1365–1371.

[107] K. Bergstad, J. J. N. Piet, J. E. Bäckvall, *J. Org. Chem.* **1999**, *64*, 2545–2548.

[108] S. Y. Jonsson, H. Adolfsson, J.-E. Bäckvall, *Org. Lett.* **2001**, *3*, 3463–3466.

[109] S. Y. Jonsson, H. Adolfsson, J. E. Bäckvall, *Chem. Eur. J.* **2003**, *9*, 2783–2788.

[110] A. H. Éll, A. Closson, H. Adolfsson, J. E. Bäckvall, *Adv. Synth. Catal.* **2003**, *345*, 1012–1016.

[111] A. Closson, M. Johansson, J. E. Bäckvall, *Chem. Commun.* **2004**, 1494–1495.

[112] B. M. Choudary, N. S. Showdari, S. Madhi, M. L. Kantam, *Angew. Chem., Int. Ed. Engl.* **2001**, *40*, 4619–4623

8
Liquid Phase Oxidation Reactions Catalyzed by Polyoxometalates
Ronny Neumann

8.1
Introduction

Environmentally benign and sustainable transformations in organic chemistry are now considered to be basic goals and requirements in the development of modern organic syntheses. In order to meet these aims, reactions should be free of dangerous waste, have high atom economy and use solvents that are as environmentally friendly as possible. These requirements for the preparation of organic compounds, as dictated by society, has led to the introduction of mainly homogeneous liquid phase catalysis into the arena of organic chemistry. Therefore, it is not surprising that in the area of oxidative transformations one major goal is the replacement of stoichiometric procedures, using classical toxic waste-producing oxidants, with catalytic procedures using environmentally benign oxidant. Oxidants whose use is being contemplated include molecular oxygen, hydrogen peroxide, nitrous oxide and several others where there is either no byproduct, the byproduct is environmentally benign, for example, water or nitrogen, or the byproduct can be easily recovered and recycled. For synthetic utility, where high conversion and selectivity are desirable, these oxidants will require activation by appropriate, usually metal-based catalysts. Furthermore, it would be preferable, if possible, to carry out these reactions in aqueous media or organic media with a low environmental load.

As always in research involving catalysis, attention must also be paid to the vital issue of catalyst integrity, recovery and recycling. In order to succeed in carrying out the required and necessary catalytic reactions, new catalysts must obviously be developed. Beyond the description of the practical utility of the new catalysts, significant in-roads into the mechanism of oxidant and substrate activation is key to the understanding of catalytic activity and selectivity. Such an understanding will extend our possibilities of finding yet better catalysts.

Therefore, in this chapter we will survey in detail, with some emphasis on our own research, the study and use of an interesting class of oxidation catalysts, polyoxometalates, and describe their utility for oxidative transformations. Emphasis will be put on both mechanistic and synthetic aspects of their use, together with a discussion of catalytic systems designed to aid in catalyst recovery and recycling. After a

Modern Oxidation Methods. Edited by Jan-Erling Bäckvall
Copyright © 2004 WILEY-VCH Verlag GmbH & Co. KGaA, Weinheim
ISBN: 3-527-30642-0

short introduction to polyoxometalates, their structure and general properties, the major parts of this chapter will deal with three different types of oxidants: (1) monooxygen donors, (2) peroxides and (3) molecular oxygen. Finally, various systems for catalyst "engineering" designed to aid in catalyst recovery and recycling will be discussed.

8.2
Polyoxometalates (POMs)

Research applications of polyoxometalates have over the past two decades become very apparent and important, as reflected by the recent publication of a special volume of *Chemical Reviews* [1] devoted to these compounds. The diversity of research in the polyoxometalate area is significant and includes their application in many fields, including structural chemistry, analytical chemistry, surface chemistry, medicine, electrochemistry and photochemistry. However, the most extensive research on the application of polyoxometalates appears to have been in the area of catalysis, where their use as Brönsted acid catalysts and oxidation catalysts has been active since the late 1970s. Research published over the past decade or two has firmly established the significant potential of polyoxometalates as homogeneous oxidation catalysts. Through development of novel ideas and concepts, polyoxometalates have been shown to have significant diversity of activity and mechanism that in the future may lead to important practical applications. In recent years, a number of excellent and general reviews on the subject of catalysis by polyoxometalates have already been published [2] and while some repetition is inevitable we will attempt to keep the redundancy in the present chapter to a minimum.

A basic premise behind the use of polyoxometalates in oxidation chemistry is the fact that polyoxometalates are oxidatively stable. In fact as a class of compounds they are thermally stable generally to at least 350–450 °C in the presence of molecular oxygen. This, *a priori*, leads to the conclusion that for practical purposes polyoxometalates would have distinct advantages over the widely investigated organometallic compounds, which are vulnerable to decomposition due to oxidation of the ligand bound to the metal center. Polyoxometalates, previously also called heteropolyanions, are oligooxide clusters of discrete structure with a general formula $[X_xM_mO_y]^{q-}$ ($x \leq m$) where X is defined as the heteroatom and M are the addenda atoms. The addenda atoms are usually either molybdenum or tungsten in their highest oxidation state, 6^+, while the heteroatom can be any number of elements, both transition metals and main group elements; phosphorous and silicon are the most common heteroatoms. A most basic polyoxometalate, $[XM_{12}O_{40}]^{q-}$ where (X = P, Si, etc; M = W, Mo), is that of a Keggin structure, Figure 8.1(a). Such Keggin type polyoxometalates, commonly available in their protic form, are significant for catalysis only as Brönsted acid catalysts.

However, since polyoxometalate synthesis is normally carried out in water, by mixing the stoichiometrically required amounts of monomeric metal salts and adjusting the pH to a specific acidic value, many other structure types are accessible by varia-

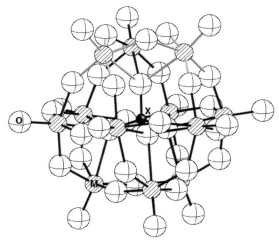

Figure 8.1 Polyoxometalates with the Keggin structure (a) $[XM_{12}O_{40}]^{9-}$ and (b) $[PM_{12}V_2O_{40}]^{5-}$

tion of the reaction stoichiometry, replacement of one or more addenda atoms with other transition or main group metals, and pH control. Oxygen atoms may also be partially substituted by fluorine, nitrogen or sulfur. In the discussion below, additional polyoxometalate structures used in catalytic applications will be described, but already at this juncture it is important to note that there are important polyoxometalate structure–oxidation reactivity and selectivity relationships, which represent a significant attraction for the use of polyoxometalates as oxidation catalysts. As may be noted from the general formula of polyoxometalates, they are also polyanionic species. This property enables fine control of polyoxometalate solubility by simple variation of the cation. A summary of this solution chemistry of polyoxometalates is given in Figure 8.2.

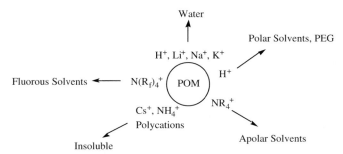

Figure 8.2 Solubility of polyoxometalates as a function of the counter cation

8.3
Oxidation with Mono-oxygen Donors

One class of polyoxometalate compounds is that where transition or main group metals, often in a lower oxidation state, substitute for a tungsten- or molybdenum-oxo group at the polyoxometalate surface. The substituting metal center is thus penta-coordinated by the "parent" polyoxometalate. The octahedral coordination sphere is completed by an additional sixth labile ligand, L (usually L = H_2O). This lability of the sixth ligand allows the interaction of the substituting transition metal atom with a reaction substrate and/or oxidant leading to reaction at the transition metal center. In analogy with organometallic chemistry the "pentadentate" polyoxometalate acts as an inorganic ligand. This analogy, in earlier years led to such transition metal-substituted polyoxometalates being termed "inorganic metalloporphyrins". Many structural variants of such transition metal-substituted polyoxometalates are known. For example, some of those used in catalytic applications are: (a) the transition metal-substituted "Keggin" type compounds, $[XTM(L)M_{11}O_{39}]^{q-}$ (X = P, Si; TM = $Co^{II,III}$, Zn^{II}, $Mn^{II,III}$, $Fe^{II,III}$, $Cu^{I,II}$, Ni^{II}, Ru^{III}, etc.; M = Mo, W), Figure 8.3(a); (b) the so-called "sandwich" type polyoxometalates, $\{[(WZnTM_2(H_2O)_2][(ZnW_9O_{34})_2]\}^{q-}$, Figure 8.3(b), having a ring of transition metals between two truncated Keggin "inorganic ligands"; and (c) the polyfluorooxometalates, Figure 8.3(c), of a quasi Wells-Dawson structure. In particular, these last two compound classes often have superior catalytic activity and stability.

One of the first uses of transition metal-substituted polyoxometalates (TMSPs) was in the context of a comparison of the catalytic activity of such TMSPs with their metalloporphyrin counterparts. Thus, it seemed natural at the time to evaluate the activity of such TMSPs with iodosobenzene and pentafluoroiodosobenzene as oxidant [3]. In particular the manganese(II/III)-substituted Keggin, $[PMn(H_2O)W_{11}O_{39}]^{5-}$ [see Figure 8.3(a)] and Wells-Dawson $[\alpha\text{-}P_2Mn(H_2O)W_{17}O_{61}]^{7-}$ polyoxometalates showed good activity and very high selectivity for the epoxidation of alkenes, and some activity for the hydroxylation of alkanes that usually compared favorably with the activity of the manganese(III) tetra-2,6-dichlorophenyporphyrin. The stereoselectivities and regioselectivities observed in epoxidation of probe substrates such as limonene, cis-stilbene, naphthalene and others, lead to the hypothesis, yet to be proven, that a reactive manganese(V) oxo intermediate was the catalytically active species. More recently a comparison was made between Wells-Dawson type polyoxometalates incorporating a metal center, e.g., $[\alpha\text{-}P_2Mn(H_2O)W_{17}O_{61}]^{7-}$, and a polyoxometalate supporting a manganese center, $[Mn^{II}(CH_3CN)_x/(P_2W_{15}Nb_3O_{62})^{9-}]$ and it was found that the incorporating analogue was about twice as active [4]. Other manganese non-polyoxometalate supporting compounds showed very significantly reduced activity.

High valent chromium(V) oxo species, $[XCr^V = OW_{11}O_{39}]^{(9-n)-}$ (X = P^V, SiI^V), which were thought to serve as analogues of the proposed manganese intermediate, were prepared and evaluated as stoichiometric oxidants for alkene epoxidation [5]. Unfortunately, the selectivity to the epoxide product was relatively low, usually at best only ~10%, indicating that the chromium-oxo species reacted via a pathway different to that proposed for the manganese compound. Through the use of a dif-

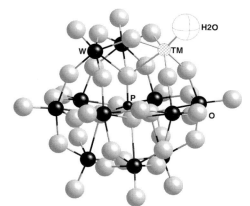

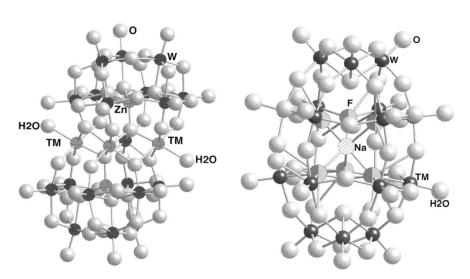

Figure 8.3 Various transition metal substituting polyoxometalates

ferent mono-oxygen donor, p-cyano-N,N-dimethylaniline-N-oxide and other N-oxides, it was later observed that the family of tetrametal substituted "sandwich" polyoxometalates, $[M_4(H_2O)_2P_2W_{18}O_{68}]^{10-}$, catalyzed epoxidation of alkenes [6]. Surprisingly, the cobalt substituted polyoxometalate was much more reactive than the corresponding manganese, iron or nickel analogues, however the $[PCo(H_2O)W_{11}O_{39}]^{5-}$ polyoxometalate was as reactive as $[Co_4(H_2O)_2P_2W_{18}O_{68}]^{10-}$. It was concluded, mainly from kinetic studies, that the active catalytic intermediate was a relatively rarely observed $Co^{IV} = O$ species. This conclusion, if correct, also leaves the identity of the active species in the manganese/iodosobenzene systems an open question since the manganese-substituted compounds are inferior catalysts with N-oxides by two orders of magnitude.

The initial observation, discussed above, that transition metal-substituted polyoxometalates could be used to catalyze oxygen transfer from iodosobenzene to organic substrates, raised the possibility of activation of other mono-oxygen donors. This conceptual advance was then coupled with the ability to incorporate noble metals into the polyoxometalate framework to utilize periodate as an oxidant. The $[Ru^{III}(H_2O)SiW_{11}O_{39}]^{5-}$ [see Figure 8.3(a)] compound catalyzed the oxidation of alkenes and alkanes using various oxygen donors [7]. In particular, sodium periodate proved to be a mild and selective oxidant enabling the oxidative cleavage of carbon–carbon double bonds in styrene derivatives to yield the corresponding benzaldehyde products in practically quantitative yields. A further kinetic and spectroscopic investigation led to the proposal of a reaction mechanism that included the formation of a metallocyclooxetane intermediate, which is transformed into a Ru^V cyclic diester in the rate determining step, requiring water. The diester is then rearranged via carbon–carbon bond cleavage to form the benzaldehyde product [8]. The periodate-$[Ru^{III}(H_2O)SiW_{11}O_{39}]^{5-}$ system was also nicely adapted in an electrochemical two phase system where iodate, formed from spent periodate, is reoxidized at a lead oxide electrode [9]. Highly selective and efficient synthesis of aldehydes by carbon–carbon bond cleavage was possible. These ruthenium-substituted Keggin compounds were also used for the oxidation of alkanes to alcohols and ketones using sodium hypochlorite [10], and the oxidation of alcohols and aldehydes to carboxylic acids with potassium chlorate as oxidant [11]. The stability of Keggin type compounds to basic sodium hypochlorite conditions is, however, suspect and research should be carried out to validate the catalyst integrity under such conditions.

Ozone is a highly electrophilic oxidant that is attractive as an environmentally benign oxidant (33% active oxygen, O_2 byproduct). Although its reactivity with nucleophilic substrates such as alkenes is well known to be very high, non-catalytic reactions at saturated carbon–hydrogen bonds are of limited synthetic value. The catalytic activation of ozone in aqueous solution was shown to be possible with the manganese-substituted polyoxometalate, $\{[WZnMn^{II}_2(H_2O)_2][(ZnW_9O_{34})_2]\}^{12-}$ [12]. Good conversions of alkanes into ketones, e.g., cyclohexane to cyclohexanone, were observed at 2 °C within 45 min. Low temperature, in situ observation of an emerald green intermediate by UV-VIS and EPR led to the suggestion that a high valent manganese ozonide was the catalytically active species. The manganese ozonide complex was formulated as POM–Mn^{IV}–O–O–O·, although other canonical or tautomeric forms may be envisaged, e.g., POM–Mn^{III}–O–O–O^+ or POM–Mn^V–O–O–O^-. In the absence of a substrate the ozonide intermediate very quickly decays by reduction or disproportionation to a brown manganese(IV) oxo or hydroxy species. In the presence of alkanes, reactivity was observed that is typical of radical intermediates, while with alkenes stereoselective epoxidation was noted. The possible reaction pathways are compiled in Scheme 8.1.

Nitrous oxide is also potentially attractive as an environmentally benign oxidant (36% active oxygen, N_2 byproduct) but it is usually considered to be an inert [13] and poor ligand towards transition metals [14]. In practice, there are only a few catalytic systems that have been shown to be efficient for the activation of N_2O for selective hydrocarbon oxidation. Notable among them are the iron containing acidic zeolites [15], which at elevated temperatures are thought to yield surface activated iron-oxo species (α-oxygen) [16], capable of transferring an oxygen atom to inert hydrocarbons

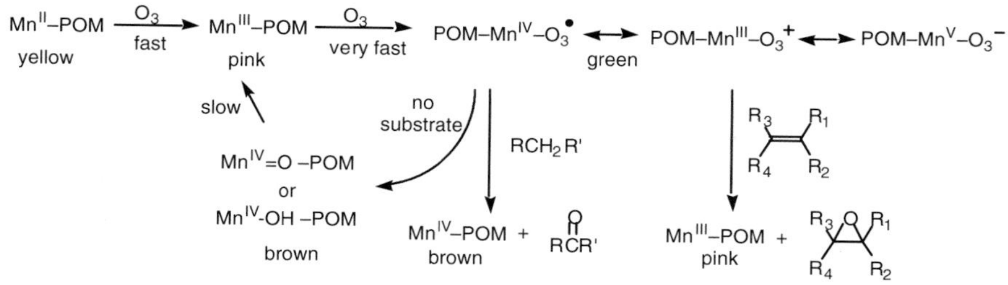

8.3 Oxidation with Mono-oxygen Donors

Scheme 8.1 Activation of ozone and proposed catalytic pathways in the presence of $\{[WZnMn^{II}_2(H_2O)_2][(ZnW_9O_{34})_2]\}^{12-}$

such as benzene [17]. Iron oxide on basic silica has been shown to catalyze, albeit non-selectively, propene epoxidation [18]. Also, stoichiometric oxygen transfer from nitrous oxide to a ruthenium porphyrin [19] yielded a high valent ruthenium-dioxo species capable of oxygen transfer to nucleophiles such as alkenes and sulfides. Under more extreme conditions (140 °C, 10 atm N_2O), the ruthenium porphyrin has recently been shown to catalyze oxygen transfer to trisubstituted alkenes [20], and the oxidation of alcohols to ketones [21]. We have just shown that a manganese-substituted polyoxometalate of the "sandwich" structure, $^8Q_{10}[Mn^{III}_2ZnW(ZnW_9O_{34})_2]$ [see Figure 8.3(b), $^8Q = (C_8H_{17})_3CH_3N^+$], is capable of selectively (>99.9%) catalyzing the epoxidation of alkenes, Scheme 8.2 [22]. It appears that there is only little correlation between the catalytic activity and the nucleophilicity of the alkene. In addition, reactions were stereoselective.

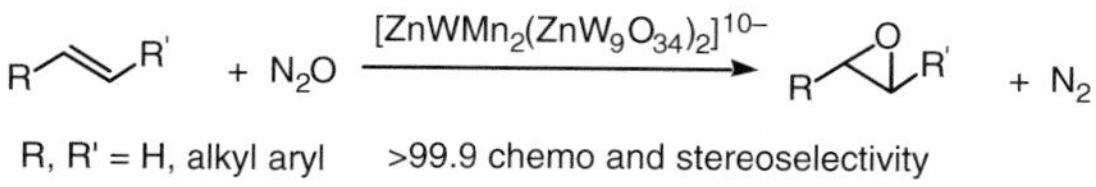

Substrate	TON
1-octene	10
E-2-octene	14
cyclooctene	19
1-decene	8
cyclohexene	9
Z-2-hexen-1-ol	21
E-2-hexen-1-ol	19
Z-stilbene	15
E-stilbene	23

Scheme 8.2 Results for epoxidation of alkenes with N_2O catalyzed by $Q_{10}[Mn^{III}_2ZnW(ZnW_9O_{34})_2]$

Also the vanadium substituted polyoxomolybdate, $^4Q_5[PV_2Mo_{10}O_{40}]$, [see Figure 1(b), $^4Q = (C_4H_9)_4N^+$], was catalytically active and highly selective in the oxidation of primary and secondary alcohols with nitrous oxide, to yield aldehydes and ketones, respectively, Scheme 8.3 [23]. In addition the same catalyst under similar conditions catalyzed the oxidation of alkylarenes, Scheme 8.3. In the oxidation of alkylarenes with hydrogen atoms at β-positions, the substrate is dehydrogenated, e.g., cumene to α-methylstyrene, whereas in the absence of β-hydrogen atoms ketones are formed, e.g., diphenylmethane to benzophenone.

It is important to note that the catalysts are orthogonal in their catalytic activity, that is $^8Q_{10}[Mn^{III}_2ZnW(ZnW_9O_{34})_2]$ is active only for alkene epoxidation but not for

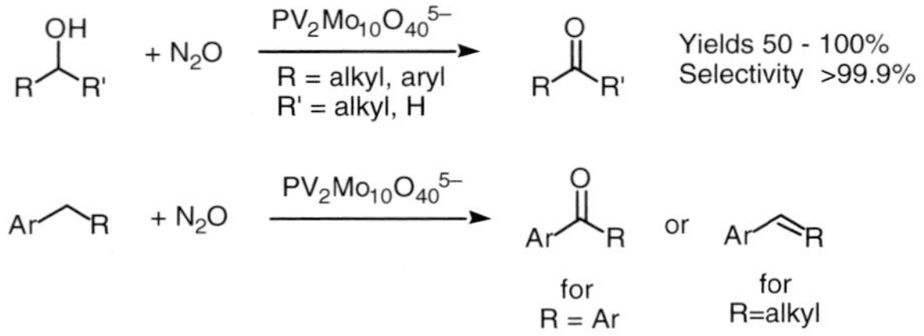

Scheme 8.3 Oxidation of alkylarenes and alcohols with N_2O catalyzed $Q_5[PV_2Mo_{10}O_{40}]$

alcohol or alkylarene oxidation, while for $^4Q_5[PV_2Mo_{10}O_{40}]$ the opposite is true. The mechanistic understanding of the catalytic activation of nitrous oxide is still quite rudimentary; however, one very notable observation is that nitrous oxide appears to be oxidized in the presence of both polyoxometalates, as was measured by the reduction (UV-VIS and EPR) of the polyoxometalates in its presence under reaction conditions without substrate. Considering the high oxidation potential of nitrous oxide as measured by its ionization potential in the gas phase, this result is surprising. A further important observation in reactions catalyzed by $^4Q_5[PV_2Mo_{10}O_{40}]$, and from correlation of rates with homolytic benzylic carbon–hydrogen bond dissociation energies and Hammett plots, indicate that perhaps an intermediate polyoxometalate–nitrous oxide complex leads to carbon–hydrogen bond cleavage. Additional experiments are required to further understand the mechanistic picture.

Sulfoxides are potentially interesting oxidants and/or oxygen donors, notably used in numerous variants of Swern type oxidations of alcohols in the presence of a stoichiometric amount of an electrophilic activating agent [24]. The deoxygenation of sulfoxides to sulfides catalyzed by metal complexes with oxygen transfer to the metal complex or to reduced species, such as hydrohalic acids, phosphines, carbenes and carbon monoxide, is also well established [25]. In this context it has recently been demonstrated in our group that sulfoxides can be used as oxygen donors/oxidants in polyoxometalate catalyzed reactions. For the first time an oxygen transfer from a sulfoxide to an alkylarene hydrocarbon, to yield sulfide and a carbonyl product, was demonstrated; in certain cases oxidative dehydrogenation was observed, Scheme 8.4 [26].

Further research on the reaction mechanism revealed that the reaction rate was correlated with the electron structure of the sulfoxide; the more electropositive sulf-

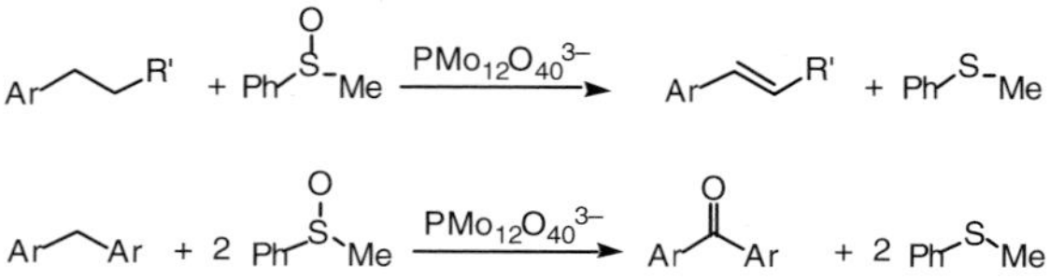

Substrate	TON
xanthene	145
diphenylmethane	142
fluorene	67
triphenylmethane	70
Isochroman	61
bibenzyl	26
dihydroanthracene	300
dihydrophenathrene	55

Scheme 8.4 Oxidation of alkylarenes with phenylmethylsulfoxide catalyzed by $Q_3[PMo_{12}O_{40}]$

oxides were better oxygen donors. Excellent correlation of the reaction rates with the heterolytic benzylic carbon–hydrogen bond dissociation energies indicated a hydride abstraction mechanism in the rate determining step to yield a carbocation intermediate. The formation of 9-phenylfluorene as a byproduct in the oxidation of triphenylmethane supports this suggestion. Further kinetic experiments and ^{17}O NMR showed the formation of a polyoxometalate–sulfoxide complex before the oxidation reaction, this complex being the active oxidant in these systems. Subsequently, in a similar reaction system, sulfoxides were used to facilitate the aerobic oxidation of alcohols [27]. In this manner, benzylic, allylic and aliphatic alcohols were all oxidized to aldehydes and ketones in a reaction catalyzed by Keggin type polyoxomolybdates, $[PV_xMo_{12-x}O_{40}]^{-(3+x)}$ ($x = 0, 2$), with DMSO as solvent. The oxidation of benzylic alcohols was quantitative within hours and selective to the corresponding benzaldehydes, but the oxidation of allylic alcohols was less selective. The oxidation of aliphatic alcohols was slower but selective. In mechanistic studies considering the oxidation of benzylic alcohols, similar to the oxidation of alkylarenes, a polyoxometalate–sulfoxide complex appears to be the active oxidant. Further isotope labeling experiments, kinetic isotope effects, and especially Hammett plots, showed that oxidation occurs by oxygen transfer from the activated sulfoxide and elimination of water from the alcohol. However, the exact nature of the reaction pathway is dependent on the identity of substituents on the phenyl ring.

Summarizing the information discussed in the section above, it can be noted that polyoxometalates appear to be versatile oxidation catalysts capable of activating various mono-oxygen donors such as iodosobenzene, periodate, ozone, nitrous oxide and sulfoxides. Some of these reactions are completely new from both a synthetic and mechanistic point of view. The various reaction pathways expressed are also rather unique and point to the many options and reaction pathways available for oxidation catalyzed by polyoxometalates.

8.4
Oxidation with Peroxygen Compounds

Before specifically discussing oxidation by peroxygen compounds using polyoxometalates as catalysts, a few general comments concerning peroxygen compounds as oxidants or oxygen donors are worth noting. Firstly, from a practical point of view, hydrogen peroxide is certainly the most sustainable oxidant of this class, since it has a high percentage of active oxygen, 47%, it is inexpensive, and the byproduct of oxidation is water. On the down-side, its use as an aqueous solution presents problems of compatibility and reactivity with hydrophobic organic substrates or solvents and some precautions must be taken, such as working under reasonably low concentrations (usually < 20 wt% in a polar organic solvent) to prevent safety hazards. Various "solid" forms of hydrogen peroxide such as urea-hydroperoxide, sodium perborate and sodium percarbonate are also available. Alkyl hydroperoxides, notably *tert*-butylhydroperoxide, have the advantage that they are freely soluble in organic media and can thus be used in strictly non-aqueous solvents. The alcohol byproduct resulting from the use of alkyl hydroperoxides as oxygen donors can often be easily recovered, for example by distillation, and at least in principle the alkyl hydroperoxide can be re-synthesized from the alcohol. There are also other peroxygen oxidants readily available; one notable inorganic compound is monoperoxosulfate, HSO_5^-, normally available as a triple salt, Oxone™.

From a mechanistic point of view, it is important to realize that polyoxometalates may interact with peroxygen oxidants in several different ways depending on the composition, structure and redox potential of the polyoxometalate compounds. On the one hand, one may expect reaction pathways typical for any oxotungstate or oxomolybdate compounds with formation of peroxo or hydroperoxy (alkylperoxy) intermediates, capable of oxygen transfer reactions with nucleophilic substrates such as alkenes to yield epoxides. On the other hand, depending on the redox potential of the polyoxometalate, a varying degree of homolytic cleavage of oxygen–oxygen and hydrogen–oxygen bonds will lead to hydroxy (alkoxy) and peroxy (peralkoxy) intermediate radical species. The trend of increased formation of radical species as a function of increasing oxidation potential is clearly evident in the series of Keggin type heteropoly acids: $H_5PV_2Mo_{10}O_{40} > H_3PMo_{12}O_{40} > H_3PW_{12}O_{40}$. In particular, hydroxy or alkoxy radicals will lead to further hydrogen abstraction from the substrate molecules and formation of additional radical species. The rate of formation and fate of these latter radical species will determine the conversion, selectivity and identity of the products formed in the reaction. This tendency for homolytic cleavage in the peroxygen compounds can also be expected to be strongly influenced by the presence of substituting transition metals into the polyoxometalate structure. A high oxidation potential of the polyoxometalate and/or presence of redox active transition metals will also lead to dismutation reactions and thus non-productive decomposition of the peroxygen oxidant and low yields based on the oxidant. There is also a more remote possibility that intermediate hydroperoxide species of a transition metal substituted into a polyoxometalate structure, for example a Fe^{III}–OOH intermediate, will lead to an Fe^V=O species or equivalent. To date, there has been no observation of such a biomimetic transformation in polyoxometalate catalytic chemistry.

Originally, *tert*-butylhydroperoxide was used together with transition metal-substituted Keggin type compounds and then later on more effectively with transition metal substituted "sandwich" compounds for the oxidation of alkanes to alcohols and ketones [28]. The oxidation of alkenes proceeded with low selectivity. Although not rigorously studied, it would seem quite certain that these reactions proceed by a radical mechanism *via* hydrogen abstraction by alkoxy radicals from the substrate. Interestingly, it has recently been observed that oxidation of alkanes with *tert*-butylhydroperoxide catalyzed by a polyoxomolybdate, $H_3PMo_{12}O_{40}$, may be redirected from oxygenation to oxydehydrogenation yielding alkenes as the major products [29]. Thus, both acyclic and cyclic alkanes were oxidized to alkenes by *tert*-butylhydroperoxide in acetic acid with $H_3PMo_{12}O_{40}$ as catalyst with reaction selectivity generally 90%. Some minor amounts of alcohols, ketones and hydroperoxides products formed *via* oxygenation with molecular oxygen were also obtained as were some acetate esters. The alkene product tended selectively towards the kinetically favored product rather than the thermodynamically more stable alkenes. Therefore, oxidation of 1-methylcyclohexane yielded mostly 3- and 4-methylcyclohexene rather than 1-methylcyclohexene. Similarly, in the oxidation of 2,2,4-trimethylpentane, the terminal alkene, 2,2,4-trimethyl-4-pentene, was formed in four-fold excess relative to the internal alkene, 2,2,4-trimethyl-3-pentene. A reaction scheme to explain the reaction selectivity is presented in Scheme 8.5.

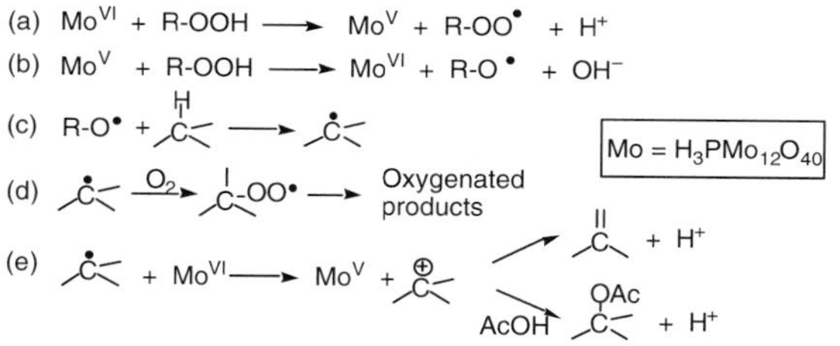

(a) $Mo^{VI} + R\text{-}OOH \longrightarrow Mo^{V} + R\text{-}OO^{\bullet} + H^{+}$
(b) $Mo^{V} + R\text{-}OOH \longrightarrow Mo^{VI} + R\text{-}O^{\bullet} + OH^{-}$
(c) $R\text{-}O^{\bullet} + \overset{H}{\underset{}{>}C<} \longrightarrow \overset{\bullet}{>}C<$
(d) $\overset{\bullet}{>}C< \xrightarrow{O_2} \overset{|}{>}C\text{-}OO^{\bullet} \longrightarrow$ Oxygenated products
(e) $\overset{\bullet}{>}C< + Mo^{VI} \longrightarrow Mo^{V} + \overset{\oplus}{>}C<$

$Mo = H_3PMo_{12}O_{40}$

$\longrightarrow >C=C< + H^{+}$

$\xrightarrow{AcOH} >\underset{OAc}{C<} + H^{+}$

Scheme 8.5 Oxidation of alkanes with tert-butylhydroperoxide catalyzed by $H_3PMo_{12}O_{40}$

tert-Butylhydroperoxide reacts with the $H_3PMo_{12}O_{40}$ catalyst to yield alkoxy and alkylperoxy radicals [reactions (a) and (b)]. The alkoxy radical, which can be trapped by spin traps and observed by EPR, homolytically abstracts hydrogen from a reactive carbon–hydrogen moiety [reaction (c)]. Instead of the usual diffusion rate-controlled oxygenation with molecular oxygen [reaction (d)], oxidative electron transfer occurs yielding a carbocation, which in turn is dehydrogenated to yield an alkene or is attacked by acetic acid to give the acetate ester as a byproduct [reaction (e)]. tert-Butylhydroperoxide has also been used for the highly selective oxidation of thioethers, e. g., tetrahydrothiophene, to the corresponding sulfoxides without further oxidation to sulfones using $H_5PV_2Mo_{10}O_{40}$ as the catalyst [30]. Although one may automatically assume that such an oxidation would take place by an oxygen transfer reaction from a polyoxometalate-alkylperoxy intermediate to the sulfide, the evidence presented indicates that in fact oxidation occurs via electron transfer from the thioether to the polyoxometalate, where the role of the tert-butylhydroperoxide is to re-oxidize the reduced polyoxometalate. This type of mechanism is in line with what is known about oxidation catalyzed by $H_5PV_2Mo_{10}O_{40}$ with oxygen as the terminal oxidant, as will be discussed in Section 8.5 below (see Scheme 8.9).

Enantioselective oxidation catalysis to yield chiral products from prochiral substrates using polyoxometalate catalysts has not been observed until recently. However, in a combined effort of several research groups it has been shown that the racemic vanadium-substituted "sandwich" type polyoxometalate, $[(V^{IV}O)_2ZnW(ZnW_9O_{34})_2]^{12-}$, is an extremely effective catalyst (up to 40 000 turnovers) at near ambient temperatures, for the enantioselective epoxidation of allylic alcohols to the 2R,3R-epoxyalcohol with the sterically crowded chiral hydroperoxide, TADOOH, as oxygen donor [31], Scheme 8.6.

The enantiomeric excesses, ee, attained using aryl substituted allylic alcohols was quite high, generally 70–90%, at high conversions, >95%. However, less sterically hindered allylic alcohols such as geraniol gave a low enantiomeric excess, 20%, of

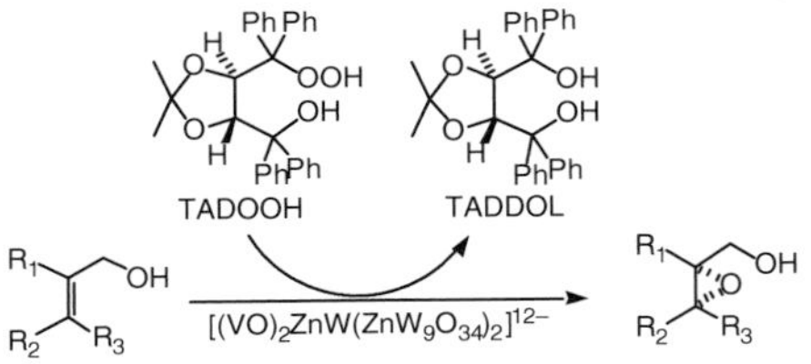

8 Liquid Phase Oxidation Reactions Catalyzed by Polyoxometalates

R_1	R_2	R_3	ee (%)
Ph	Ph	H	82
Me	Ph	H	84
Me	4-MeOPh	H	70
H	Ph	H	50
Ph	H	H	44

Scheme 8.6 Enantioselective epoxidation of allylic alcohols with a chiral hydroperoxide, TADOOH, catalyzed by $[(V^{IV}O)_2ZnW(ZnW_9O_{34})_2]^{12-}$

chiral 2R,3R-epoxygeraniol. The chiral induction observed in the reaction is thought to be due to the presence of a vanadium template for the chiral hydroperoxide and the allylic alcohol. Thus, non-functionalized alkenes, e.g., 1-phenylcyclohexene showed essentially negligible enantioselectivity. Also, substitution of vanadium with other transition metals yielded significantly lower enantioselectivity and low conversion of allylic alcohols.

As noted above the oxotungstate or oxomolybdate nature of polyoxometalate compounds boded well for their activation of hydrogen peroxide. Thus, Ishii and coworkers described the first use of polyoxometalates with 30–35% aqueous hydrogen peroxide as the oxidant. They used the commercially available phosphotungstic acid, $H_3PW_{12}O_{40}$ as catalyst. In order to utilize a biphasic reaction medium (organic substrate/aqueous oxidant) they added a quaternary ammonium salt, hexadecylpyridinium bromide, to dissolve the $[PW_{12}O_{40}]^{3-}$ in the organic apolar solvent reaction phase. Ishii's group, in addition to others, gave numerous examples of oxidation reactions typical for the use of reactions with hydrogen peroxide in the presence of tungsten-based catalysts. The first examples dealt with the epoxidation of allylic alcohols [32] and alkenes [33]. Generally high epoxide yields, >90%, were obtained with only a relatively small excess of hydrogen peroxide.

An evaluation of the catalytic activity for epoxidation reveals turnover frequencies of 5–15 per hour per tungsten atom. Under more acidic conditions and at higher temperatures the epoxides are sensitive to hydrolysis leading to the formation of vicinal diols, which are subsequently oxidized to keto-alcohols, α,β-diketones [34] or, at longer reaction times, undergo oxidative carbon–carbon bond cleavage to yield carboxylic acids and ketones. The phosphotungstate polyoxometalate was also effective for oxidation of secondary alcohols to ketones, while primary alcohols were not reactive, allowing for the high yield regioselective oxidation of non-vicinal diols to the corresponding keto-alcohols; αω-diols did, however, react to give lactones (e.g., γ-butyrolactone from 1,4-butanediol) in high yields [35]. Other research showed that alkynes [36], amines [37], and sulfides [37], could be oxidized efficiently to ketones, N-oxides, and sulfoxides and sulfones, respectively. Various quinones were also synthesized from active arene precursors [38].

While the synthetic applications involving oxidation of the various substrate types was being investigated mainly by Ishii's group [32–38] using $[PW_{12}O_{40}]^{3-}$ as the catalyst, other researchers have actively pursued studies aimed at an understanding of the identity of the true catalyst in these reactions [39–42]. In this context it should be noted that the isostructural compound $[SiW_{12}O_{40}]^{4-}$ showed almost no catalytic activity compared with $[PW_{12}O_{40}]^{3-}$ under identical conditions [32], although most recently a similar γ-$SiW_{10}O_{34}(H_2O)_2^{4-}$ polyoxometalate has been shown to have similar activity to $\{PO_4[WO(O_2)_2]\}^{3-}$ (normalized per tungsten atom) [43]. At practically the same time, Csanyi and Jaky [39], and the groups of Brégault [40], Griffith [41] and Hill [42] suggested, and convincingly proved, using various spectroscopic and kinetic probes, that the $[PW_{12}O_{40}]^{3-}$ and even more so the lacunary $[PW_{11}O_{39}]^{7-}$ polyoxometalate formed at pH ~3–4 was unstable in the presence of aqueous hydrogen peroxide, leading mainly to the formation of the peroxophosphotungstate, $\{PO_4[WO(O_2)_2]\}^{3-}$. This compound had been previously synthesized and characterized by Venturello et al. and had been shown to have very similar catalytic activity in various oxidation reactions with hydrogen peroxide [44]. A general conclusion resulting from these studies of the groups of Ishii, Venturello, Csanyi and Jaky, Brégault, Griffith and Hill is that the $\{PO_4[WO(O_2)_2]\}^{3-}$ peroxophosphotungstate compound is one on the best catalysts, especially from the point of view of synthetic versatility, amongst all of the many peroxotungstates that have been studied. A more extensive review of the complex phosphotungstate solution chemistry in the presence of hydrogen peroxide is beyond the scope of this present chapter.

In more recent years, it has been shown by Xi and coworkers that by the careful choice of the quaternary ammonium counter cation and reaction solvent, a possibly technologically practical process for the epoxidation of propene to propene oxide could be envisiged using $\{PO_4[WO(O_2)_2]\}^{3-}$ as catalyst [45]. For example, in the presence of hydrogen peroxide using a combination of toluene and tributylphosphate as solvent, a soluble $\{PO_4[WO(O_2)_2]\}^{3-}$ compound was obtained. Once the hydrogen peroxide is used up a $\{PO_4[WO_3]\}^{3-}$ compound is formed that is insoluble in the reaction medium, allowing simple recovery for recycling of the phosphotungstate species. Importantly, it was claimed that the system could be coupled with the synthesis of hydrogen peroxide from hydrogen and oxygen using the classic ethylanthraquinone process for hydrogen peroxide preparation.

The hydrolytic instability of the simple and lacunary Keggin type polyoxometalates, $[PW_{12}O_{40}]^{3-}$ and $[PW_{11}O_{39}]^{7-}$, in the presence of aqueous hydrogen peroxide, leading to formation in solution of various peroxotungstate species of varying catalytic activity, led to two intertwined issues. The first issue that came up was the necessity to carefully analyze the stability of polyoxometalates under hydrogen peroxide/hydrolytic conditions. For example, it had been claimed that lanthanide-containing polyoxometalates, $[LnW_{10}O_{36}]^{q-}$, were active catalysts for alcohol oxidation [46], however, subsequent research showed that they in fact decomposed to smaller and known peroxotungstate species, which were the catalytically active species [47]. On the other hand, other Keggin compounds appeared to be stable in the presence of aqueous hydrogen peroxide. For example, a stable peroxo species based on the Keggin structure, $[SiW_9(NbO_2)_3O_{37}]^{7-}$ was synthesized, characterized and used for epoxidation of reactive allylic alcohols but not alkenes [48]. The $Q_5[PV_2Mo_{10}O_{40}]$ (Q = quaternary ammonium cation) in aqueous hydrogen peroxide/acetic acid was stable and catalyzed the oxidation of alkylaromatic substrates at the benzylic position [49], while $Q_5[PV_2W_{10}O_{40}]$, used for the oxidation of benzene to phenol, also remained intact during the reaction [50]. Likewise, titanium-substituted Keggin type phosphotungstates are apparently stable in the presence of hydrogen peroxide [51]. In addition, it would appear that various iron-substituted Keggin compounds reported by Mizuno and coworkers for alkene and alkane oxidation are also stable in the presence of hydrogen peroxide, although the study was not completely definitive [52]. From these examples and others not noted, it is clear that certain Keggin type polyoxometalates can be stable under certain reaction conditions. Parameters to consider in this context are pH, the relative stability of the specific polyoxometalate at such a pH, solvent and temperature.

A second issue is whether there are polyoxometalate structures that are intrinsically more stable towards the hydrolytic conditions of aqueous hydrogen peroxide. It was observed that in general, larger polyoxometalates, specifically polyoxotungstates of various "sandwich" type structures were solvolytically stable towards hydrogen peroxide. Unfortunately, often the substituting transition metal catalyzes the fast decomposition of hydrogen peroxide leading to low reaction yields and non-selective reactions of little synthetic value. However, there is now a considerable body of research into several types of transition metal-substituted polyoxometalates that are synthetically useful. Various iron containing polyoxometalates of "sandwich" type structures have been investigated by the Hill group and found to have good activity for alkene oxidation with only moderate non-productive decomposition of hydrogen peroxide [53]. A relatively new class of transition metal-substituted compounds, polyfluorooxometalates, $[TM(H_2O)H_2W_{17}O_{55}F_6]^{q-}$, which have a quasi Wells-Dawson structure [see Figure 8.3(c)], and where there is partial replacement of oxygen by fluorine, proved to be very active and stable oxidation catalysts that can be monitored by ^{19}F NMR for epoxidation of alkenes and allylic alcohols with hydrogen peroxide [54]. The nickel substituted compound was the most active of the series studied. Previously, our group observed that a far more catalytically active class of compounds, which were also stable in oxidation reactions using aqueous hydrogen peroxides, were the $\{[WZnTM_2(H_2O)_2][(ZnW_9O_{34})_2]\}^{q-}$ "sandwich" type polyoxometalates. Originally we observed that among this class of compounds, the manganese and analogous rhodium derivatives dissolved in the organic phase were uniquely active when reactions were carried out in biphasic systems, preferably 1,2-dichloroethane/water [55]. At lower temperatures, highly selective epoxidation could be carried out even with cyclohexene, which is normally highly susceptible to allylic oxidation. Non-productive decomposition of hydrogen peroxide at low temperatures was minimal, but increased with temperature and was also dependent on the reactivity of the substrate. The rhodium compound was preferable in terms of minimization of hydrogen peroxide decomposition, but of course it is more expensive. Up to tens of thousands of turnovers could be attained for reactive hydrocarbon substrates [56].

The synthetic utility of the $\{[WZnMn^{II}_2(H_2O)_2][(ZnW_9O_{34})_2]\}^{12-}$ polyoxometalate as catalyst for hydrogen peroxide activation, was then extended to additional substrate classes having various functional units [57]. Thus, allylic primary alcohols were oxidized selectively to the corresponding epoxides in high yields and >90% selectivity. Allylic secondary alcohols were oxidized to a mixture of αβ-unsaturated ketones (the major product) and epoxides (the minor product). Secondary alcohols were oxidized to ketones and sulfides to a mixture of sulfoxides and sulfones. The reactivity of simple alkenes is inordinately affected by the steric bulk of the substrate. For example, the general reactivity scale for the epoxidation of alkenes indicates a strong correlation between the rate of the epoxidation and the nucleophilicity of the alkene, which is in turn correlated with the degree of substitution at the double bond. Thus, it was expected and observed that 2,3-dimethyl-2-butene would be more reactive than 2-methyl-2-heptene, however, other more bulky substrates such as 1-methycyclohexene were found to be less reactive than cyclohexene, in contrast to what would normally be expected. Furthermore, α-pinene did not react at all. This led, for example, to unusual reaction selectivity in limonene epoxidation where both epoxides were formed in equal amounts, in contrast to the usual situation where epoxidation at the endo double bond is highly preferred [58]. In these catalytic systems high turnover conditions can be easily achieved and high conversions are attained for reactive substrates, but sometimes for less reactive substrates, such as terminal alkenes, conversions and yields are low. The conversion can be increased by continuous or semi-continuous addition of hydrogen peroxide and removal of spent aqueous phases.

After the original studies on the activity of the $\{[WZnMn^{II}_2(H_2O)_2][(ZnW_9O_{34})_2]\}^{12-}$ polyoxometalates in the mid 1990s, recent industrial interest revived research in this area. Originally, the large size of the "sandwich" type structure was thought to be disadvantageous for large scale and practical applications, because even at low molar percent loads of catalyst, relatively large amounts of polyoxometalate would be required. However, the large molecular sizes (high molecular weights) have an under appreciated advantage in that they significantly simplify catalyst recovery from homogeneous solutions *via* easily applied nano-filtration techniques. This reverses some of the previous thinking in this area.

The newly initiated reinvestigation on the use of "sandwich" type polyoxometalates, $\{[WZnTM_2(H_2O)_2][(ZnW_9O_{34})_2]\}^{q-}$, showed that for a significant series of transition metals, they were exceptionally active catalysts for epoxidation of allylic alcohols using toluene or ethyl acetate as solvent [59]. The identity of the transition metal did not affect the reactivity, chemoselectivity, or stereoselectivity of the allylic alcohol epoxidation by hydrogen peroxide. These selectivity features support a conclusion that a tungsten peroxo complex rather than a high-valent transition-metal-oxo species operates as the key intermediate in the sandwich-type POMs-catalyzed epoxidations. The marked enhancement of reactivity and selectivity of allylic alcohols *versus* simple alkenes was explained by a template formation in which the allylic alcohol is coordinated through metal–alcoholate bonding and the hydrogen-peroxide oxygen source is activated in the form of a peroxo tungsten complex. 1,3-Allylic strain expresses a high preference for the formation of the *threo* epoxy alcohol, whereas in substrates with 1,2-allylic strain the *erythro* diastereomer was favored. In contrast to acyclic allylic alcohols the $\{[WZnTM_2(H_2O)_2][(ZnW_9O_{34})_2]\}^{q-}$ catalyzed oxidation of the cyclic allylic alcohols by hydrogen peroxide yielded significant amounts of enone rather than epoxides.

In the present section we have highlighted research that has been carried out using polyoxometalates as catalysts for oxidation with peroxygen compounds. Not all of the synthetic applications have been noted, but those missing have been reviewed previously [2]. It is important to stress that from a synthetic point of view, various substrates with varying functional groups can be effectively transformed into desired products. In this sense polyoxometalates constitute one class of compounds, among others, that may be considered for such transformations. In general, the often simple preparation of catalytically significant polyoxometalates, along with the conceivable recovery from solution by nano-filtration, presents a conceptual advantage in the use of polyoxometalates. From a mechanistic point of view, the wide range of properties available in the various classes of polyoxometalate compounds allows one to express reactivity in a number of ways, ranging from nucleophilic–electrophilic reactions between peroxo or hydroperoxy intermediates and organic substrates, to radical and radical chain reactions via alkoxy or hydroxy radicals formed by homolytic cleavage of peroxygen compounds by polyoxometalates.

8.5
Oxidation with Molecular Oxygen

The basic ecological and economic advantage and impetus for the use of oxygen from air as the primary oxidant for catalytic oxidative transformations are eminently clear. Yet, the chemical properties of ground state molecular oxygen limit its usefulness as an oxidant for broad synthetic applications. The limiting properties are the radical nature of molecular oxygen, the strong oxygen–oxygen bond and the fact that one electron reduction of oxygen is generally not thermodynamically favored, $\Delta G > 0$. The ground state properties of molecular oxygen lead to the situation that under typical liquid phase conditions, reactions proceed by the well-known autooxidation pathways, Scheme 8.7.

(a) $RH + M^{n+} \longrightarrow RH^{+\bullet} + M^{(n-1)+} \longrightarrow R^\bullet + M^{(n-1)+} + H^+$ electron and proton transfer
or
(b) $RH + M^{n+} \longrightarrow R^\bullet + M^{(n-1)+} + H^+$ hydrogen abstraction

(c) $R^\bullet + O_2 \longrightarrow ROO^\bullet$ propagation

(d) $ROO^\bullet + RH \longrightarrow ROOH + R^\bullet$

(e) $ROOH + M^{(n-1)+} \longrightarrow RO^\bullet + M^{n+} + OH^-$
  hydroperoxide decomposition
(f) $ROOH + M^{n+} \longrightarrow ROO^\bullet + M^{(n-1)+} + H^+$

Scheme 8.7 Metal catalyzed autooxidation pathways

Metal based catalysts may affect such pathways in various ways, but most notably they have an influence in initiating the radical chain propagation and decomposing intermediate alkylhydroperoxide species to alkoxy and peraalkoxy radicals, as discussed in Section 8.4 above. It is also very instructive to note that in nature, common monooxygenase enzymes, such as cytochrome P-450 and methane monooxygenase, use reducing agents to activate molecular oxygen, Scheme 8.8.

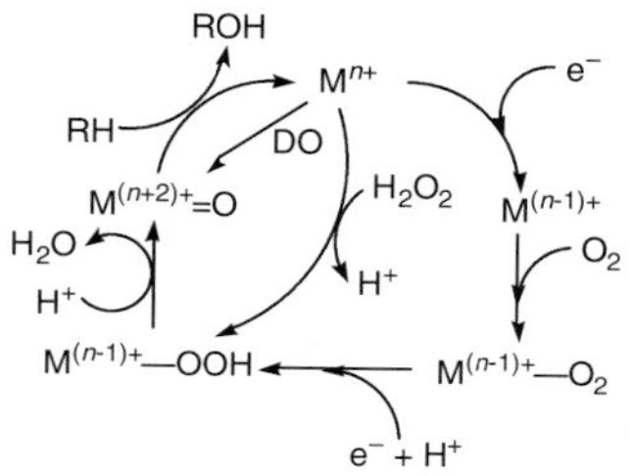

Scheme 8.8 Oxidation under reducing conditions – monooxygenase type reactions

The scheme depicted is not presented as an exact mechanistic representation, but rather to illustrate several basic points. Firstly, one may observe that oxygen is a unique oxidant compared with other oxygen donors; the oxygen donors being in principle reduced relative to molecular oxygen. In fact, even the active oxidizing intermediate in metal-catalyzed autooxidation pathways is the reduced peroxo intermediate [Scheme 8.7, reaction (d)]. In addition, only one oxygen atom of molecular oxygen in both schemes is incorporated in the product; the other atom is reduced, coupled with the formation of water. Secondly, the requirement of a reducing agent for activation negates the basic ecological and economic impetus for the use of molecular oxygen since the reducing agent becomes in fact a limiting or sacrificial reagent. These observations lead to the conclusion that newer and preferred methods of molecular oxygen activation should employ a superbiotic or abiotic approach. Polyoxometalates have played a part in such approaches to oxidative transformations.

Polyoxometalates have been investigated as catalysts for aerobic oxidation reactions that are based on various mechanistic motifs. As indicated above, one way to utilize molecular oxygen is to oxidize a hydrocarbon in the presence of a reducing agent in a reaction that proceeds *via* an autooxidation type mechanism, with the appropriate radical species as intermediates. In the most synthetically interesting case, a polyoxometalate may initiate a radical chain reaction between oxygen and an aldehyde as the reducing and sacrificial reagent. Aldehydes are practical, sacrificial reagents because the relatively low carbon–hydrogen homolytic bond energy allows easy formation of the initial intermediates, the acylperoxo radical or an acylhydroperoxide (peracid). Also some aldehydes, such as isobutyraldehye, are readily available and inexpensive. As for all peroxygen species, these active intermediates may then be used for the epoxidation of alkenes, the oxidation of alkanes to ketones and alcohols, and for the Baeyer-Villiger oxidation of ketones to esters. This has been demonstrated using both vanadium ($H_5PV_2Mo_{10}O_{40}$) and cobalt ($Co^{II}PW_{11}O_{39}^{5-}$) containing Keggin type polyoxometalates as catalysts, with isobutyraldehyde as the preferred acylperoxo/peracid precursor [60]. Significant yields at very high selectivities were obtained in most examples.

Polyoxometalates with the required redox properties can also be used in a straightforward manner as autooxidation catalysts. In this way the trisubstituted Keggin compound, $[M_3(H_2O)_3PW_9O_{37}]^{6-}$ (M = Fe^{III} and Cr^{III}) and $[Fe_2M(H_2O)_3PW_9O_{37}]^{7-}$ (M = Ni^{II}, Co^{II}, Mn^{II} and Zn^{II}) were used in the autooxidation of alkanes such as propane and isobutane to acetone and *tert*-butyl alcohol [61]. Later $[Fe_2Ni(OAc)_3PW_9O_{37}]^{10-}$ was prepared and used to oxidize alkanes such as adamantane, cyclohexane, ethylbenzene and n-decane where the reaction products (alcohol and ketone) and regioselectivities were typical for metal catalyzed autooxidations [62]. An interesting recent application of such an autooxidation is the oxidation of 3,5-di-*tert*-catechol by iron and/or vanadium substituted polyoxometalates [63]. In this reaction there is a very high turnover number, >100 000. Here the polyoxometalates are excellent mimics of catechol dioxygenase. Finally, another use of a polyoxometalate, mainly $(PCo^{II}Mo_{11}O_{39})^{5-}$, was to catalyze autooxidation of cumene to the hydroperoxo/peroxo intermediate by the Co^{II} component of the polyoxometalate, followed by oxygen transfer to an alkene such as 1-octene, to yield epoxide catalyzed by the molybdate component [64]. With the analogous tungsten polyoxometalate there was negligible oxygen transfer. In these reactions the cumene acts as a sacrificial reducing agent.

As indicated above, other mechanistic motifs have been utilized in aerobic oxidation catalyzed by polyoxometalates. Perhaps the oldest and possibly most developed of all the mechanistic motifs considered is an abiotic approach whereby, the polyoxometalate activates the reaction substrate, both organic and inorganic, rather than the oxygen that serves as the ultimate oxidant. In such catalytic reactions the polyoxometalate undergoes a redox type interaction, involving electron transfer with the reaction substrate leading to its oxidation and concomitant reduction of the polyoxometalate. Generally, the initial electron transfer is rate determining, but exceptions are known. Molecular oxygen is used to reoxidize the reduced polyoxometalate. The mechanism is summarized in Scheme 8.9.

(a) $RH_2 + POM^{n-} \xrightarrow{slow} RH_2^{+\bullet} + POM^{(n+1)-}$

(b) $RH_2^{+\bullet} + POM^{(n+1)-} \xrightarrow{fast} R + POM^{(n+2)-} + 2H^+$

RH_2 = substrate; R = product

(c) $POM^{(n+2)-} + 2H^+ + 1/2\,O_2 \longrightarrow POM^{n-} + H_2O$

Scheme 8.9 Redox type mechanism for oxidation with polyoxometalates

The basic requirement for a catalyst for such a reaction is that the oxidation potential be sufficient for oxidation of organic substrates. Yet a too high oxidation potential is also not desirable because then it will not be possible to reoxidize the polyoxometalate with molecular oxygen. For example, $(Co^{III}W_{12}O_{40})^{5-}$ has a high oxidation potential, facilitiaing oxidation of substrates such as xylene, but the resulting $(Co^{II}W_{12}O_{40}]^{6-}$ is not oxidized by molecular oxygen and thus can be used only as a stoichiometric oxidant [65]. It turns out that most commonly used catalysts for the reaction sequence described in Scheme 8.9 are the phosphovanandomolybdates, $(PV_xMo_{12-x}O_{40})^{(3+x)-}$, particularly but not exclusively when $x = 2$. This compound in its acid form has an oxidation potential of ~0.7 V as measured by cyclic voltammetry.

The use of $H_5PV_2Mo_{10}O_{40}$ was first described as a co-catalyst in the Wacker reaction [66]. The Wacker reaction oxidation of terminal alkenes is a reaction that epito-

8.5 Oxidation with Molecular Oxygen

mizes the mechanistic motif as expressed in Scheme 8.9. The $H_5PV_2Mo_{10}O_{40}$ polyoxometalate acts to reoxidize the palladium species, which is in fact, in the absence of a co-catalyst, a stoichiometric oxidant of alkenes. The use of $H_5PV_2Mo_{10}O_{40}$ replaces the classic $CuCl_2$ system, which because of the high chloride concentration is both corrosive and forms chlorinated side-products. In the 1990s, the Wacker type oxidation of ethylene to acetaldehyde was significantly improved by Grate and coworkers at Catalytica [67]. An interesting extension of the use of $H_5PV_2Mo_{10}O_{40}$ in palladium-catalyzed reactions has been to add benzoquinone as an additional co-catalyst to reoxidize the primary palladium catalyst; the resulting hydroquinone is, in turn, reoxidized by the polyoxometalate. This catalytic sequence has been used for the palladium-catalyzed oxidation of alkenes [68] and conjugated dienes [69].

$H_5PV_2Mo_{10}O_{40}$ was also used to oxidize gaseous hydrogen bromide to molecular bromine, which was utilized *in situ* for the selective bromination of phenol to 4-bromophenol [70]. More recently, $H_5PV_2Mo_{10}O_{40}$ has been used in a similar way with molecular iodine to carry out catalytic quantitative iodination of a wide range of aromatic substrates without formation of any hydrogen iodide as byproduct [71]. Other early interest in the catalytic chemistry of $H_{3+x}PV_xMo_{12-x}O_{40}$ was in the oxidation of sulfur-containing compounds of interest in the purification of industrial waste and natural gas. Oxidations included that of H_2S to elemental sulfur, sulfur dioxide to sulfur trioxide (sulfuric acid), mercaptans to disulfides and sulfides to sulfoxides and sulfones [72]. Hill and his group have continued the investigation of the oxidation chemistry of sulfur compounds, and shown that for H_2S oxidation catalysts of low oxidation potential are sufficient for these reactions, because the oxidation of H_2S to elemental sulfur is thermodynamically favored ($\Delta G < 0$) [73].

In our opinion, a significant challenge for the use of the mechanistic motif indicated in Scheme 8.9 is the use of $[PV_2Mo_{10}O_{40}]^{5-}$ for the direct oxidation of hydrocarbon substrates coupled with the suppression of autooxidation pathways. Perhaps an early use of $(PV_2Mo_{10}O_{40})^{5-}$ in this context was the reaction described by Brégault and coworkers, where $H_5PV_2Mo_{10}O_{40}$ was used in combination with dioxygen to oxidatively cleave vicinal diols [74] and ketones [75]. For example, 1-phenyl-2-propanone can be cleaved to benzaldehyde (benzoic acid) and acetic acid ostensibly through the αβ-diketone intermediate, 1-phenyl-1,2-propane dione. Similarly, cycloalkanones can be cleaved to keto-acids and di-acids. In general, the conversions and selectivities are very high. Both vanadium centers and acidic sites appeared to be a requisite for the reaction.

It would be interesting to carry out the oxidative cleavage of diols also under non-acidic conditions, as a possible pathway to the formation of a chiral pool from natural carbohydrate sources. In this context, nearly neutral forms iodomolybdates, $[IMo_6O_{24}]^{5-}$, have been found to show some activity for aerobic carbon–carbon bond cleavage reactions of diols with phenyl substituents, but unfortunately aliphatic diols are less reactive [76]. In the late 1980s to early 1990s the $[PV_2Mo_{10}O_{40}]^{5-}$ polyoxometalate was shown to be active in a series of oxidative dehydrogenation reactions, such as the oxydehydrogenation of cyclic dienes to the corresponding aromatic derivatives [77], and the selective oxydehydrogenation of alcohol compounds to aldehydes with no over-oxidation to the carboxylic acids [78]. Significantly, autooxidation of the aldehyde to the carboxylic acid was strongly inhibited, in fact particularly at higher concentrations 0.1–1 mol%, $[PV_2Mo_{10}O_{40}]^{5-}$ can be considered to be an excellent autoxidation inhibitor. Similar to alcohol dehydrogenation to aldehydes, amines may be dehydrogenated to intermediate and unstable imines [78]. In the presence of water, aldehyde is formed, which immediately may undergo further reaction with the initial amine to yield a Schiff base. Since the Schiff base is formed under equilibrium conditions, aldehydes are eventually the sole products. Under the careful exclusion of water, the intermediate imine was efficiently dehydrogenated to the corresponding nitrile.

During this period, the oxydehydrogenation of activated phenols to quinones was also demonstrated. In this way, oxidation of activated phenols in alcohol solvents yielded only oxidative dimerization products, diphenoquinones. Unfortunately under these mild conditions, the less reactive phenols did not react. It was observed that there was a clear correlation of the reaction rate with the oxidation potential of the phenol, indicating clearly that an electron transfer step was rate determining. An interesting extension of this work is the oxidation of 2-methyl-1-naphthol to 2-methyl-1,4-naphthaquinone (Vitamin K_3, menadione) in fairly high selectivities, ~83% at atmospheric O_2 [79]. This work could lead to a new environmentally favorable process to replace the stoichiometric CrO_3 oxidation of 2-methylnaphthalene used today. More recently, the finding that $[PV_2Mo_{10}O_{40}]^{5-}$ could catalyze the oxydehydrogenation of hydroxylamine to nitrosium cations led to an effective and general method for aerobic selective oxidation of alcohols to aldehydes or ketones by the use of nitroxide radicals and $[PV_2Mo_{10}O_{40}]^{5-}$ as cocatalysts. Typically, quantitative yields were obtained for oxidation of aliphatic, allylic and benzylic alcohols to the corresponding ketones or aldehydes with very high selectivity [80]. Based mostly on kinetic evidence and some spectroscopic support, a reaction scheme was formulated, Scheme 8.10. The results indicated that the polyoxometalate oxidizes the nitroxyl radical to the nitrosium cation. The latter oxidized the alcohol to the ketone/aldehyde and is reduced to the hydroxylamine, which in turn is reoxidized by $[PV_2Mo_{10}O_{40}]^{5-}$.

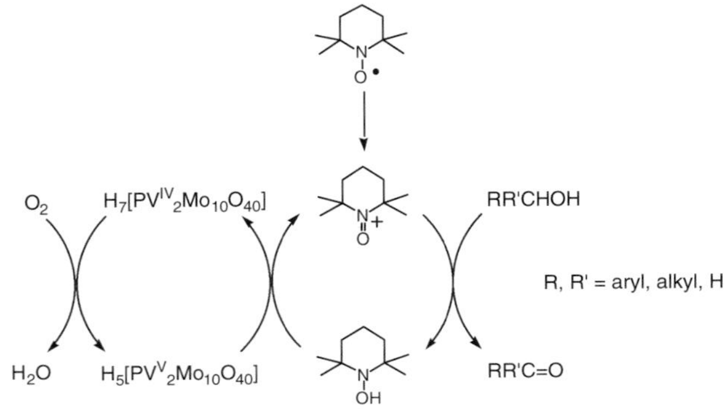

Scheme 8.10 Aerobic oxidation of alcohols with TEMPO and $H_5PV_2Mo_{10}O_{40}$

Another very important example of the use of polyoxometalates for oxydehydrogenation is the technology proposed by Hill and Weinstock for the delignification of wood pulp [81]. In the first step, lignin is oxidized selectively in the presence of cellulose and the polyoxometalate is reduced. The now oxidized and water soluble lignin component is separated from the whitened pulp and mineralized at high temperature with oxygen to CO_2 and H_2O. During the mineralization process the polyoxometalate is reoxidized by molecular oxygen (air) and can be used for an additional process cycle.

A mechanistic exploration of $(PV_2Mo_{10}O_{40})^{5-}$ catalyzed oxydehydrogenations utilizing kinetic and spectroscopic tools was also carried out [82]. The room temperature oxydehydrogenation of α-terpinene to p-cymene was chosen as a model reaction. Dehydrogenation was explained by a series of fast electron and proton transfers leading to the oxidized or dehydrogenated product and the reduced polyoxometalate. Interestingly, there were clear indications that the reoxidation of the reduced polyoxometalate by molecular oxygen went through an inner sphere mechanism, presumably via formation of a μ-peroxo intermediate. Subsequent research has given conflicting but still inconclusive evidence that the reoxidation might occur via an outer sphere mechanism [83].

In the reactions reviewed in the paragraphs immediately above, the oxidation of the hydrocarbon substrate by the polyoxometalate catalyst is purely a dehydrogenation reaction and no oxygenation of the substrate was observed, as is implicit in Scheme 8.9. An important extension of this mechanistic theme would be to couple electron transfer from the hydrocarbon to the polyoxometalate with oxygen transfer from the polyoxometalate to the reduced hydrocarbon substrate. This type of reactivity is known in an important area of gas phase heterogeneous oxidation reactions, whereby a metal oxide compound at high temperature, ~450 °C, transfers oxygen from the lattice of the oxide to a hydrocarbon substrate. This type of mechanism was originally proposed by Mars and van Krevelen, and the reaction is important in several industrial applications, such as oxidation of propene to acrolein and butane to maleic anhydride. Recently, it was shown by us that with the $[PV_2Mo_{10}O_{40}]^{5-}$ catalyst, electron transfer–oxygenation reactions were possible for oxidation of hydrocarbons at moderate temperatures, <80 °C [84]. Substrates oxygenated in this manner include polycyclic aromatic compounds and alkyl aromatic compounds. Thus, anthracene was oxidized to anthraquinone and active secondary alkyl arenes were oxidized to ketones. Use of $^{18}O_2$ and isotopically labeled polyoxometalates, as well as carrying out stoichiometric reactions under anaerobic conditions, provided strong evidence for a homogeneous Mars–van Krevelen type mechanism and evidence against autooxidation and oxidative nucleophilic substitution as alternative possibilities, Scheme 8.11.

Evidence for the activation of the hydrocarbon by electron transfer was inferred from the excellent correlation of the reaction rate with the oxidation potential of the substrate. For anthracene the intermediate cation radical was observed by ESR spectroscopy, whereas for xanthene the cation radical quickly underwent additional electron and proton transfer, yielding a benzylic cation species observed by ^1H NMR. Comparison of the oxidation potentials of the organic substrates, 1.35–1.50 V, and that of the catalyst, ~0.7 V, and analysis of the reaction rates led to the conclusion

(a) $RH_2 + H_5[PV_2Mo_{10}O_{40}]^{5-} \xrightarrow{slow} RH_2^{+\bullet} + H_5[PV_2Mo_{10}O_{40}]^{6-}$

for RH = anthracene

(b) $RH_2^{+\bullet} + H_5[PV_2Mo_{10}O_{40}]^{6-} \longrightarrow R{=}O + H_7[PV_2Mo_{10}O_{39}]^{5-}$

for RH = xanthene

(b') $RH_2^{+\bullet} + H_5[PV_2Mo_{10}O_{40}]^{6-} \longrightarrow R^+ + H_7[PV_2Mo_{10}O_{40}]^{7-}$

$R^+ + H_7[PV_2Mo_{10}O_{40}]^{7-} \longrightarrow R{=}O + H_7[PV_2Mo_{10}O_{39}]^{5-}$

(c) $H_7[PV_2Mo_{10}O_{39}]^{5-} + O_2 \longrightarrow H_5[PV_2Mo_{10}O_{40}]^{5-} + H_2O$

Scheme 8.11 Mars–van Krevelen type oxygenation of anthracene and xanthene

that the electron transfer step from the hydrocarbon to the polyoxometalate occurs through an outer-sphere mechanism. The reactions are thermodynamically feasible because of the high negative charge of the polyoxometalate catalyst. As shown by Marcus theory, this introduces a large electrostatic work function and lowers the free energy of the reaction.

Beyond, the two mechanistic themes presented above, i.e., autooxidation and redox type reactions involving electron transfer, a ruthenium-substituted "sandwich" type polyoxometalate was shown to be a catalyst for oxidation by a "dioxygenase" type mechanism, as outlined in Scheme 8.12 [85].

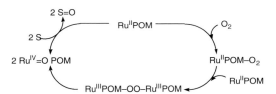

Scheme 8.12 Oxygen activation by a Ru-polyoxometalate: a dioxygenase mechanism

A range of supporting evidence for such a mechanism in the hydroxylation of adamantane and for alkene epoxidation was obtained by providing evidence against autooxidation reactions (radical traps, isotope effects and other reaction probes), and by substantiating the "dioxygenase" mechanism by confirming the reaction stoichiometry and isolating and characterizing a ruthenium-oxo intermediate. The intermediate was also shown to be viable for oxygen transfer in a quantitative and stereoselective manner. The catalytic cycle was also supported by kinetic data.

As can be concluded from the details presented in this section of the review, the variety of properties available in polyoxometalate compounds enables them to be used for aerobic oxidation, which may proceed by a number of mechanistic schemes. In some cases, practical synthetic techniques are already available, especially for aerobic alcohol oxidation and other oxidative dehydrogenation reactions. In other cases it is hoped that mechanistic possibilities that have been put forward will stimulate future synthetic developments.

8.6
Heterogenization of Homogeneous Reaction Systems

Beyond questions of catalytic activation of oxidants and/or substrates by catalysts in general and by polyoxometalates for oxidation, an important part of catalysis research is connected with questions of catalyst recovery and recycling. In general one can discern between two broad approaches. The first basic approach is to immobilize a catalyst with proven catalytic properties onto a solid support, leading to a catalytic system that may be filtered and reused. Such approaches include concepts such as simple use of catalysts as insoluble bulk material, impregnation of a catalyst onto a solid and usually inert matrix, attachment through covalent or ionic bonds of a catalyst to a support, inclusion of a catalyst in a membrane or other porous material and several others. The second basic approach is to use biphasic liquid/liquid systems, such that at separation temperatures, which are usually ambient, the catalyst and product phases may be separated by phase separation; the catalysts phase is reused and the product is worked up in the usual manner. Numerous biphasic media have been discussed in the literature, which include using catalysts in aqueous, fluorous, ionic liquid, super critical fluid and other liquid phases. Some research in this general area of catalyst recovery has also been carried out using polyoxometalates as catalysts, with emphasis naturally being placed on reactions with oxygen or hydrogen peroxide as the most attractive oxidants for large-scale applications.

In the area of solid/liquid reactions, the first application of liquid phase oxidation involving heterogenization of the homogeneous catalyst was impregnation onto a solid support. One set of research reports in this area was to impregnate phosphovanadomolybdate catalysts, $[PV_xMo_{12-x}O_{40}]^{(3+x)-}$ (x = 2, 3, 4), onto active carbon as the best catalyst for aerobic oxidation. In this way, first $[PV_2Mo_{10}O_{40}]^{5-}$ and then $[PV_6Mo_6O_{40}]^{9-}$ on carbon were used to catalyze oxidation of alcohols, amines and phenols [78, 86]. Recently, a ruthenium-containing polyoxometalate has also been used for alcohol oxidation [87]. Toluene is a good solvent for many reactions and does not lead to measurable leaching. On the other hand, polar solvents tend to dissolve the catalysts into solution. Similarly, $[PV_2Mo_{10}O_{40}]^{5-}$ on several supports, such as carbon or textile fibers, was active for oxidation of various odorous volatile organics such as acetaldehdye, 1-propanethiol and thiolane [88]. The impetus of such research was not preparative (synthetic) but rather to deodorize air.

Very recently, an iron substituted polyoxometalate supported on cationic silica was found to be active for oxidation of sulfides and aldehydes [89]. For similar reactions, $H_5PV_2Mo_{10}O_{40}$ supported on a mesoporous molecular sieve, both by adsorption onto MCM-41 and by electrostatic binding to MCM-41 modified with amino groups, was active in the aerobic oxidation of alkenes and alkanes in the presence of isobutyraldehyde as sacrificial reagent [90]. It was originally thought that the catalyst support was inert and was useful for increasing the surface area for the heterogeneous reaction. However, it was noticeable from the combined research efforts that for autooxidation of aldehydes and oxidation of sulfides, silica supports were suitable, but for the oxidation of alcohols, amines and phenols, carbon supports were far superior. In fact, the unique and high activity of active carbon *versus* other supports such as silica or alumina, led to the suggestion that the support may be actively involved in the catalysis. A subsequent study led to the formulation that quinones, probably formed on the active carbon surface through the presence of the polyoxometalate and oxygen, might play a role as an intermediate oxidant [91]. Thus, a catalytic cycle may be considered, whereby a surface quinone oxidizes the alcohol to the aldehyde and is reduced to a hydroquinone, which in turn is reoxidized in the presence of the catalyst and molecular oxygen.

Since heteropoly acids can form complexes with crown ether type complexes [92], an interesting twist, which is especially useful in oxidation with the acidic $H_5PV_2Mo_{10}O_{40}$, was to employ the inexpensive polyethylene glycol as solvent [93]. Upon cooling the reaction mixture the $H_5PV_2Mo_{10}O_{40}$–polyethylene glycol phases separate from the product. In this way, previously known reactions with $H_5PV_2Mo_{10}O_{40}$, such as aerobic oxidation of alcohols, dienes and sulfides, and Wacker type oxidations, were demonstrated. Beyond the simple use of polyethylene glycol as solvent, the attachment of both hydrophilic polyethylene glycol and hydrophobic polypropylene glycol to silica by the sol-gel synthesis leads to solid particles, which upon dispersion in organic solvents lead to liquid-like phases, Scheme 8.13. Addition of $H_5PV_2Mo_{10}O_{40}$ leads to what we have termed solvent anchored supported liquid phase catalysis and reactivity typical for this catalyst [94]. The balance of hydrophilicity–hydrophobicity of the surface is important for tweaking the catalytic activity.

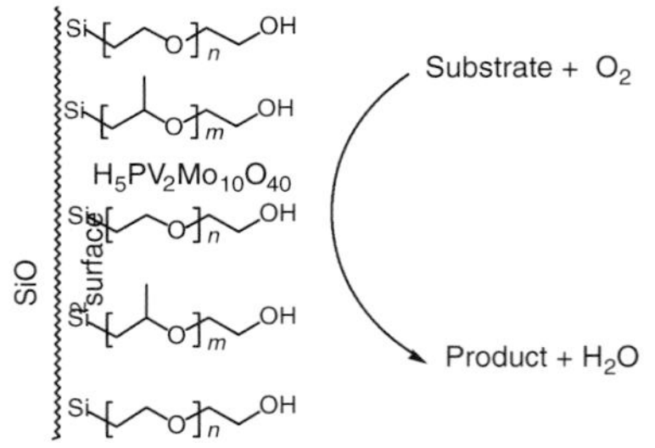

Scheme 8.13 Solvent anchored supported liquid phase catalysis: silica–PEG/PPG–$H_5PV_2Mo_{10}O_{40}$

Catalysts useful for reactions with hydrogen peroxide have also been heterogenized on a solid support. Since polyoxometalates are anionic, preparation of silica particles with quaternary ammonium moieties on the surface led to a useful catalytic assembly with $\{[WZnMn_2(H_2O)_2][(ZnW_9O_{34})_2]\}^{12-}$ as the active species. Importantly, using the sol-gel synthesis for the preparation of silica, the surface hydrophobicity could be controlled by choice of the organosilicate precursors [95]. This control of surface hydrophobicity led to the tuning of the catalytic activity and gave essentially the same reactivity as in the previously reported biphasic liquid/liquid reaction medium. No organic solvent was needed. Reactions can be carried out by mixing aqueous hydrogen peroxide and the organic substrate with a solid catalyst particle, which is easily recoverable. Although, polyoxometalates are commonly synthesized as water-soluble alkali salts, the idea of carrying out reactions in biphasic media, polyoxometalate-water/organic substrate, has not been realized until recently. It has now been demonstrated that $\{[WZn_3(H_2O)_2][(ZnW_9O_{34})_2]\}^{12-}$ in water catalyzes the oxidation of alcohols with hydrogen peroxide, Scheme 8.14 [96]. The catalytic system is quite effective for oxidation of secondary and primary alcohols to ketones and carboxylic acids, respectively. An important and key characteristic of this catalytic system is that the catalyst, $Na_{12}[WZn_3(H_2O)_2][(ZnW_9O_{34})_2]$ does not have to be prepared beforehand. Assembly, *in situ*, of the polyoxometalate by mixing sodium tungstate, zinc nitrate and nitric acid in water is sufficient to attain a fully catalytically active system. After completion of the reaction and phase separation, the catalyst water solution may be reused without loss of activity.

Concepts and techniques to utilize polyoxometalates in an efficient way are just in

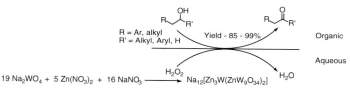

Scheme 8.14 Aqueous biphasic reactions *via* self-assembly for oxidation of alcohols with hydrogen peroxide

their infancy. As the number of synthetic uses of polyoxometalates increases and as the practical potential of polyoxometalate oxidation catalysis becomes a reality, one might expect a number of new methods for catalyst "engineering" to aid in recovery and recycling.

8.7
Conclusion

Liquid phase oxidation catalysis by polyoxometalates became a research topic only about 25 years ago. Since then various applications of polyoxometalates as practical oxidation catalysts useful for synthesis have been demonstrated. Additional synthetic procedures are not far away, due to the wide variety of polyoxometalates that can be prepared and the important structure–activity relationships that have been shown. In fact, it is the mechanistic research that has been carried out which points to many new and possibly exciting synthetic applications. Although polyoxometalates are of high molecular weight, efficient methods of catalyst recycling such as nanofiltration and some use of supports and biphasic media are already available. This bodes well for the eventual use of polyoxometalate catalysts, along with benign oxidants, as an attractive platform for replacing the still common use of environmentally damaging stoichiometric oxidants.

References

[1] *Chemical Reviews*, **1998**, *98* (Volume 1).
[2] (a) M. T. POPE, *Isopoly and Heteropoly Anions*, Springer, Berlin, Germany, **1983**; (b) A. MÜLLER, *Polyoxometalate Chemistry*, Kluwer Academic, Dordrecht, The Netherlands, **2001**; (c) I. V. KOZHEVNIKOV, *Catalysis by Polyoxometalates*, Wiley, Chichester, **2002**; (d) C. L. HILL, C. M. PROSSER-MCCARTHA, *Coord. Chem. Rev*, **1995**, *143*, 407; (e) N. MIZUNO, M. MISONO, *Chem. Rev.* **1998**, *98*, 171; (f) R. NEUMANN, *Prog. Inorg. Chem.* **1998**, *47*, 317.
[3] (a) C. L. HILL, R. B. BROWN, *J. Am. Chem. Soc.* **1986**, *108*, 536; (b) D. MANSUY, J. F. BARTOLI, P. BATTIONI, D. K. LYON, R. G. FINKE, *J. Am. Chem. Soc.* **1991**, *113*, 7222.
[4] H. WEINER, Y. HAYASHI, R. G. FINKE, *Inorg. Chem.* **1999**, *38*, 2579.
[5] (a) D. E. KATSOULIS, M. T. POPE, *J. Chem. Soc., Chem. Commun.* **1986**, 1186; (b) A. M. KHENKIN, C. L. HILL, *J. Am. Chem. Soc.* **1993**, *115*, 8178–8186.
[6] X. ZHANG, K. SASAKI, C. L. HILL, *J. Am. Chem. Soc.* **1996**, *118*, 4809.
[7] R. NEUMANN, C. ABU-GNIM, *J. Chem. Soc., Chem. Commun.* **1989**, 1324.
[8] R. NEUMANN, C. ABU-GNIM, *J. Am. Chem. Soc.* **1990**, *112*, 6025
[9] E. STECKHAN, C. KANDZIA, *Synlett* **1992**, 139.
[10] M. BRESSAN, A. MORVILLO, G. ROMANELLO, *J. Mol. Catal.* **1992**, *77*, 283.
[11] L. I. KUZNETSOVA, V. A. LIKHOLOBOV, L. G. DETUSHEVA, *Kinet. Catal.* **1993**, *34*, 914.
[12] R. NEUMANN, A. M. KHENKIN, *Chem. Commun.* **1998**, 1967.
[13] R. G. S. BANKS, R. J. HENDERSON, J. M. PRATT, *J. Chem. Soc. A* **1968**, 2886–2889.
[14] F. BOTTOMLY, I. J. B. LIN, M. MUKAIDA, *J. Am. Chem. Soc.* **1980**, *102*, 5238.
[15] (a) G. I. PANOV, *CATTECH* **2000**, *4*, 18; (b) G. I. PANOV, A. URIATE, M. A. RODKIN, V. I. SOBOLEV, V. I. *Catal. Today* **1998**, *41*, 365.
[16] N. P. NOTTE, *Top. Catal.* **2000**, *13*, 387.
[17] M. A. RHODKIN, V. I. SOBOLEV, K. A. DUBKOV, N. H. WATKINS, G. I. PANOV, G. I. *Stud. Surf. Sci. Catal.* **2000**, *130A*, 881.
[18] V. DUMA, D. HÖNICKE, *J. Catal.* **2000**, *191*, 93.
[19] J. T. GROVES, J. S. ROMAN, *J. Am. Chem. Soc.* **1995**, *117*, 5594.
[20] T. YAMADA, K. HASHIMOTO, Y. KITAICHI, K. SUZUKI, T. IKENO, *Chem. Lett.* **2001**, 268.
[21] K. HASHIMOTO, Y. KITAICHI, H. TANAKA, T. IKENO, T. YAMADA, *Chem. Lett.* **2001**, 922.
[22] R. BEN-DANIEL, L. WEINER, R. NEUMANN, *J. Am. Chem. Soc.* **2002**, *124*, 8788.
[23] R. BEN-DANIEL, R. NEUMANN, *Angew. Chem., Int. Ed. Engl.* **2003**, *42*, 92.
[24] (a) A. J. MANCUSO, D. SWERN, *Synthesis* **1981**, 165; (b) T. T. TIDWELL, *Org. React.* **1990**, *39*, 297; (c) T. T. TIDWELL, *Synthesis* **1990**, 857.
[25] V. Y. KUKUSHKIN, *Coord. Chem. Rev.* **1995**, *139*, 375–407.
[26] A. M. KHENKIN, R. NEUMANN, *J. Am. Chem. Soc.* **2002**, *124*, 4198.
[27] A. M. KHENKIN, R. NEUMANN, *J. Org. Chem.* **2002**, *67*, 7075.
[28] (a) M. FARAJ, C. L. HILL, *J. Chem. Soc., Chem. Commun.* **1987**, 1487; (b) R. NEUMANN, A. M. KHENKIN, *Inorg. Chem.* **1995**, *34*, 5753; (c) M. R. CRAMAROSSA, L. FORTI, M. A. FEDOTOV, L. G. DETUSHEVA, V. A. LIKHOLOBOV, L. I. KUZNETSOVA, G. L. SEMIN, F. CAVANI, F. TRIFIRÓ, *J. Mol. Catal.* **1997**, *127*, 85; (d) Y. MATSUMOTO, M. ASAMI, M. HASHIMOTO, M. MISONO, *J. Mol. Catal.* **1996**, *114*, 161.
[29] A. M. KHENKIN, R. NEUMANN, *J. Am. Chem. Soc.* **2001**, *123*, 6437.
[30] R. D. GALL, M. FARAJ, C. L. HILL, *Inorg. Chem.* **1994**, *33*, 5015.
[31] (a) W. ADAM, P. L. ALSTERS, R. NEUMANN, C. R. SAHA-MÖLLER, D. SEEBACH, R. ZHANG, *Org. Lett.* **2003**, *5*, 725; (b) W. ADAM, P. L. ALSTERS, R. NEUMANN, C. R. SAHA-MÖLLER, D. SEEBACH, R. ZHANG, *J. Org. Chem.* in the press.
[32] Y. MATOBA, Y. ISHII, M. OGAWA, *Synth. Commun.* **1984**, *14*, 865.
[33] (a) Y. ISHII, K. YAMAWAKI, T. URA, H. YA-

MADA, T. YOSHIDA, M. OGAWA, *J. Org. Chem.* **1988**, *53*, 3587; (b) T. OGUCHI, Y. SAKATA, N. TAKEUCHI, K. KANEDA, Y. ISHII, M. OGAWA, *Chem. Lett.* **1989**, 2053; (c) M. SCHWEGLER, M. FLOOR, H. VAN BEKKUM, *Tetrahedron Lett.* **1988**, *29*, 823.

[34] (a) Y. SAKATA, Y. KATAYAMA, Y. ISHII, *Chem. Lett.* **1992**, 671; (b) Y. SAKATA, Y. ISHII, *J. Org. Chem.* **1991**, *56*, 6233; (c) T. IWAHAMA, S. SAKAGUCHI, Y. NISHIYAMA, Y. ISHII, *Tetrahedron Lett.* **1995**, *36*, 1523.

[35] Y. ISHII, K. YAMAWAKI, T. YOSHIDA, M. OGAWA, *J. Org. Chem.* **1988**, *53*, 5549.

[36] F. P. BALLISTRERI, S. FAILLA, E. SPINA, G. A. TAMASELLI, *J. Org. Chem.* **1989**, *54*, 947.

[37] S. SAKAUE, Y. SAKATA, Y. NISHIYAMA, Y. ISHII, *Chem. Lett.* **1992**, 289.

[38] (a) H. ORITA, M. SHIMIZU, T. HAYKAWA, K. TAKEHIRA, *React. Kinet. Catal. Lett.* **1991**, *44*, 209; (b) L. A. PETROV, N. P. LOBANOVA, V. L. VOLKOV, G. S. ZAKHAROVA, I. P. KOLENKO, L. YU. BULDAKOVA, *Izv. Akad. Nauk SSSR, Ser. Khim.* **1989**, 1967; (c) M. SHIMIZU, H. ORITA, T. HAYAKAWA, K. TAKEHIRA, *Tetrahedron Lett.* **1989**, *30*, 471.

[39] (a) L. J. CSANYI, K. JAKY, *J. Mol. Catal.* **1990**, *61*, 75; (b) L. J. CSANYI, K. JAKY, *J. Catal.* **1991**, *127*, 42.

[40] (a) L. SALLES, C. AUBRY, F. ROBERT, G. CHOTTARD, R. THOUVENOT, H. LEDON, J.-M. BRÉGAULT, *New J. Chem.* **1993**, *17*, 367; (b) C. AUBRY, G. CHOTTARD, N. PLATZER, J.-M. BRÉGAULT, R. THOUVENOT, F. CHAUVEAU, C. HUET, H. LEDON, *Inorg. Chem.* **1991**, *30*, 4409; (c) L. SALLES, C. AUBRY, R. THOUVENOT, F. ROBERT, C. DORÉMIEUX-MORIN, G. CHOTTARD, H. LEDON, Y. JEANNIN, J.-M. BRÉGAULT, *Inorg. Chem.* **1994**, *33*, 871.

[41] (a) A. C. DENGEL, W. P. GRIFFITH, B. C. PARKIN, *J. Chem. Soc., Dalton Trans.* **1993**, 2683; (b) A. J. BAILEY, W. P. GRIFFITH, B. C. PARKIN, *J. Chem. Soc., Dalton Trans.* **1995**, 1833.

[42] D. C. DUNCAN, R. C. CHAMBERS, E. HECHT, C. L. HILL, *J. Am. Chem. Soc.* **1995**, *117*, 681.

[43] K. KAMATA, K. YONEHARA, Y. SUMIDA, K. YAMAGUCHI, S. HIKICHI, N. MIZUNO, *Science* **2003**, *300*, 964.

[44] C. VENTURELLO, R. D'ALOISO, J. C. BART, M. RICCI, *J. Mol. Catal.* **1985**, *32*, 107.

[45] (a) Z. XI, N. ZHOU, Y. SUN, K. LI, *Science*, **2001**, *292*, 1139; (b) Y. SUN, Z. XI, G. CAO, *J. Mol. Catal. A* **2001**, *166*, 219; (c) Z. XI, H. WANG, Y. SUN, N. ZHOU, G. CAO, M. LI, *J. Mol. Catal. A* **2001**, *168*, 299.

[46] (a) R. SHIOZAKI, H. GOTO, Y. KERA, *Bull. Chem. Soc. Jpn.* **1993**, *66*, 2790; (b) R. SHIOZAKI, H. KOMINAMI, Y. KERA, *Synth. Commun.* **1996**, *26*, 1663.

[47] W. P. GRIFFITH, N. MORLEY-SMITH, H. I. S. NOGUEIRA, A. G. F. SHOAIR, M. SURIAATMAJA, A. J. P. WHITE, D. J. WILLIAMS, *J. Organomet. Chem.* **2000**, *607*, 146.

[48] M. W. DROEGE, R. G. FINKE, *J. Mol. Catal.* **1991**, *69*, 323.

[49] R. NEUMANN, M. DE LA VEGA, *J. Mol. Catal.* **1993**, *84*, 93.

[50] K. NOMIYA, H. YANAGIBAYASHI, C. NOZAKI, K. KONDOH, E. HIRAMATSU, Y. SHIMIZU, *J. Mol. Catal. A* **1996**, *114*, 181.

[51] (a) T. YAMASE, T. OZEKI, S. MOTOMURA, *Bull. Chem. Soc. Jpn.* **1992**, *65*, 1453; (b) O. A. KHOLDEEVA, G. M. MAKSIMOV, R. I. MAKSIMOVSKYAYA, L. A. KOVALEVA, M. A. FEDETOV, V. A. GROGORIEV, C. L. HILL, *Inorg. Chem.* **2000**, *39*, 3828.

[52] (a) Y. SEKI, J. S. MIN, M. MISONO, N. MIZUNO, *J. Phys. Chem. B* **2000**, *104*, 5940; (b) N. MIZUNO, Y. SEKI, Y. NISHIYAMA, I. KIYOTO, M. MISONO, *J. Catal.* **1999**, *184*, 550; (c) N. MIZUNO, I. KIYOTO, C. NOZAKI, M. MISONO, *J. Catal.* **1999**, *181*, 171; (d) N. MIZUNO, C. NOZAKI, L. KIYOTO, M. MISONO, *J. Am. Chem. Soc.* **1998**, *120*, 9267.

[53] (a) A. M. KHENKIN, C. L. HILL, *Mendeleev Commun.* **1993**, 140; (b) X. ZHANG, Q. CHEN, D. C. DUNCAN, R. J. LACHICOTTE, C. L. HILL, C. L. *Inorg. Chem.* **1997**, *36*, 4381; (c) X. ZHANG, Q. CHEN, D. C. DUNCAN, C. F. CAMPANA, C. L. HILL, *Inorg. Chem.* **1997**, *36*, 4208; (d) X. ZHANG, T. M. ANDERSON, Q. CHEN, C. L. HILL, *Inorg. Chem.* **2001**, *40*, 418.

[54] R. BEN-DANIEL, A. M. KHENKIN,

R. Neumann, *Chem. Eur. J.* **2000**, *6*, 3722.

[55] (a) R. Neumann, M. Gara, *J. Am. Chem. Soc.* **1994**, *116*, 5509; (b) R. Neumann, A. M. Khenkin, *J. Mol. Catal.* **1996**, *114*, 169.

[56] R. Neumann, M. Gara, *J. Am. Chem. Soc.* **1995**, *117*, 5066.

[57] (a) R. Neumann, D. Juwiler, *Tetrahedron* **1996**, *47*, 8781; (b) R. Neumann, A. M. Khenkin, D. Juwiler, H. Miller, M. Gara, *J. Mol. Catal.* **1997**, *117*, 169.

[58] M. Bösing, A. Nöh, I. Loose, B. Krebs, *J. Am. Chem. Soc.* **1998**, *120*, 7252.

[59] (a) W. Adam, P. L. Alsters, R. Neumann, C. R. Saha-Möller, D. Sloboda-Rozner, R. Zhang, *Synlett* **2002**, 2011; (b) W. Adam, P. L. Alsters, R. Neumann, C. R. Saha-Möller, D. Sloboda-Rozner, R. Zhang, *J. Org. Chem.* **2003**, *68*, 1721.

[60] (a) M. Hamamoto, K. Nakayama, Y. Nishiyama, Y. Ishii, *J. Org. Chem.* **1993**, *58*, 6421; (b) N. Mizuno, T. Hirose, M. Tateishi, M. Iwamoto, *Chem. Lett.* **1993**, 1839; (c) N. Mizuno, M. Tateishi, T. Hirose, M. Iwamoto, *Chem. Lett.* **1993**, 1985; (d) N. Mizuno, T. Hirose, M. Tateishi, M. Iwamoto, *Stud. Surf. Sci. Catal.* **1994**, *82*, 593; (e) A. M. Khenkin, A. Rosenberger, R. Neumann, *J. Catal.*, **1999**, *182*, 82.

[61] J. E. Lyons, P. E. Ellis, V. A. Durante, *Stud. Surf. Sci. Catal.* **1991**, *67*, 99.

[62] (a) N. Mizuno, T. Hirose, M. Tateishi, M. Iwamoto, *J. Mol. Catal.* **1994**, *88*, L125; (b) N. Mizuno, M. Tateishi, T. Hirose, M. Iwamoto, *Chem. Lett.* **1993**, 2137.

[63] H. Weiner, R. G. Finke, R.G. *J. Am. Chem. Soc.* **1999**, *121*, 9831.

[64] R. Neumann, M. Dahan, *J. Chem. Soc., Chem. Commun.* **1995**, 2277.

[65] (a) A. W. Chester, *J. Org. Chem.* **1970**, *35*, 1797; (b) L. Eberson, L.-G. Wistrand, *Acta Chem. Scand. B* **1980**, *34*, 349; (c) L. Eberson, *J. Am. Chem. Soc.* **1983**, *105*, 3192.

[66] (a) K. I. Matveev, *Kinet. Catal.* **1977**, *18*, 716; (b) K. I. Matveev, I. V. Kozhevnikov, *Kinet. Catal.* **1980**, *21*, 855.

[67] J. R. Grate, D. R. Mamm, S. Mohajan, *Mol. Eng.* **1993**, *3*, 205; J. R. Grate, D. R. Mamm, S. Mohajan in *Polyoxometalates: From Platonic Solids to Anti-Retroviral Activity*: M. T. Pope, A. Müller, Eds., Kluwer: The Netherlands, **1993**, p. 27.

[68] (a) H. Grennberg, K. Bergstad, J.-E. Bäckvall, *J. Mol. Catal.* **1996**, *113*, 355; (b) T. Yokota, S. Fujibayashi, Y. Nishyama, S. Sakaguchi, Y. Ishii, *J. Mol. Catal.* **1996**, *114*, 113.

[69] K. Bergstad, H. Grennberg, J.-E. Bäckvall, *Organometallics* **1998**, *17*, 45.

[70] R. Neumann, I. Assael, *J. Chem. Soc., Chem. Commun.* **1998**, 1285.

[71] O. V. Branytska, R. Neumann, *J. Org. Chem.* **2003**, *68*, 9510.

[72] (a) I. V. Kozhevnikov, V. I. Simagina, G. V. Varnakova, K. I. Matveev, *Kinet. Catal.* **1979**, *20*, 506; (b) B. S. Dzhumakaeva, V. A. Golodov, *J. Mol. Catal.* **1986**, *35*, 303; (c) V. E. Karbanko, V. N. Sidelnikov, I. V. Kozhevnikov, K. I. Matveev, *React. Kinet. Catal. Lett.* **1982**, *21*, 209.

[73] (a) M. K. Harrup, C. L. Hill, *Inorg. Chem.* **1994**, *33*, 5448; (b) M. K. Harrup, C. L. Hill, *J. Mol. Catal. A* **1996**, *106*, 57; (c) C. L. Hill, R. D. Gall, *J. Mol. Catal. A* **1996**, *114*, 103.

[74] J.-M. Brégault, B. El Ali, J. Mercier, J. Martin, C. Martin, *C. R. Acad. Sci. II* **1989**, *309*, 459.

[75] (a) B. El Ali, J.-M. Brégault, J. Martin, C. Martin, *New J. Chem.* **1989**, *13*, 173; (b) B. El Ali, J.-M. Brégault, J. Mercier, J Martin, C. Martin, O. Convert, *J. Chem. Soc., Chem. Commun.* **1989**, 825; (c) A. Atlamsani, M. Ziyad, J.-M. Brégault, *J. Chim. Phys., Phys.-Chim. Biol.* **1995**, *92*, 1344.

[76] A. M. Khenkin, R. Neumann, *Adv. Synth. Catal.* **2002**, *344*, 1017.

[77] R. Neumann, M. Lissel, *J. Org. Chem.* **1989**, *54*, 4607–4610.

[78] R. Neumann, M. Levin, *J. Org. Chem.* **1991**, *56*, 5707–5710.

[79] K. I. Matveev, E. G. Zhizhina, V. F. Odyakov, *React. Kinet. Catal. Lett.* **1995**, *55*, 47.

[80] R. Ben-Daniel, P. L. Alsters, R. Neumann, *J. Org. Chem.* **2001**, *66*, 8650.

[81] (a) I. A. Weinstock, R. H. Atalla, R. S. Reiner, M. A. Moen, K. E. Ham-

MEL, C. J. HOUTMAN, C. L. HILL, *New J. Chem.* **1996**, *20*, 269; (b) I. A. WEINSTOCK, R. H. ATALLA, R. S. REINER, M. A. MOEN, K. E. HAMMEL, C. J. HOUTMAN, C. L. HILL, M. K. HARRUP, *J. Mol. Catal. A-Chem.* **1997**, *116*, 59; (c) I.A. WEINSTOCK, R. H. ATALLA, R. S. REINER, C. J. HOUTMAN, C. L. HILL, *Holzforshung* **1998**, *52*, 304.

[82] R. NEUMANN, M. LEVIN, *J. Am. Chem. Soc.* **1992**, *114*, 7278.

[83] D. C. DUNCAN, C. L. HILL, *J. Am. Chem. Soc.* **1997**, *119*, 243.

[84] (a) A. M. KHENKIN, R. NEUMANN, *Angew. Chem., Int. Ed. Engl.* **2000**, *39*, 4088; (b) A. M. KHENKIN, L. WEINER, Y. WANG, R. NEUMANN, *J. Am. Chem. Soc.* **2001**, *123*, 8531.

[85] (a) R. NEUMANN, M. DAHAN, *Nature* **1997**, *388*, 353; (b) R. NEUMANN, M. DAHAN, *J. Am. Chem. Soc.* **1998**, *120*, 11969.

[86] (a) S. FUJIBAYASHI, K. NAKAYAMA, M. HAMAMOTO, S. SAKAGUCHI, Y. NISHIYAMA, Y. ISHII, *J. Mol. Catal. A* **1996**, *110*, 105; (b) K. NAKAYAMA, M. HAMAMOTO, Y. NISHIYAMA, Y. ISHII, *Chem. Lett.* **1993**, 1699.

[87] K. YAMAGUCHI, N. MIZUNO, *New J. Chem.* **2002**, *26*, 972.

[88] L. XU, E. BORING, C. L. HILL, *J. Catal.* **2000**, *195*, 394.

[89] (a) N. M. OKUN, T. M. ANDERSON, C. L. HILL, *J. Am. Chem. Soc.* **2003**, *125*, 3194; (b) N. M. OKUN, T. M. ANDERSON, C. L. HILL, *J. Mol. Catal. A* **2003**, *197*, 283.

[90] A. M. KHENKIN, R. NEUMANN, A. B. SOROKIN, A. TUEL, *Catal. Lett.* **1999**, *63*, 189.

[91] A. M. KHENKIN, I. VIGDERGAUZ, R. NEUMANN, *Chem. Eur. J.* **2000**, *6*, 875–882.

[92] R. NEUMANN, I. ASSAEL, *J. Chem. Soc., Chem. Commun.* **1989**, 547.

[93] A. HAIMOV, R. NEUMANN, *Chem. Commun.* **2002**, 876.

[94] (a) R. NEUMANN, M. COHEN, *Angew. Chem., Int. Ed. Engl.* **1997**, *36*, 1738; (b) M. COHEN, R. NEUMANN, *J. Mol. Catal. A* **1999**, *146*, 293.

[95] R. NEUMANN, H. MILLER, *J. Chem. Soc., Chem. Commun.* **1995**, 2277.

[96] D. SLOBODA-ROZNER, P. L. ALSTERS, R. NEUMANN, *J. Am. Chem. Soc.* **2003**, *125*, 5280.

9
Oxidation of Carbonyl Compounds
Jacques Le Paih, Jean-Cédric Frison and Carsten Bolm

9.1
Introduction

The oxidation of aldehydes (alkanals) and ketones (alkanones) has been reviewed extensively [1–3], and there are compilations based on reagent types [4–8] and oxidation methods for most functionalized compounds including those having carbonyl groups. Books [9, 10] and comprehensive review articles [11–16] on carboxylic acids and their derivatives also provide important background information on the oxidation of carbonyl compounds. This account will focus exclusively on the synthesis of carboxylic acid derivatives. After a brief summary of the well-established methods, new directions in oxidative transformations of carbonyl compounds will be described. Among these, in particular, catalytic [17, 18] and asymmetric versions will be emphasized.

9.2
Oxidations of Aldehydes

In addition to the information given in the general literature cited above, the oxidation of aldehydes has specifically been reviewed [19, 20]. The presentation here will begin with an overview of reagents that have been used for the conversion of aldehydes into carboxylic acids and derivatives thereof. Subsequently, more specific oxidation reactions such as dismutations and oxidative rearrangements will be described. In the final part, oxidations of aldehyde derivatives such as acetals, oximes and hydrazones will be presented.

9.2.1
Conversions of Aldehydes to Carboxylic Acid Derivatives by Direct Oxidations

Since aldehydes are at an intermediate oxidation level between alcohols and carboxylic acids, reagents that are capable of oxidizing alcohols to carboxylic acid derivatives can generally also be applied for aldehyde oxidations. The various oxidants in-

Modern Oxidation Methods. Edited by Jan-Erling Bäckvall
Copyright © 2004 WILEY-VCH Verlag GmbH & Co. KGaA, Weinheim
ISBN: 3-527-30642-0

troduced for this transformation can be grouped into different classes with distinct properties.

9.2.1.1 Metal-free Oxidants

Oxygen

On contact with air, aldehydes are readily oxidized to carboxylic acids. Generally, this autoxidation proceeds by a radical chain mechanism [21], and it can be accelerated by irradiation with ultraviolet light or the addition of catalysts [22]. In the oxidation of aromatic aldehydes with oxygen polyoxometalates [23], supported $Fe(acac)_3$ [24], $Ni(acac)_2$ [25], $CoCl_2$ [26], a Ru(diaminodiphosphine)Cl_2 derivative [27], and $Rh(CO)(PPh_3)_2Cl$ [28] have been used. Base catalysts are also effective. For example, sodium pyrazolide catalyzes the air oxidation of aromatic substrates even at room temperature [29]. Aliphatic aldehydes are converted into the corresponding carboxylic acids with air [30] or O_2/O_3 mixtures [31] as oxidants. Again, metal catalysts [32–36] such as platinum [34], Mn(stearate)$_2$ [35], $CeO_2/RuCl_3$ [36] or others [23, 24] have been applied.

Peroxides

Organic peroxides [37] are particularly attractive oxidants that have frequently been applied in the preparation of carboxylic acids starting from aldehydes. Hydrogen peroxide [38] oxidizes under both basic [39] and acidic conditions [40]. Catalytic processes for oxidations of aldehydes with hydrogen peroxide involve the use of molybdenum oxides [41], vanadium oxides [42], tungsten oxometalates [43] and titanium silicates [44] as catalysts. Esters can be obtained when the reaction is performed in the presence of an alcohol [42, 44]. Recently, a two-phase system applying a phase transfer catalyst has been introduced, which affords carboxylic acids in high yield [45]. The system is free of both organic solvent and metal catalyst and allows the selective oxidation of functionalized aldehydes such as **1** and **3**. The corresponding carboxylic acids **2**, possessing an olefinic double bond, and **4**, bearing an oxidation sensitive secondary hydroxyl group, are obtained in 85% and 79% yield, respectively [Eqs. (1) and (2)].

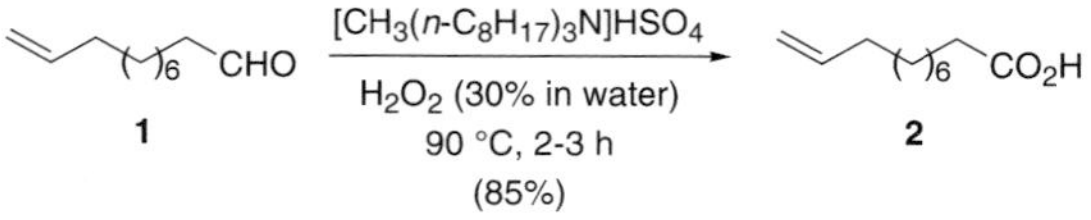

(1)

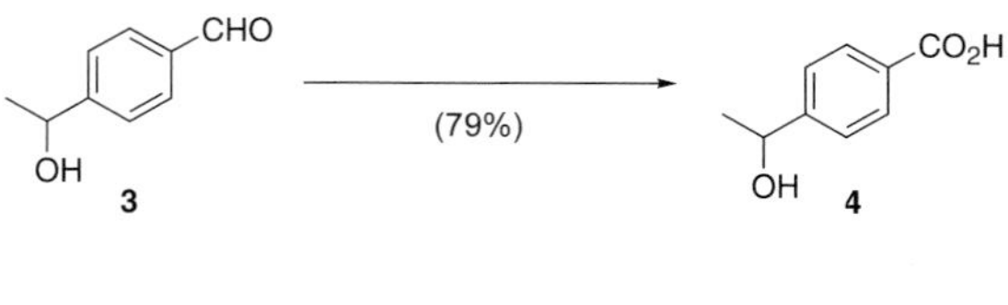

(2)

A common oxidant is *tert*-butyl hydroperoxide [46–48], which can either be activated by metal salts [47] or used in an aqueous, metal-free process [48]. In this latter case, it is possible to oxidize aldehydes to carboxylic acids selectively, while other potentially reactive functional groups in the molecule remain untouched. An example is provided by the selective oxidation of p-thiomethoxybenzaldehyde (5) with tert-butyl hydroperoxide in the presence of cetyltrimethylammonium sulfate {[CTA]$_2$SO$_4$} to give carboxylic acid 6 in 98% yield [Eq. (3)]. By increasing the reaction temperature to 70 °C the corresponding p-sulfonyl benzoic acid can be obtained quantitatively [48].

$$\text{5} \xrightarrow[\substack{t\text{-BuOOH (30\%),} \\ 20\,°\text{C, 20 h} \\ (98\%)}]{[\text{CTA}]_2\text{SO}_4\ \text{pH} >13} \text{6} \tag{3}$$

Other peroxides such as 2-hydroperoxyhexafluoro-2-propanol [49], cumene hydroperoxide [50] or dimethyldioxirane [51] have also been used. Peracids, which can be formed *in situ* by reacting hydrogen peroxide with the corresponding acid [52], are also effective. Perbenzoic acids are the most commonly used reagents for such purpose [53–55]. In some cases, they have been activated by metal catalysts [54, 55].

Hydroxide and alkoxide
Aldehydes are oxidized to the corresponding acids in molten sodium hydroxide. The reaction conditions are very harsh, and fused alkali is required for this transformation. By this process vanillic acid can be obtained in very high yield starting from vanillin [56]. Other aldehyde oxidations involving hydroxy or alkoxy derivatives (Cannizzaro and Tishchenko reactions) will be discussed in Section 9.2.2.1.

9.2.1.2 Metal-based Oxidants

Manganese
Potassium permanganate has been widely used for the oxidation of aldehydes to carboxylic acids under acidic, neutral or basic conditions [57, 58]. In reactions with enolizable aldehydes, competitive enol cleavage reactions can occur, when bases or strong acids are used (see Section 9.3.1). The best yields are achieved under neutral conditions, albeit at a reduced reaction rate. If the solubility of the substrate is low, an organic solvent such as acetone can be added [59]. Phase transfer-assisted permanganate oxidations have been developed using tetraalkylammonium salts [60, 61] or crown ethers [62] as phase transfer catalysts. These reagents solubilize permanganate in the organic layer or utilize it directly from the solid state. Even though oxidative cleavage reactions are common under these conditions, use of benzyltriethylammonium permanganate allows clean oxidations of aldehydes to carboxylic acids [61]. An optimized system involves potassium permanganate in a mixture of *tert*-butanol and aqueous sodium dihydrogenphosphate [63]. Sodium permanganate [64] and copper [65] permanganate can also be used, but the substrate scope is limited. Recently, metal permanganates have been introduced for the oxidation of organometallic compounds of palladium [66] and ruthenium [67] bearing an aldehyde group on

the ligand. No oxidation of the metal occurred. For example, ruthenium complex **8** was obtained in 64% yield by oxidation of aldehyde **7** with potassium permanganate in water and acetone as the co-solvent [Eq. (4)].

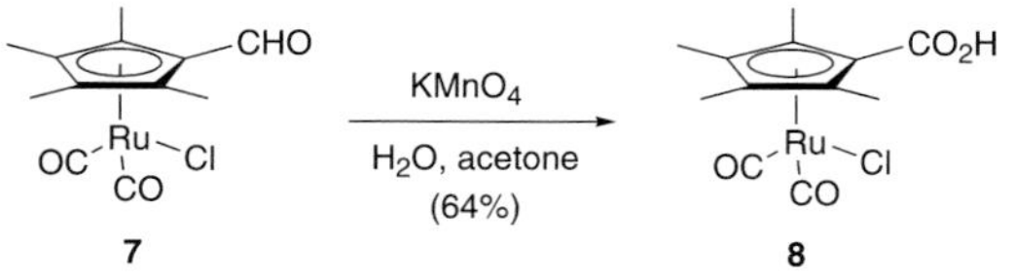

$$\text{7} \xrightarrow[\text{(64\%)}]{\text{KMnO}_4, \text{H}_2\text{O, acetone}} \text{8} \quad (4)$$

The oxidation of aldehydes to carboxylic acids by the lower valent manganese dioxide [68] is a very slow process, which requires a higher temperature [69]. Corey developed the most straightforward application of this reagent [70], which is particularly recommended for the conversion of α,β-unsaturated aldehydes [70, 71], since it proceeds without double bond isomerization. Aromatic aldehydes can be converted into the corresponding esters [72]. The reaction involves the use of cyanide ions and proceeds *via* cyanohydrins. The latter are oxidized to α-ketonitriles, which undergo a cyanide substitution reaction with alkoxide as the nucleophile to give the desired esters.

Chromium

In general the oxidation of aldehydes to carboxylic acids proceeds smoothly with chromium(IV) reagents under acidic conditions [57, 73]. As with metal permanganates, cleavage reactions can occur with enolizable aldehydes. As a result, the corresponding carbonyl derivatives with one carbon less than the starting material are formed (see Section 9.3.1). The most common chromium reagent for the aldehyde oxidation is the Jones reagent [74], which is a solution of chromium trioxide in dilute sulfuric acid [75]. Other chromium compounds, such as pyridinium halochromate [76, 77] or pyridinium dichromate in N,N-dimethylformamide [78, 79], have also been applied. In these cases, esters can be prepared directly when the reaction is performed in the presence of an alcohol [79]. Some kinetic studies have been performed in oxidations with pyridium fluorochromate [77] and quinolinium dichromate [80]. Recently, the oxidation of aldehydes using catalytic amounts of chromium salts [73] in the presence of an excess of sodium periodate [81] in an acidic medium [82] has been described.

Silver

Both oxides of silver, Ag_2O and AgO, have been used for aldehyde oxidations. The silver(I) reagent has mainly been applied in transformations of aliphatic [83] and aromatic aldehydes [84], less so in oxidations of organometallic complexes bearing an aldehyde function [85] or α,β-unsaturated compounds [86]. Use of the silver(II) oxide is less common [70, 87], probably due to its limited availability and high cost. Cyanide ions catalyze this oxidation in methanol leading to carboxylic acids (and not to esters; as compared with manganese) [70]. In reactions with α,β-unsaturated aldehydes, double bond isomerizations have been observed [70]. Less common silver reagents such as silver picolinate [88], tetrakis(pyridine)silver peroxodisulfate [89] and silver carbonate supported on celite [90] have also been used. Oxidations with the last reagent in the presence of alcohol affords esters instead of acids [90], which is in contrast with the reactions with other silver oxides.

Other metals
Aldehydes can be oxidized by nickel [91] and ruthenium oxides. Whereas ruthenium tetroxide [92] gives poor conversions to the corresponding acids, catalytic amounts of ruthenate [93] in the presence of a secondary oxidant, such as $NaBrO_3$ or $K_2S_2O_8$, give better yields. Other ruthenium species for the catalytic oxidation of aldehydes to carboxylic acids are tris(triphenylphosphine) ruthenium dichloride [94] in the presence of hypervalent iodine [95] and ruthenium trichloride [96] in combination with periodate [81].

Finally, oxidations of aldehydes in the presence of alcohols or amines lead to esters or amides, respectively. The reactions require ruthenium [97], rhodium [98, 99] or palladium [100] catalysts. Formally they are oxidative dehydrogenations of the corresponding hemiacetals or hemiaminals, generated *in situ* by addition of the alcohol or the amine to the aldehyde. Hydrogen is formed during the reaction and needs to be trapped by a hydrogen acceptor such as an unsaturated compound (bearing a double or a triple bond) [97], an aryl halide [100] or another oxidant [99]. These scavengers and oxidants also avoid the reduction of the aldehyde, which would lead to a dismutation process (see Section 9.2.2.1).

9.2.1.3 Halogen-based Oxidants

Molecular halogens
Aromatic and aliphatic aldehydes are oxidized smoothly by bromine in aqueous solution [101]. The reaction can also be performed in alcohol as the solvent, but in this case, the corresponding esters are formed in good yields [102, 103]. Iodine in the presence of alkali is also effective for this transformation [103].

Halo amines, amides, and imides
N-Bromo- [104–106] or N-iodo- [107] succinimide are efficient reagents for the oxidation of aldehydes to carboxylic acid derivatives. The corresponding acyl halides are formed as intermediates, which are easily converted into acids [104], esters [105, 107], or amides [106] by reaction with water, alcohol or amine, respectively. Fluoride-containing reagents such as fluoro oxysulfate [108] and selectfluor {1-chloromethyl-4-fluoro-1,4-diazoniabicyclo[2.2.2] octane bis(tetrafluoroborate)} [109] are also able to oxidize aldehydes to carboxylic acid derivatives. In these cases, acyl fluorides are the intermediates.

Hypochlorite
Hypochlorites are very convenient oxidizing agents, with sodium [110, 111] and calcium [112, 113] hypochlorite being the most useful ones. Oxidations of aliphatic and aromatic aldehydes afford the corresponding carboxylic acids [112] or esters

[111, 113]. Other hypochlorites, and in particular *tert*-butyl hypochlorite [114], have also been used. For example, the latter reagent was applied in the oxidation of a (cyclobutadiencarboxaldehyde) iron tricarbonyl complex (**9**), which afforded ester **10** in 90% yield [Eq. (5)]. No oxidation of the metal center occurred. The reaction path involves the aldehyde oxidation by *tert*-butyl hypochlorite leading to the corresponding acyl chloride, which is then converted into the methyl ester by reaction with methanol.

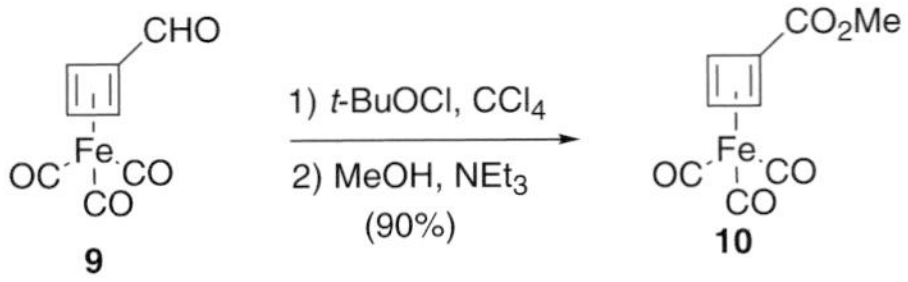

$$\mathbf{9} \xrightarrow[\text{2) MeOH, NEt}_3]{\text{1) } t\text{-BuOCl, CCl}_4} \mathbf{10} \quad (90\%) \tag{5}$$

The success of reactions with hypochlorites is dependent on the solvent and the starting material. Thus, if an aromatic aldehyde bears donor substituents, such as methoxy groups, on the aromatic ring, electrophilic chlorinations can occur to give undesired byproducts [114].

Chlorite
A very useful oxidant for the conversion of aldehydes into carboxylic acids is sodium chlorite [115]. However, since other oxidants such as hypochlorite and chlorine dioxide are formed in the course of this oxidation, the product yield can be diminished. In order to avoid this effect, scavengers such as hydrogen peroxide [116–118], dimethyl sulfoxide [116], sulfamic acid [116, 119–121], resorcinol [119] or 2-methylbutene [122–124] have been added. Chlorites are particularly appropriate for the oxidation of functionalized aliphatic [116–118, 122], α,β-unsaturated [116, 117, 120, 123] and aromatic aldehydes bearing electron-withdrawing groups [116, 117, 121, 124]. In the case of donor substituents on the aromatic ring, chlorinated compounds are formed as byproducts [116].

9.2.1.4 Sulfur- and Selenium-based Oxidants

Sulfur
Oxidations of aldehydes to derivatives of carboxylic acids can be performed by use of Caro's acid (peroxomonosulfuric acid) [125], which is prepared by treatment of sulfuric acid with hydrogen peroxide or obtained from ammonium peroxodisulfate. In this manner, 100% of metacrolein is converted into ethylmetacrylate in the presence of ethanol [125]. Acrolein is oxidized to acrylic acid (99% conversion) at neutral pH using Oxone [126], which is a stable water-soluble oxidant having the approximate composition $K_2SO_4 \cdot 2KHSO_5 \cdot KHSO_4$.

Selenium
Oxidations with selenium as the oxidizing agent have recently been reviewed [127]. The combination of a catalytic amount of selenium dioxide and a secondary oxidant

is a convenient system for aldehyde oxidation [128–130]. Acrolein is selectively oxidized to acrylic acid in 90% yield using a 15% aqueous solution of hydrogen peroxide [128]. With a 30% hydrogen peroxide solution, a wide range of aldehydes has been oxidized to afford products in high yield [129]. Arylselenic acid derivatives are also known to activate hydrogen peroxide and *tert*-butyl hydroperoxide [130–132]. The latter combination is very selective. Thus, even with methoxybenzaldehyde derivatives only small quantities of the Dakin product (see Dakin reactions, under Section 9.2.2.2) are formed [130]. This reaction has also been performed in a triphasic system using a perfluorinated selenium catalyst [132].

9.2.1.5 Nitrogen-based Oxidants

Aldehydes are readily oxidized to carboxylic acids using nitric acid of various concentrations. Aliphatic substrates can be applied, however the selectivity is low and in the case of hydroxyaldehydes [133], the corresponding diacids are obtained. Aromatic aldehydes are smoothly converted into the corresponding carboxylic acids [134]. Optimization studies revealed that a 5.5 molar solution of nitric acid was optimal for the conversion of benzaldehyde into benzoic acid giving a 94% yield of the desired product [134]. Nitrogen heterocycles, and in particular phenanthrolines containing aldehyde functions, are also readily oxidized by nitric acid to the corresponding carboxylic acids [135].

Other nitrogen-containing reagents that are able to oxidize aldehydes, are nitrobenzene in the presence of cyanide ions [136], peroxyacetyl nitrate [137] and Angeli's salt (sodium trioxodinitrate) [138]. However, the scope and efficiency of these reagents remain limited.

9.2.1.6 Miscellaneous

Sodium perborate is a convenient oxidant for the oxidation of aromatic aldehydes [139]. In contrast, aliphatic substrates are inert. Recently it was shown that aldehydes can also be oxidized to carboxylic acids by IBX (*o*-iodoxybenzoic acid) [140].

3-Benzylthiazolium bromide, in the presence of a base and a primary alcohol, catalyzes a redox reaction in which the aldehyde is oxidized to the corresponding methyl ester. Organic compounds such as acridine [141] or flavine [142] are reduced.

Electrochemical [143, 144] and biochemical processes [145] are also effective for the synthesis of carboxylic acid derivatives from aldehydes.

9.2.2
Conversions of Aldehydes into Carboxylic Acid Derivatives by Aldehyde Specific Reactions

9.2.2.1 Dismutations and Dehydrogenations

Cannizzaro reactions

On treatment with aqueous or alcoholic alkali solutions aromatic and aliphatic aldehydes **11** lacking α-hydrogens undergo dismutations (Cannizzaro reactions) to give the corresponding carboxylic acid salts **12** and alcohols **13** [Eq. (6)] [146,147].

$$2 \text{ RCHO} \xrightarrow{\text{HO}^-} \text{RCO}_2^- + \text{RCH}_2\text{OH} \qquad (6)$$

    11   **12**   **13**

Improvements to the reactions involve the use of microwave irradiation [148] and transformations without solvent [149]. The dismutation mechanism of benzaldehyde and pivaldehyde has been studied by *ab initio* calculations [150].

Intramolecular Cannizzaro reactions can occur when two aldehyde functionalities are in close proximity to each other. Then, dismutation followed by ring-closure leads to lactones [151].

On an alternative pathway, internal Cannizzaro reactions afford mandelic acid-type compounds from phenylglyoxal derivatives [152–156]. Copper complexes [153, 154], chromium perchlorate [154], cobalt Schiff's bases [155] and yttrium chloride [156] have been applied as catalysts. An asymmetric version [Eq. (7)] has been developed using phenylglyoxal (**14**) as substrate and a combination of Cu(OTf)$_2$ and (*S,S*)-Ph-box **16** as the chiral catalyst [154]. After 24 h at room temperature isopropyl mandelate (**15**) was obtained with an enantioselectivity of 28% ee.

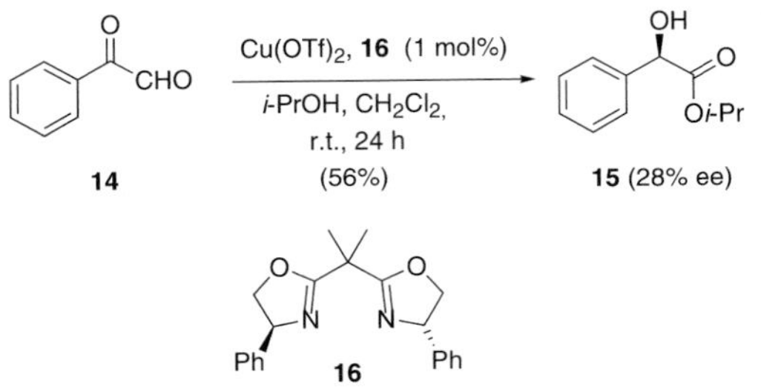

$$(7)$$

Finally, cross-Cannizzaro reactions allow alcohols to be synthesized in good yields (>50%) from aromatic aldehydes, using an excess of sacrificial paraformaldehyde, which is oxidized to formic acid [157].

Tishchenko reactions

A modification of the Cannizzaro reaction was discovered by Claisen [158] and later extended by Tishchenko [159]. It involves the use of sodium or aluminum alkoxides in the conversion of aliphatic and aromatic aldehydes into the corresponding esters [160]. Other catalysts, such as boric acid [161], superoxide ion/crown ether [162] and metal catalysts [163] have also been applied in this oxidation reaction.

Dialdehydes allow an intramolecular Tishchenko reaction [164, 165]. In the case of δ-keto aldehydes the reaction leads to lactones [164]. Other substrates such as terephthaldehyde or dodecanedial behave differently and result in the formation of polymers [165].

A related intramolecular reaction of this type is the so-called Evans-Tishchenko reaction [166]. Here, a β-hydroxy ketone is reduced in the presence of an aldehyde yielding 1,3-diol monoesters. Several metal catalysts such as samarium iodide [166, 167] and zirconocene complexes [168] are effective. The reaction is highly diastereoselective. For example, starting from chiral β-hydroxy ketone **17** and an excess of aldehyde in the presence of a catalytic amount of SmI_2, 1,3-diol monoester **18** is obtained as a single compound in 99% yield [166] [Eq. (8)].

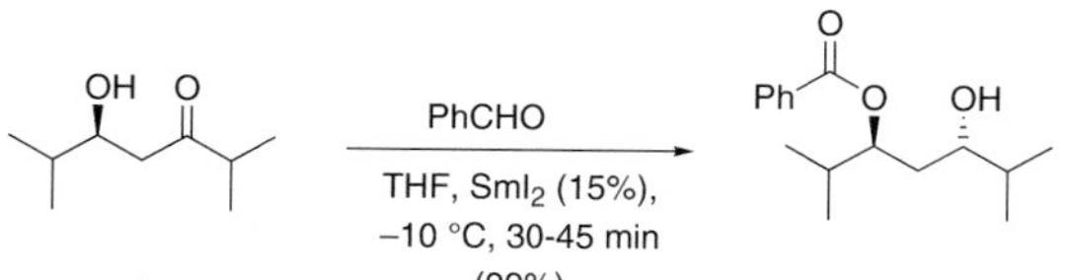

$$\underset{\mathbf{17}}{\text{OH O structure}} \quad \xrightarrow[\substack{\text{THF, SmI}_2 \text{ (15\%),} \\ -10\ °C,\ 30\text{-}45\ \text{min} \\ (99\%)}]{\text{PhCHO}} \quad \underset{\mathbf{18}\ (anti{:}syn > 99{:}1)}{\text{Ph-O-CO structure OH}} \qquad (8)$$

The excellent level of *anti* selectivity in the formation of the 1,3-diol monoester can be explained by transition state **A** shown in Scheme 9.1.

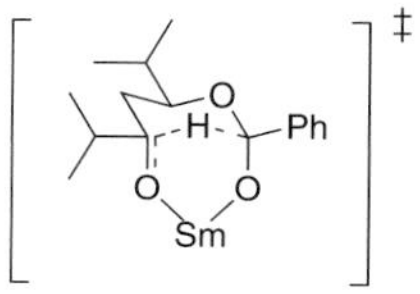

Scheme 9.1 Transition state of the intramolecular Evans-Tishchenko reaction

Other modifications of the Tishchenko reaction are also known, and catalytic cascade reactions are particularly interesting. Examples include combinations of Tishchenko with aldol reactions in the presence of achiral [169] and chiral catalysts [170], Tishchenko/esterification sequences [171], tandem semipinacol rearrangement/Tishchenko reactions [172] and couplings of vinyl esters with aldehydes [173].

9.2.2.2 Oxidative Aldehyde Rearrangements

Dakin reactions

On treatment with peroxides, aromatic aldehydes **19**, and especially those containing hydroxyl or alkoxy groups, undergo oxidations to yield aryl formates **20** (Scheme 9.2, path a). This process, usually known as the Dakin reaction [174], occurs with substrates having electron-donating substituents on the aryl group (see also Section 9.3.2.1). Since the phenol formyl esters **20** are easily hydrolyzed, they are often not isolated but converted directly into the corresponding phenols **21**. The mechanism of the Dakin reaction involves a nucleophilc attack by the oxygen of a peroxy compound at the carbonyl carbon of the aldehyde to form Criegee intermediate **B**. Subsequent migration of the aryl substituent affords products **20**. If instead of the aryl group the hydride migrates (Scheme 9.2, path b), carboxylic acids **22** are formed.

Peroxides are usually used to oxidize aldehydes to aryl formates. In the case of hydroxy aldehydes, hydrogen peroxide under basic conditions is the most suitable oxidant [175]. With benzaldehydes and aromatic aldehydes bearing alkoxy groups, use of hydrogen peroxide catalyzed by selenium derivatives [176] or performance of the reaction under acidic conditions [177] gives better results. Peracids, and in particular

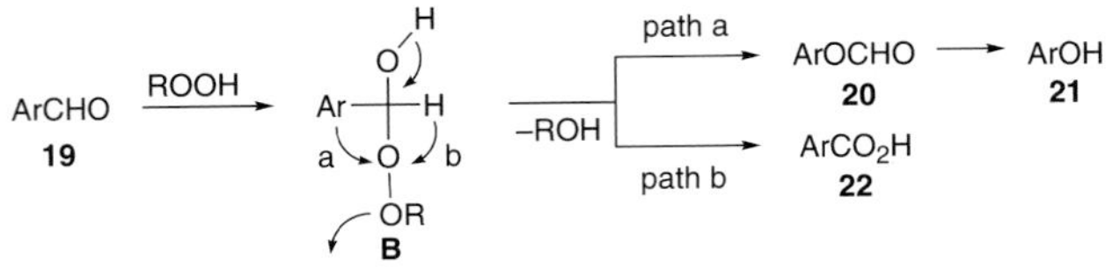
ArCHO
19
ROOH
Ar
H
O
H
a
O
b
OR
B
−ROH
path a
path b
ArOCHO
20
ArOH
21
$ArCO_2H$
22

Scheme 9.2 Mechanism of the Dakin reaction

m-chloroperbenzoic acid (*m*-CPBA) [178], have also been applied in the Dakin reaction [178, 179]. In addition, other oxidants such as percarbonate [180] and UHP (urea-hydrogen peroxide complex) [181] generate formate derivatives.

Electron-rich heterocyclic aldehydes [182, 183] undergo the same type of reaction. Thus, 2-formylfuran **23** is converted into the corresponding unsaturated lactone **24** in 69% yield by treatment with hydrogen peroxide and formic acid as shown in Eq. (9) [182].

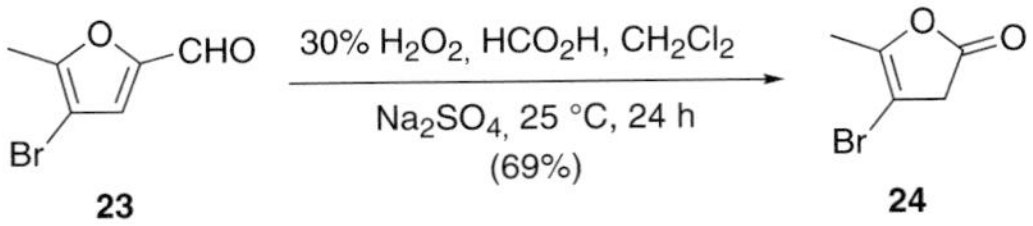
CHO
Br
23
30% H_2O_2, HCO_2H, CH_2Cl_2
Na_2SO_4, 25 °C, 24 h
(69%)
Br
24

(9)

When reacted with peracids, primary aliphatic aldehydes are mainly oxidized to carboxylic acids. However, if the α-carbon is benzylic or branched, formate esters are also formed [184]. Furthermore, a heteroatom in the α-position of the aldehyde function facilitates the formate ester synthesis [185].

Finally, α,β-unsaturated aldehydes are converted into vinyl formates, when treated with hydroperoxides [186]. Catalytic quantities of aryl diselenides catalyze this oxidation [Eq. (10)].

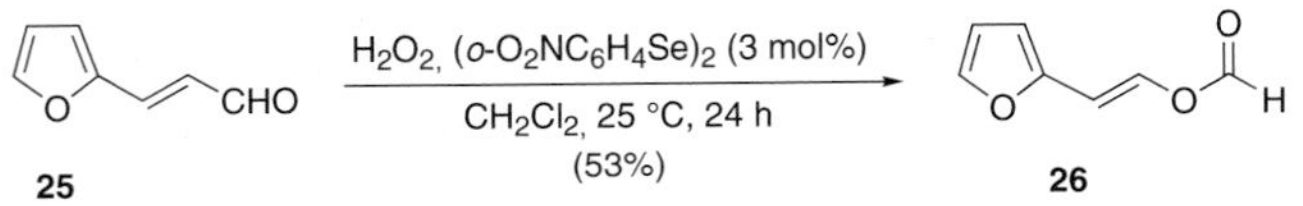
CHO
25
H_2O_2, (o-$O_2NC_6H_4Se)_2$ (3 mol%)
CH_2Cl_2, 25 °C, 24 h
(53%)
H
26

(10)

Thus, with bis-*o*-nitrophenyl diselenide as the catalyst, vinyl formate **26** is obtained in 53% yield starting from furane **25**. No oxidation of the heterocycle or the double bond was observed [186].

Miscellaneous rearrangements

Aldehydes can be converted into amides by a Schmidt rearrangement (see Schmidt reaction, under Section 9.3.2.2). This reaction is particularly interesting for the formation of sugar lactams [187]. When aldehydes are treated with α-azido alcohols in the presence of acid, the rearrangement yields cyclic imidates [188]. Nitriles can be obtained directly in high yields by reacting aldehydes with triazidochlorosilane [189].

9.2.3
Conversions of Aldehyde Derivatives into Carboxylic Acid Derivatives

9.2.3.1 Acetals
As described in Section 9.2.1, esters can be prepared from aldehydes by oxidation in the presence of alcohol. Although oxidative deprotections of acetals to give ketones are known [190, 191], acetals usually afford esters upon oxidation. Non-cyclic acetals are transformed into esters by several oxidants, such as trichloroisocyanuric acid [192], dioxirane [193], ozone [194], peracids [195] and Caro's acid [196]. In the case of cyclic acetals, the corresponding haloalkyl carboxylic ester [197] can be obtained by treatment of haloform derivatives and AIBN. Similarly, monoprotected diols are generated by oxidative ring cleavage. This reaction has been performed with ozone [194], hydrogen peroxide [198], oxygen [199], nitrogen dioxide [200], electrophilic halogens [201], chromium derivatives [202], hypervalent iodide [95, 203], Oxone [204, 205], and triphenylmethyl carbenium tetrafluoroborate [206, 207]. The resulting diol monoesters are particularly interesting in sugar chemistry. For example, triphenylmethyl carbenium tetrafluoroborate has been used in the synthesis of hydroxy benzoate **28** from benzylidene **27** as shown in Eq. (11) [207].

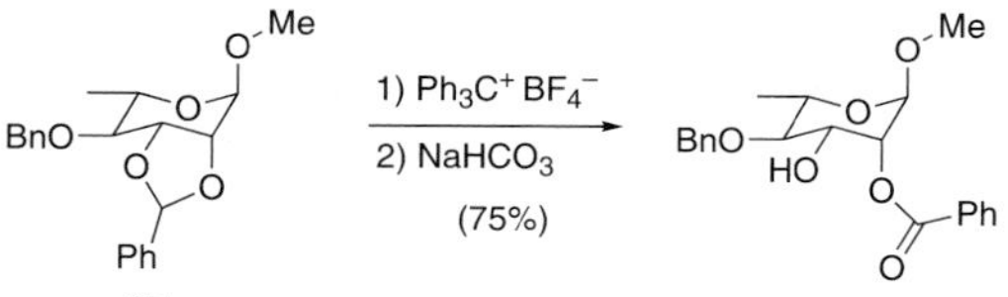

$$\text{27} \xrightarrow[\text{2) NaHCO}_3]{\text{1) Ph}_3\text{C}^+ \text{BF}_4^-} \text{28} \quad (75\%) \tag{11}$$

Catalytic systems based on palladium [208], ruthenium [209] and vanadium [210] complexes with *tert*-butyl hydroperoxide as oxidant or a cobalt catalyst under aerobic conditions [211] have also been applied in this reaction. In carbohydrate chemistry in particular [212] other cyclic acetals and cyclic hemi-acetals have been oxidized to the corresponding lactones with a plethora of oxidants.

9.2.3.2 Nitrogen Derivatives
Oxidations of nitrogen-containing compounds [213] derived from aldehydes and ketones have been reviewed. The most common products from aldimine oxidations are nitriles [214]. However, other compounds such as nitrile oxides [215], oxaziridines [205, 216] and formamides or amides stemming from a Beckmann rearrangement (see Beckmann reaction, under Section 9.3.2.2.) can also be formed. Furthermore, the initial aldehyde can be regenerated by oxidative deprotection [190, 217].

Imines
Aldimines are oxidized to nitriles by a copper(II) catalyst and oxygen as oxidant [218] or by manganese dioxide [219]. Recently, the catalytic oxidation with nickel/copper formate in the presence of ammonium peroxodisulfate has been reported to give

nitriles in high yields [220]. Finally, formamides can be obtained by oxidative rearrangement of imines [221].

Oxime derivatives

Aldoximes are among the most suitable starting materials for the synthesis of nitriles. A plethora of reagents [222–227] has been introduced for this Beckmann-type fragmentation (see Beckmann reaction, under Section 9.3.2.2). Among these are acetic anhydride [223], metal salts [224] or other metal catalysts [225], selenium dioxide [226] and Burgess' reagent [227]. In the presence of halide derivatives [228] or metal salts [229] aldoximes can also be oxidized to nitrile oxides.

Alternatively, the Beckmann rearrangement of aldoximes affords amides (see Beckmann reaction, under Section 9.3.2.2). Recent developments along these lines have been focused on utilizing SiO_2 at high temperature under microwave irradiation [230]. Furthermore, a one-pot amide synthesis using hydroxylamine in the presence of Al_2O_3 and sulfonic acid has been introduced [231].

Hydrazone derivatives

Hydrazones stemming from aldehydes are easily converted into nitriles. For example, when treated with alkylating agents hydrazones form hydrazonium salts [232], which upon elimination give nitriles as the formally oxidized products. Alternatively, hydrazones can be oxidized directly, and for this reaction oxidants such as peracids [233], oxone [234] and dioxirane [235] were found to be effective. Finally, catalytic systems based on combinations of hydrogen peroxide with metal catalysts [236] or selenium compounds [237] have been developed for this process. Other reactions along these lines include oxidations of monohydrazones giving diazo compounds [238] or nitrilimines [239] and conversions of vicinal dihydrazones into alkynes [240].

9.2.3.3 Miscellaneous Substrates

Bisulfite adducts of aldehydes, $RCH(OH)SO_3Na$, are conveniently oxidized to carboxylic acid derivatives by treatment with a mixture of dimethylsulfoxide and acetic anhydride [241]. The important intermediate in this reaction is an α-ketosulfonate. Its sulfonate fragment can easily be displaced by nucleophiles such as water, alcohol or amine, giving the corresponding acids, esters, or amides, respectively.

Oxidations of acyl anion equivalents allow a mild transformation of aldehydes into carboxylic acid derivatives [242–244]. 1,3-Dithianes, formed from aldehydes and 1,3-propane-dithiol, are readily converted into their 2-lithio salts by treatment with an appropriate base. The reaction of these salts with dimethyl disulfide affords the corresponding *ortho*-thioformates [242]. Their alcoholysis in the presence of a mixture of mercuric chloride and mercuric oxide leads to the formation of esters in good yields [242]. Analogously, O-trimethylsilyl-protected cyanohydrins derived from aldehydes can easily be deprotonated to give acyl anion equivalents. Upon treatment with N,N-dimethyl-O-(diphenylphosphinyl)-hydroxylamine they form tetrahedral intermediates, which are readily hydrolyzed to give the corresponding amides [243]. This reaction is particularly valuable in the conversion of aromatic aldehydes, where good yields of amides have been achieved. In the same manner, α-aminonitriles are oxidized directly in the presence of potassium *tert*-butylate and oxygen to afford amides in excellent yields at room temperature [244].

9.2.4
Oxidative Decarboxylations of Aldehydes

Aldehydes with branching at the α-position can easily be cleaved into ketones by decarboxylation in the presence of copper complexes and air [245, 246]. For example, in a pyridine/water medium at 70 °C for 4.5 h in the presence of a catalytic amount of $CuCl_2$, isobutyraldehyde gives acetone in 75% yield [245]. This oxidative cleavage reaction has successfully been applied in the modification of lateral chains of steroids using a copper complex bearing a phenanthroline or bipyridine ligand [246]. Furthermore, in an oxygen atmosphere α,β-unsaturated aldehydes undergo an oxidative deformylation, generating allylic peroxides, which are reduced by phosphines to afford alcohols [247].

9.3
Oxidations of Ketones

The focus of this section is on ketone oxidations leading to carboxylic acid derivatives. Firstly, cleavage reactions of ketones followed by oxidative rearrangements and more specific transformations will be discussed. Oxidations of ketones leading to compounds other than carboxylic acids such as oxidative deprotections [190] and oxidations of groups next to the carbonyl moiety [248] are not included.

9.3.1
Ketone Cleavage Reactions

The direct oxidation of ketones to carboxylic acids usually implies a carbon–carbon bond cleavage [249]. Product mixtures are often obtained. However, for the preparation of dicarboxylic acids from cyclic ketones this reaction has proven to be synthetically very useful.

9.3.1.1 Simple Acyclic Ketones
Open-chain alkanones and in particular those with methyl substituents are transformed into their corresponding carboxylic acids by treatment with hypohalite [250]. In this so-called haloform reaction, α-halo ketones are important intermediates. For example, by reaction with hypohalite, methyl ketone **29** is converted into α,α,α-trihalo ketone **30**, which is easily hydrolyzed to give carboxylate **31** and haloform **32** [Eq. (12)].

$$R-CO-CH_3 \xrightarrow{OX^-} R-CO-CX_3 \xrightarrow{OH^-} R-CO-O^- + CHX_3 \quad (12)$$

29   **30**   **31**   **32**

Haloform reactions are generally performed with halogens in the presence of hydroxide [251] or directly with hypohalites [252]. Alternative methods affording carboxylic acids from methyl ketones (or other enolizable substrates) include the aerobic oxidation in the presence of a catalytic amount of dinitrobenzene [253] with a base in a dipolar aprotic solvent {such as DMF [254] or HMPT (hexamethylphosphoric triamide) [255, 256]} and the use of stoichiometric quantities of hypervalent iodide derivatives [95, 257] or nitrosylpentacyanoferrate [258]. Furthermore, metal catalysts can be used, and systems such as *tert*-butyl hydroperoxide in the presence of rhenium oxide [259], oxygen in combination with a copper complex [260], heteropolyacids [261] and Mn^{II}/Co^{II} systems [262] were found to be applicable. Finally, aryl ketones are selectively oxidized to aliphatic carboxylic acids by treatment with periodate [81] in the presence of ruthenium trichloride [263].

9.3.1.2 Simple Cyclic Ketones

The oxidation of cyclic ketones is particularly interesting, especially for the synthesis of dicarboxylic acids [264]. An industrially important process is the conversion of cyclohexanone into adipic acid. When substituents are present in the α-position of the ketone, the cleavage generally occurs at the more hindered side. The carbon–carbon bond-breaking process of cyclic ketones can be mediated by a wide range of chromium [265] (see Chromium, under Section 9.2.1.2) and manganese reagents [266] (see Manganese, under Section 9.2.1.2). Other methods involve cerium salts [267] and potassium superoxide [268]. Reactions under haloform conditions [269] or electrochemical processes [270] have also been studied. Moreover, aerobic oxidations of cyclic ketones have been performed in alkali solution [254] or in the presence of metal catalysts [271–278] such as copper(II) derivatives [272], heteropolyanions [273], oxovanadium complexes [274], iron catalysts [275, 276] and carbonyl complexes of both rhodium [277] and rhenium [278]. The cleavage of cyclohexanones leads to keto acids [267, 272–275] or aldehydo acids [276] depending on the substrate and the reaction conditions. Furthermore, oximinocarboxylic esters are obtained by treatment of cyclic ketones with NOCl in the presence of an alcohol in liquid SO_2 [279]. A modification of this procedure using LDA (lithium diisopropyl amide) in the presence of ethyl nitrite has been employed in the carbon–carbon bond-breaking process of bicyclic ketones [Eq. (13)] [279]. For example, 2-norbonanone (**33**) can be converted into cyclopentane derivative **34** in 46 % yield.

$$\underset{33}{\text{[2-norbornanone]}} \xrightarrow[\text{THF, petroleum ether}]{\text{LDA, EtONO}} \underset{34}{\text{EtO}_2\text{C}-\text{[cyclopentane]}-\text{NOH}} \qquad (13)$$
(46%)

In an alternative approach alkyl nitrite in the presence of a Lewis acid is utilized [280].

Halo carboxylic acids and esters are either obtained by cleavage of cyclic ketones by electrochemical methods using halide electrolytes [281] or by treatment of the substrates with an oxovanadium complex $VO(OEt)Cl_2$ in the presence of a halide donor

[282] [Eq. (14)]. Thus, according to the latter sequence, ethyl-δ-bromo butanoate (**36**) was obtained in 60% yield from cyclobutanone **35** at room temperature after 13 h.

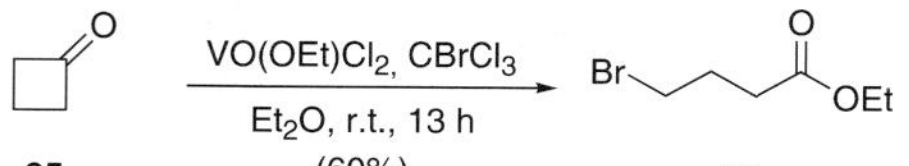

$$\underset{\textbf{35}}{\text{cyclobutanone}} \xrightarrow[\text{Et}_2\text{O, r.t., 13 h}]{\text{VO(OEt)Cl}_2,\ \text{CBrCl}_3} \underset{\textbf{36}}{\text{Br}\sim\sim\text{C(O)OEt}} \quad (60\%) \tag{14}$$

9.3.1.3 Functionalized Ketones

Cyclic ketones with functional groups at the α-position of the carbonyl moiety are easily cleaved into the corresponding dicarbonyl compounds. The most commonly used substrates are α-hydroxy ketones [283], α-diketones [284], α-halo [285], and α-nitro ketones [286]. 1,3-Diketones are oxidized to dicarboxylic acids, and in this case loss of one carbon atom has been observed [287]. Ketone derivatives are also cleaved. For example, ketoximes afford unsaturated nitrile derivatives by Beckmann fragmentation (see Beckmann reaction, under Section 9.3.2.2).

9.3.2
Oxidative Rearrangements of Ketones

9.3.2.1 Baeyer-Villiger Reactions

Introduction and mechanism

More than a century ago Baeyer and Villiger discovered the rearrangement of ketones into esters or lactones [288]. The reaction has been used widely in organic synthesis, and many reagents and conditions have been discovered providing a solid basis for predictable applications [174] (see Dakin reactions, under Section 9.2.2.2). Criegee proposed a two-step mechanism, which is illustrated in Scheme 9.3 [289]. Firstly, peroxide **38** adds to the carbonyl moiety of ketone **37**. This step is reversible and is followed by the rearrangement of the tetrahedral intermediate, the so-called Criegee adduct **B** (Scheme 9.3). The subsequent irreversible migration of one of the

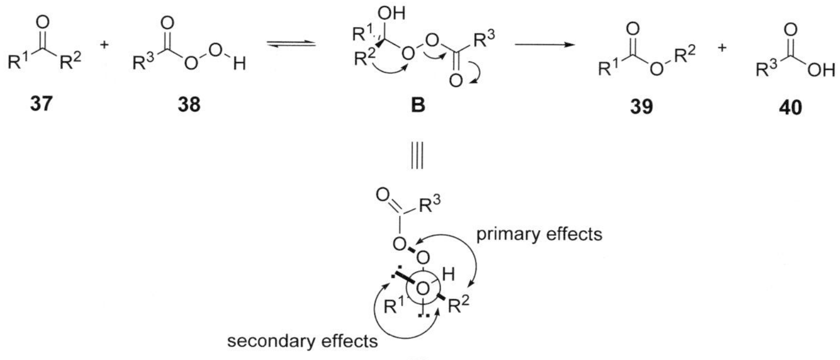

Scheme 9.3 Criegee adduct **B** as an intermediate of the Baeyer-Villiger reaction

two ketonic substituents and the simultaneous cleavage of the O–O bond leads to ester **39** and acid **40** in a concerted manner.

The rearrangement is regioselective with migration of the group that is best able to stabilize the developing positive charge [290]. Thus, the migration rate of ketonic groups normally decreases in the following order: quaternary > tertiary > secondary > cyclopropyl ≫ methyl. Other parameters, such as donor heteroatoms in the α-position [291], β-silyl or β-stannyl groups [292], increase the migration rate. In contrast, electron-withdrawing groups, e.g., halogens, retard it. In transformations of some particular compounds, long-range substituent effects have been observed [293]. Since the migration of the substituent proceeds with retention of the configuration, the Baeyer-Villiger reaction is particularly attractive for the synthesis of optically active products [294].

The stereoelectronic components of the reaction have already been discussed above. For a long time it was assumed that the migrating group occupied an antiperiplanar position with respect to the dissociating oxygen–oxygen bond of the peroxide (see representation **B′** in Scheme 9.3). Recent experimental studies confirmed this primary stereoelectronic effect (**B′**) and revealed that it is indeed, at least in part, responsible for the selectivity of the migration step [295].

Reagents
The choice of the oxidant is primordial, as it acts both as the nucleophile and the leaving group. A reactivity order of oxidants has been established [174]. Peracids and especially m-CPBA [296] are the most efficient oxidants for the Baeyer-Villiger reaction, followed by alkyl peroxides. The reaction with peracids proceeds faster under acid catalysis [297], but buffered solutions can also be used. Some ketones are unreactive under standard conditions, and need to be transformed into their hemiketals [298]. Safer peracids such as MMPP (magnesium monoperphthalate) [299, 300] and sodium perborate [301] have also been applied giving similar results as m-CPBA. Recently, processes with solid-supported peracids [302] and transformations in the solid state [303] have been developed.

Alkyl hydroperoxides and hydrogen peroxides, which are not reactive enough to promote the Baeyer-Villiger reactions, can only be used in combination with catalysts or in the presence of carboxylic acids to form the peracids *in situ* [304].

The Baeyer-Villiger reaction with molecular oxygen as the terminal oxidant in the presence of a sacrificial aldehyde is a synthetically interesting variant of the original protocol [305, 306]. Recently, it has been shown that this process can also be performed in supercritical carbon dioxide [307].

Acidic solvents, and particularly fluorinated alcohols, activate hydrogen peroxide, and even in the absence of a catalyst the rearrangement proceeds faster [308]. Other more uncommon non-metallic oxidants such as perhydrates [309] and TEMPO (2,2,6,6-tetramethylpiperidinyl-1-oxyl) in combination with sodium hypochlorite [309] have also been applied.

Activation by catalysts

Metal catalysis can be used to increase the efficiency of Baeyer-Villiger reactions [310]. Such catalysis is of interest, since both steps of the Baeyer-Villiger reaction (cf. Scheme 9.3) can be influenced. Either the carbonyl compound or the oxidant is activated to lead to a more efficient formation of **B** (first step), or the rearrangement (second step) is enhanced by supporting the decomposition of the Criegee adduct [311].

Along these lines, Mukaiyama and coworkers developed various catalytic systems, which utilize molecular oxygen as the oxidant and show high activity in Baeyer-Villiger reactions. Further components of these systems are a sacrificial aldehyde and a metal complex. The first generation catalysts were nickel complexes [312]. Later, other metal reagents based on iron [313], ruthenium and manganese oxides [314] or heteropolyoxometalates [315] were introduced. Furthermore, mesoporous catalysts have been applied [316]. Redox molecular sieves [317], hydrotalcites and titanosilicates [318] were also found to catalyze the reaction by activation of hydrogen peroxide or oxygen. Corma followed another approach, in which the ketone and not the oxidant is activated by tin containing zeolites [319]. DFT calculations showed, however, that the rate enhancement could not be explained by a simple carbonyl activation provided by such catalysts alone [320]. Currently, Sn-MCM-41 is the most efficient heterogeneous catalyst in terms of TONs (turn over numbers). Alternatively, other Sn^{IV}-containing catalysts can be used, some of which show a similar behaviour to Sn-MCM-41 [321].

Activation of hydrogen peroxide has been achieved by the use of methyltrioxorhenium (MTO) [322]. Strukul and coworkers employed cationic platinum complexes as catalysts and hydrogen peroxide as the oxidant in the conversion of cyclohexanones into caprolactones [323]. A niobiocene complex has been applied giving esters with a regioselectivity opposite to that generally observed [324]. Some supported platinum [325], nickel [326] and methyltrioxorhenium [327] catalysts have also been used in reactions with hydrogen peroxide.

Through the activation of the ketone, Baeyer-Villiger reactions can be accelerated by Lewis acids. A $SnCl_4$/diamine system in combination with trimethylsilyl peroxide as oxidant gave good results with cyclobutanones [328]. By using a tin bis(perfluoroalkanesulfonyl)amide as catalyst, the reaction can be performed under fluorous biphasic conditions, allowing a complete recovery and reuse of the catalyst [329]. Metal triflates have also been applied in the presence of m-CPBA. While scandium triflate is not more efficient than trifluoromethane sulfonic acid [330], bismuth triflate has the advantage of being recyclable and reusable in three catalytic runs without loss of activity [331].

Among the non-metallic catalytic systems for Baeyer-Villiger reactions [174], those with flavine-type catalysts are particularly interesting. Use of catalytic amounts (5 mol%) of flavine analogue **42** in the presence of hydrogen peroxide gave lactone **41** with a yield of up to 90% from ketone **43** (Scheme 9.4) [332].

Aryl diselenides (see Selenium, under Section 9.2.1.4) in combination with hydrogen peroxide are also attractive catalysts for Baeyer-Villiger reactions [333].

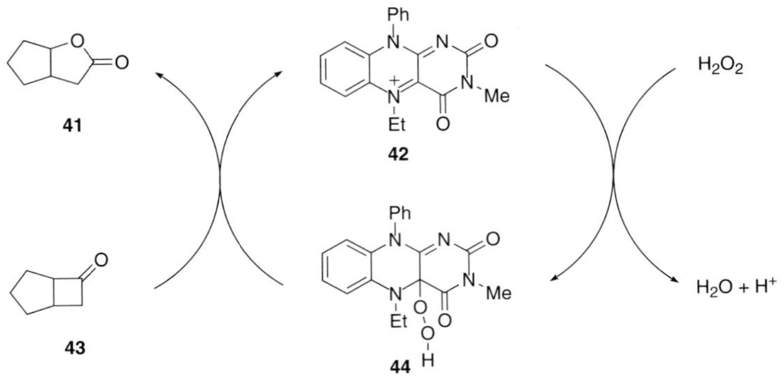

9 Oxidation of Carbonyl Compounds

Scheme 9.4 Oxidation of a bicyclic ketone by hydrogen peroxide catalyzed by a flavine analogue

Asymmetric Baeyer-Villiger reactions
Stoichiometric versions

Two major substrate types have successfully been converted in asymmetric Baeyer-Villiger reactions employing stoichiometric quantities of chiral auxiliaries. Racemic bicyclic ketones lead to two regioisomeric optically active lactones by a stereodivergent process [334], and monocyclic *meso*-substrates are desymmetrized affording enantioenriched lactones [335].

Thus, with regard to the reaction itself, three different approaches can be distinguished. In the first, optically active lactones are obtained using chiral oxidants [336]. In the second, chiral ketone derivatives are applied [337]. An example of the latter type is shown in Eq. (15). The oxidation of ketal **45** in the presence of an excess of $SnCl_4$ and *m*-CPBA as oxidant gives lactone **46** in quantitative yield with 89% enantiomeric excess (ee) [Eq. (15)] [337].

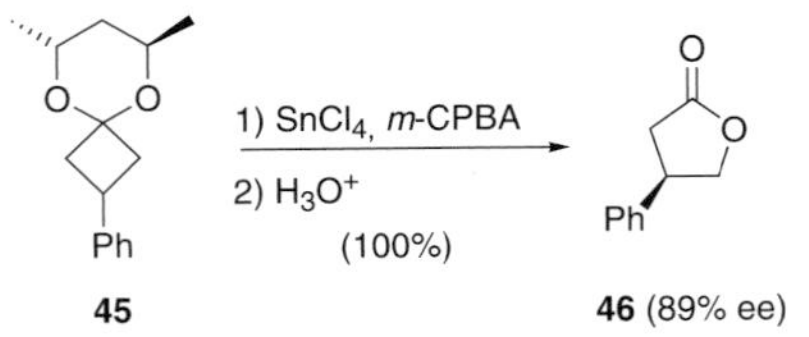

$$\text{45} \xrightarrow[\text{2) } H_3O^+]{\text{1) } SnCl_4, \text{ } m\text{-CPBA}} \text{46 (89\% ee)} \quad (100\%) \tag{15}$$

The third approach involves the use of a chiral Lewis acid. For this purpose, a chirally modified zirconium complex has been developed. By using *tert*-butylhydroperoxide as the oxidant and with a ligand combination of enantiopure BINOL and (achiral) 2,2'-biphenol a bicyclic cyclobutanone derivative was converted into the corresponding lactones with an enantioselectivity up to 84% ee [338].

Catalytic versions

In 1993, Bolm introduced chiral copper complex **49** and applied it in the first metal-catalyzed asymmetric Baeyer-Villiger reaction. Under Mukaiyama-type reaction conditions using molecular oxygen as the oxidant and pivaldehyde as the co-reductant, the transformation of 2-aryl cyclohexanones into enantiomerically enriched caprolactones occurred with enantioselectivities of up to 65% ee [339]. Bicyclic cyclobutanone derivatives and *meso*-ketones were also suitable substrates [340]. The best result was achieved with tricyclic ketone **47**, which was oxidized to lactone **48** with 91% ee in high yield [Eq. (16)].

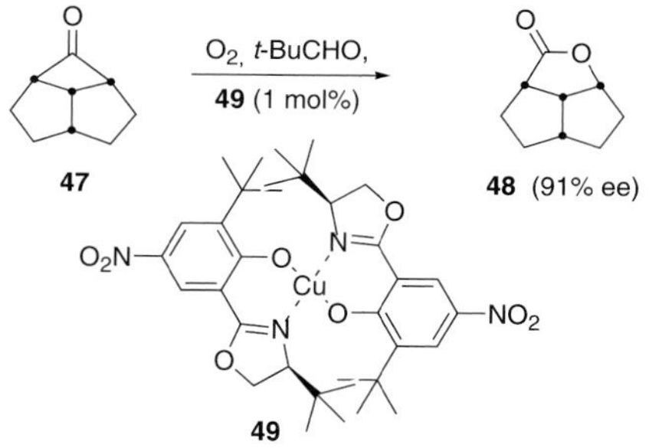

(16)

Other metal catalysts are also capable of performing asymmetric Baeyer-Villiger reactions [341]. For example, Strukul and coworkers found a chiral diphosphine/platinum system with high activity and good enantioselectivity [342]. Katsuki and coworkers investigated an analogous system and replaced platinum by palladium [343]. In this case, the reaction worked best with a chiral 2-(phosphinophenyl)-pyridine, which allowed lactones **46** and **48** to be obtained with 80 and >99% ee, respectively. It is worth noting that both Bolm's copper and Strukul's platinum catalyst are also applicable on substituted cyclic ketones other than cyclobutanones. Most systems developed later required the high reactivity of the four-membered cycloalkanones to achieve efficient conversions.

Diethyl zinc in combination with a chiral aminoalcohol and oxygen as oxidant afforded lactone **46** with up to 32% ee in good yield from 3-phenylcyclobutanone (**50**) [Eq. (17)] [344]. Applying the Sharpless/Katsuki titanium-based epoxidation system to related substrates gave products with both moderate enantioselectivities and yields [345].

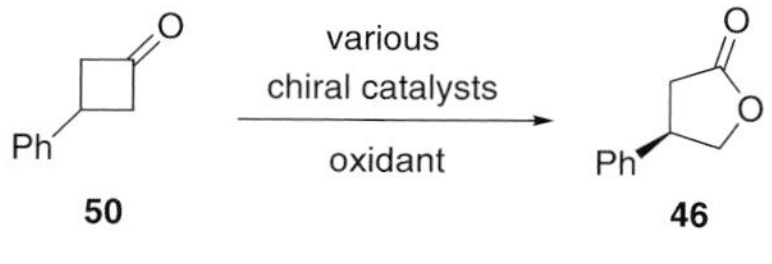

(17)

Currently, the most efficient catalysts in terms of activity and enantioselectivity are binaphthol- and salen-based systems. The former is exemplified by a combination of BINOL and magnesium iodide, which catalyzes the conversion of prochiral ketones into enantiomerically enriched lactones with up to 65% ee [346]. When BINOL derivatives are combined with organoaluminum reagents, enantioselectivities of up to

76% ee have been reached [347]. Use of a chiral oxidant did not improve the stereoselectivity in this system any further [348].

Katsuki successfully applied two different metal salen complexes in asymmetric Baeyer-Villiger reactions. Cationic cobalt complex **51** catalyzes the reaction between cyclobutanone **50** and UHP as oxidant to give lactone **46** with 77% ee in 72% yield [349]. In the same transformation zirconium salen complex **52** affords the product with higher enantioselectivity (87% ee) (Scheme 9.5). Over 99% ee could be achieved in the conversion of tricyclic ketone **47** [350].

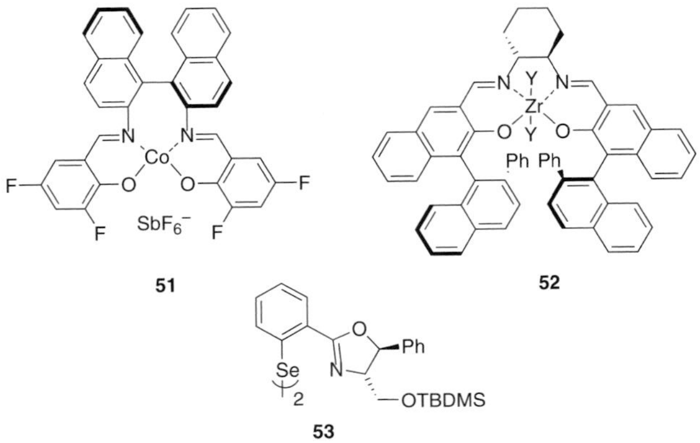

Scheme 9.5 Chiral catalysts used in asymmetric Baeyer-Villiger reactions

Another approach towards asymmetric Baeyer-Villiger reactions involves the use of chiral diselenide **53** [351]. In combination with ytterbium triflate and hydrogen peroxide (30%) it forms a catalyst, which is able to produce lactone **46** from cyclobutanone **50** with 19% ee in up to 92% yield.

Murahashi modified the flavine/hydroperoxide system developed by Furstoss and coworkers [332] and introduced a chiral bisflavine. Under appropriate reaction conditions lactone **46** was obtained with 74% ee in 70% yield [352].

Biocatalytic versions
Among the various methods for performing asymmetric Baeyer-Villiger reactions, biocatalytic processes represent an interesting alternative. In enzymatic transformations good activities and high enantioselectivities have been achieved with a number of substrates. Often, however, generalization problems still have to be dealt with [353].

9.3.2.2 Ketone Amidations

Beckmann reaction
The conversion of oximes, which are readily available from ketones, into amides is known as the Beckmann rearrangement [354]. This reaction has been used exten-

sively in organic synthesis [355] and can be considered as the nitrogen analogue of the Baeyer-Villiger rearrangement (Scheme 9.6). Firstly, the hydroxyl group of oxime **C** is transformed into a good leaving group (OA). Subsequently, migration of the *anti* substituent of the heterocarbonyl group and simultaneous cleavage of the nitrogen–oxygen bond of **D** occurs. The regioselectivity of the rearrangement is determined by the configuration of the oxime. However, under isomerization conditions (**D** and **D′**), the migratory aptitude of the substituents and the stability of the oxime govern the selectivity of the reaction. In this case, aryl moieties migrate more easily than alkyl groups and hydrogen. Certain substituents such as endocyclic vinyl groups do not migrate at all.

A nitrilium ion **E** or an imidate derivative **G** is formed as an intermediate. Their hydrolysis leads either to the formation of amide **H** or, in the case of nitrilium ion **E**, to nitrile **F** by fragmentation and formal loss of a carbocation. Under particular conditions nitrilium salts **E** derived from special substrates can be isolated [356].

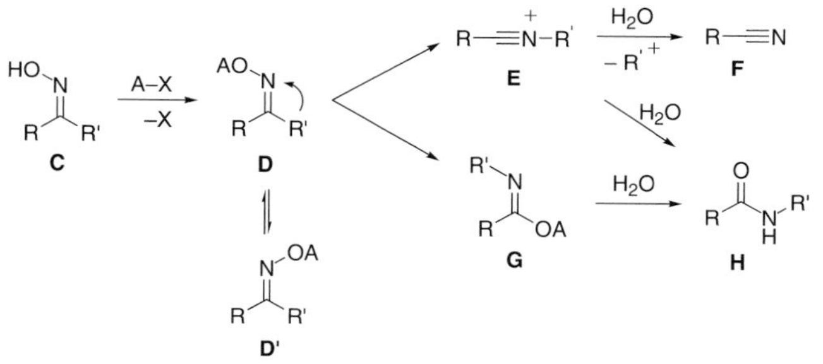

Scheme 9.6 Mechanistic path of the Beckmann rearrangement

Recently, *ab initio* calculations on the mechanism of the Beckmann rearrangement of various substrates have been performed [357].

The transformation of the oxime hydroxyl group into a better leaving group can be achieved by a plethora of reagents [355]. The most common way of achieving this reaction is to use a strong acid [355, 358]. Usually, however, this requires high temperature. Other reagents such as montmorillonite K 10 [359], iminium salts [360], thionyl chloride [361], catalytic quantities of phosphorous(V) [362–364] in ionic liquids [364], phosphine with a halide source [365] and Lewis acids [366] are also effective. Recently, new methods have been developed involving a rhodium catalyst in the presence of sulfonic acid [367], heterogeneous catalysts [368], supercritical water [369] and 2,4,6-trichloro[1,3,5]triazines (**54**) [370] in DMF at room temperature. In the last case, excellent yields have been achieved for a great variety of substrates. For example, use of **54** in the Beckmann rearrangement of acetophenone oxime (**55**) affords amide **56** in quantitative yield [Eq. (18)]. The active species in this reaction is believed to be a Vilsmeier-Haack type complex formed by reaction between triazine **54** and a molecule of DMF.

$$\underset{\underset{55}{Ph\quad Me}}{\overset{N^{\nwarrow OH}}{\|}} \xrightarrow[\substack{\text{DMF, r.t., 6 h}\\(100\%)}]{54} \underset{56}{\overset{O}{\underset{Me}{\|}}\underset{H}{\overset{}{N}}\text{-}Ph} \qquad (18)$$

where **54** is 2,4,6-trichloro-1,3,5-triazine (cyanuric chloride).

Another mild method for the activation of the oxime hydroxyl group consists in its conversion into an ester moiety or, more commonly, into a sulfonate derivative. They can be rearranged into amides under basic [371] or acidic [372] conditions also involving Lewis acids [366]. Trimethylsilyl chloride [373] and chloroformate [374] in the presence of a Lewis acid can also be used. Amidines have been obtained by Beckmann rearrangement of a tosyloxime in the presence of benzotriazol [375].

Recently, an environmentally friendly method has been developed starting from oxime carbamate derivatives [376]. They are diazotized by treatment with amyl nitrite in the presence of sulfuric acid and afford the corresponding amides with CO_2 and N_2 as the only byproducts.

Finally, it was found that ketones can be used as starting materials directly. When treated with $NH_2\text{-}O\text{-}SO_3H$ [377] or MSH (o-mesitylene sulfonyl hydroxylamine) [378] they efficiently rearrange to give amides. Other single-step procedures, which promote both the formation of the oxime and its rearrangement, involve ZnO [379], silica supported $NaHSO_4$ [380] or alumina/CH_3SO_3H [231].

Depending on the reaction conditions, fragmentation rather than rearrangement can occur. Substrates with a quaternary [381] or a heteroatom-substituted carbon [382] as the potential migrating group as well as α-ketoximes [383] undergo this reaction. Ring-contraction by Beckmann type rearrangement has also been observed and leads to α-amino acid derivatives [384].

Progress has recently been made in cascade reactions. Use of aluminum reagents leads to the formation of nitrilium species, which add nucleophiles such as hydride [385], thiol [386] or cyanide [387]. Finally, cascade Friedel-Craft/Beckmann amidation reactions of electron-rich arenes have been developed. They involve reactions of carboxylic acids and hydroxylamine in the presence of polyphosphoric acid [388].

Schmidt reaction

Treatment of ketones with azide derivatives affords amides by a Schmidt reaction [389]. Its mechanism is similar to that of the Beckmann rearrangement. The key intermediate is an iminodiazonium ion [390]. In transformations of bridged bicyclic ketones both the Schmidt and the Beckmann reaction are complementary [391] since inverse selectivities have been observed [392].

The Schmidt reaction can be performed intermolecularly using NaN_3 [393], $TMSN_3$ [394] or alkylazide [395]. With a hydroxyazide the corresponding imidate salt is obtained, which reacts with nucleophiles [396]. Reactions using alkylazide in the presence of $TiCl_4$ or a proton source have also been studied [397, 398]. Chiral molecules bearing a terminal azido group afford products with good diastereoselectivity [398]. An example of an intermolecular desymmetrization reaction [399, 400] is shown in Eq. (19). Using β-hydroxyazide **58** in the desymmetrization of ketone **57** leads to hydroxy lactames **59** and **60** in a ratio of 7:93 [399].

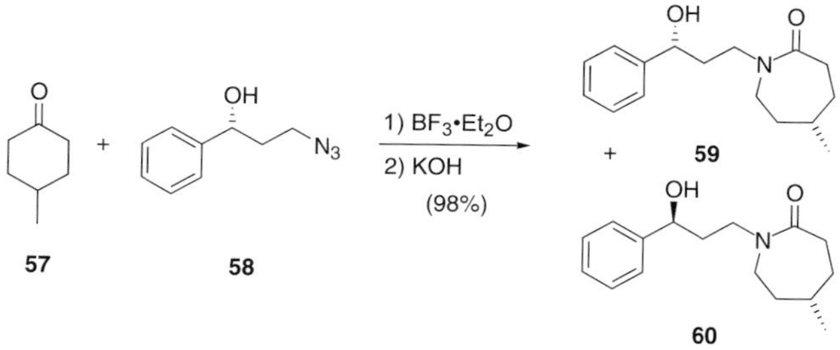

$$\text{57} + \text{58} \xrightarrow[\text{2) KOH}]{\text{1) BF}_3\cdot\text{Et}_2\text{O}} \text{59} + \text{60} \quad (98\%) \tag{19}$$

The Schmidt reaction has also been applied to silyl enol ethers, which furnish lactames on treatment with TMSN₃ and UV irradiation [401].

Cascade Diels-Alder/Schmidt reactions [402] allow the synthesis of polycyclic compounds, which lead to useful precursors for the synthesis of natural products. Finally, fragmentation reactions can occur, and in some cases the corresponding nitriles are obtained [403].

9.3.2.3 Miscellaneous Rearrangements

Aryl alkyl ketones **61** undergo oxidative rearrangement to α-arylalkanoic acids **62** [Eq. (20)] [404], which are industrially important intermediates for the synthesis of compounds with anti-inflammatory properties.

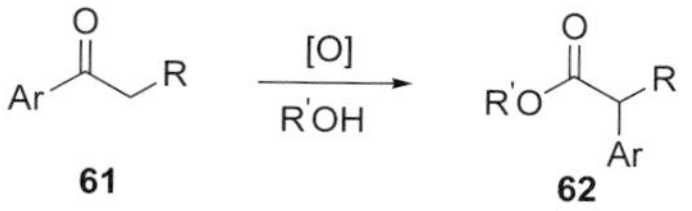

$$\text{61} \xrightarrow[\text{R'OH}]{[\text{O}]} \text{62} \tag{20}$$

The most common reagents used for this transformation are hypervalent main group oxidants such as thallium(III) [405, 406], lead(IV) [407, 408] and hypervalent iodide(III) [95, 409]. In order to avoid any α-carbonyl oxidation the reaction is usually performed in the presence of an alcohol or a combination of trialkyl orthoformate and a strong acid to ensure rapid acetalization. However, depending on the reaction conditions, hydroxyacids can also be isolated as major products [410, 411].

A wide variety of ketones with both aryl and alkyl groups can be used in this reaction. Dialkyl ketones also react, but mixtures of carboxylic acids are generally obtained due to the similarity of the migratory aptitude of the alkyl groups. Functionalized ketones behave analogously. Thus, α-diazo ketones yield α-arylalkanoic acid derivatives in the presence of metal salts through the Wolff rearrangement [412, 413]. α-Hydroxy ketones are effectively converted when treated with DAST (diethylaminosulfur trifluoride) [414]. In a similar manner, α-halo ketones are transformed into α-arylalkanoic acid derivatives in the presence of a base by the Favorskii rearrangement [415]. In this reaction zinc [416], thallium [417] and silver salts [418, 419] as well as peracids (by oxidation) [420] and photochemical processes [421] are also applicable. The use of enantioenriched α-bromoacetals leads to a stereospecific formation of α-arylalkanoic acid derivatives [418]. The required α-halo ketones can be formed *in situ* by treatment of ketones with halide in the presence of alkoxide [422, 423]. Sometimes, however, an electrochemical oxidation [423] is necessary.

Selective migrations on addition of a base have also been achieved starting from α-chloro-α-sulfoxidyl ketones [424] or α-chloro-α-sulfonyl ketones [425]. In the latter case, enantiopure substrates lead to the formation of β-sulfonyl carboxylic acids with excellent enantiocontrol. α-Selenium derivatives are oxidized to α-arylalkanoic derivatives by peracids [426].

α,β-Unsaturated ketones can either be transferred into their α,γ-unsaturated counterparts by treatment with $Pb(OAc)_4$ [407, 427] or to β-hydroxyacid derivatives when reacted with hypervalent iodine [95, 428]. Finally, α-hydroxy carboxylic acids are obtained from aryl α-diketones in a base mediated benzil–benzilic acid rearrangement [429].

9.3.3
Willgerodt Reactions

The Willgerodt reaction allows the transformation of a straight or branched aryl alkyl ketone into its corresponding amide and/or acid ammonium salt by heating the substrate in the presence of ammonium polysulfide [430]. Interestingly, the position of the carbonyl group in the ketone is irrelevant, and only terminal carboxylic acid derivatives are obtained. Unfortunately, the yield decreases dramatically with increasing chain length. The reaction can also be performed with aliphatic and unsaturated ketones, albeit the product yields are low. When sulfur, in combination with dry primary or secondary amines, is used as the reagent, the reaction is called the Kindler modification of the Willgerodt reaction [431, 432]. In this case the product is a thioamide, which can easily be hydrolyzed to the corresponding carboxylic acid. The conditions have been optimized [432], and particularly good results have been obtained with morpholine as the amine component [433]. In addition, aliphatic ketones also react, as demonstrated by the example shown in Eq. (21). The Willgerodt-Kindler reaction of pentan-3-one (**63**) gives linear thioamide **64** in 53% yield.

$$\underset{\mathbf{63}}{\text{pentan-3-one}} \xrightarrow[\substack{\text{DMF} \\ 130\,°C,\ 3\,h \\ (53\%)}]{S_8\,/\,\text{morpholine}} \underset{\mathbf{64}}{\text{thioamide}} \qquad (21)$$

Other variants of this reaction involve the use of HCl salts of a volatile secondary amine in the presence of sodium acetate in DMF [434], dimethylammonium dimethylcarbamate [435] or catalysts [436]. Microwave irradiation leads to good yields of thioamide derivatives in short reaction times [437]. Transformations of dialdehydes with secondary diamines in the presence of sulfur can be used for the synthesis of polymers [438].

The mechanism of the Willgerodt reaction is not entirely clear [439]. Most likely it does not proceed by a skeletal rearrangement, but involves a series of consecutive oxidations and reductions along the carbon chain. A thorough understanding of this reaction is still to be established.

9.4
Conclusions

Oxidations of carbonyl compounds to carboxylic acid derivatives are essential and important tools in organic synthesis. Many methods are already mature and allow the desired transformations to be performed in a highly predictable and selective manner. Nevertheless, further improvements are required, in particular with respect to environmental issues and the use of oxidative transformations on a large industrial scale. Catalytic and asymmetric carbonyl oxidations have emerged but most of them still appear to be at a very basic stage. Particularly interesting discoveries have been made in reinvestigations of "old reactions" such as the Baeyer-Villiger, Cannizzaro and Tishchenko reactions. They now take on a different appearance, which will make them even more attractive for application in the synthesis of complex organic molecules. For sure, the future will bring more of those exciting developments into oxidation chemistry.

References

[1] H. S. Verter in *The Chemistry of Functional Groups. The Chemistry of the Carbonyl Group* (Ed.: J. Zabicky), Wiley-Interscience, New York, **1970**, p. 71.

[2] *Methods for the Oxidation of Organic Compounds* (Ed.: A. H. Haines), Academic Press, London, **1988**, p. 241.

[3] *Oxidation* in Organic Compounds (Ed.: M. Hudlicky) ACS Monograph Series, **1990**, pp. 174 and 186.

[4] *Advanced Organic Chemistry* (Eds.: M. B. Smith, J. March), Wiley-Interscience, New York, **2001**, p. 1507.

[5] (a) *Methoden der Chemie (Houben-Weyl)*, Vol. 4/1a (Ed.: H. Kropf) 4th edn., Thieme, Stuttgart, **1981**; (b) *Methoden der Chemie (Houben-Weyl)*, Vol. 4/1b (Ed.: E. Müller) 4th edn., Thieme, Stuttgart, **1975**.

[6] (a) *Oxidation* in Organic Chemistry (Ed.: K. E. Wiberg), Part A, Academic Press, New York, **1965**; (b) *Oxidation* in Organic Chemistry (Ed.: K. E. Trahanovsky), Part B, Academic Press, New York, **1973**; (c) *Oxidation* in Organic Chemistry (Ed.: K. E. Trahanovsky), Part C Academic Press, New York, **1978**; (d) *Oxidation* in Organic Chemistry (Ed.: K. E. Trahanovsky), Part D, Academic Press, New York, **1982**.

[7] (a) *Oxidation, Vol. 1* (Ed.: R. L. Augustine), Dekker, New York, **1969**; (b) *Oxidation, Vol. 2* (Eds.: R. L. Augustine, D. J. Trecker), Dekker, New York, **1970**.

[8] *Modern Synthetic Reactions* (Ed.: H. O. House), 2nd edn., Benjamin, Menlo Park, **1972**.

[9] *Methoden der Chemie (Houben-Weyl)*, Vol. E5 (Ed.: J. Falbe), 4th edn., Thieme, Stuttgart, **1985**, p. 209.

[10] *Methoden der Chemie (Houben-Weyl)*, Vol. 8 (Ed.: E. Müller), 4th edn., Thieme, Stuttgart, **1952**, pp. 407 and 557.

[11] *Comprehensive Organic Transformations* (Ed.: R. C. Larock), 2nd edn., Wiley, New York, **1999**, p. 1621.

[12] M. A. OGLIARUSO, J. F. WOLFE in *The Chemistry of Functional Groups. The Chemistry of Acid Derivatives* (Ed.: S. PATAI), Part 1, Wiley-Interscience, New York, **1979**, p. 267.

[13] M. A. OGLIARUSO, J. F. WOLFE in *Synthesis of Carboxylic Acids, Esters and their Derivatives* (Eds.: S. PATAI, Z. RAPPOPORT), Wiley-Interscience, New York, **1991**.

[14] V. F. KUCHEROV, L. A. YANOVSKAYA in *The Chemistry of Carboxylic Acids and Esters* (Ed.: S. PATAI), Wiley-Interscience, New York, **1969**, p. 175.

[15] M. A. OGLIARUSO, J. F. WOLFE in *Comprehensive Organic Functional Group Transformations*, Vol. 5 (Eds.: A. R. KATRITZKY, O. METH-COHN, C. W. REES), Elsevier Science, Oxford, **1995**, p. 23.

[16] (a) *Survey of Organic Synthesis*, Vol. 1 (Eds.: C. A. BUEHLER, D. E. PEARSON), Wiley-Interscience, New York, **1970**; (b) *Survey of Organic Synthesis*, Vol. 2 (Eds.: C. A. BUEHLER, D. E. PEARSON), Wiley-Interscience, New York, **1977**.

[17] *Metal Catalyzed Oxidation of Organic Compounds* (Eds.: R. A. SHELDON, J. K. KOCHI), Academic Press, New York, **1981**.

[18] R. A. SHELDON, J. K. KOCHI, *Adv. Catal.* **1976**, *25*, 272.

[19] H.-G. PADEKEN in *Methoden der Chemie (Houben-Weyl)*, Vol. E3 (Ed.: J. FALBE), 4th edn., Thieme, Stuttgart, **1983**, p. 634.

[20] J. ROCEK in *The Chemistry of Functional Groups. The Chemistry of the Carbonyl Group* (Ed.: S. PATAI), Wiley-Interscience, New York, **1966**, p. 461.

[21] (a) M. NICLAUSE, J. LEMAIRE, M. LETORT, *Adv. Photochem.* **1966**, *4*, 25; (b) *Mechanisms of Oxidation of Organic Compounds* (Ed.: W. A. WATERS), Wiley, New York, **1964**; (c) C. LEHTINEN, V. NEVALAINEN, G. BRUNOW, *Tetrahedron* **2001**, *57*, 4741.

[22] *Catalytic Activation of Dioxygen by Metal Complexes* (Ed.: L. I. SIMANDI), Kluwer Academic Publishers, Dordrecht, **1992**, p. 318.

[23] (a) M. HAMAMOTO, K. NAKAYAMA, Y. NISHIYAMA, Y. ISHII, *J. Org. Chem.* **1993**, *58*, 6421; (b) A. N. KHARAT, P. PENDLETON, A. BADALYAN, M. ABEDINI, M. MOHAMMAD, *J. Mol. Catal. A* **2001**, *175*, 277.

[24] P. MASTRORILLI, C. F. NOBILE, *Tetrahedron Lett.* **1994**, *35*, 4193.

[25] J. HOWARTH, *Tetrahedron Lett.* **2000**, *41*, 6627.

[26] T. PUNNIYAMURTHY, S. J. SINGH KALRA, J. IQBAL, *Tetrahedron Lett.* **1994**, *35*, 2959.

[27] W.-K. WONG, X.-P. CHEN, J.-P. GUO, Y.-G. CHI, W.-X. PAN, W.-Y. WONG, *J. Chem. Soc., Dalton Trans.* **2002**, 1139.

[28] J.-I. HOJO, S. YUASA, N. YAMAZOE, I. MOCHIDA, T. SEIYAMA, *J. Catal.* **1975**, *36*, 93.

[29] S. OHTA, T. TACHI, M. OKAMOTO, *Synthesis* **1983**, 291.

[30] P. J. GARRAT, C. W. DOECKE, J. C. WEBER, L. A. PAQUETTE, *J. Org. Chem.* **1986**, *51*, 449.

[31] C. K. LEONDING, M. SCHÖFTNER, J. FRIEDHUBER, (for DSM Chemie) US 5.686.638 (**1997**).

[32] F. KOCH in *Applied Homogeneous Catalysis with Organometallic Compounds* (Eds.: B. CORNILS, W. A. HERRMANN), Wiley-VCH, Weinheim, **2002**, p. 427.

[33] M. BESSON, P. Gazellot in *Fine Chemicals through Heterogeneous Catalysis* (Eds.: R. A. SHELDON, H. van BEKKUM), Wiley-VCH, Weinheim, **2001**, p. 491.

[34] (a) A. ABAD, M. ARNO, A. C. CUNAT, M. L. MARIN, R. J. ZARAGOZA, *J. Org. Chem.* **1992**, *57*, 6861; (b) J. FRIED, J. C. SIH, *Tetrahedron Lett.* **1973**, *14*, 3899.

[35] D. P. RILEY, D. P. GETMAN, G. R. BECK, R. M. HEINTZ, *J. Org. Chem.* **1987**, *52*, 287.

[36] F. VONCANSON, Y. P. GUO, J. L. NAMY, H. B. KAGAN, *Synth. Commun.* **1998**, *28*, 2577.

[37] (a) J. MEIJER, A. H. HOGT, B. FISCHER in *Acros Organics, Chemistry Review Prints*, No. 6; (b) *Organic Peroxides* (Ed.: W. ANDO), Wiley, Chichester, **1992**; (c) H. HEANEY, *Top. Curr. Chem.* **1993**, *164*, 1; (d) R. SHELDON, *Top. Curr. Chem.* **1993**, *164*, 23; (e) *Peroxide Chemistry Mechanistic and Preparative Aspects of Oxygen Transfer* (Ed.: W. ADAM), Research Report-DFG, Wiley-VCH, Weinheim, **2000**.

[38] (a) A. S. Rao, H. R. Mohan in *Handbook of Reagents for Organic Synthesis, Oxidizing and Reducing Agents* (Eds.: S. D. Burke, R. L. Danheiser), Wiley, Chichester, **1999**, p. 174; (b) *Catalytic Oxidations with Hydrogen Peroxide as Oxidant* (Ed.: G. Strukul), Kluwer, Dordrecht, **1992**.

[39] (a) H. Koyama, T. Kamikawa, *J. Chem. Soc., Perkin Trans. 1* **1998**, 203; (b) C. González, E. Guitián, L. Castedo, *Tetrahedron* **1999**, 55, 5195; (c) R. Ballini, M. Petrini, *Synthesis* **1986**, 1024; (d) A. J. Kirby, J. M. Percy, *Tetrahedron* **1988**, 44, 6903.

[40] R. Longeray, P. Lanteri, X. Lu, C. Huet, (for Air Liquide) EP 424242 (**1991**).

[41] B. M. Trost, Y. Masuyama, *Tetrahedron Lett.* **1984**, 25, 173.

[42] R. Gopinath, B. K. Patel, *Org. Lett.* **2000**, 2, 577.

[43] C. Venturello, M. Gambaro, *J. Org. Chem.* **1991**, 56, 5924.

[44] S. P. Chavan, S. W. Dantale, C. A. Govande, M. S. Venkatraman, C. Praveen, *Synlett* **2002**, 267.

[45] K. Sato, M. Hyodo, J. Takagi, M. Aoki, R. Noyori, *Tetrahedron Lett.* **2000**, 41, 1439.

[46] A. K. Jones, T. E. Wilson, S. S. Nikam in *Handbook of Reagents for Organic Synthesis, Oxidizing an Reducing Agents* (Eds.: S. D. Burke, R. L. Danheiser), Wiley, Chichester, **1999**, p. 61.

[47] (a) Y. Masuyama, M. Takahashi, Y. Kurusu, *Tetrahedron Lett.* **1984**, 25, 4417; (b) S. Zhang, R. E. Shepherd, *Inorg. Chim. Acta* **1992**, 193, 217.

[48] F. Fringuelli, R. Pellegrino, O. Piermatti, F. Pizzo, *Synth. Commun.* **1994**, 24, 2665.

[49] B. Ganem, R. P. Heggs, A. J. Biloski, *Tetrahedron Lett.* **1980**, 21, 685.

[50] E. G. E. Hawkins, *J. Chem. Soc.* **1950**, 2169.

[51] A. L. Baumstark, M. Beeson, P. C. Vasquez, *Tetrahedron Lett.* **1989**, 30, 5567.

[52] (a) R. Balicki, *Synth. Commun.* **2001**, 31, 2195; (b) R. H. Dodd, M. Le Hyaric, *Synthesis* **1993**, 295.

[53] (a) J. W. Patterson, *J. Org. Chem.* **1990**, 55, 5528; (b) Y. Ogata, Y. Sawaki, *J. Org. Chem.* **1969**, 34, 3985; (c) A. Adejare, J. Shen, A. M. Ogunbadeniyi, *J. Fluorine Chem.* **2000**, 105, 107.

[54] Y. Watanabe, K. Takehira, M. Shimizu, T. Hayakawa, H. Orita, *J. Chem. Soc., Chem. Commun.* **1990**, 927.

[55] T.-C. Zheng, D. E. Richardson, *Tetrahedron Lett.* **1995**, 36, 837.

[56] I. A. Pearl, *Org. Synth. Coll. Vol. 4*, **1963**, 974.

[57] (a) *The Oxidation of Organic Compounds by Permanganate Ion and Hexavalent Chromium* (Ed.: D. G. Lee), Open Court, La Salle, **1980**; (b) *Organic Synthesis by Oxidation with Metal Compounds* (Eds.: W. J. Mijs, C. R. J. I. de Jonge), Plenum, New York, **1986**.

[58] (a) *Manganese Compounds as Oxidizing Agent* in Organic Chemistry (Ed.: D. Arndt), Open Court, La Salle, **1981**; (b) A. J. Fatiadi, *Synthesis* **1987**, 85.

[59] (a) J.-I. Kim, G. B. Schuster, *J. Am. Chem. Soc.* **1992**, 114, 9303; (b) G. Parinello, J. K. Stille, *J. Am. Chem. Soc.* **1987**, 109, 7122; (c) S. G. Jagadeesh, G. L. Krupadanam, G. Srimannarayana, *Synth. Commun.* **2001**, 31, 1547; (d) A. Tsuge, T. Moriguchi, S. Mataka, M. Tashiro, *Liebigs Ann. Chem.* **1996**, 769; (e) B. Wang, S. Gangwar, G. M. Pauletti, T. J. Siahaan, R. T. Borchardt, *J. Org. Chem.* **1997**, 62, 1363; (f) A. G. Schultz, E. G. Antoulinakis, *J. Org. Chem.* **1996**, 61, 4555; (g) P. Kutschy, M. Suchy, A. Andreani, M. Dzurilla, M. Rossi, *Tetrahedron Lett.* **2001**, 42, 9281; (h) A. J. Kirby, I. V. Komarov, N. Feeder *J. Chem. Soc., Perkin Trans. 2* **2001**, 522.

[60] (a) T. Sala, M. V. Sargent, *J. Chem. Soc., Chem. Commun.* **1978**, 253; (b) T. L. Capson, M. D. Thompson, V. M. Dixit, R. G. Gaughan, C. D. Poulter, *J. Org. Chem.* **1988**, 53, 5903; (c) F. M. Menger, J. U. Rhee, H. K. Rhee, *J. Org. Chem.* **1975**, 40, 3803.

[61] D. Scholz, *Monatsh. Chem.* **1979**, 110, 1471.

[62] D. J. Sam, H. E. Simmons, *J. Am. Chem. Soc.* **1972**, 94, 4024.

[63] (a) R. J. Heffner, J. Jiang, M. M. Jouillé, *J. Am. Chem. Soc.* **1992**, 114, 10181; (b) A. Abiko, J. C. Roberts, T. Takemasa, S. Masmune, *Tetrahedron*

Lett. **1986**, *27*, 4537; (c) Y. CHEN, Z. XIONG, J. YANG, Y. LI, *J. Chem. Res. (S)* **1997**, 472; (d) B. LIU, S. DUAN, A. C. SUTTERER, K. D. MOELLER, *J. Am. Chem. Soc.* **2002**, *124*, 10101; (e) F. BENEDETTI, P. MAMAN, S. NORDEBO, *Tetrahedron Lett.* **2000**, *41*, 10075.

[64] F. MENGER, C. LEE, *Tetrahedron Lett.* **1981**, *22*, 1655.

[65] N. A. NOURELDIN, D. G. LEE, *J. Org. Chem.* **1982**, *47*, 2790.

[66] J. VICENTE, J.-A. ABAD, B. RINK, F.-S. HERNANDEZ, M. C. RAMÍREZ DE ARELLANO, *Organometallics* **1997**, *16*, 5269.

[67] D. R. T. KNOWLES, H. ADAMS, P. M. MAITLIS, *Organometallics* **1998**, *17*, 1741.

[68] (a) G. CAHIEZ, M. ALAMI in *Handbook of Reagents for Organic Synthesis, Oxidizing and Reducing Agents* (Eds.: S. D. BURKE, R. L. DANHEISER), Wiley, Chichester, **1999**, p. 231; (b) A. J. FATIADI, *Synthesis* **1976**, 65; (c) A. J. FATIADI, *Synthesis* **1976**, 133.

[69] M. Z. BARAKAT, M. F. ABDEL-WAHAB, M. M. EL-SADR, *J. Chem. Soc.* **1956**, 4685.

[70] (a) E. J. COREY, N. W. GILMAN, B. E. GANEM, *J. Am. Chem. Soc.* **1968**, *90*, 5616; (b) E. J. COREY, J. A. KATZENELLENBOGEN, N. W. GILMAN, S. A. ROMAN, B. W. ERICKSON, *J. Am. Chem. Soc.* **1968**, *90*, 5618.

[71] (a) R. BAUBOUY, J. GORE, *Synthesis* **1974**, 572; (b) R. H. SCLESSINGER, E. J. IWANOWICZ, J. P. SPRINGER, *J. Org. Chem.* **1986**, *51*, 3070; (c) M. G. CONSTANTINO, P. M. DONATE, N. PETRAGNANI, *J. Org. Chem.* **1986**, *51*, 253; (d) E. P. WOO, F. SONDHEIMER, *Tetrahedron* **1970**, *26*, 3933; (e) B. S. BAL, W. E. CHILDERS, H. W. PINNICK, *Tetrahedron* **1981**, *37*, 2091; (f) H. YAMAMOTO, T. ORITANI, *Tetrahedron Lett.* **1995**, *36*, 5797; (g) P. ANGERS, P. CANONNE, *Tetrahedron Lett.* **1995**, *36*, 2397; (i) G. LAI, W. K. ANDERSON, *Synth. Commun.* **1997**, *27*, 1281; (h) M. HOHMANN, N. KRAUSE, *Chem. Ber.* **1995**, *128*, 851.

[72] (a) C.-W. KUO, J.-M. FANG, *Synth. Commun.* **2001**, *31*, 877; (b) G. LAI, W. K. ANDERSON, *Synth. Commun.* **1997**, *27*, 1281; (c) M. KOREEDA, Y. WANG, *J. Org. Chem.* **1997**, *62*, 446; (d) P. D. JONES, T. E. GLASS, *Tetrahedron Lett.* **2001**, *42*, 2265; (e) D. L. BOGER, S. E. WOLKENBERG, *J. Org. Chem.* **2000**, *65*, 9120.

[73] (a) *Chromium Oxidations* in Organic Chemistry (Eds.: G. CAINELLI, G. CARDILLO), Springer, New York, **1984**; (b) J. MUZART, *Chem. Rev.* **1992**, *92*, 113.

[74] K. BOWDEN, I. M. HEILBRON, E. R. H. JONES, B. C. L. WEEDON, *J. Chem. Soc.* **1946**, 39.

[75] (a) J. WROBEL, A. DIETRICH, B. J. GORHAM, K. SESTANJ, *J. Org. Chem.* **1990**, *55*, 2694; (b) R. L. HALTERMAN, M. A. MCEVOY, *J. Am. Chem. Soc.* **1990**, *112*, 6690; (c) M.-T. LAI, L.-D. LIU, H.-W. LIU, *J. Am. Chem. Soc.* **1991**, *113*, 7388; (d) S. D. DAVIES, S. W. EPSTEIN, A. C. GARNER, O. ICHIHARA, A. D. SMITH, *Tetrahedron: Asymmetry* **2002**, *13*, 1555; (e) F. A. LUZZIO, R. W. FICHT, *J. Prakt. Chem.* **2000**, *342*, 498; (f) M. GHOSH, R. G. DULINA, R. KAKARLA, M. J. SOFIA, *J. Org. Chem.* **2000**, *65*, 8387; (h) P. C. SILVA, J. S. COSTA, V. L. P. PEREIRA, *Synth. Commun.* **2001**, *31*, 595; (i) N. PERLMAN, A. ALBECK, *Synth. Commun.* **2000**, *30*, 4443; (j) P. GIZECKI, R. DHAL, L. TOUPET, G. DUJARDIN, *Org. Lett.* **2000**, *2*, 585; (k) S. RÁDL, J. STACH, J. HAJICEK, *Tetrahedron Lett.* **2002**, *43*, 2087; (l) C. B. DE KONING, J. P. MICHAEL, A. L. ROUSSEAU, *J. Chem. Soc., Perkin Trans. 1* **2000**, 787.

[76] P. SALEHI, H. FIROUZABADI, A. FARROKHI, M. GOLIZADEH, *Synthesis* **2001**, 2273.

[77] S. AGARWAL, K. CHOWDHURY, K. K. BANERJI, *J. Org. Chem.* **1991**, *56*, 5111.

[78] (a) E. J. COREY, G. SCHMIDT, *Tetrahedron Lett.* **1979**, *20*, 399; (b) C. H. HEATHCOCK, S. D. YOUNG, J. P. HAGEN, R. PILLI, U. BADERTSCHER, *J. Org. Chem.* **1985**, *50*, 2095; (c) D. ENDERS, M. BARTSCH, J. RUNSINK, *Synthesis* **1999**, 243.

[79] (a) P. J. GAREGG, L. OLSSON, S. OSCARSON, *J. Org. Chem.* **1995**, *60*, 2200; (b) G. T. NADOLSKI, B. S. DAVIDSON, *Tetrahedron Lett.* **2001**, *42*, 797.

[80] G. S. CHAUBEY, S. DAS, M. K. MAHANTI, *Bull. Chem. Soc. Jpn.* **2002**, *75*, 2215.

[81] (a) A. J. Fatiadi, *Synthesis* **1974**, 229; (b) A. G. Wee, J. Sobodian in *Handbook of Reagents for Organic Synthesis, Oxidizing and Reducing Agents* (Eds.: S. D. Burke, R. L. Danheiser), Wiley, Chichester, **1999**, p. 423.

[82] (a) P. Alsters, E. Schmleder-van de Vondervoort, (for DSM Fine Chemicals) EP 1245556 (**2002**); (b) See also: M. Zhang, J. Li, Z. Somg, R. Desmond, D. M. Tschaen, E. J. J. Grabowski, P. J. Reider, *Tetrahedron Lett.* **1998**, *39*, 5323.

[83] (a) J. E. McMurry, R. G. Dushin, *J. Am. Chem. Soc.* **1990**, *112*, 6942; (b) C. Kuroda, P. Theramobgkol, J. R. Engebrecht, J. D. White, *J. Org. Chem.* **1986**, *51*, 956; (c) T. Nishiyama, J. F. Woodhall, J. F. Lawson, E. N. Lawson, W. Kitching, *J. Org. Chem.* **1989**, *54*, 2183; (d) S. W. Pelletier, D. L. Herald, *J. Chem. Soc., Chem. Commun.* **1971**, 10; (e) J. B. Lambert, D. E. Markó, *J. Am. Chem. Soc.* **1985**, *107*, 7978; (f) J.-F. Hoeffler, C. Grosdemange-Billiard, M. Rohmer, *Tetrahedron Lett.* **2001**, *42*, 3065; (g) P. Coutrot, C. Grison, C. Bomont, *J. Organomet. Chem.* **1999**, *586*, 208; (h) S. Flock, H. Frauenrath, *Synlett* **2001**, 839; (i) A. Dondoni, A. Marra, A. Boscarata, *Chem. Eur. J.* **1999**, *5*, 3562; (j) H. Fraunerath, D. Brethauer, S. Reim, M. Maurer, G. Raabe, *Angew. Chem.* **2001**, *113*, 176, *Angew. Chem., Int. Ed. Engl.* **2001**, *40*, 177.

[84] (a) T. Kataoka, T. Iwamura, H. Tsutsui, Y. Kato, Y. Banno, Y. Aoyama, H. Shimizu, *Het. Commun.* **2001**, *12*, 317; (b) G. V. M. Sharma, A. Llangovan, B. Lavanya, *Synth. Commun.* **2000**, *30*, 397; (c) D. J. Chadwick, J. Chambers, G. D. Meakins, R. L. Snowden, *J. Chem. Soc., Perkin Trans. 1* **1973**, 1766; (d) G. Tamagnan, Y. Gao, V. Bakthavachalam, W. L. White, J. L. Neumeyer, *Tetrahedron Lett.* **1995**, *56*, 5861; (e) I. Sasaki, J. C. Daran, G. G. A. Balavoine, *Synthesis* **1999**, 815; (f) M. Inoue, M. W. Carson, A. J. Frontier, S. J. Danishefsky, *J. Am. Chem. Soc.* **2001**, *123*, 1878; (g) Y. L. Janin, E. Roulland, A. Beurdeley-Thomas, D. Decaudin, C. Moneret, M.-F. Poupon, *J. Chem. Soc., Perkin Trans. 1* **2002**, 529; (h) E. J. Corey, J. Das, *J. Am. Chem. Soc.* **1982**, *104*, 5551.

[85] A. F. Neto, J. Miller, V. F. de Andrade, S. Y. Fujimoto, M. M. de Freitas Afonso, F. C. Archanjo, V. A. Darin, M. L. Andrade e Silva, A. D. L. Borges, G. Z. Del Ponte, *Anorg. Allg. Chem.* **2002**, *628*, 209.

[86] (a) Y. Chen, G. Zhou, L. Liu, Z. Xiong, Y. Li, *Synthesis* **2001**, 1305; (b) H.-M. Tai, M.-Y. Chang, A.-Y. Lee, N.-C. Chang, *J. Org. Chem.* **1999**, *64*, 659; (c) A. B. Pepermann, *J. Org. Chem.* **1981**, *46*, 5039.

[87] (a) S. C. Thomason, D. G. Kubler, *J. Educ. Chem.* **1968**, *45*, 546; (b) M. Jung, K. Lee, H. Jung, *Tetrahedron Lett.* **2001**, *42*, 3997.

[88] T. G. Clarke, N. A. Hampson, J. B. Lee, J. R. Morley, B. Scanlon, *Can. J. Chem.* **1969**, *47*, 1649.

[89] H. Firouzabadi, P. Salehi, I. Mohammadpour-Baltork, *Bull. Chem. Soc. Jpn.* **1992**, *65*, 2878.

[90] (a) M. Fétizon in *Handbook of Reagents for Organic Synthesis, Oxidizing and Reducing Agents* (Eds.: S. D. Burke, R. L. Danheiser), Wiley, Chichester, **1999**, p. 361; (b) S. Morgenlie, *Acta Chem. Scand.* **1973**, *27*, 3009.

[91] (a) K. Nakagawa, S. Mineo, S. Kawamura, *Chem. Pharm. Bull.* **1978**, *26*, 299; (b) A. K. Banerjee, E. V. Cabrera, *J. Chem. Res. (S)* **1998**, 380.

[92] L. M. Berkowitz, P. N. Rylander, *J. Am. Chem. Soc.* **1958**, *80*, 6682.

[93] (a) S. V. Ley, R. Leslie in *Reagents for Organic Synthesis*, Vol. 6 (Ed.: L. A. Paquette), Wiley, Chichester, **1995**, p. 4282; (b) A. J. Bailey, W. P. Griffith, S. I. Mostafa, P. A. Sherwood, *Inorg. Chem.* **1993**, *32*, 268; (c) G. Green, W. P. Griffith, D. M. Hollinshead, S. V. Ley, M. Schröder, *J. Chem. Soc., Perkin Trans. 1* **1984**, 681.

[94] P. Müller, J. Godoy, *Tetrahedron Lett.* **1981**, 2361.

[95] (a) R. M. Moriarty, O. Prakash, *Org. React.* **1999**, *54*, 273; (b) *Hypervalent Iodide* in *Organic Synthesis* (Ed.: A. Varvoglis), Academic Press, Cornwall, **1997**; (c) H. Togo, K. Sakuratani, *Synlett* **2002**, 1966; (d) T. Wirth, *Top. Curr. Chem.* **2003**, *224*, 185.

[96] J.-P. Brière, R. H. Blaauw, J. C. J. Benningshof, A. E. van Ginkel, J. H. van Maarseveen, H. Hiemstra, *Eur. J. Org. Chem.* **2001**, 2371.

[97] (a) T. Naota, S.-I. Murahashi, *Synlett*, **1991**, 693; (b) Y. Blum, D. Reshef, Y. Shvo, *Tetrahedron Lett.* **1981**, 22, 1541; (c) S.-I. Murahashi, T. Naota, K. Ito, Y. Maeda, H. Taki, *J. Org. Chem.* **1987**, 52, 4319.

[98] R. Grigg, T. R. B. Mitchell, S. Sutthivaiyakit, *Tetrahedron* **1981**, 37, 4313.

[99] A. Tillack, I. Rudloff, M. Beller, *Eur. J. Org. Chem.* **2001**, 523.

[100] Y. Tamaru, Y. Yamada, Z.-I. Yoshida, *Synthesis* **1983**, 474.

[101] J. Palou, *Chem. Soc. Rev.* **1994**, 357.

[102] M. Al Neirabeyeh, M. D. Pujol, *Tetrahedron Lett.* **1990**, 31, 2273.

[103] (a) E. L. Eliel, L. Clawson, D. E. Knox, *J. Org. Chem.* **1985**, 50, 2707; (b) S. Yamada, D. Morizono, K. Yamamoto, *Tetrahedron Lett.* **1992**, 33, 4329.

[104] H. Gilman, C. G. Brannen, R. K. Ingham, *J. Am. Chem. Soc.* **1956**, 78, 1689.

[105] I. E. Markó, A. Mekhalfia, W. D. Ollis, *Synlett* **1990**, 347.

[106] (a) Y.-F. Cheung, *Tetrahedron Lett.* **1979**, 40, 3809; (b) C. P. Decicco, D. J. Nelson, *Tetrahedron Lett.* **1993**, 51, 8213; (c) I. E. Markó, A. Mekhalfia, *Tetrahedron Lett.* **1990**, 49, 7237.

[107] C. McDonald, H. Holcomb, K. Kennedy, E. Kirkpatrick, T. Leathers, P. Vanemon, *J. Org. Chem.* **1989**, 54, 1213.

[108] S. Stavber, Z. Planinsek, M. Zupan, *J. Org. Chem.* **1992**, 57, 5334.

[109] R. E. Banks, N. J. Lawrence, A. L. Popplewell, *Synlett* **1994**, 831.

[110] J. M. Galvin, E. N. Jacobsen in *Handbook of Reagents for Organic Synthesis, Oxidizing and Reducing Agents* (Eds.: S. D. Burke, R. L. Danheiser), Wiley, Chichester, **1999**, p. 407.

[111] R. V. Stevens, K. T. Chapman, C. A. Stubbs, W. W. Tam, K. F. Albizati, *Tetrahedron Lett.* **1982**, 23, 4647.

[112] S. O. Nwaukwa, P. M. Keehn, *Tetrahedron Lett.* **1982**, 23, 3131.

[113] C. E. McDonald, L. E. Nice, A. W. Shaw, N. B. Nestor, *Tetrahedron Lett.* **1993**, 34, 2741.

[114] S. R. Wilson, S. Tofigh, R. N. Misra, *J. Org. Chem.* **1982**, 47, 1360.

[115] T. Hase, K. Wähälä in *Reagents for Organic Synthesis*, Vol. 7 (Ed.: L. A. Paquette), Wiley, Chichester, **1995**, p. 4533.

[116] E. Dalcanale, F. Montanari, *J. Org. Chem.* **1986**, 51, 567.

[117] A. Raach, O. Reiser, *J. Prakt. Chem.* **2000**, 342, 605.

[118] H. Tabuchi, T. Hamamoto, S. Miki, T. Tejima, A. Ichihara, *J. Org. Chem.* **1994**, 59, 4749.

[119] B. O. Lindgren, T. Nilsson, *Acta Chem. Scand.* **1973**, 27, 888.

[120] K. Shishido, K. Goto, S. Miyoshi, Y. Takaishi, M. Shibuya, *J. Org. Chem.* **1994**, 59, 406.

[121] G. Bringmann, J. Hollenz, R. Weirich, M. Rübenacker, C. Funke, M. R. Boyd, R. J. Gulakowski, G. François, *Tetrahedron* **1998**, 54, 497.

[122] (a) J. Mann, A. Thomas, *Tetrahedron Lett.* **1986**, 30, 3533; (b) A. B. Smith III, T. L. Leenay, *J. Am. Chem. Soc.* **1989**, 111, 5761; (c) G. A. Kraus, M. Taschner, *J. Org. Chem.* **1980**, 45, 1175; (d) X. Teng, Y. Takayama, S. Okamoto, F. Sato, *J. Am. Chem. Soc.* **1999**, 121, 11916.

[123] (a) G. A. Kraus, B. Roth, *J. Org. Chem.* **1980**, 45, 4825; (b) W.-W. Lee, H. J. Shin, S. Chang, *Tetrahedron: Asymmetry* **2001**, 12, 29; (c) S. Kuwahara, S. Hamade, W. S. Leal, J. Ishikawa, O. Kodama, *Tetrahedron* **2000**, 56, 8111; (d) Y. Tobe, D. Yamashita, T. Takahashi, M. Inata, J.-I. Sato, K. Kakiuchi, K. Kobiro, Y. Odaira, *J. Am. Chem. Soc.* **1990**, 112, 775; (e) Y.-S. Hon, W.-C. Lin, *Tetrahedron Lett.* **1995**, 42, 7693.

[124] (a) H. D. H. Showalter, A. D. Sercel, B. M. Leja, C. D. Wolfangel, L. A. Ambroso, W. L. Elliot, D. W. Fry, A. J. Kraker, C. T. Howard, G. H. Lu, C. W. Moore, J. M. Nelson, B. J. Roberts, P. W. Vincent, W. A. Denny, A. M. Thompson, *J. Med. Chem.* **1997**, 40, 413; (b) A. Ahmed, R. A. Bragg, J. Clayden, K. Tchabanenko, *Tetrahedron Lett.* **2001**, 42, 3407.

[125] (a) A. Nishihara, I. Kubota, *J. Org. Chem.* **1968**, 33, 2525; (b) R. J. Ken-

NEDY, A. M. STOCK, *J. Org. Chem.* **1960**, *25*, 1901.

[126] T.-C. ZHENG, D. E. RICHARDSON, *Tetrahedron Lett.* **1995**, *36*, 833.

[127] (a) N. RABJOHN, *Org. React.* **1976**, *24*, 261; (b) J. DRABOWICZ, M. MIKOLAJCZYK, *Top. Curr. Chem.* **2000**, *208*, 143.

[128] C. W. SMITH, R. T. HOLM, *J. Org. Chem.* **1957**, *22*, 746.

[129] M. BRZASZCZ, K. KLOC, M. MAPOSAH, J. MLOCHOWSKI, *Synth. Commun.* **2000**, *30*, 4425.

[130] H. WÓJTOWICZ, M. BRZASZCZ, K. KLOC, J. MLOCHOWSKI, *Tetrahedron* **2001**, *57*, 9743.

[131] J.-K. CHOI, Y.-K. CHANG, S. Y. HONG, *Tetrahedron Lett.* **1988**, *29*, 1967.

[132] G.-J. TEN BRINK, J. M. VIS, I. W. C. E. ARENDS, R. A. SHELDON, *Tetrahedron* **2002**, *58*, 3977.

[133] C. E. CANTRELL, D. E. KIELY, G. J. ABRUSCATO, J. M. RIORDAN, *J. Org. Chem.* **1977**, *42*, 3562.

[134] Y. OGATA, H. TEZUKA, Y. SAWAKI, *Tetrahedron* **1967**, *23*, 1007.

[135] (a) M. YANAGIDA, L. P. SINGH, K. SAYAMA, K. HARA, R. KATOH, A. ISLAM, K. SUGIHARA, H. ARAKAWA, M. K. NAZEERUDDIN, M. GRÄTZEL, *J. Chem. Soc., Dalton Trans.* **2000**, 2817; (b) P. T. KAYE, K. W. WELLINGTON, *Synth. Commun.* **2001**, *31*, 799.

[136] J. CASTELLS, F. PUJOL, H. LLITJOS, M. MORENO-MAÑAS, *Tetrahedron* **1982**, *38*, 337.

[137] P. H. WENDSCHUH, C. T. PATE, J. N. PITTS, *Tetrahedron Lett.* **1973**, 2931.

[138] L. TORUN, T. MOHAMMAD, H. MORRISON, *Tetrahedron Lett.* **1999**, *40*, 5279.

[139] (a) A. McKILLOP, D. KEMP, *Tetrahedron* **1989**, *45*, 3299; (b) A. BANERJEE, B. HAZRA, A. BHATTACHARYA, S. BANERJEE, G. C. BANERJEE, S. SENGUPTA, *Synthesis* **1989**, 765; (c) A. McKILLOP, W. R. SANDERSON, *Tetrahedron* **1995**, *51*, 6145; (d) J. MUZART, *Synthesis* **1995**, 1325.

[140] (a) R. MAZITSCHEK, M. MÜLBAIER, A. GIANNIS, *Angew. Chem.* **2002**, *114*, 4216, *Angew. Chem., Int. Ed. Engl.* **2002**, *41*, 4059; (b) For other oxidations with IBX see K. C. NICOLAOU, T. MONTAGNON, P. S. BARAN, *Angew. Chem.* **2002**, *114*, 1035, *Angew. Chem., Int. Ed. Engl.* **2002**, *41*, 993 and references therein.

[141] H. INOUE, K. HIGASHIURA, *J. Chem. Soc., Chem. Commun.* **1980**, 549.

[142] (a) S.-W. TAM, L. JIMENEZ, F. DIEDERICH, *J. Am. Chem. Soc.* **1992**, *114*, 1503; (b) P. MATTEI, F. DIEDERICH, *Helv. Chim. Acta* **1997**, *80*, 1555.

[143] T. SHONO in *Comprehensive Organic Chemistry*, Vol. 7 (Ed.: B. M. TROST), Wiley, **1991**, p. 789.

[144] (a) S. RAJENDRAN, D. C. TRIVEDI, *Synthesis* **1995**, 153; (b) S. TORII, T. INOKUCHI, T. SUGIURA, *J. Org. Chem.* **1986**, *51*, 155; (c) M. OKIMOTO, T. CHIBA, *J. Org. Chem.* **1988**, *53*, 218.

[145] A. SCHMID, F. HOLLMANN, B. BÜHLER in *Enzyme Catalysis* in Organic Synthesis, Vol. 3 (Eds.: K. DRAUZ, H. WALDMANN), Wiley-VCH, Weinheim **2002**, p. 1194.

[146] S. CANNIZZARO, *Ann.* **1853**, *88*, 129.

[147] (a) T. A. GIESSMAN, *Org. React.* **1944**, *2*, 94; (b) R. M. KELLOG in *Comprehensive Organic Synthesis*, Vol. 8 (Ed.: B. M. TROST), Pergamon Press, Oxford, **1991**, p. 79.

[148] A. SHARIFI, M. M. MOJTAHEDI, M. R. SAISI, *Tetrahedron Lett.* **1999**, *40*, 1179.

[149] K. YOSHIZAWA, S. TOYOTA, F. TODA, *Tetrahedron Lett.* **2001**, *42*, 7983.

[150] J. R. SHELDON, J. H. BOWIE, S. DUA, J. D. SMITH, R. A. J. O'HAIR, *J. Org. Chem.* **1997**, *62*, 3931.

[151] K. BOWDEN, A. M. BUTT, M. STREATER, *J. Chem. Soc., Perkin Trans. 2* **1992**, 567.

[152] J. HINE, G. F. KOSER, *J. Org. Chem.* **1971**, *36*, 3591.

[153] S.-J. JIN, P. K. ARORA, L. M. SAYRE, *J. Org. Chem.* **1990**, *55*, 3011.

[154] A. E. RUSSELL, S. P. MILLER, J. P. MORKEN, *J. Org. Chem.* **2000**, *65*, 8381.

[155] K. MARUYAMA, Y. MURAKAMI, K. YODA, T. MASHINO, A. NISHINAGA, *J. Chem. Soc., Chem. Commun.* **1992**, 1617.

[156] P. R. LIKHAR, A. K. BANYOPADHYAY, *Synlett* **2000**, 538.

[157] (a) J. L. THAKURIA, M. BARUAH, J. S. SANDHU, *Chem. Lett.* **1999**, 995; (b) R. S. VARMA, K. P. NAICKER, P. J. LIESEN, *Tetrahedron Lett.* **1998**, *39*, 8437.

[158] L. CLAISEN, *Ber.* **1887**, *20*, 646a.

[159] W. Tishchenko, *Chem. Zentr.* **1906**, *77*, 1309

[160] (a) Y. Ogata, A. Kawasaki, *Tetrahedron* **1969**, *25*, 929; (b) T. Saegusa, T. Ueshima, *J. Org. Chem.* **1968**, *33*, 3310; (c) T. Saegusa, T. Ueshima, S. Kitagawa, *Bull. Chem. Soc. Jpn.* **1969**, *42*, 248.

[161] P. R. Stapp, *J. Org. Chem.* **1973**, *38*, 1433.

[162] F. Le Bideau, T. Coradin, D. Gourier, J. Henique, E. Samuel, *Tetrahedron Lett.* **2000**, *41*, 5215.

[163] (a) N. Menasche, Y. Shvo, *Organometallics* **1991**, *10*, 3885; (b) D. Bankston, *J. Org. Chem.* **1989**, *54*, 2003; (c) T. Seki, K. Akutsu, H. Hattori, *Chem. Commun.* **2001**, 1000; (d) M. R. Bürgstein, H. Berberich, P. W. Roesky, *Chem. Eur. J.* **2001**, *7*, 3078.

[164] (a) M. Adinolfi, G. Barone, F. De Lorenzo, A. Iadonisi, *Synthesis* **1999**, 336; (b) J.-L. Hsu, J.-M. Fang, *J. Org. Chem.* **2001**, *66*, 8573.

[165] I. Yamaguchi, T. Kimishima, K. Osakada, T. Yamamoto, *J. Polym. Sci. A* **1997**, 1265.

[166] D. A. Evans, A. H. Hoveyda, *J. Am. Chem. Soc.* **1990**, *112*, 6447.

[167] (a) D. Romo, S. D. Meyer, D. D. Johnson, S. L. Schreiber, *J. Am. Chem. Soc.* **1993**, *115*, 7906; (b) A. B. Smith, L. Dongjoo, C. M. Adams, M. C. Kozlowski, *Org. Lett.* **2002**, *4*, 4539; (c) M. Lautens, J. T. Colucci, S. Hiebert, N. D. Smith, G. Bouchain, *Org. Lett.* **2002**, *4*, 1879; (d) K.-U. Schöning, R. K. Hayashi, D. R. Powell, A. Kirschning, *Tetrahedron: Asymmetry* **1999**, *10*, 817.

[168] Y. Umekawa, S. Sakaguchi, Y. Nishiyama, Y. Ishii, *J. Org. Chem.* **1997**, *62*, 3409.

[169] (a) C. Schneider, M. Hansch, *Chem. Commun.* **2001**, 1218; (b) C. M. Mascarenhas, M. O. Duffey, S.-Y. Liu, J. P. Morken, *Org. Lett.* **1999**, *1*, 1427; (c) L. Lu, H.-Y. Chang, J.-M. Fang, *J. Org. Chem.* **1999**, *64*, 843; (d) I. Simpura, V. Nevalainen, *Tetrahedron Lett.* **2001**, *42*, 3905; (e) R. Mahrwald, B. Costisella, *Synthesis* **1996**, 1087; (f) P. M. Bodnar, J. T. Shaw, K. A. Woerpel, *J. Org. Chem.* **1997**, *62*, 5674; (g) T. Ooi, T. Miura, Y. Itagaki, H. Ichikawa, K. Maruoka, *Synthesis* **2002**, 279.

[170] (a) C. M. Mascarenhas, S. P. Miller, P. S. White, J. P. Morken, *Angew. Chem.* **2001**, *113*, 621, *Angew. Chem., Int. Ed. Engl.* **2001**, *40*, 601; (b) O. Loog, U. Mäerog, *Tetrahedron: Asymmetry* **1999**, *10*, 2411.

[171] O. P. Törmäkangas, A. M. P. Koskinen, *Tetrahedron Lett.* **2001**, *42*, 2743.

[172] C.-A. Fan, B.-M. Wang, Y.-Q. Tu, Z.-L. Song, *Angew. Chem.* **2001**, *113*, 3995, *Angew. Chem., Int. Ed. Engl.* **2001**, *40*, 3877.

[173] M. Takeno, S. Kikuchi, K.-I. Morita, Y. Nishiyama, Y. Ishii, *J. Org. Chem.* **1995**, *60*, 4974.

[174] (a) G. R. Krow, *Org. React.* **1993**, *43*, 251; (b) C. H. Hassal, *Org. React.* **1957**, *9*, 73; (c) G. R. Krow in *Comprehensive Organic Chemistry, Vol. 7* (Ed.: B. M. Trost), Pergamon Press, **1991**, p. 671.

[175] (a) A. V. R. Rao, N. Sreenivasan, D. R. Reddy, V. H. Deshpande, *Tetrahedron Lett.* **1987**, *28*, 455; (b) L. S. Kiong, J. H. P. Tyman, *J. Chem. Soc., Perkin Trans. 1* **1981**, 1942; (c) J. Zhu, R. Beugelmans, A. Bigot, G. P. Singh, M. Bois-Choussy, *Tetrahedron Lett.* **1993**, *34*, 7401; (d) J. A. Elix, Y. Jin, M. T. Alder, *Aust. J. Chem.* **1989**, *42*, 765; (e) K. van Laak, H.-D. Scharf, *Tetrahedron* **1989**, *45*, 5511.

[176] (a) L. Syper, *Synthesis* **1989**, 167; (b) L. Syper, J. Mlochowski, *Tetrahedron* **1987**, *43*, 207; (b) M. E. Jung, T. I. Lazarova, *J. Org. Chem.* **1997**, *62*, 1553; (d) T. Fukuyama, L. Yang, *Tetrahedron Lett.* **1986**, *27*, 6299; (e) S. A. Kulkarni, M. V. Paradkar, *Synth. Commun.* **1992**, *22*, 1555; (f) J. A. Guzmán, V. Mendoza, E. García, C. F. Garibay, L. Z. Olivares, L. A. Maldonado, *Synth. Commun.* **1995**, *25*, 2121.

[177] (a) M. Matsumoto, H. Kobayashi, Y. Hotta, *J. Org. Chem.* **1984**, *49*, 4740; (b) A. Roy, K. R. Reddy, P. K. Mohanta, H. Ila, H. Junjappa, *Synth. Commun.* **1999**, *29*, 3781.

[178] (a) C. R. Broka, S. Chan, B. Peterson, *J. Org. Chem.* **1988**, *53*, 1584; (b) S. V. Kolotuchin, A. I. Meyers, *J. Org. Chem.* **1999**, *64*, 7921;

(c) J. A. ELIX, U. M. JENIE, *Aust. J. Chem.* **1989**, *42*, 987; (d) C. PULGARIN, R. TABACCHI, *Helv. Chim. Acta* **1988**, *71*, 876; (e) S. M. MAJERUS, N. ALIBHAI, S. TRIPATHY, T. DURST, *Can. J. Chem.* **2000**, *78*, 1345; (f) G. B. JONES, C. J. MOODY, *J. Chem. Soc., Perkin Trans. 1* **1989**, 2455.

[179] (a) J. ROYER, M. BEUGELMANS-VERRIER, *C. R. Hebd. Seances Acad. Sci., Ser. C* **1971**, *272*, 1818; (b) H. ISHII, K.-I. HARADA, T. ISHIDA, E. UEDA, T. NAKAJIMA, *Tetrahedron Lett.* **1975**, 319; (c) L.-H. KUO, C.-D. LIN, S.-S. JWO, L.-L. LIN, S.-L. LEE, S.-J. TSAI, (for Sinon Corporation) US 2002123655 (**2002**).

[180] G. W. KABALKA, N. K. REDDY, C. NARAYANA, *Tetrahedron Lett.* **1992**, *33*, 865.

[181] (a) R. S. VARMA, K. P. NAICKER, *Org. Lett.* **1999**, *1*, 189; (b) W. E. NOLAND, B. L. KEDROWSKI, *Org. Lett.* **2000**, *2*, 2109.

[182] (a) J.-A. H. NÄSMAN, K. G. PENSAR, *Synthesis* **1985**, 786; (b) C. W. JEFFORD, D. JAGGI, J. BOUKOUVALAS, *Tetrahedron Lett.* **1989**, *30*, 1237.

[183] Z. L. HICKMAN, C. F. STURINO, N. LACHANCE, *Tetrahedron Lett.* **2000**, *41*, 8217.

[184] (a) F. NICOTRA, F. RONCHETTI, G. RUSSO, L. TOMA, P. GARIBOLDI, B. M. RANZI, *J. Chem. Soc., Chem. Commun.* **1984**, 383; (b) G. BUCHBAUER, V. M. HENEIS, V. KREJCI, C. TALSKY, H. WUNDERER, *Monatsh. Chem.* **1985**, *116*, 1345; (c) A. F. BARRERO, E. J. Alvarez-MANZANEDA, R. ALVAREZ-MANZANEDA, R. CHAHBOUN, R. MENESES, M. B. APARICIO, *Synlett* **1999**, 713.

[185] (a) A. DEBOER, R. E. ELLWANGER, *J. Org. Chem.* **1974**, *39*, 77; (b) E. ALVAREZ, R. PÉREZ, M. RICO, R. M. RODRÍGUEZ, J. D. MARTÍN, *J. Org. Chem.* **1996**, *61*, 3003; (c) B. ALCAIDE, M. F. ALY, M. A. SIERRA, *J. Org. Chem.* **1996**, *61*, 8819.

[186] L. SYPER, *Tetrahedron* **1987**, *43*, 2853.

[187] P. NORRIS, D. HORTON, B. R. LEVINE, *Tetrahedron Lett.* **1995**, *36*, 7811.

[188] A. HAJNAL, J. WÖLFLING, G. SCHNEIDER, *Synlett* **2002**, 1077.

[189] S. S. ELSMORY, A.-A. S. EL-AHL, H. SOLIMAN, F. A. AMER, *Tetrahedron Lett.* **1995**, *36*, 2639.

[190] *Protective Groups in Organic Synthesis* (Eds.: T. W. GREENE, P. G. M. WUTZ), 3rd edn., Wiley, New York, **1999**.

[191] (a) M. M. HERAVI, M. TAJBAKHSH, S. HABIBZADEH, M. GHASSEMZADEH, *Monatsh. Chem.* **2001**, *132*, 985; (b) N. KOMATSU, A. TANIGUCHI, S. WADA, H. SUZUKI, *Adv. Synth. Catal.* **2001**, *343*, 473.

[192] M. BONI, F. GHELFI, U. M. PAGNONI, A. PINETTI, *Bull. Chem. Soc. Jpn.* **1994**, *67*, 156.

[193] R. CURCI, L. D'ACCOLTI, M. FIORENTINO, C. FUSCO, W. ADAM, M. E. GONZALEZ-NUNEZ, R. MELLO, *Tetrahedron Lett.* **1992**, *33*, 4225.

[194] P. DESLONGCHAMPS, C. MOREAU, D. FRÉHEL, R. CHÊNEVERT, *Can. J. Chem.* **1975**, *53*, 1204.

[195] H. RHEE, J. Y. KIM, *Tetrahedron Lett.* **1998**, *39*, 1365.

[196] (a) S. PENG, F.-L. QING, Y.-Q. LI, C.-M. HU, *J. Org. Chem.* **2000**, *65*, 694; (b) C. WAKSELMAN, H. MOLINES, M. TORDEUX, *J. Fluorine Chem.* **2000**, *102*, 211.

[197] H.-X. LIN, L.-H. XU, N.-J. HUANG, *Synth. Commun.* **1997**, *27*, 303.

[198] (a) S. S. ZLOTSKY, M. N. NAZAROV, L. G. KULAK, D. L. RAKHMANKULOV, *J. Prakt. Chem.* **1992**, *334*, 441; (b) T. TAKEDA, H. WATANABE, T. KITAHARA, *Synlett* **1997**, 1149.

[199] E. M. KURAMSHIN, L. G. KULAK, M. N. NAZAROV, S. S. ZLOTSKY, D. L. RAKHMANKULOV, *J. Prakt. Chem.* **1989**, *331*, 591.

[200] T. NISHIGUTCHI, T. OHOSIMA, A. NISHIDA, S. FUJISAKI, *J. Chem. Soc., Chem. Commun.* **1995**, 1121.

[201] (a) A. F. MINGOTAUD, D. FLORENTIN, A. MARQUET, *Synth. Commun.* **1992**, *22*, 2401; (b) S. SUGAI, T. KODAMA, S. AKABOSHI, S. IKEGAMI, *Chem. Pharm. Bull.* **1984**, *32*, 99.

[202] (a) C. J. BARNETT, T. M. WILSON, D. A. EVANS, T. C. SOMERS, *Tetrahedron Lett.* **1997**, *38*, 735; (b) N. CHIDAMBARAM, S. BHAT, S. CHANDRASEKARAN, *J. Org. Chem.* **1992**, *57*, 5013.

[203] T. SUEDA, S. FUKUDA, M. OCHIAI, *Org. Lett.* **2001**, *3*, 2387.

[204] M. CURINI, F. EPIFANO, M. C. MARCOTULLIO, O. ROSATI, *Synlett* **1999**, 777.

[205] W. Adam, C. Saha-Möller, P. A. Ganeshpure, *Chem. Rev.* **2001**, *101*, 3499.

[206] M. E. Jung, *Reagents for Organic Synthesis, Vol. 8* (Ed.: L. A. Paquette), Wiley, Chichester, **1995**, p. 5348.

[207] H. P. Wessel, D. R. Bundle, *J. Chem. Soc., Perkin Trans. 1* **1985**, 2251.

[208] T. Hosokawa, Y. Imada, S.-I. Murahashi, *J. Chem. Soc., Chem. Commun.* **1983**, 1245.

[209] S.-I. Murahashi, Y. Oda, T. Naota, *Chem. Lett.* **1992**, 2237.

[210] B. M. Choudary, P. N. Reddy, *Synlett* **1995**, 959.

[211] Y. Chen, P. G. Wang, *Tetrahedron Lett.* **2001**, *42*, 4955.

[212] (a) M. Besson, P. Gallezot in *Fine Chemicals through Heterogeneous Catalysis* (Eds.: R. A. Sheldon, H. van Bekkum), Wiley-VCH, Weinheim, **2001**, p. 507; (b) S. H. J. Arts, E. J. M. Mombarg, H. van Bekkum, R. A. Sheldon, *Synthesis* **1997**, 597; (c) I. Isaac, I. Stasik, D. Beaupère, R. Uzan, *Tetrahedron Lett.* **1995**, *36*, 383.

[213] (a) J. H. Boyer, *Chem. Rev.* **1980**, *80*, 495; (b) R. N. Butler, *Chem. Rev.* **1984**, *84*, 249.

[214] (a) M. North in *Comprehensive Organic Functional Group Transformations, Vol. 3* (Eds.: A. R. Katritzky, O. Meth-Cohn, C. W. Rees), Elsevier Science, Oxford, **1995**, p. 611; (b) K. Friedrich, K. Wallenfels in *The Chemistry of the Cyano Group* (Ed. Z. Rappoport), Wiley-Interscience, New York, **1970**, p. 67.

[215] (a) *Nitriles Oxides, Nitrones, and Nitronates in Organic Synthesis* (Ed.: K. G. B. Torsell), Wiley-VCH, New York, **1988**; (b) C. Grundmann in *The Chemistry of the Cyano Group* (Ed. Z. Rappoport), Wiley-Interscience, New York, **1970**, p. 791.

[216] (a) F. A. Davis, A. C. Sheppard, *Tetrahedron* **1989**, *45*, 5703; (b) L. Martiny, K. A. Jørgensen, *J. Chem. Soc., Perkin Trans. 1* **1995**, 699.

[217] (a) S. S. Chaudhari, K. G. Akamanchi, *Synthesis* **1999**, 760; (b) K. Aghapoor, M. M. Heravi, M. A. Nooshabadi, M. M. Ghassemzadeh, *Monatsh. Chem.* **2002**, *133*, 107; (c) M. H. Hashemi, Y. A. Beni, *Synth. Commun.* **2001**, *31*, 295; (d) P. Salehi, M. M. Khodaei, M. Goodarzi, *Synth. Commun.* **2002**, *32*, 1259; (e) G. Blay, E. Benach, I. Fernández, S. Galletero, J. R. Pedro, R. Ruiz, *Synthesis* **2000**, 403.

[218] P. K. Arora, L. M. Sayre, *Tetrahedron Lett.* **1991**, *32*, 1007.

[219] G. Lai, N. K. Bhamare, W. K. Anderson, *Synlett* **2001**, 230.

[220] F.-E. Chen, H. Fu, Y. Cheng, Y.-X. Lü, *Synthesis* **2000**, 1519.

[221] P. Nongkunsarn, C. A. Ramsden, *Tetrahedron* **1997**, *53*, 3805.

[222] (a) T. Isobe, T. Ishikawa, *J. Org. Chem.* **1999**, *64*, 6984; (b) J.-P. Dulcere, *Tetrahedron Lett.* **1981**, *22*, 1599; (c) K. Mai, G. Patil, *Synthesis* **1986**, 1037; (d) C. Botteghi, G. Chelucci, G. Del Ponte, M. Marchetti, S. Paganelli, *J. Org. Chem.* **1994**, *59*, 7125; (e) M. Ghiaci, K. Bakhtiari, *Synth. Commun.* **2001**, *31*, 1803; (f) B. Tamami, A. R. Kiasat, *Synth. Commun.* **2000**, *30*, 235; (g) R. Hekmatshoar, M. M. Heravi, Y. S. Beheshtiha, K. Asadolah, *Monatsh. Chem.* **2002**, *133*, 111.

[223] (a) M. Koós, *Tetrahedron Lett.* **2000**, *41*, 5403; (b) M. E. Pierce, D. J. Carini, G. F. Huhn, G. J. Wells, J. F. Arnett, *J. Org. Chem.* **1993**, *58*, 4642.

[224] D. G. Desai, S. S. Swami, G. D. Mahale, *Synth. Commun.* **2000**, *30*, 1623.

[225] (a) E. Choi, C. Lee, Y. Na, S. Chang, *Org. Lett.* **2002**, *4*, 2369; (b) J. W. Grissom, D. Klingberg, S. Meyenburg, B. L. Stallman, *J. Org. Chem.* **1994**, *59*, 7876; (c) O. Attanasi, P. Palma, F. Serra-Zanetti, *Synthesis* **1983**, 741.

[226] G. Sonovsky, J. A. Krogh, S. G. Umhoefer, *Synthesis* **1979**, 722.

[227] (a) C. P. Miller, D. H. Kaufman, *Synlett* **2000**, 1169; (b) B. Jose, M. S. Sulatha, P. M. Pillai, S. Prathapan, *Synth. Commun.* **2000**, *30*, 1509.

[228] (a) H. Irngartinger, A. Weber, T. Escher, P. W. Fettel, F. Gassner, *Eur. J. Org. Chem.* **1999**, 2087; (b) F. Foti, G. Grassi, F. Risitano, S. La Rosa, *J. Heterocycl. Chem.* **2001**, 539; (c) X. Han, N. R. Natale, *J. Heterocycl. Chem.* **2001**, 415.

[229] N. Arai, M. Iwakoshi, K. Tanabe, K. Narasaka, *Bull. Chem. Soc. Jpn.* **1999**, *72*, 2277.

[230] A. Loupy, S. Régnier, *Tetrahedron Lett.* **1999**, *40*, 6221.

[231] H. Sharghi, M. H. Sarvari, *J. Chem. Res. (S)* **2001**, 446.

[232] (a) J. F. Le Borgne, T. Cuvigny, M. Larcheveque, H. Normant, *Synthesis* **1976**, 238; (b) M. T. Nguyen, L. F. Clarke, A. F. Hegarty, *J. Org. Chem.* **1990**, *55*, 6177; (c) A. Kamal, M. Arifuddin, N. V. Rao, *Synth. Commun.* **1998**, *28*, 4507; (d) Z. Xiao, J. W. Timberlake, *Tetrahedron* **1998**, *54*, 12715.

[233] (a) R. Fernández, C. Gash, J.-M. Lassaletta, J.-M. Llera, J. Vásquez, *Tetrahedron Lett.* **1993**, *34*, 141; (b) D. Enders, A. Plant, *Synlett* **1994**, 1054; (c) D. Enders, C. F. Janeck, G. Raabe, *Eur. J. Org. Chem.* **2000**, 3337; (d) R. Fernández, E. Martín-Zamora, C. Pareja, J. M. Lassaletta, *J. Org. Chem.* **2001**, *66*, 5201.

[234] T. Ramaligan, B. V. S. Reddy, R. Srinivas, J. S. Yadav, *Synth. Commun.* **2000**, *30*, 4507.

[235] A. Altamura, L. D'Accoli, A. Detomaso, A. Dinoi, M. Fiorentino, C. Fusco, R. Curci, *Tetrahedron Lett.* **1998**, *39*, 2009.

[236] (a) S. Stankovik, J. H. Espenson, *Chem. Commun.* **1998**, 1579; (b) H. Rudler, B. Denise, *Chem. Commun.* **1998**, 2145; (c) J. Mlochowski, K. Kloc, E. Bubicz, *J. Prakt. Chem.* **1994**, *336*, 467.

[237] (a) S. B. Said, J. Skarzewski, J. Mlochowski, *Synthesis* **1989**, 223; (b) J. Mlochowski, M. Giurg, E. Kubicz, S. B. Said, *Synth. Commun.* **1996**, *26*, 291.

[238] (a) *Modern Catalytic Methods for Organic Synthesis with Diazo Compounds* (Eds.: M. P. Doyle, M. A. McKervey, T. Ye), Wiley-Interscience, New York, **1998**; (b) T. Nakamura, T. Momose, T. Shida, T. Kinoshita, T. Takui, Y. Teki, K. Itoh, *J. Am. Chem. Soc.* **1995**, *117*, 11292; (c) K. Matsuda, T. Yamagata, T. Seta, H. Iwamura, K. Hori, *J. Am. Chem. Soc.* **1997**, *119*, 8058; (d) W. Nagai, Y. Hirata, *J. Org. Chem.* **1989**, *54*, 635; (e) E. K. Moltzen, A. Senning, H. Lütjens, A. Krebs, *J. Org. Chem.* **1991**, *56*, 1317; (f) R. N. Butler, P. D. O'Shea, A. L. Burke, *J. Chem. Soc., Chem. Commun.* **1987**, 1210.

[239] D.-W. Chen, Z.-C. Chen, *Synth. Commun.* **1995**, *25*, 1671.

[240] (a) R. Weiss, J. Seubert, *Angew. Chem.* **1994**, *106*, 900, *Angew. Chem., Int. Ed. Engl.* **1994**, *33*, 891; (b) J. Tsuji, H. Takahashi, T. Kajimoto, *Tetrahedron Lett.* **1973**, 4573; (c) G. G. Bargamov, E. M. Kagramanova, M. D. Bargamova, *Russ. Chem. Bull.* **1998**, *47*, 656; (d) I. Oprean, H. Ciupe, L. Gansca, F. Hodosan, *J. Prakt. Chem.* **1987**, *329*, 283; (e) J. Bestmann, K. Kumar, L. Kisielowski, *Chem. Ber.* **1983**, *116*, 2378.

[241] P. G. M. Wuts, C. L. Bergh, *Tetrahedron Lett.* **1986**, *27*, 3995.

[242] R. A. Ellison, W. D. Woessner, C. C. Williams, *J. Org. Chem.* **1972**, *37*, 2757.

[243] G. Boche, F. Bosold, M. Niessner, *Tetrahedron Lett.* **1982**, *23*, 3255.

[244] T.-H. Chuang, C.-C. Yang, C.-J. Chang, J.-M. Fang, *Synlett* **1990**, 733.

[245] H. B. Charman, (Imperial Chemical Industries) GB 1133882 (**1968**).

[246] (a) K. L. Perlman, H. M. Darwish, H. F. Deluca, *Tetrahedron Lett.* **1994**, *35*, 2295; (b) V. van Rheenen, *Tetrahedron Lett.* **1969**, 985.

[247] H. Kigoshi, Y. Imamura, K. Mizuta, H. Niwa, K. Yamada, *J. Am. Chem. Soc.* **1993**, *115*, 3056.

[248] (a) A. B. Jones in *Comprehensive Organic Chemistry*, Vol. 7 (Ed.: B. M. Trost), Wiley, **1991**, p. 151; (b) D. R. Buckle, I. L. Pinto in *Comprehensive Organic Chemistry*, Vol. 7 (Ed.: B. M. Trost), Wiley, **1991**, p. 119.

[249] *Heterocyclic Fragmentation of Organic Molecules* (Ed.: T.-L. Ho), Wiley-Interscience, New York, **1993**.

[250] R. C. Fuson, B. A. Bull, *Chem. Rev.* **1934**, *15*, 275.

[251] N. A. Porter, D. M. Scott, I. J. Rosenstein, B. Giese, A. Veit, H. G. Zeitz, *J. Am. Chem. Soc.* **1991**, *113*, 1791.

[252] S. Kajigaeshi, T. Nakagawa, N. Nagasaki, S. Fujisaki, *Synthesis* **1985**, 674.

[253] (a) H.-R. Bjørsvik, L. Liguori, J. A. V. Merinero, *J. Org. Chem.* **2002**, *67*, 7493; (b) H.-R. Bjørsvik, L. Liguori,

R. R. Gonzalez, J. A. V. Merinero, *Tetrahedron Lett.* **2002**, *43*, 4985.

[254] A. Zabjek, A. Petric, *Tetrahedron Lett.* **1999**, *40*, 6077.

[255] T. J. Wallace, H. Pobiner, A. Schriesheim, *J. Org. Chem.* **1965**, *30*, 3768.

[256] Y. Zhang, G. B. Schuster, *J. Org. Chem.* **1995**, *60*, 7192.

[257] (a) R. M. Moriarty, I. Prakash, R. Penmasta, *J. Chem. Soc., Chem. Commun.* **1987**, 202; (b) J. C. Lee, J.-H. Choi, Y. C. Lee, *Synlett* **2001**, 1563.

[258] A. Katho, M. T. Beck, *Synlett* **1992**, 165.

[259] S. Gurunath, A. Sudalai, *Synlett* **1999**, 559.

[260] A. Atlamsani, J.-M. Brégeault, *New J. Chem.* **1991**, *15*, 671.

[261] B. El Ali, J.-M. Brégeault, J. Mercier, J. Martin, C. Martin, O. Convert, *J. Chem. Soc., Chem. Commun.* **1989**, 825.

[262] F. Minisci, F. Recupero, F. Fontana, H.-R. Bjørsvik, L. Liguori, *Synlett* **2002**, 610.

[263] D. A. Oare, C. H. Heathcock, *J. Org. Chem.* **1990**, *55*, 157.

[264] A. Cox in *Comprehensive Organic Chemistry, Vol. 2* (Eds.: D. Barton, W. D. Ollis), Pergamon Press, **1979**, p. 661.

[265] J. Rocek, A. Riehl, *J. Org. Chem.* **1967**, *32*, 3569.

[266] (a) M. Jáky, J. Szammer, E. Simon-Trompler, *J. Chem. Soc., Perkin Trans. 2* **2000**, 1597; (b) M. N. Greco, B. E. Maryanoff, *Tetrahedron Lett.* **1992**, *33*, 5009.

[267] (a) L. He, C. A. Horiuchi, *Bull. Chem. Soc. Jpn.* **1999**, *72*, 2515; (b) L. He, M. Kanamori, C. A. Horiuchi, *J. Chem.Res. (S)* **1999**, 122.

[268] C. Sotiriou, W. Lee, R. W. Giese, *J. Org. Chem.* **1990**, *55*, 2159.

[269] G. Rothenberg, Y. Sasson, *Tetrahedron Lett.* **1996**, *37*, 13641.

[270] A. Kunai, K. Hatoh, Y. Hirano, J. Harada, K. Sasaki, *Bull. Chem. Soc. Jpn.* **1985**, *58*, 1717.

[271] (a) J.-M. Brégeault, F. Launay, A. Atlamsani, *C. R. Acad. Sci. Paris, Série IIc* **2001**, 11; (b) O. Hamed, A. El-Qisairi, P. M. Henry, *J. Org. Chem.* **2001**, *66*, 180; (c) K. Tanaka, (Mitsubishi Gas Chemical) JP 2002030027 (**2002**).

[272] A. Atlamsani, J.-M. Brégeault, *Synthesis* **1993**, 79.

[273] A. Altamsani, J.-M. Brégeault, M. Ziyad, *J. Org. Chem.* **1993**, *58*, 5663.

[274] T. Hirao, M. Mori, Y. Ohshiro, *Bull. Chem. Soc. Jpn.* **1989**, *62*, 2399.

[275] S. Ito, M. Matsumoto, *J. Org. Chem.* **1983**, *48*, 1133.

[276] (a) Y. Tanihara, T. Setoyama, (Mitsubishi Chemical Industries) JP 2000178225 (**2000**); (b) H. Mori, T. Suzuki, (Mitsubishi Chemical) JP 2000239217 (**2000**); (c) M. Fujii, T. Takewaki, T. Setoyama, (Mitsubishi Chemical) JP 2002003439 (**2002**).

[277] G. D. Mercer, J. S. Shu, T. B. Rauchfuss, D. M. Roundhill, *J. Am. Chem. Soc.* **1975**, *97*, 1966.

[278] K. Osowska-Pacewicka, H. Alper, *J. Org. Chem.* **1988**, *53*, 808.

[279] (a) C. M. Moorhoff, L. A. Paquette, *J. Org. Chem.* **1991**, *56*, 703; (b) C. M. Moorhoff, L. A. Paquette, *J. Org. Chem.* **1991**, *56*, 6728.

[280] N. Kakeya, T. Takai, (Ube Industries) JP09020722 (**1997**).

[281] S. Hara, S.-Q. Chen, T. Hatakeyama, T. Fukuhara, M. Sekiguchi, N. Yoneda, *Tetrahedron Lett.* **1995**, *36*, 6511.

[282] T. Hirao, T. Fujii, S.-I. Miyata, Y. Ohshiro, *J. Org. Chem.* **1991**, *56*, 2264.

[283] (a) M. Ohno, I. Oguri, S. Eguchi, *J. Org. Chem.* **1999**, *64*, 8995; (b) M. Golinski, S. Vasudevan, R. Floresca, C. P. Brock, D. S. Watt, *Tetrahedron Lett.* **1993**, *34*, 55; (c) E. Hata, T. Takai, T. Yamada, T. Mukaiyama, *Chem. Lett.* **1994**, 535; (d) L. El Aakel, F. Launay, A. Atlamsani, J.-M. Brégeault, *Chem. Commun.* **2001**, 2218; (e) D. T. C. Yang, C. J. Zhang, P. P. Fu, G. W. Kabalka, *Synth. Commun.* **1997**, *27*, 1601; (f) C. Coin, V. Le Boisselier, I. Favier, M. Postel, E. Dunach, *Eur. J. Org. Chem.* **2001**, 735; (g) M. Kirihara, S. Takizawa, T. Momose, *J. Chem. Soc., Perkin Trans. 1* **1998**, 7.

[284] (a) D. T. C. Yang, T. T. Evans, F. Yamazaki, C. Narayanna, G. W. Kabalka, *Synth. Commun.* **1993**, *23*, 1183; (b) M. Utaka, M. Nakatani, A. Takeda, *Tetrahedron* **1985**, *41*, 2167; (c) E. K. Starostin, A. A. Mazurchik, A. V. Igna-

TENKO, G. I. NIKISHIN, *Synthesis* **1992**, 917.

[285] D. T. C. YANG, Y. H. CAO, G. W. KABALKA, *Synth. Commun.* **1995**, *25*, 3695.

[286] (a) K. KOSTOVA, M. HESSE, *Helv. Chim. Acta* **1984**, *67*, 1725; (b) R. BALLINI, G. BOSICA, F. GIGLI, *Tetrahedron* **1998**, *54*, 7573.

[287] (a) J. COSSY, D. BELOTTI, V. BELLOSTA, D. BROCCA, *Tetrahedron Lett.* **1994**, *35*, 6089; (b) M. ABU-OMAR, J. H. ESPENSON, *Organometallics* **1996**, *15*, 3543; (c) H. H. WASSERMAN, J. E. PICKETT, *Tetrahedron* **1985**, *41*, 2155; (d) S. W. ASHFORD, K. C. GREGA, *J. Org. Chem.* **2001**, *66*, 1523.

[288] A. BAEYER, V. VILLIGER, *Ber. Dtsch. Chem. Ges.* **1899**, *32*, 3625.

[289] (a) R. CRIEGEE, *Liebigs Ann. Chem.* **1948**, *560*, 127; (b) M. RENZ, B. MEUNIER, *Eur. J. Org. Chem.* **1999**, 737.

[290] M. F. HAWTHORNE, W. D. EMMONS, K. S. MCCALLUM, *J. Am. Chem. Soc.* **1958**, *80*, 6393.

[291] D. W. YOUNG, P. J. CORRINGER, G. BURTIN, *J. Chem. Soc., Perkin Trans. 1* **2000**, 3451.

[292] (a) P. F. HUDRLIK, A. M. HUDRLIK, G. NAGENDRAPPA, T. YIMENU, E. T. ZELLERS, E. CHIN, *J. Am. Chem. Soc.* **1980**, *102*, 6894; (b) T. AIDA, M. ASAOKA, S. SONODA, H. TAKEI, *Heterocycles* **1993**, *36*, 427; (c) S. HORVAT, P. KARALLAS, J. M. WHITE, *J. Chem. Soc., Perkin Trans. 2* **1998**, 2151; (d) R. P. BAKALE, M. A. SCIALDONE, C. R. JOHNSON, *J. Am. Chem. Soc.* **1990**, *112*, 6729.

[293] G. MEHTA, N. MOHAL, *J. Chem. Soc., Perkin Trans. 1* **1998**, 505.

[294] (a) K. MISLOW, J. BRENNER, *J. Am. Chem. Soc.* **1953**, *75*, 2318; (b) J. A. BERSON, S. SUZUKI, *J. Am. Chem. Soc.* **1959**, *81*, 4088.

[295] (a) R. M. GOODMAN, Y. KISHI, *J. Am. Chem. Soc.* **1998**, *120*, 9392; (b) C. M. CRUDDEN, A. C. CHEN, L. A. CALHOUN, *Angew. Chem.* **2000**, *112*, 2973; *Angew. Chem., Int. Ed. Engl.* **2000**, *39*, 2851. For early studies on this subject, see: (c) R. NOYORI, T. SATO, H. KOBAYASHI, *Tetrahedron Lett.* **1980**, *21*, 2569; (d) R. NOYORI, H. KOBAYASHI, T. SATO, *Tetrahedron Lett.* **1980**, *21*, 2573;

(e) R. NOYORI, T. SATO, H. KOBAYASHI, *Bull. Chem. Soc. Jpn.* **1983**, *56*, 2661.

[296] (a) K. S. FELDMAN, M. J. WU, D. P. ROTELLA, *J. Am. Chem. Soc.* **1990**, *112*, 8490; (b) W. M. PANKAU, W. KREISER, *Helv. Chim. Acta* **1998**, *81*, 1997; (c) D. L. VARIE, J. BRENNAN, B. BRIGGS, J. S. CRONIN, D. A. HAY, J. A. RIECK, M. J. ZMIJEWSKI, *Tetrahedron Lett.* **1998**, *39*, 8405; (d) J. D. FELL, C. H. HEATHCOCK, *J. Org. Chem.* **2002**, *67*, 4742.

[297] Y. OKUNO, *Chem. Eur. J.* **1997**, *3*, 212.

[298] P. A. GRIECO, K. W. HUNT, *Org. Lett.* **2000**, *2*, 1717.

[299] H. HEANEY, *Aldrichim. Acta* **1993**, *26*, 35.

[300] (a) T. HOSAYA, Y. KURIYAMA, K. SUZUKI, *Synlett* **1995**, 635; (b) T. MINO, S. MASUDA, M. NISHIO, M. YAMASHITA, *J. Org. Chem.* **1997**, *62*, 2633.

[301] M. ESPIRITU, P. N. HANDLEY, R. NEUMANN, *Adv. Synth. Catal.* **2003**, *345*, 325.

[302] A. LAMBERT, J. A. ELINGS, D. J. MACQUARRIE, G. CARR, J. H. CLARK, *Synlett* **2000**, 1052.

[303] (a) G. ROTHENBERG, A. D. DOWNIE, C. L. RASTON, J. L. SCOTT, *J. Am. Chem. Soc.* **2001**, *123*, 8701; (b) S. R. GHORPADE, U. R. KALKOTE, S. P. CHAVAN, S. R. BHIDE, T. RAVINDRANATHAN, V. G. PURANIK, *J. Org. Chem.* **2001**, *66*, 6803.

[304] (a) M. S. COOPER, H. HEANEY, A. J. NEWBOLD, W. R. SANDERSON, *Synlett* **1990**, 533; (b) G. ALLAN, A. J. CARNELL, M. L. ESCUDERO HERNANDEZ, A. PETTMAN, *Tetrahedron* **2001**, *57*, 8193; (c) K. I. KUCHKOVA, Y. M. CHUMAKOV, Y. A. SIMONOV, G. BOCELLI, A. PANASENKO, P. F. VLAD, *Synthesis* **1997**, 1045; (d) K. B. WIBERG, C. ÖSTERLE, *J. Org. Chem.* **1999**, *64*, 7763.

[305] K. KANEDA, S. UENO, T. IMANAKA, E. SHIMOTSUMA, Y. NISHIYAMA, Y. ISHII, *J. Org. Chem.* **1994**, *59*, 2915.

[306] C. BOLM, C. PALAZZI, G. FRANCIO, W. LEITNER, *Chem. Commun.* **2002**, 1588.

[307] (a) R. NEUMANN, K. NEIMANN, *Org. Lett.* **2000**, *2*, 2861; (b) A. BERKESSEL, M. R. M. ANDREAE, *Tetrahedron Lett.* **2001**, *42*, 2293; (c) A. BERKESSEL, M. R. M. ANDREAE, H. SCHMICKLER,

J. Lex, *Angew. Chem.* **2002**, *114*, 4661; *Angew. Chem., Int. Ed. Engl.* **2002**, *41*, 4481.

[308] (a) R. D. Chambers, M. Clark, *Tetrahedron Lett.* **1970**, 2741; (b) P. A. Ganeshpure, W. Adam, *Synthesis* **1996**, 179.

[309] C. Palomo, J. M. Aizpurua, C. Cuevas, R. Urchegui, A. Linden, *J. Org. Chem.* **1996**, *61*, 4400.

[310] (a) C. Bolm, O. Beckmann, T. K. K. Luong, *Transition Metals for Organic Synthesis*, Vol. 2 (Eds.: M. Beller, C. Bolm), Wiley-VCH, **1998**, p. 213; (b) C. Bolm in *Advances in Catalytic Processes*, Vol. 2 (Ed.: M. P. Doyle), JAI Press, Greenwich, **1997**, p. 43; (c) G. Strukul, *Angew. Chem.* **1998**, *110*, 1256, *Angew. Chem., Int. Ed. Engl.* **1998**, *37*, 1199.

[311] P. Carlqvist, R. Eklund, T. Brinck, *J. Org. Chem.* **2001**, *66*, 1193.

[312] (a) T. Yamada, K. Takahashi, K. Kato, T. Takai, S. Inoki, T. Mukaiyama, *Chem. Lett.* **1991**, 641; (b) C. Bolm, G. Schlinghoff, K. Weikkardt, *Tetrahedron Lett.* **1993**, *34*, 3405.

[313] (a) S.-I. Murahashi, Y. Oda, T. Naota, *Tetrahedron Lett.* **1992**, *33*, 7557; (b) S.-I. Murahashi, *Angew. Chem.* **1995**, *107*, 2670, *Angew. Chem., Int. Ed. Engl.* **1995**, *34*, 2443.

[314] T. Inokushi, M. Kanazaki, T. Sugimoto, S. Torii, *Synlett* **1994**, 1037.

[315] M. Hamamoto, K. Nakayama, Y. Nishiyama, Y. Ishii, *J. Org. Chem.* **1993**, *58*, 6421.

[316] R. Raja, J. M. Thomas, *Chem. Commun.* **2001**, 675.

[317] R. Raja, J. M. Thomas, G. Sankar, *Chem. Commun.* **1999**, 525.

[318] (a) K. Kaneda, T. Yamashita, *Tetrahedron Lett.* **1996**, *37*, 4555; (b) K. Kaneda, S. Ueno, T. Imanaka, *J. Chem. Soc., Chem. Commun.* **1994**, 797.

[319] (a) A. Corma, M. T. Navarro, L. Nemeth, M. Renz, *Nature* **2001**, *412*, 423; (b) M. Renz, T. Blasco, A. Corma, V. Fornes, R. Jensen, L. Nemeth, *Chem. Eur. J.* **2002**, *8*, 4708.

[320] R. B. Sever, T. W. Root, *J. Phys. Chem B* **2003**, *107*, 10848. For a comparative study on the mechanism of tin and titanium catalyzed Baeyer-Villiger oxidation see also R. B. Sever, T. W. Root, *J. Phys. Chem B* **2003**, *107*, 10521.

[321] A. Corma, M. T. Navarro, M. Renz, *J. Catal.* **2003**, *219*, 242.

[322] (a) W. A. Herrmann, R. W. Fischer, J. D. G. Correia, *J. Mol. Catal. A* **1994**, *94*, 213; (b) A. M. Faisca Phillips, C. Romao, *Eur. J. Org. Chem.* **1999**, 1767; (c) R. Bernini, E. Mincione, M. Cortese, G. Aliotta, A. Oliva, R. Saladino, *Tetrahedron Lett.* **2001**, *42*, 5401.

[323] M. Del Todesco Frisone, R. Giovanetti, F. Pinna, G. Strukul, *Stud. Surf. Sci. Catal.* **1991**, *66*, 405.

[324] M. C. Fermin, J. W. Bruno, *Tetrahedron Lett.* **1993**, *34*, 7545.

[325] C. Palazzi, F. Pinna, G. Strukul, *J. Mol. Catal. A* **2000**, *151*, 245.

[326] I. C. Chisem, J. Chisem, J. H. Clark, *New J. Chem.* **1998**, *22*, 81.

[327] R. Bernini, E. Mincione, M. Cortese, R. Saladino, G. Gualandi, M. C. Belfiore, *Tetrahedron Lett.* **2003**, *44*, 4823.

[328] (a) R. Gottlich, K. Yamakoshi, H. Sasai, M. Shibasaki, *Synlett* **1997**, 971; (b) F. Velazquez, H. F. Olivo, *Org. Lett.* **2000**, *2*, 1931; (c) See also M. Suzuki, H. Takada, R. Noyori, *J. Org. Chem.* **1982**, *47*, 902.

[329] X. Hao, O. Yamazaki, A. Yoshida, J. Nishikido, *Tetrahedron Lett.* **2003**, *44*, 4977.

[330] H. Kotsuki, K. Arimura, T. Araki, T. Shinohara, *Synlett* **1999**, 462.

[331] M. Mujahid Alam, R. Varala, S. R. Adapa, *Synth. Commun.* **2003**, *33*, 3035.

[332] C. Mazzini, J. Lebreton, R. Furstoss, *J. Org. Chem.* **1996**, *61*, 8.

[333] R. A. Sheldon, G. J. ten Brink, J. M. Vis, I. W. C. E. Arens, *J. Org. Chem.* **2001**, *66*, 2429.

[334] H. B. Kagan, *Croat. Chem. Acta* **1996**, *69*, 669.

[335] M. C. Willis, *J. Chem. Soc., Perkin Trans. 1* **1999**, 1765.

[336] M. Aoki, D. Seebach, *Helv. Chim. Acta* **2001**, *84*, 187.

[337] T. Sugimura, Y. Fujiwara, A. Tai, *Tetrahedron Lett.* **1997**, *38*, 6019.

[338] C. Bolm, O. Beckmann, *Chirality* **2000**, *12*, 523.

[339] (a) C. Bolm, G. Schlinghoff, K. Weick-

HARDT, *Angew. Chem.* **1994**, *106*, 1944, *Angew. Chem., Int. Ed. Engl.* **1994**, *33*, 1848; (b) SEE ALSO: Y. PENG, X. FENG, K. YU, Z. LI, Y. JIANG, C.-H. YEUNG, *J. Organomet. Chem.* **2001**, *619*, 204.

[340] (a) C. BOLM, G. SCHLINGHOFF, *J. Chem. Soc., Chem. Commun.* **1995**, 1247; (b) C. BOLM, G. SCHLINGHOFF, *J. Mol. Catal. A* **1997**, *117*, 347.

[341] (a) C. BOLM, T. K. K. LUONG, O. BECKMANN in *Asymmetric Oxidation Reactions* (Ed.: T. KATSUKI), Oxford University Press, **2001**, p. 147; (b) C. BOLM in *Peroxide Chemistry* (Ed.: W. ADAM), Wiley-VCH, **2000**, 494; (c) D. R. KELLY, *Chim. Oggi* **2000**, *18*, 33 and 52; (d) C. Bolm, O. Beckmann in *Comprehensive Asymmetric Catalysis, Vol. 2* (Eds.: E. N. JACOBSEN, A. PFALTZ, H. YAMAMOTO), Springer-Verlag, **1999**, p. 803; (d) C. BOLM, *Med. Res. Rev.* **1999**, *19*, 348.

[342] (a) A. GUSSO, R. BACCIN, F. PINNA, G. STRUKUL, *Organometallics* **1994**, *13*, 3442; (b) A. VARAGNOLO, F. PINNA, G. STRUKUL, *J. Mol. Catal. A* **1997**, *117*, 413; (c) C. PANEGHETTI, R. GAVAGNIN, F. PINNA, G. STRUKUL, *Organometallics* **1999**, *18*, 5057.

[343] T. SHINOHARA, S. FUJIOKA, H. KATSUKI, *Heterocycles* **2001**, *55*, 237.

[344] K. ITO, A. ISHII, T. KURODA, T. KATSUKI, *Synlett* **2003**, 643.

[345] (a) M. LOPP, A. PAJU, T. KANGER, T. PEHK, *Tetrahedron Lett.* **1996**, *37*, 7583; (b) T. KANGER, K. KRIIS, A. PAJU, T. PEHK, M. LOPP, *Tetrahedron: Asymmetry* **1998**, *9*, 4475.

[346] C. BOLM, O. BECKMANN, A. COSP, C. PALAZZI, *Synlett* **2001**, 1461.

[347] C. BOLM, O. BECKMANN, C. PALAZZI, *Can. J. Chem.* **2001**, *79*, 1593.

[348] C. BOLM, O. BECKMANN, T. KÜHN, C. PALAZZI, W. ADAM, P. B. RAO, C. R. SAHA-MÖLLER, *Tetrahedron: Asymmetry* **2001**, *12*, 2441.

[349] (a) T. UCHIDA, T. KATSUKI, *Tetrahedron Lett.* **2001**, *42*, 6911; (b) T. UCHIDA, T. KATSUKI, K. ITO, S. AKASHI, A. ISHII, T. KURODA, *Helv. Chim. Acta* **2002**, *85*, 3078.

[350] A. WATANABE, T. UCHIDA, K. ITO, T. KATSUKI, *Tetrahedron Lett.* **2002**, *43*, 4481.

[351] Y. MIYAKE, Y. NISHIBAYASHI, S. UEMURA, *Bull. Chem. Soc. Jpn.* **2002**, *75*, 2233.

[352] S.-I. MURAHASHI, S. ONO, Y. IMADA, *Angew. Chem.* **2002**, *114*, 2472, *Angew. Chem., Int. Ed. Engl.* **2002**, *41*, 2366.

[353] (a) V. ALPHAND, R. FURSTOSS in *Asymmetric Oxidation Reaction* (Ed.: T. KATSUKI), Oxford University Press, **2001**, p. 214; (b) V. ALPHAND, R. FURSTOSS in *Handbook of Enzyme Catalysis in Organic Synthesis, Vol. 2* (Eds.: K. DRAUZ, H. WALDMANN), Wiley-VCH, **1995**, p. 744; (c) S. M. ROBERTS, P. W. H. WAN, *J. Mol. Catal. B* **1998**, *4*, 111; (d) M. D. MIHOVILOVIC, B. MÜLLER, P. STANETTY, *Eur. J. Org. Chem.* **2002**, 3711.

[354] E. BECKMANN, *Chem. Ber.* **1886**, *19*, 988.

[355] (a) R. E. GAWLEY, *Org. React.* **1988**, *35*, 1; (b) D. CRAIG in *Comprehensive Organic Chemistry, Vol. 7* (Ed.: B. M. TROST), Pergamon Press, **1991**, p. 689.

[356] J. C. JOCHIMS, S. HEHL, S. HERZBERGER, *Synthesis* **1990**, 1128.

[357] (a) M. T. NGUYEN, G. RASPOET, L. G. VANQUICKENBORNE, *J. Am. Chem. Soc.* **1997**, *119*, 2552; (b) M. T. NGUYEN, G. RASPOET, L. G. VANQUICKENBORNE, *J. Chem. Soc., Perkin Trans. 2* **1997**, 821; (c) V. SIMUNIC-MEZNARIC, Z. MIHALIC, H. VANCIK, *J. Chem. Soc., Perkin Trans. 2* **2002**, 2154.

[358] (a) K. GOPALAIAH, S. CHANDRASKHAR, *Tetrahedron Lett.* **2001**, *42*, 8123; (b) S. CHANDRASEKHAR, K. GOPALAIAH, *Tetrahedron Lett.* **2002**, *43*, 2455.

[359] A. I. BOSCH, P. DE LA CRUZ, E. DIEZ-BARRA, A. LOUPY, F. LANGA, *Synlett* **1995**, 1259.

[360] Y. IZUMI, T. FUJITA, *J. Mol. Catal. A* **1996**, *106*, 43.

[361] J. SKARZEWSKI, M. ZIELINSKA-BLAJET, I. TUROWSKA-TYRK, *Tetrahedron: Asymmetry* **2001**, *12*, 1923.

[362] A. PADWA, D. L. HERZOG, W. R. NADLER, M. H. OSTERHOUT, A. T. PRICE, *J. Org. Chem.* **1994**, *59*, 1418.

[363] H. SATO, H. YOSHIOKA, Y. IZUMI, *J. Mol. Catal. A* **1999**, *149*, 25.

[364] (a) J. PENG, Y. DENG, *Tetrahedron Lett.* **2001**, *42*, 403; (b) R. X. REN, L. D. ZUEVA, W. OU, *Tetrahedron Lett.* **2001**, *42*, 8441.

[365] (a) N. IRANPOOR, H. FIROUZABADI, G. AGHAPOUR, *Synth. Commun.* **2002**,

32, 2535; (b) S. Thiebaut, C. Gerardin-Charbonnier, C. Selve, Tetrahedron **1999**, 55, 1329.

[366] (a) M. Boruah, D. Konwar, *J. Org. Chem.* **2002**, 67, 7138; (b) M. M. Khodaei, F. A. Meybodi, N. Rezai, P. Salehi, *Synth. Commun.* **2001**, 31, 2047; (c) D. C. Barman, A. J. Thakur, D. Prajapati, J. S. Sandhu, *Chem. Lett.* **2000**, 1196; (d) A. J. Thakur, A. Boruah, D. Prajapati, J. S. Sandhu, *Synth. Commun.* **2000**, 30, 2105; (e) J. S. Yadav, B. V. S. Reddy, A. V. Madhavi, Y. S. S. Ganesh, *J. Chem. Res. (S)* **2002**, 236; (f) B. M. Khadilkar, D. J. Upadhyaya, *Synth. Commun.* **2002**, 32, 1867.

[367] (a) M. Arisawa, M. Yamagushi, *Org. Lett.* **2001**, 3, 311; (b) H. Kusama, Y. Yamashita, K. Naradaka, *Bull. Chem. Soc. Jpn.* **1995**, 68, 373.

[368] (a) T. Tatsumi in *Fine Chemicals through Heterogeneous Catalysis* (Eds.: R. A. Sheldon, H. van Bekkum), Wiley-VCH, Weinheim, 2001, p. 185; (b) B.-Q. Xu, S.-B. Cheng, X. Zhang, S.-F. Ying, Q.-M. Zhu, *Chem. Commun.* **2000**, 1121; (c) Y. Ko, M. H. Kim, S. J. Kim, G. Seo, M.-Y. Kim, Y. S. Uh, *Chem. Commun.* **2000**, 829.

[369] Y. Ikushima, K. Hatakeda, O. Sato, T. Yokoyama, M. Arai, *J. Am. Chem. Soc.* **2000**, 122, 1908.

[370] L. De Luca, G. Giacomelli, A. Porcheddu, *J. Org. Chem.* **2002**, 67, 6272.

[371] (a) O. Muraoka, B.-Z. Zheng, K. Okamura, E. Tabata, G. Tanabe, M. Kubo, *J. Chem. Soc., Perkin Trans. 1* **1997**, 113; (b) E. Albertini, A. Barco, S. Benetti, P. Pollini, V. Zanirato, *Tetrahedron* **1997**, 53, 17177; (c) R. P. Frutos, D. M. Spero, *Tetrahedron Lett.* **1998**, 38, 2475.

[372] (a) B. Westermann, I. Gedrath, *Synlett* **1996**, 665; (b) J. D. White, P. Hrnciar, F. Stappenbeck, *J. Org. Chem.* **1999**, 64, 7871; (c) W. S. Murphy, B. Sarsam, G. Ferguson, J. F. Gallaher, *J. Chem. Soc., Perkin Trans. 1* **1998**, 4121.

[373] M. H. Wu, E. N. Jacobsen, *Tetrahedron Lett.* **1998**, 38, 1693.

[374] T. Harada, T. Ohno, S. Kobayashi, T. Mukaiyama, *Synthesis* **1991**, 1216.

[375] R. Anilkumar, S. Chandrasekhar, *Tetrahedron Lett.* **2000**, 41, 5427.

[376] A. R. Kratritzky, D. A. R. Monteux, D. O. Tymoshenko, *Org. Lett.* **1999**, 1, 577.

[377] R. Anilkumar, S. Chandraskhar, *Tetrahedron Lett.* **2000**, 41, 7235.

[378] (a) G. A. Olah, A. P. Fung, *Synthesis* **1979**, 537; (b) A. Laurent, P. Jacquault, J. L. Di Martino, J. J. Hamelin, *J. Chem. Soc., Chem. Commun.* **1995**, 1101.

[379] A. Kanasawa, S. Gillet, P. Delair, A. E. Greene, *J. Org. Chem.* **1998**, 63, 4660.

[380] H. Sharghi, M. Hosseini, *Synthesis* **2002**, 1057.

[381] B. Das, N. Ravindranath, B. Venkataiah, P. Madhusudhan, *J. Chem. Res. (S)* **2000**, 482.

[382] (a) A. García Martínez, E. Teso Vilar, A. Garcia Fraile, S. de la Moya Cerero, B. Lora Maroto, *Eur. J. Org. Chem.* **2002**, 781; (b) X. Jiang, C. Wang, Y. Hu, H. Hu, D. F. Covey, *J. Org. Chem.* **2000**, 65, 3555; (c) K. Takatori, Y. Takeuchi, Y. Shinohara, K. Yamaguchi, M. Nakamura, T. Hirosawa, T. Shimizu, M. Saito, S. Aizawa, O. Sugiyama, Y. Ohtsuka, M. Kajiwara, *Synlett* **1999**, 975; (d) M. S. Laxmischa, G. S. R. Subba Rao, *Tetrahedron Lett.* **2000**, 41, 3759.

[383] (a) P. A. Petukhov, A. V. Tkachev, *Tetrahedron* **1997**, 53, 2535; (b) Z. Tokic-Vujosevic, Z. Cekovic, *Synthesis* **2001**, 2028; (c) J. Kehler, E. Breuer, *J. Chem. Soc., Chem. Commun.* **1997**, 1751; (d) V. M. Pejanovic, J. A. Petrovic, J. J. Csanadi, S. M. Stankovic, D. A. Miljkovic, *Tetrahedron* **1995**, 51, 13379; (e) H. Fujioka, M. Miyazaki, H. Kitagawa, T. Yamanaka, H. Yamamoto, K. Takuma, Y. Kita, *J. Chem. Soc., Chem. Commun.* **1993**, 1634; (f) A. Garcia Martinez, E. Teso Vilar, A. Garcia Fraile, S. de la Moya Cerero, B. Lora Maroto, *Tetrahedron: Asymmetry* **1996**, 7, 2177.

[384] (a) T.-L. Ho, E. Gorobets, *Tetrahedron* **2002**, 58, 4969; (b) M. J. Figueira, J. M. Blanco, O. Caamaño, F. Fernández, X. García-Mera, C. López, *Synthesis* **2000**, 1459; (c) I. P. Andrews,

R. J. J. Dorgan, J. F. Hudner, T. Harvey, N. Hussein, D. C. Lathbury, N. J. Lewis, G. S. Macaulay, D. O. Morgan, R. Stockmann, C. R. White, *Tetrahedron Lett.* **1996**, *37*, 4811.

[385] A. Fredenhagen, H. H. Peter, *Tetrahedron* **1996**, *52*, 1235.

[386] (a) N. S. Mani, M. Wu, *Tetrahedron: Asymmetry* **2000**, *11*, 4687; (b) D. Schinzer, U. Abel, P. G. Jones, *Synlett* **1997**, 632.

[387] K. Maruoka, T. Miyazaki, M. Ando, Y. Matsumura, S. Sakane, K. Hattori, H. Yamamoto, *J. Am. Chem. Soc.* **1983**, *105*, 2831.

[388] T. Cablewski, P. A. Gurr, K. D. Raner, C. R. Strauss, *J. Org. Chem.* **1994**, *59*, 5814.

[389] H. Wolff, *Org. React.* **1946**, *3*, 307.

[390] J. P. Richard, T. L. Amyes, Y.-G. Lee, V. Jagannadham, *J. Am. Chem. Soc.* **1994**, *116*, 10833.

[391] (a) B. T. Smith, J. A. Wendt, J. Aubé, *Org. Lett.* **2002**, *4*, 2577; (b) M. Tanaka, M. Oba, K. Tamai, H. Suemune, *J. Org. Chem.* **2001**, *66*, 2667; (b) G. R. Krow, O. H. Cheung, Z. Hu, Y. B. Lee, *J. Org. Chem.* **1996**, *61*, 5574.

[392] G. R. Krow, S. W. Szczepanski, J. Y. Kim, N. Liu, A. Sheikh, Y. Xiao, J. Yuan, *J. Org. Chem.* **1999**, *64*, 1254.

[393] (a) M. Moreno-Mañas, E. Trepat, R. M. Sebastián, A. Vallribera, *Tetrahedron: Asymmetry* **1999**, *10*, 4211; (b) R. W. Baker, R. V. Kyasnoor, M. V. Sargent, *Tetrahedron Lett.* **1999**, *40*, 3475.

[394] M. J. Mphahlele, *J. Chem. Soc., Perkin Trans. 1* **1999**, 3477.

[395] J. Aubé, K. Schildknegt, K. A. Agrios, C. Mossman, G. L. Milligan, P. Desai, *J. Am. Chem. Soc.* **2000**, *122*, 7226.

[396] V. Gracias, G. L. Milligan, J. Aubé, *J. Org. Chem.* **1996**, *61*, 10.

[397] A. Wrobleski, J. Aubé, *J. Org. Chem.* **2001**, *66*, 886.

[398] J. Aubé, P. Desai, K. Schildknegt, R. Iyengar, *Org. Lett.* **2000**, *2*, 1625.

[399] J. Aubé, K. Furness, *Org. Lett.* **1999**, *1*, 495.

[400] G. Vidari, M. Tripolini, P. Novella, P. Allegrucci, L. Garlaschelli, *Tetrahedron: Asymmetry* **1997**, *8*, 2893.

[401] P. A. Evans, D. P. Modi, *J. Org. Chem.* **1995**, *60*, 6662.

[402] J. E. Golden, J. Aubé, *Angew. Chem.* **2002**, *114*, 4492, *Angew. Chem., Int. Ed. Engl.* **2002**, *41*, 4316.

[403] J.-Y. Mérour, S. J. Piroëlle, *J. Heterocyclic Chem.* **1994**, *31*, 141.

[404] (a) M. F. Schlecht in *Comprehensive Organic Chemistry, Vol. 7* (Ed.: B. M. Trost), Pergamon Press, **1991**, p. 815; (b) C. Giordano, G. Castaldi, F. Uggeri, *Angew. Chem.* **1984**, *96*, 413, *Angew. Chem., Int. Ed. Engl.* **1984**, *23*, 413.

[405] M. P. Sibi in *Handbook of Reagents for Organic Synthesis, Oxidizing and Reducing Agents* (Eds.: S. D. Burke, R. L. Danheiser), Wiley, Chichester, **1999**, p. 448.

[406] (a) E. C. Taylor, R. A. Conley, A. H. Katz, A. McKillop, *J. Org. Chem.* **1984**, *49*, 3840; (b) P. Camps, S. Giménez, X. Farrés, D. Mauléon, G. Carganico, *Liebigs Ann. Chem.* **1993**, 641; (c) T. G. van Aardt, P. S. van Heerden, D. Ferreira, *Tetrahedron Lett.* **1998**, *39*, 3881.

[407] M. L. Mihailovic, Z. Cekovic in *Handbook of Reagents for Organic Synthesis, Oxidizing and Reducing Agents* (Eds.: S. D. Burke, R. L. Danheiser), Wiley, Chichester, **1999**, p. 190.

[408] (a) T. Yamauchi, K. Nakao, K. Fujii, *J. Chem. Soc., Perkin Trans. 1* **1987**, 1433.

[409] (a) H. Togo, S. Abe, G. Nogami, M. Yokoyama, *Bull. Chem. Soc. Jpn.* **1999**, *72*, 2351; (b) H. Togo, S. Abe, G. Nogami, M. Yokoyama, *Synlett* **1998**, 534.

[410] O. V. Singh, *Tetrahedron Lett.* **1990**, *31*, 3055.

[411] (a) J. P. Guthrie, J. Cossar, J. Lu, *Can. J. Chem.* **1991**, *69*, 1904; (b) J. D. Andre, A. Heymes, P. J. Grossi, M. Venerucci, (SANOFI, Industria Chimica Produtti Francis) EP 184572 (**1986**).

[412] G. B. Gill in *Comprehensive Organic Chemistry, Vol. 3* (Ed.: B. M. Trost), Pergamon Press, **1991**, p. 887.

[413] (a) N. Shibata, B. K. Das, H. Honjo, Y. Takeuchi, *J. Chem. Soc., Perkin Trans. 1* **2001**, 1605; (b) J.-Y. Winum, M. Kamal, A. Leydet, J.-P. Roque, J.-L. Montero, *Tetrahedron Lett.* **1996**, *37*, 1781; (c) A. R. Katritzky, S. Zhang, Y. Fang,

[414] T. Yamauchi, K. Hattori, S.-I. Ikeda, K. Tamaki, *J. Chem. Soc., Perkin Trans. 1* **1990**, 1683.

[415] J. Mann in *Comprehensive Organic Chemistry*, Vol. 3 (Ed.: B. M. Trost), Pergamon Press, **1991**, p. 839.

[416] (a) H. R. Sonawane, D. G. Kulkarni, N. R. Ayyangar, *Tetrahedron Lett.* **1990**, *31*, 7495; (b) O. Piccolo, F. Spreafico, G. Visentin, *J. Org. Chem.* **1987**, *52*, 10; (c) C. Giordano, G. Castaldi, F. Uggeri, F. Gurzoni, *Synthesis* **1985**, 436.

[417] T. Yamauchi, K. Nakao, K. Fujii, *J. Chem. Soc., Perkin Trans. 1* **1987**, 1255.

[418] G. Castaldi, S. Cavicchioli, C. Giordano, F. Uggeri, *J. Org. Chem.* **1987**, *52*, 3018.

[419] A. Srikishna, G. Sunderbabu, *Tetrahedron Lett.* **1989**, *30*, 3561.

[420] (a) M. N. Elinson, S. K. Feducovich, A. S. Dorofeev, A. N. Vereshchagin, G. I. Nikishin, *Tetrahedron* **2000**, *56*, 9999; (b) T. Shono, Y. Matsumura, S. Katoh, T. Fujita, T. Kamada, *Tetrahedron Lett.* **1989**, *30*, 371.

[421] W. Oppolzer, S. Rosset, J. De Brahander, *Tetrahedron Lett.* **1997**, *38*, 1539.

[422] S. D. Higgins, C. B. Thomas, *J. Chem. Soc., Perkin Trans. 1* **1983**, 1483.

[423] (a) H. R. Sonawane, N. S. Bellur, D. G. Kulkarni, N. R. Ayyangar, *Tetrahedron* **1994**, *50*, 1243; (b) H. R. Sonawane, D. G. Kulkarni, N. R. Ayyangar, (Council of Scientific and Industrial Research) EP 336031 (**1989**).

[424] T. Satoh, H. Unno, *Tetrahedron* **1997**, *53*, 7843.

[425] T. Satoh, S. Motohashi, S. Kimura, N. Tokutake, K. Yamakawa, *Tetrahedron Lett.* **1993**, *34*, 4823.

[426] S. Uemura, S.-I. Fukuzawa, T. Yamauchi, K. Hattori, S. Mizutaki, K. Tamaki, *J. Chem. Soc., Chem. Commun.* **1984**, 426.

[427] F. Mathew, B. Myrboh, *Tetrahedron Lett.* **1990**, *31*, 3757.

[428] O. V. Singh, C. P. Grag, R. P. Kapoor, *Synthesis* **1990**, 1025.

[429] G. B. Gill in *Comprehensive Organic Chemistry*, Vol. 3 (Ed. B. M. Trost), Pergamon Press, **1991**, p. 821.

[430] (a) M. Carmack, M. A. Spielman, *Org. React.* **1946**, *3*, 83; (b) E. V. Brown, *Synthesis* **1975**, 358.

[431] M. R. Kanyonyo, A. Gozzo, D. M. Lambert, D. Lesieur, J. H. Poupaert, *Bull. Chem. Soc. Belg.* **1997**, *106*, 39.

[432] (a) T. Lundstedt, R. Carlson, R. Shabana, *Acta Chem. Scand., Ser. B.* **1987**, *41*, 157; (b) R. Carlson, T. Lundstedt, *Acta Chem. Scand., Ser. B.* **1987**, *41*, 164.

[433] F. Dutron-Woitrin, R. Merenyl, H. G. Viehe, *Synthesis* **1985**, 77.

[434] J. O. Amupitan, *Synthesis* **1983**, 730.

[435] W. Schroth, J. Andersch, *Synthesis* **1989**, 202.

[436] H. S. Chiou, M. R. Rubino, S. W. Jahoda, D. Lindley, J. R. Battler, (Hoechst Celanese) US 5149866 (**1992**).

[437] (a) M. F. Moghaddam, M. Ghaffarzadeh, *Synth. Commun.* **2001**, *31*, 317; (b) K. D. Raner, C. R. Strauss, R. W. Trainor, J. S. Thorn, *J. Org. Chem.* **1995**, *60*, 2456; (c) M. Nooshabadi, K. Aghapoor, H. R. Darabi, M. M. Mojtahedi, *Tetrahedron Lett.* **1999**, *40*, 7549.

[438] T. Kanbara, Y. Kawai, K. Hasegawa, H. Morita, T. Yamamoto, *J. Polym. Sci. A* **2001**, *39*, 3739.

[439] M. Carmack, *J. Heterocyclic Chem.* **1989**, *26*, 1319.

Org. Lett. **2000**, *2*, 3789; (d) A. R. Katritzky, S. Zhang, A. H. M. Hussein, Y. Fang, *J. Org. Chem.* **2001**, *66*, 5606; (e) A. D. Allen, B. Cheng, M. H. Fenwick, B. Givehchi, H. Henry-Riyad, V. A. Nikolaev, E. A. Shikhova, D. Tahmassedi, T. T. Tidwell, S. Wang, *J. Org. Chem.* **2001**, *66*, 2611.

10
Manganese-based Oxidation with Hydrogen Peroxide
Jelle Brinksma, Johannes W. de Boer, Ronald Hage, and Ben L. Feringa

10.1
Introduction

Oxidation reactions are among the most important transformations in synthetic chemistry [1] and offer important methodology for the introduction and modification of functional groups [2]. Currently used stoichiometric oxidations based on, for example, nitric acid, chromic acid and derivatives, alkyl hydroperoxides, permanganate, osmium tetraoxide, bleach and peracids frequently suffer from high costs, formation of toxic waste or low atom efficiency, providing a strong incentive to develop sustainable catalytic alternatives [3, 4].

Oxidations with molecular oxygen [5], the primary oxidant in biological systems [6], are desirable and are already used in large scale industrial processes [7], but practicability is often hampered by low conversion, modest selectivity or safety issues associated with peroxide build-up. Although both oxygen atoms of O_2 can be transferred in dioxygenation or singlet oxygen reactions, often stoichiometric amounts of a reducing agent are required to convert one of the oxygen atoms of O_2 into water. Hydrogen peroxide is particularly attractive as it has a high active oxygen content (47%), gives H_2O as the only waste product and is relatively cheap. One of the problems frequently encountered in metal-catalyzed oxidations with hydrogen peroxide is the concomitant decomposition of H_2O_2 (catalase activity), which makes the use of a large excess of H_2O_2 necessary to reach full conversion. Applications of H_2O_2, in particular metal-catalyzed [8] epoxidations [9], and green oxidations [4] have recently been reviewed. In this chapter an overview of Mn-catalyzed oxidations with H_2O_2 is presented.

$KMnO_4$ still occupies a position as an oxidant in nearly every introductory course on organic chemistry [10, 11]. In sharp contrast is the limited use of this oxidizing agent in the practice of synthetic chemistry and it is only in recent years that several discoveries have revealed the potential of manganese catalysts for selective oxidations. This evolved in part from studies on the structure and function of Mn-based redox enzymes [12]. Following a brief summary of biomimetic manganese systems, oxidative transformations with H_2O_2 catalyzed by Mn-complexes are outlined.

Modern Oxidation Methods. Edited by Jan-Erling Bäckvall
Copyright © 2004 WILEY-VCH Verlag GmbH & Co. KGaA, Weinheim
ISBN: 3-527-30642-0

10.2
Biomimetic Manganese Oxidation Catalysis

Manganese can frequently be found in the catalytic redox center of several enzymes [12] including superoxide dismutase [13], catalase [14] and the oxygen evolving complex photosystem II (PS II) [15]. Superoxide (O_2^-), a harmful radical for living organisms, is the product of single electron reduction of oxygen [16]. Owing to the high toxicity it needs to be converted into less reactive species. Superoxide dismutases are metalloenzymes which catalyze the dismutation of the superoxide (O_2^-) to oxygen (O_2) and hydrogen peroxide (H_2O_2) [17]. The latter product can be degraded by catalase enzymes to water and oxygen (*vide supra*). Superoxide dismutase (SOD) enzymes can be classified into two major structural families: copper-zinc SOD and manganese or iron SOD [16, 18]. The active site of manganese SOD contains a mononuclear five-coordinate Mn^{III}-ion bound to three histidines, one aspartate residue and one water or hydroxide ligand. The mechanism of the catalytic conversion of superoxide into oxygen starts by binding of the superoxide radical anion to the Mn^{III}-monomer leading to the reduction to Mn^{II} and oxidation of superoxide into oxygen [13, 19]. Subsequently, the catalytic cycle is closed by binding of a second superoxide to the Mn^{II}-ion resulting in the oxidation of Mn^{II} and reduction of the superoxide anion to H_2O_2.

In photosystem II (PS II), located in the thylakoid membrane of chloroplasts in green plants, algae and a number of cyanobacteria, two water molecules are oxidized to dioxygen [15]. PS II consists of light harvesting pigments, a water oxidation center (WOC), and electron transfer components [15]. Based on detailed spectroscopic analyses it has been recognized that a tetranuclear Mn-cluster is the active catalyst for the oxygen evolution, and this structure has recently been confirmed by X-ray analysis of PS II [20]. However, the exact mechanism of the water oxidation has not been elucidated so far.

Catalases decompose H_2O_2 to water and oxygen and these manganese enzymes have been isolated from three different bacteria: *Lactobacillus plantarum* [21], *Thermus thermophilus* [22], and *Thermoleophilum album* [14]. X-ray crystallographic structure analysis [23] elucidated that these catalases contain a dinuclear manganese center. During the catalytic process the dinuclear manganese active site cycles between the Mn_2^{II}- and Mn_2^{III}-oxidation states [24]. EPR [25], NMR [26] and UV-Vis [26a] spectroscopic studies revealed that for the H_2O_2 disproportionation both Mn_2^{II}- and Mn_2^{III}-oxidation states are involved [27]. The proposed catalase mechanism is depicted in Scheme 10.1. Hydrogen peroxide decomposition is initiated by (a) the binding of H_2O_2 to the Mn^{III}-Mn^{III} dinuclear center, followed by (b) reduction to the Mn^{II}-Mn^{II} intermediate and concomitant oxidation of the peroxide to O_2 [27, 28]. Subsequent binding of a second molecule of H_2O_2 to the Mn^{II}-Mn^{II} species (c) effects the reduction of H_2O_2 to H_2O and results in the oxidation of the Mn^{II}-Mn^{II} species (d), the step that closes the catalytic cycle [13].

To gain insight into the mechanisms of these enzymes, a variety of Mn complexes that mimic the active site have been developed [28]. Dismukes and coworkers reported the first functional catalase model that exhibits high activity towards H_2O_2 decomposition; even after turnover numbers of 1000, no loss of activity towards H_2O_2 decomposition was observed [29]. The dinuclear Mn^{II}-complex is based on ligand 1

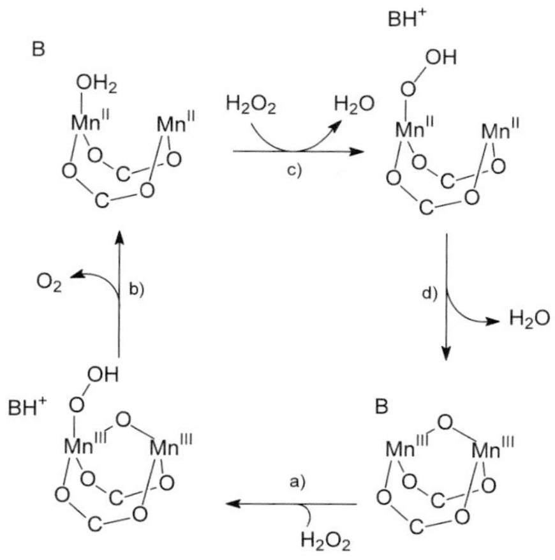

Scheme 10.1 Proposed mechanism for H_2O_2 decomposition by manganese catalase

(Figure 10.1). EPR and UV-Vis spectroscopic investigations revealed that under conditions of H_2O_2 decomposition, both Mn^{III}-Mn^{III} and Mn^{II}-Mn^{II} oxidation states are present, similar to those observed for the natural manganese catalase enzymes [28].

Sakiyama explored various dinuclear manganese complexes as catalase mimics derived from 2,6-bis{N-[(2-dimethylamino)ethyl]iminomethyl-4-methylphenolate} (**2**, Figure 10.1) and related ligands. Employing UV-Vis and MS techniques both mono- and di-Mn^{IV}-oxo intermediates could be detected [30]. Notably, the proposed mechanism (Scheme 10.2) is different from that reported for the manganese catalases and model compounds containing ligand **1** [30].

Manganese complexes of 1,4,7-triazacyclononane (tacn) or 1,4,7-trimethyl-1,4,7-triazacyclononane (**3**, tmtacn, Figure 10.1) were studied by Wieghardt as models for the oxygen evolving center of photosystem II as well as manganese catalase [31]. Turnover numbers of the H_2O_2 decomposition as high as 1300 are readily reached [31d]. More recently Krebs and Pecoraro used the tripodal bpia ligand (bpia = bis-(picolyl)(N-methylimidazol-2-yl)amine) as a Mn-catalase model system. Several Mn-complexes based on this ligand were found as structural mimics of the catalase enzyme. Remarkably, the catalytic activity was found to be within 2–3 orders of magnitude relative to the catalase enzyme [32]. Various manganese oxidation catalysts, which evolved from these systems, will be discussed in the following paragraphs.

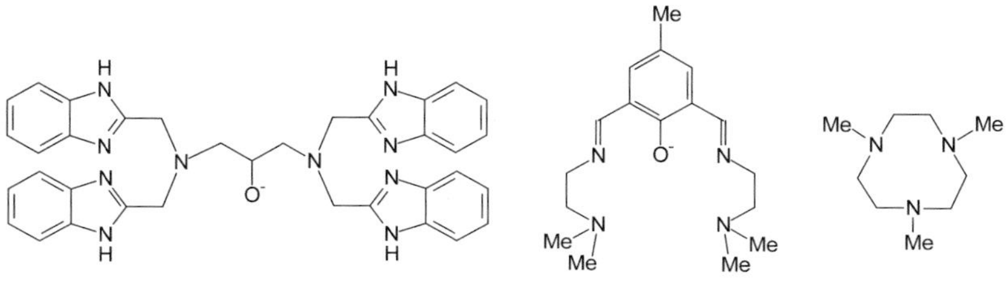

Fig. 10.1 Ligands studied in manganese catalase mimics

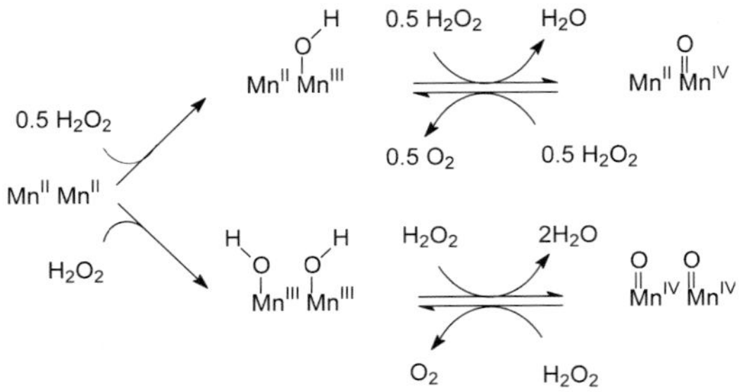

Scheme 10.2 Proposed mechanism of H_2O_2 decomposition catalyzed by Mn-complexes based on ligand 2 [30]

10.3
Bleaching Catalysis

Bleaching processes in the paper industry and bleaching of stains on textiles through the use of detergents have been studied intensively and the oldest bleaching procedures for laundry cleaning employ H_2O_2 and high temperatures [28]. Several catalysts have been investigated to attain lower bleaching temperatures of 40–60 °C or to achieve effective bleaching under ambient conditions [28, 33]. Manganese complexes based on 1,4,7-trimethyl-1,4,7-triazacyclononane, that is $[Mn_2O_3(tmtacn)_2](PF_6)_2$ **4** (Mn-tmtacn, Figure 10.2) were studied extensively by Unilever Research as bleach catalysts for stain removal at ambient temperatures [34]. Unfortunately, due to textile damage as a result of high oxidation activity, commercialization for laundry applications was ceased [34].

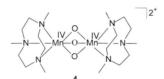

Fig. 10.2 Mn–tmtacn bleaching catalyst

10.4
Catalytic Epoxidation

Epoxides are an important and extremely versatile class of organic compounds and the development of new methods for the selective epoxidation of alkenes continues to be a major challenge [2, 9, 35]. The epoxidation of olefins can be achieved by applying a variety of oxidants including peroxycarboxylic acids [36], dioxiranes [37], alkylhydroperoxides [38], hypochlorite [39], iodosylbenzene [39], oxygen [40] and hydrogen peroxide [9, 35c, 38]. With a few exceptions, most of the oxidants have the disadvantage that in addition to the oxidized products, stoichiometric amounts of waste products are formed, which have to be separated from the often sensitive epoxides. The use of H_2O_2 in combination with Mn-complexes offers several advantages, including high reactivity of the catalytic system, although the oxidant is often partially destroyed due to catalase type activity typically associated with Mn-catalysis [28]. It should also be noted that unselective side reactions might occur after the homolytic cleavage of H_2O_2 leading to hydroxyl radicals [41].

10.4.1
Manganese Porphyrin Catalysts

Manganese porphyrins and several other metal porphyrin complexes, in particular Fe and Cr systems, have been studied intensively as catalysts in epoxidation reactions of alkenes [39, 40]. A variety of oxidants such as iodosylarenes, alkylhydroperoxides, peracids and hypochlorite in addition to H_2O_2 were employed [39, 40]. The early porphyrin-based catalysts often showed rapid deactivation, due to oxidative degradation of the ligand. Spectacular improvements in robustness and activity of catalysts for olefin epoxidation and hydroxylation of alkanes were obtained after the introduction of halogen substituents into the porphyrin ligands [42]. General disadvantages of manganese porphyrin-based epoxidation catalysts are the difficulty in preparing the ligands and the often tedious purification that is required.

Initial attempts to use H_2O_2 as oxidant for alkene epoxidation with porphyrin-based catalysts were unsuccessful due to dismutation of H_2O_2 into H_2O and O_2, leading to a fast depletion of the oxidant. Introduction of bulky groups on the porphyrin ligand allowed the use of aqueous H_2O_2, although only low conversions were obtained. Mansuy and coworkers demonstrated that the catalytic system could be improved greatly by performing the oxidation reaction in the presence of large amounts of imidazole [43, 44]. The role of imidazole is proposed to be two-fold: (a) acting as a stabilizing axial ligand and (b) to promote the formation of the $Mn^V=O$ intermediate (the actual epoxidizing species) by heterolysis of Mn^{III}–OOH. This catalytic system provides epoxides in yields up to 99%. The amount of axial ligand could be significantly reduced by the addition of a catalytic amount of carboxylic acid [45]. Under two-phase reaction conditions with a small amount of benzoic acid, the oxidation reaction was dramatically accelerated and high conversions in less than 10 min at 0 °C could be obtained (Scheme 10.3, Table 10.1) [45].

Carboxylic acids and nitrogen containing additives are considered to facilitate the heterolytic cleavage of the O–O bond in the manganese hydroperoxy intermediate, resulting in a catalytically active manganese(V)-oxo intermediate [46]. However, competing homolytic cleavage of the O–O bond leads to the formation of hydroxyl radicals and an unselective oxidation reaction, which is a serious problem when using H_2O_2 in numerous metal-catalyzed oxidations [40]. The proposed catalytic epoxidation cycle of manganese porphyrin **5** starts with the conversion into the well established Mn^V-oxo species (Scheme 10.4) [41a, 47]. Subsequently, the oxygen atom is transferred to the olefin *via* a concerted- (route **a**) or stepwise- (route **b**) pathway followed by release of the Mn^{III}-species and formation of the epoxide. In the stepwise route **b**, which involves a neutral carbon radical intermediate, rotation around the former double bond results in *cis/trans* isomerization leading to *trans*-epoxides when starting from *cis*-alkenes, as is observed experimentally [47].

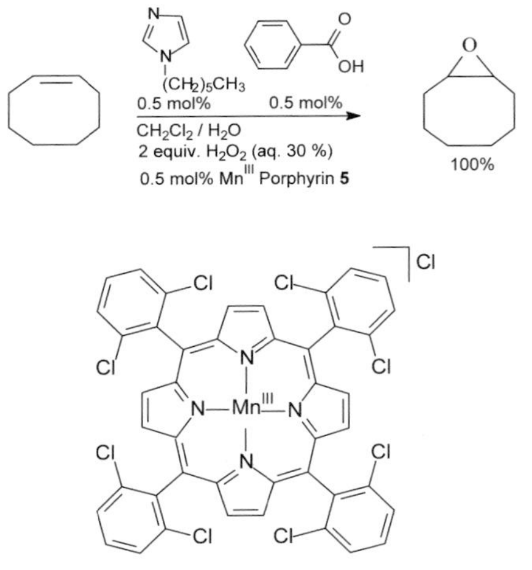

Scheme 10.3 Manganese porphyrin complex **5**; an effective catalyst for epoxidation reactions with H_2O_2 [45]

Tab. 10.1 Oxidation of olefins with Mn^{III} porphyrin complex **5** [45]

Substrate	Conversion (%)	Epoxide (%)	Reaction time (min)
Cyclooctene	100	100	10
Dodec-1-ene	96	92	15
α-Methylstyrene	100	100	7
cis-Stilbene	90	85 (cis)	20
trans-Stilbene	0	0	300
trans-4-Octene	75	54 (trans)	15

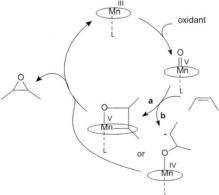

Scheme 10.4 The proposed catalytic epoxidation cycle: **a**, concerted pathway, **b**, stepwise pathway

10.4 Catalytic Epoxidation

Gradual improvement in the stereoselectivity of the oxidation of *cis*-stilbene was observed by increasing the number of β-halogen atoms at the porphyrin ligand, pointing to an enhanced preference for a concerted pathway. *trans*-Alkenes are poor substrates for these catalysts [45, 48].

Enhanced epoxidation rates were observed by using a modified Mn-porphyrin complex **6** in which the carboxylic acid and imidazole groups are both covalently linked to the ligand (Scheme 10.5) [49]. Employing 0.1 mol% of the Mn complex and 2 equiv. of H_2O_2, cyclooctene was converted in only 3 min to the corresponding epoxide with 100% conversion and selectivity. Similar results were obtained for alkenes such as α-methylstyrene, *p*-chlorostyrene, α-pinene and camphene with turnover numbers up to 1000. The proposed mechanism is similar to oxidation reactions with porphyrin-based catalysts in the presence of the external co-catalysts imidazole and carboxylic acid, with a prominent role for the pending carboxylic acid group in the peroxide heterolysis (Scheme 10.5) [49].

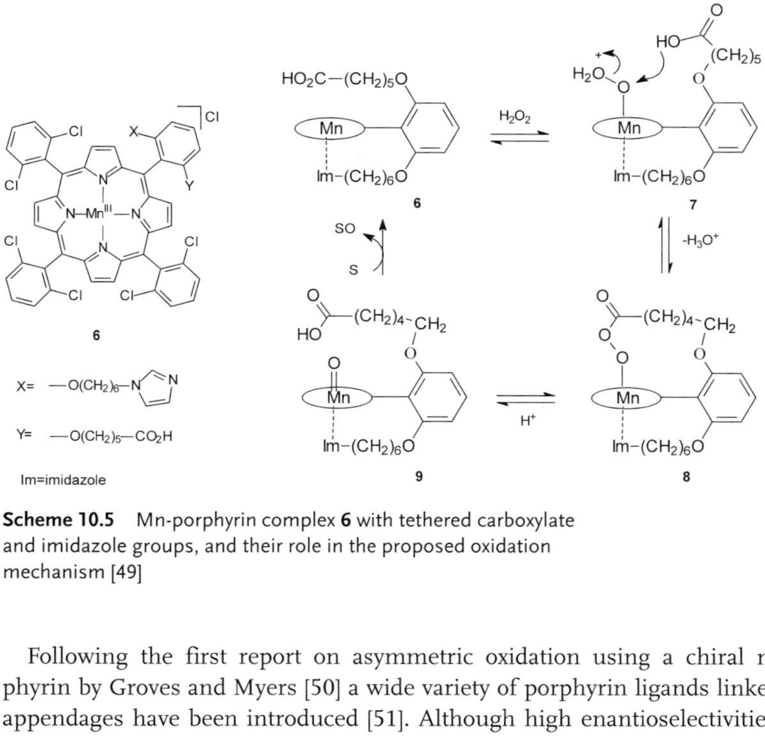

Scheme 10.5 Mn-porphyrin complex **6** with tethered carboxylate and imidazole groups, and their role in the proposed oxidation mechanism [49]

Following the first report on asymmetric oxidation using a chiral metalloporphyrin by Groves and Myers [50] a wide variety of porphyrin ligands linked to chiral appendages have been introduced [51]. Although high enantioselectivities were observed with iodosylbenzene as oxidant, the use of H_2O_2 only resulted in moderate ee so far [52]. For a discussion of the ingenious chiral ligand designs and the stereochemical issues involved the reader is referred to a number of excellent reviews [51, 53].

The immobilization of homogeneous Mn-porphyrin epoxidation catalysts on silica to achieve easy catalyst recovery has been realized through anchoring of the porphyrin ligand A [54] or the axial imidazole ligand B (Figure 10.3) [55]. The advan-

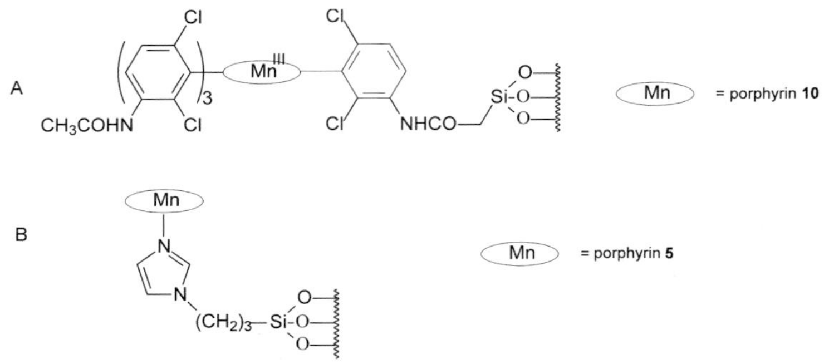

Fig. 10.3 Immobilized Mn-porphyrin epoxidation catalysts

tages of the supported catalysts are lost to some extent due to reduced epoxidation activity compared with the homogeneous system.

10.4.2
Manganese–salen Catalysts

Following the seminal report by Kochi and coworkers on the use of Mn-salen complexes as epoxidation catalysts [56], the groups of Jacobsen [57a] and Katsuki [57b] described a major breakthrough in Mn-catalyzed olefin epoxidation by the introduction of a chiral diamine functionality in the salen ligand (Figure 10.4).

Compared with chiral manganese porphyrin complexes [50], the use of the Mn-salen catalysts results generally in ee's higher than 90% and yields exceeding 80% [39, 58]. A wide range of oxidants including hypochlorite [58], iodosylbenzene [58], or m-chloroperbenzoic acid (m-CPBA) can be applied [59]. Excellent ee's are observed for epoxidation reactions of cis-disubstituted alkenes and trisubstituted alkenes catalyzed by the Mn-salen complexes **11** and **12**, employing iodosylbenzene as the oxidant. In sharp contrast, the epoxidation of trans-olefins showed moderate selectiv-

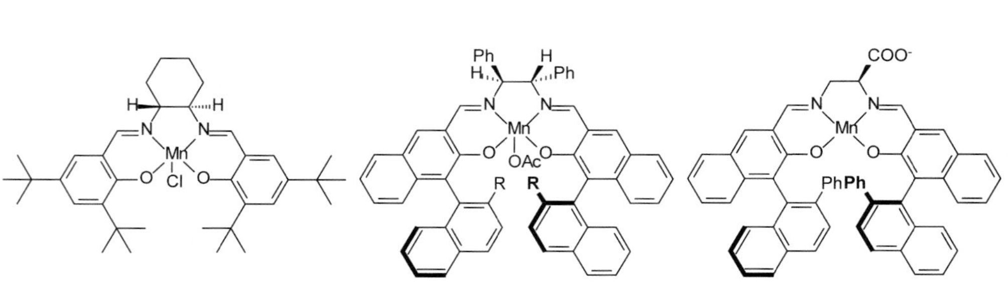

Fig. 10.4 Chiral manganese complexes introduced by Jacobsen (**11**) and Katsuki (**12, 13**) for asymmetric epoxidation of unfunctionalized olefins

ities (ee <60%), however, the enantioselectivities could be improved by the introduction of additional chiral groups at the 3,3'-positions of the phenolate part of the ligand. For the conversion of *trans*-stilbene ee up to 80% have been reported using these modified salen ligands [58].

The oxidizing species in this catalytic epoxidation is proposed to be an Mn^V-oxo intermediate [59d, e], similar to that in the Mn-porphyrin catalyzed epoxidation, as was confirmed by electrospray ionization mass spectrometry [60]. An extensive discussion of the stereoselectivity, mechanism and scope of this asymmetric epoxidation [58] using the preferred oxidant iodosylbenzene is beyond the scope of this chapter, although several mechanistic features apply to the epoxidation with H_2O_2. Despite the fact that there is consensus on the nature of the active species (i.e., an Mn^V-oxo intermediate), some controversy remains as to the exact way the enantioselection takes place. Three key issues can be distinguished: (1) the catalyst structure (i.e., if the salen ligand is planar, bent or twisted), (2) the trajectory of approach of the reacting alkene, and (3) the mode of oxygen transfer from the salen $Mn^V=O$ to the alkene (involving a concerted pathway, a stepwise radical pathway or a metallaoxetane intermediate) [39, 58, 61]. Cumulative experimental evidence indicates that, in addition to the catalyst structure being either planar or twisted, the substituents at the C_2-symmetric diimine bridge and bulky substituents at the 3,3'-positions play an important role in governing the trajectory of the side-on approach of the olefin, and as a consequence the asymmetric induction. With the five-membered chelate ring, comprising the ethylenediamine and the Mn^V-ion, being non-planar, the approach of the olefin over the downwardly bent benzene ring of the salen ligand along one of the Mn–N bonds can be envisaged (Scheme 10.6). The largest substituent of the alkene is then pointing away from the 3,3'-substituents and this governs the stereochemical outcome of the reaction between the Mn^V-oxo intermediate and the alkene [58].

Although high ee's are obtained for a wide range of substrates, the stability of the Mn-salen complexes is often a severe problem and turnover numbers are usually in the range of 40–200. More recently, a robust salen catalyst was introduced by Katsuki [62] based on ligand **13** with a carboxylic acid functionality attached to the diamine bridge (Figure 10.4). With this new catalyst, 2,2-dimethylchromene was converted into the corresponding epoxide in 99% ee with iodosylbenzene as oxidant. Turnover numbers as high as 9200 after a 6 h reaction time were reached but results with H_2O_2 as terminal oxidant have not been reported yet [62].

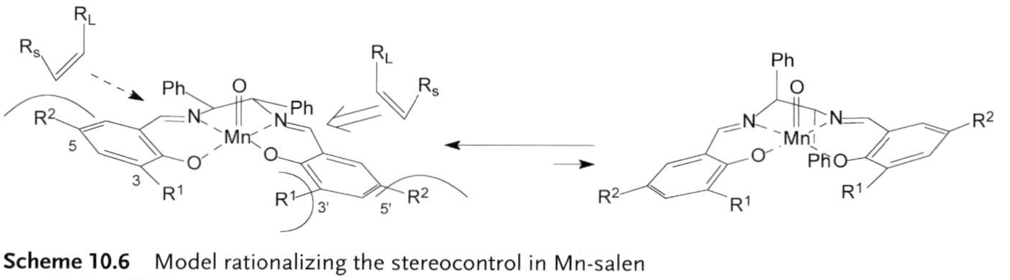

Scheme 10.6 Model rationalizing the stereocontrol in Mn-salen epoxidation [58c]

While iodosylbenzene and hypochlorite are the most common oxidants, considerable effort has been devoted to the use of hydrogen peroxide in epoxidations with Mn-salen catalysts. Promising results have been reported for certain substrates (*vide infra*), although low turnover numbers (generally up to 20–50) were obtained with H_2O_2 as the terminal oxidant for a limited range of substrates. Employing H_2O_2 as oxidant, the manganese-salen systems were found to be only catalytically active in the presence of additives such as imidazole, or derivatives thereof, or carboxylates [41b, 63–65]. The role of these additives is considered to prevent O–O bond homolysis leading to radical pathways and destruction of the catalyst, as has been discussed for the Mn-porphyrin based catalysts (*vide supra*).

Berkessel designed a chiral dihydrosalen ligand with a covalently attached imidazole group. With this new salen complex (**14**) (Figure 10.5), 1,2-dihydronaphthalene was converted into the corresponding epoxide in 72% yield and with moderate ee (up to 64%; Table 10.2) using a dilute (1%) aqueous solution of H_2O_2 as oxidant. An important feature of this system is that epoxidation reactions can be performed without the need for further additives [41b].

Using Mn-salen **15** together with *N*-methylimidazole as an axial ligand, Katsuki obtained up to 96% ee in the epoxidation of substituted chromene with 30% aqueous H_2O_2 as oxidant, although the yield of the epoxide was only 17%. With an excess of H_2O_2 (10 equiv.) and an enhanced concentration of the reactants, the yield was increased to 98%, with only a slight drop in ee to 95% (Table 10.2) [64]. It should be noted that, despite these excellent results, only a very limited number of substrates were tested (ee ranging from 88 to 98%).

Pietikäinen reported that in the presence of carboxylate salts, 30% aqueous H_2O_2 could be used as an oxidant for the asymmetric epoxidation with chiral Mn-salen catalysts (ee ranging from 64 to 96%, Table 10.2) [65b]. Furthermore, it was shown that the use of *in situ* prepared peroxycarboxylic acids, from the corresponding anhydrides and anhydrous H_2O_2, gives improved enantioselectivity in the epoxidation of alkenes if compared with the use of aqueous H_2O_2 in the presence of a carboxylate salt [66]. In particular, good results are obtained with maleic anhydride and UHP (urea-H_2O_2) in combination with the Mn^{III}-salen complex **16a** and NMO as additive. Although the number of substrates tested is again limited, in general 3–5% higher enantioselectivities were obtained and the reaction time was shortened under these conditions. The use of urea-H_2O_2 for Mn^{III}-salen catalyzed epoxidation of alkenes

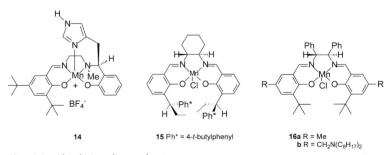

Fig. 10.5 Chiral Mn-salen catalysts

Tab. 10.2 Epoxidation of olefins by Mn^{III}-salen complexes employing H_2O_2 as oxidant

Substrate	Mn-salen	Oxidant (equiv.)	Epoxide yield (%)	ee (%)	Ref.
naphthalene-fused alkene	14	1% H_2O_2 (10)	72	64	[41b]
naphthalene-fused alkene	16a	30% H_2O_2 (1.5)	74	69	[66]
naphthalene-fused alkene	16a	urea.H_2O_2/maleic anhydride (1.5)	70	73	[66]
Ph, Ph-alkene	11	30% H_2O_2 (4)	84	96	[65b]
AcNH, O_2N-chromene	15	30% H_2O_2 (1)	17	96	[64]
AcNH, O_2N-chromene	15	30% H_2O_2 (10)	98	95	[64]
O_2N-chromene	16b	urea.H_2O_2 (2)	>99	>99	[67b]

has also been described by Kureshy [67]. Moderate to excellent ee were reported for chromene derivatives (55–99%) using ammonium acetate as additive, although for styrene only 39% ee was obtained.

10.4.3
Mn-1,4,7-triazacyclononane Catalysts

The tridentate macrocycle ligand 1,4,7-triazacyclononane (tacn) and in particular complexes of 1,4,7-trimethyl-1,4,7-triazacyclononane (3, Figure 10.1) such as $[Mn_2O_3(tmtacn)_2](PF_6)_2$ (4, Figure 10.2) have been studied extensively in oxidation chemistry [28, 31]. In combination with H_2O_2 the dinuclear manganese complex 4 is a highly active and selective epoxidation catalyst [33, 68]. High turnover numbers (>400) were obtained using styrene and vinyl benzoic acid as substrates. In methanol-carbonate buffer solutions, conversions of 99% were reached without notable catalyst degradation [68]. The scope of the $[Mn_2O_3(tmtacn)_2](PF_6)_2$ complex for epoxidation reactions was considerably enlarged by De Vos and Bein by performing the reactions in acetone as the solvent [69]. Although the procedure is not suitable for the epoxidation of electron-deficient olefins, high turnover numbers of up to 1000 have been reported for the conversion of various alkenes into the corresponding epoxides by an *in situ* prepared Mn-tmtacn complex using $MnSO_4$ as the manganese source (Table 10.3). For styrene, complete conversion with 98% epoxide selectivity is reached at 0 °C in acetone with 2 equiv. of H_2O_2 as oxidant (Scheme 10.7) [69].

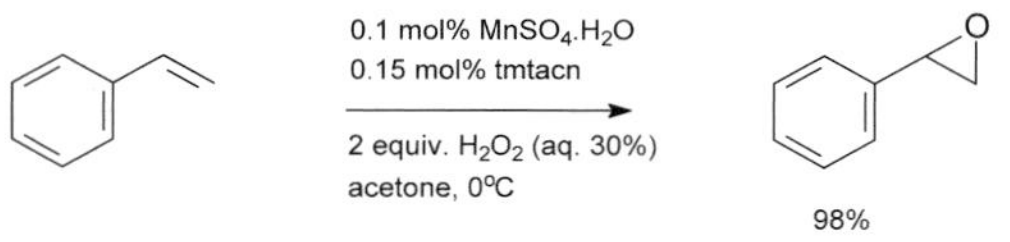

Scheme 10.7 Oxidation of styrene catalyzed by the manganese complex formed *in situ* in acetone [69]

Tab. 10.3 Oxidation of selected olefins with Mn-tmtacn

Substrate	Turnover number[a]	Selectivity (%)[b]
Cyclohexene	290	87
Styrene	1000	>98
cis-2-Hexene	540	>98
1-Hexene	270	>98
trans-β-Methylstyrene	850	90

[a] Turnover number in mole product/mole catalyst (after 3 h). [b] Selectivity: moles of epoxide/moles of converted substrate.

It has to be emphasized that no cleavage of the double bond is observed although some *cis/trans* isomerization occurs during oxidation of *cis*-alkenes. Furthermore, with alkenes such as cyclohexene only minor amounts of allylic oxidation products are found.

Apart from high turnover, there is a need to develop catalytic systems which employ H_2O_2 very efficiently. As many Mn- or Fe-catalysts are known to induce decomposition of H_2O_2 [28], often a large excess of H_2O_2 is needed to reach full conversion. A reduction in catalase activity is indeed possible by performing the oxidation reactions in acetone at sub-ambient temperatures. In contrast, the use of other solvents results in severe decomposition of the oxidant. The oxidation characteristics of the Mn-tmtacn complex in acetone were explained by a mechanism involving the nucleophilic addition of H_2O_2 to acetone, resulting in the formation of 2-hydroperoxy-2-hydroxypropane (hhpp, **17**) as depicted in Scheme 10.8 [70]. Most probably, due to the reduction of the H_2O_2 concentration in acetone, the epoxidation reaction is favored over oxidant decomposition. It is proposed that at low temperature hhpp is serving as an oxidant reservoir, which gradually releases H_2O_2 maintaining a low oxidant concentration [69b], although direct involvement of **17** in the epoxidation pathway cannot be excluded.

Unfortunately the combination of acetone and H_2O_2 can also result in the formation of explosive cyclic peroxides and therefore this solvent is not acceptable for in-

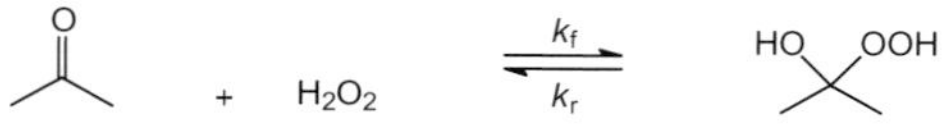

Scheme 10.8 Reaction of acetone with H_2O_2

dustrial applications involving H_2O_2. Hydrogen peroxide decomposition by Mn-tmtacn complexes in CH_3CN as the solvent can also effectively be suppressed by addition of oxalate [71] or ascorbic acid [72] as co-catalysts. Greatly enhanced epoxidation activity of the *in situ* prepared Mn-tmtacn complex is observed after addition of a catalytic amount of oxalate buffer [71]. In general, full conversion is reached with less than 1 mol% of the catalysts in 1 h. Besides oxalic acid, several other bi- or polydentate ligands such as diketones or diacids in combination with Mn-tmtacn and H_2O_2 favor olefin epoxidation over oxidant decomposition [71]. Employing this mixed catalytic system, allylic olefins (e.g., allyl acetate) and particularly terminal olefins (e.g., 1-hexene, see Scheme 10.9 and Table 10.4) are converted into the corresponding epoxides in high yields with only 1.5 equiv. of H_2O_2 [71].

In addition to the efficient use of oxidant, the isomerization of *cis*- and *trans*-alkenes is greatly reduced in the presence of the oxalate buffer. The epoxidation of 2-hexene was found to be completely stereospecific (>98%) using only 1.5 equiv. of the oxidant. Compared with the method using acetone as solvent [69], in which the Mn-tmtacn catalyst produced as much as 34% *trans*-epoxide starting from *cis*-2-hexene, the system based on the oxalate buffer represents a significant improvement.

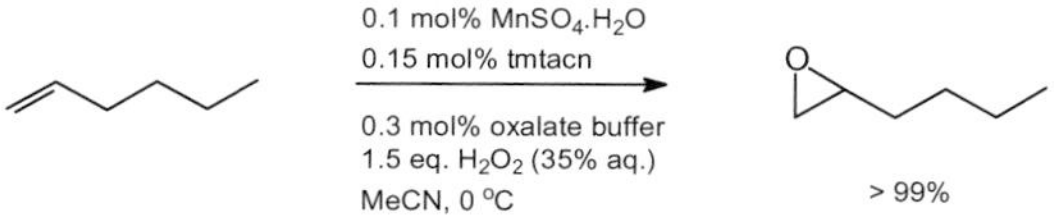

Scheme 10.9 Selective epoxidation of 1-hexene by Mn-tmtacn using H_2O_2 in the presence of an oxalate buffer

Tab. 10.4 Representative examples of epoxidation of terminal and deactivated olefins with the Mn-tmtacn/oxalate system [71]

Substrate	Epoxide yield (%)	Epoxide selectivity (%)[a]
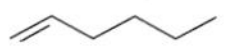	>99	>99
	35	>98 (*trans*)
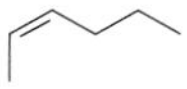	72	>98 (*cis*)
	83	92
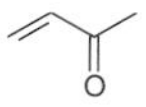	66	96
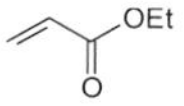	55	94
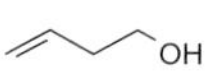	88	91
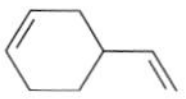	89	91 (diepoxide) 8 (mono)

[a] Selectivity: moles of epoxide/moles of converted substrate.

Furthermore, various functional groups, for example, CH_2OH, CH_2OR, COR, CO_2R, including electron withdrawing groups, are tolerated. Despite the high reactivity of the catalytic system, epoxidation is preferred over alcohol oxidation in the case of olefins bearing alcohol moieties. This procedure is also suitable for the oxidation of dienes resulting in bis-epoxidation. For example, 4-vinylcyclohexene gives the corresponding diepoxide.

Although the precise role of the oxalate co-catalyst is not known to date, the formation of a Mn-tmtacn/oxalate species (**18**, Figure 10.6) [73], related to known Cu^{2+}- and Cr^{3+}-structures [74], was proposed. It has been suggested that the addition of a catalytic amount of the bidentate oxalate impedes the formation of μ-peroxo-bridged dimers **19** and as a result the catalase type decomposition of H_2O_2, often associated with dinuclear complexes, is suppressed [28].

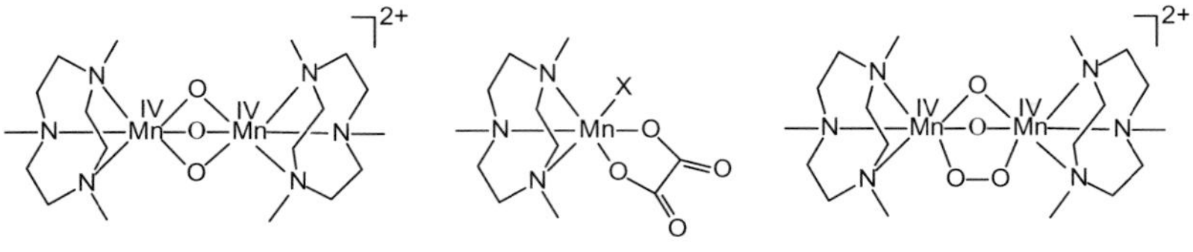

Fig. 10.6 Mn-tmtacn and proposed structures for Mn-tmtacn/oxalate oxidation catalyst (X = activated "O" to be transferred [73]

Additives such as ascorbic acid (**21**, Scheme 10.10) or squaric acid result in a further improvement of the epoxidation with H_2O_2, catalyzed by the Mn-tmtacn complex. A limited number of substrates have been studied so far but nearly quantitative yields of epoxides with retention of the olefin configuration were found employing catalyst loadings of only 0.03 mol%. Electron-deficient methyl acrylate and the terminal olefin 1-octene were also converted into the corresponding epoxides with yields of 97% and 83%, respectively. A typical oxidation is shown in Scheme 10.10. The exact role of ascorbic acid as co-catalyst remains unclear, but the H_2O_2 efficiency with this Mn catalytic system is one of the highest reported so far.

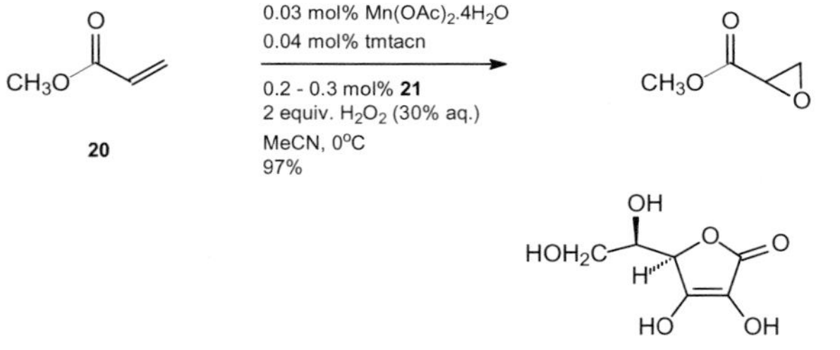

Scheme 10.10 Epoxidation in the presence of ascorbic acid (**21**)

Despite the fact that the additive **21** is chiral, enantioselectivity in the epoxidation is not observed [72]. However, enantiomerically enriched epoxides have been obtained with manganese complexes based on chiral analogues of the tmtacn ligand (Figure 10.7) [75, 76]. Beller and Bolm reported the first asymmetric epoxidation reaction catalyzed by *in situ* prepared Mn catalysts from N-substituted chiral tacn ligands [75 a]. The chirality was introduced via alkylation of the secondary amine moieties to generate the C_3-symmetric ligands depicted in Figure 10.7.

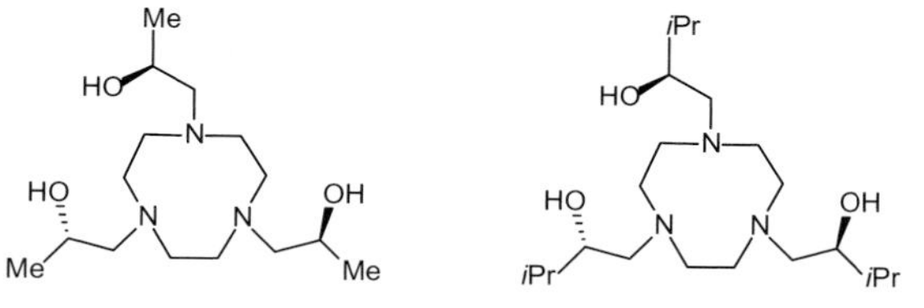

Fig. 10.7 C_3-symmetric chiral ligands

Using H_2O_2 as the oxidant and the Mn complex based on ligand **22**, styrene was converted into the corresponding epoxide with an ee of 43%, although only 15% yield was achieved after 5 h. Using longer reaction times, higher temperatures and higher catalyst loadings, the yield was increased but simultaneously the ee decreased [75 a]. Employing the sterically more demanding ligand **23**, ee's in the range of 13–38% were observed for styrene and chromene. Higher ee's were achieved with cis-β-methylstyrene as the substrate. The *trans*-epoxide was found as the major product with 55% ee whereas the *cis*-epoxide was produced as the minor product with an ee of 13% (Scheme 10.11).

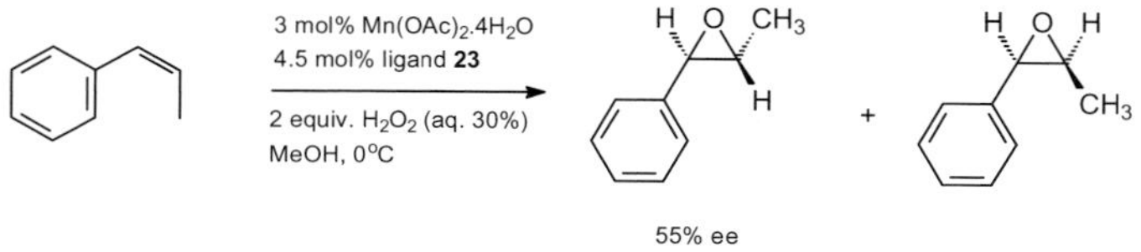

Scheme 10.11 Asymmetric epoxidation with the Mn complex of ligand **23**

An enantiopure C_3-symmetric trispyrrolidine-1,4,7-triazacyclononane ligand **24** was recently introduced (Figure 10.8) [75 d]. The tacn derivative was obtained by reduction of an L-proline derived cyclotripeptide and the corresponding dinuclear manganese complex was applied in the catalytic enantioselective epoxidation of vinylarenes with H_2O_2 as the oxidant. For the epoxidation of styrene, 3-nitrostyrene and 4-chlorostyrene, excellent conversions (up to 88%) and ee up to 30% were found [75 d].

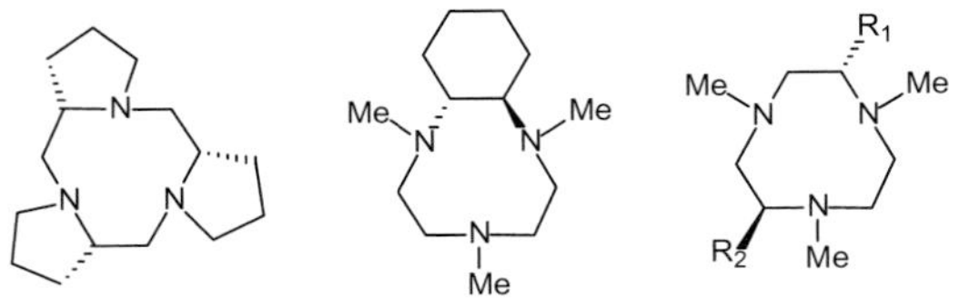

Fig. 10.8 Chiral tmtacn ligands

$R_1 = R_2 = Me$
$R_1 = R_2 = {}^iPr$

Besides the chiral C_3- and C_1-symmetric ligands, C_2-symmetric tacn analogues have been used (Figure 10.8) [75]. So far modest enantioselectivity has been obtained but the potential of these tmtacn analogues by further fine-tuning of the chiral ligand structure is evident.

Various successful attempts to improve the catalyst selectivity have been made by encapsulation of the Mn-tmtacn complex in zeolites [77]. Immobilization of the triazacyclononane ligand on an inorganic support resulted in a new class of heterogeneous manganese catalysts with increased epoxidation selectivity [78]. The conversions were usually lower than with the homogenous catalysts [78].

Little is known about the mechanism or the exact nature of the active intermediates in the oxidations with Mn-tmtacn. High-valent manganese, mono- or dinuclear manganese-oxo species as well as radical intermediates may be involved. During the oxidation reactions an induction period is often observed, indicating that the original $[Mn_2O_3(tmtacn)_2](PF_6)_2$ complex has first to be converted into a catalytically active species. It was reported that the catalytic activity of Mn-tmtacn was significantly increased when it was pre-treated with an excess of H_2O_2 prior to the addition of the substrate (in the case of benzyl alcohol oxidation) [79]. From the 16-line spectrum obtained by electron paramagnetic resonance spectroscopy (EPR) measurements it was inferred that the Mn^{IV}-Mn^{IV} dimer was instantaneously reduced by H_2O_2 to a dinuclear Mn^{III}-Mn^{IV} mixed-valent species in acetone. This mixed-valent species gradually changes to a Mn^{II}-species. EPR studies of the catalysts under comparable catalytic oxidation conditions using alkenes as substrates instead of alcohols showed again the mixed-valence Mn^{III}-Mn^{IV} dimer [33a, 69]. Based on EPR data similar manganese species were identified during related phenol oxidation experiments [80]. Barton proposed the formation of an Mn^V=O intermediate during the oxidation of 2,6-di-*tert*-butylphenol with Mn-tmtacn and H_2O_2 [81]. In electrospray mass spectrometry (ES-MS) experiments the mononuclear Mn^V=O species could indeed be assigned [82]. This species was also generated in oxidation reactions using a mononuclear Mn^{IV}-complex [68a] and from an *in situ* prepared Mn^{II}-complex using $MnSO_4$ and free tmtacn ligand. Further studies are necessary to firmly establish the actual catalyst.

10.4.4
Miscellaneous Catalysts

A drawback associated with the 1,4,7-trimethyl-1,4,7-triazacyclononane based catalyst is the often tedious procedure to achieve modifications of the ligand structure [83]. Furthermore, the sensitivity of the corresponding manganese complexes to changes in the original tmtacn structure often leads to completely inactive complexes. In view of the excellent catalytic activity of the Mn-tmtacn systems, a major challenge is therefore the design of novel dinucleating ligands, featuring the three N-donor set (as for the tmtacn ligand) for each manganese site [84], retaining the high oxidation activity.

Recently, high epoxidation activity was found for manganese complexes based on the dinucleating ligand N,N,N',N'-tetrakis(2-pyridylmethyl)-1,3-propanediamine (tptn) [85]. The ligand contains a three-carbon spacer between the three N-donor sets. This type of ligand is readily accessible and easy modification of the ligand structure is achieved. Complexes of this type have also been reported as mimics for PS II [86].

The complex **27** [$Mn_2O(OAc)_2$tptn] is able to catalyze the oxidation of various alkenes including styrene, cyclohexene and *trans*-2-octene to the corresponding epoxides in good yields and turnovers of up to 870 (Scheme 10.12). In sharp contrast, complex **28** (Scheme 10.12) based on tpen, featuring a two-carbon spacer between the three N-donor sets in the ligand, was not reactive in epoxidation reactions [85].

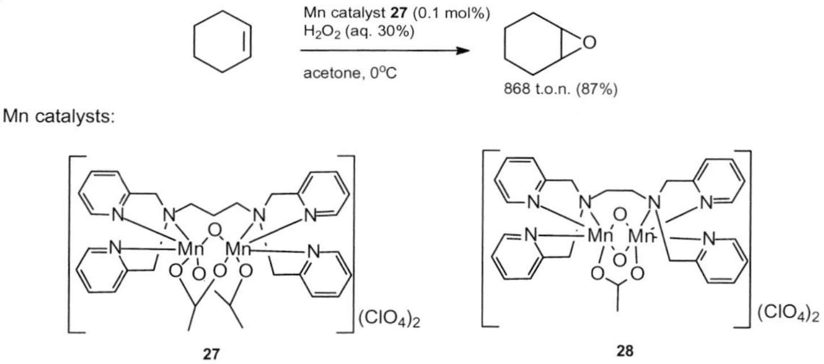

Scheme 10.12 Epoxidation with Mn-tptn catalyst and structures of manganese complexes

High selectivity was observed in the epoxidation reaction of cyclic alkenes (especially cyclohexene) with the important feature that apart from the epoxides no allylic oxidation products were found. Excellent results are also found for internal alkenes, such as, *trans*-2-octene and *trans*-4-octene, whereas terminal linear alkenes give slightly lower yields. The oxidation of *cis*-β-methylstyrene with H_2O_2 in the presence of $Mn_2O(OAc)_2$tptn catalyst **27** also provides, in addition to the corresponding *cis*-epoxide, a considerable amount of *trans*-epoxide. Cis/trans isomerization has frequently been observed in mechanistic studies using porphyrin and manganese-salen catalysts and is usually attributed to the formation of a radical intermediate (**a**, Scheme 10.13), with a lifetime sufficient for internal rotation before ring closure *via* reaction path B, providing the thermodynamically more stable *trans*-epoxide (**b**) [86]. In case of a fast collapse of the radical intermediate (*via* reaction path A), retention of configuration will be observed. The Mn-tptn based catalyst provides a viable alternative to the Mn-tmtacn and Mn-salen systems with high activity for epoxidation and the distinct advantage that ligand variation for further catalyst fine-tuning is readily accomplished.

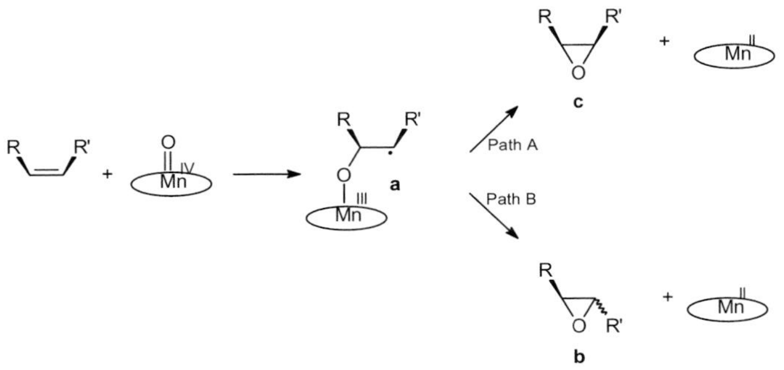

Scheme 10.13 Radical pathways to epoxides

A remarkably simple and effective Mn-based epoxidation system, using 1.0–0.1 mol% of $MnSO_4$, no ligands and 30% aqueous H_2O_2 as the oxidant in the presence of bicarbonate, was introduced by Burgess [9, 87]. Bicarbonate and H_2O_2 form the actual oxidant peroxy monocarbonate (Scheme 10.14), which is proposed to react with the Mn-ion to generate the active epoxidation complex, as was supported by EPR studies [87, 88].

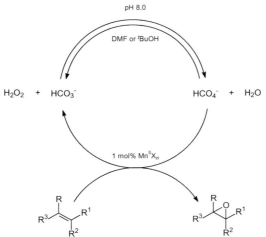

Scheme 10.14 $MnSO_4$ catalyzed epoxidation with bicarbonate/hydrogen peroxide [9]

A variety of cyclic alkenes and aryl- and trialkyl-substituted alkenes are converted into the epoxides in high yields using 10 equiv. of H_2O_2. Monoalkyl alkenes were unreactive with this system. A variety of additives were tested to increase the H_2O_2 efficiency by enhancing the activity for epoxidation and suppressing H_2O_2 decomposition. The use of 6 mol% of sodium acetate in ^tBuOH or 4 mol% of salicylic acid in DMF as the solvent resulted in an improved epoxidation system with higher epoxide yields, decreased reaction times and lower amounts of H_2O_2 (5 equiv.) required (Table 10.5) [87].

This ligand free epoxidation system has attractive features in particular in the context of the development of green oxidation procedures [3]. The Mn-salt/bicarbonate

Tab. 10.5 Epoxidation of alkenes using $MnSO_4$/salicylic acid catalyst [87b]

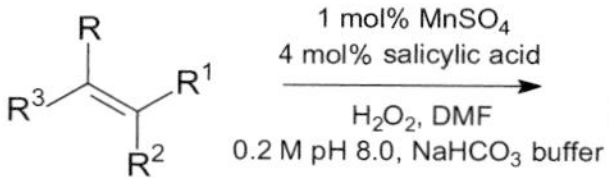

Reaction conditions: 1 mol% $MnSO_4$, 4 mol% salicylic acid, H_2O_2, DMF, 0.2 M pH 8.0, $NaHCO_3$ buffer

Alkene	Epoxide	Equiv. H_2O_2	Yield (%)
 cyclohexene	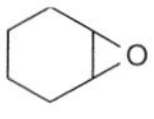 cyclohexene oxide	2.8	96
 2,3-dimethyl-2-butene type (trialkyl cyclohexene)	 corresponding epoxide	5	89
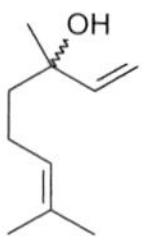 homoallyl alcohol	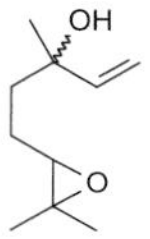 epoxy alcohol	5	91
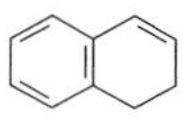 1,2-dihydronaphthalene	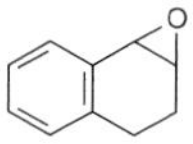 1,2-epoxy-1,2,3,4-tetrahydronaphthalene	5	97
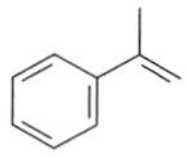 α-methylstyrene	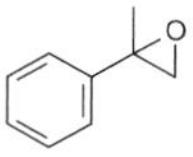 α-methylstyrene oxide	5	95
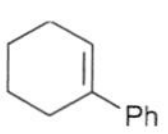 1-phenylcyclohexene (Ph)	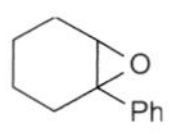 1-phenylcyclohexene oxide (Ph)	5	95
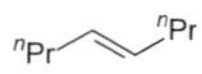 ^nPr-CH=CH-^nPr (trans)	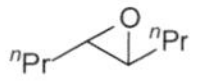 ^nPr-epoxide-^nPr	25	75
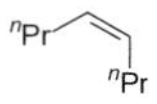 ^nPr-CH=CH-^nPr (cis)	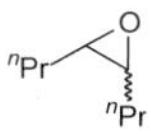 ^nPr-epoxide-^nPr	25	75 [a]

[a] Approximately 1:1 cis/trans mixture.

system is also catalytically active in ionic liquids. Epoxidation of various alkenes with 30% aqueous H_2O_2 can be accomplished with a catalytic amount of $MnSO_4$ in combination with TMAHC (tetramethylammonium hydrogen carbonate) in the ionic liquid [bmim][BF_4] (1-butyl-3-methylimidazolium tetrafluoroborate). Moderate to excellent yields are obtained for internal alkenes and the ionic liquid can be reused at least 10 times when fresh amounts of the Mn-salt and bicarbonate are added [89].

10.5
cis-Dihydroxylation

Dihydroxylation is an important synthetic transformation and several reagents can be used for the addition of two hydroxyl groups to an alkene. Both OsO_4 and alkaline $KMnO_4$ are suitable for cis-dihydroxylation [1, 11], but the catalytic version of the OsO_4 method using O_2 [90], H_2O_2 [91], or other oxidants [92], is the method of choice for this transformation [93]. The introduction of the highly enantioselective OsO_4 catalyzed dihydroxylation by Sharpless and the extensive use in synthetic chemistry in recent years has provided ample demonstration of the key role of this oxidation reaction [94]. The toxicity of OsO_4 and the stoichiometric nature of the $KMnO_4$ dihydroxylation, which usually provides only modest yields of diols, are strong incentives for the development of catalytic Mn- or Fe-based dihydroxylation reactions with H_2O_2 [95]. The high reactivity associated with Mn-systems is an important issue as over-oxidation of the diol product is frequently observed.

During alkene oxidation with a new heterogenized Mn-tmtacn system, De Vos and Jacobs [73] found that substantial amounts of cis-diol are formed in addition to the expected epoxide. The heterogenization procedure of the tacn ligand started with the conversion of dimethyl tacn (dmtacn, **29**, Figure 10.9) into the silylated compound **30** with 3-(glycidyloxy)propyltrimethoxysilane followed by immobilization on an SiO_2 surface and subsequent metalation of the new heterogenized ligand with $MnSO_4\cdot H_2O$ [73].

In oxidation reactions with **31**, H_2O_2 as the oxidant and CH_3CN as the solvent, alkenes, were converted into the corresponding cis-diols. The catalyst activity with respect to cis-diol formation is still modest (10–60 mol cis-diol per mol Mn) and epox-

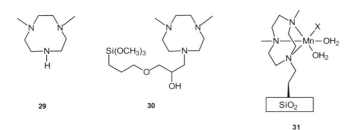

Fig. 10.9 Structures of dmtacn (**29**), heterogenized ligand **30** and the proposed active complex **31** (X = activated "O" to be transferred) [73]

10.5 cis-Dihydroxylation

ides are the major products. In the oxidation of internal alkenes, as for 2-hexene, only a slight loss of configuration was found both for epoxide and *cis*-diol. Control experiments with dmtacn **29** showed severe peroxide decomposition and no oxidation products were obtained. A sufficiently long-lived mononuclear complex **31** was postulated as the active species for both epoxidation and *cis*-dihydroxylation. This complex contains *cis*-coordination sites for labile ligands (e.g., H_2O and X), and both oxygen atoms from H_2O and X (the activated oxygen) are proposed to be transferred to the olefin to produce the *cis*-diol [73].

Recently, greatly enhanced *cis*-dihydroxylation activity was observed using $[Mn_2O_3(tmtacn)_2](PF_6)_2$ **4** (Figure 10.2) in the combination with glyoxylic acid methylester methyl hemiacetal (gmha **32**, Scheme 10.15) [96].

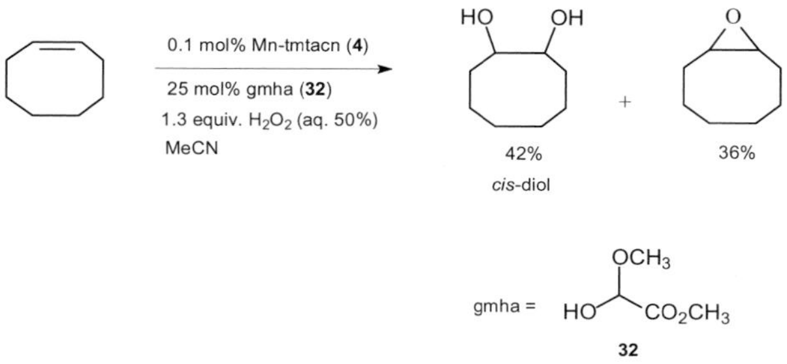

Scheme 10.15 Improved dihydroxylation in the presence of glyoxylic acid methylester methyl hemiacetals [96]

This mixed Mn-tmtacn/activated carbonyl system resulted in a highly active and H_2O_2 efficient catalyst for the epoxidation of olefins as well as the first homogeneous Mn-based catalytic *cis*-dihydroxylation system with H_2O_2. Catalytic oxidations were performed by slow addition of aqueous, 50% H_2O_2 (1.3 equiv. with respect to the substrate) to a mixture of alkene, Mn-tmtacn catalyst (0.1 mol%), and gmha (25 mol%) in CH_3CN at 0 °C. Under these reaction conditions high conversions are reached whereas only 30% excess of oxidant with respect to substrate is needed (Table 10.6). The H_2O_2 efficiency is dramatically improved compared with previous Mn based systems [85]. In most cases the conversions were also significantly higher than those obtained with oxalate as co-catalyst using 1.3 equiv. of H_2O_2. Substantial amounts of *cis*-diols are formed and the epoxide/*cis*-diol ratio depends strongly on the alkene structure. The highest amount of *cis*-diol was found for cyclooctene (Scheme 10.15), which afforded the *cis*-diol as the main product (42%, 420 TON). Minor amounts of 2-hydroxycyclooctanone were observed due to overoxidation of the diol. The ring size of cycloalkenes has a profound influence on the epoxide/*cis*-diol ratio. For cyclic olefins, almost no *trans*-diol could be detected (ratio *cis*-diol/*trans*-diol >99.5/0.5). *Cis*-diol formation is also observed for aliphatic acyclic alkenes. Yields of diol are significantly lower for *trans*-2-hexene than from *cis*-2-hexene, but the epoxide/*cis*-diol ratio was similar for both substrates. The aryl-substituted alkenes yield almost exclusively epoxide under these conditions. No diols

Tab. 10.6 cis-Dihydroxylation and epoxidation of selected olefins by H_2O_2 with Mn-tmtacn/gmha catalyst [96]

Substrate	Conversion (%)	Product	Turnover number (TON)[a]
Cyclopentene	97	epoxide	610
		cis-diol	260
		cyclopentenone	80
Cyclohexene	88	epoxide	590
		cis-diol	90
		2-cyclohexenone	80
Cyclooctene	90	epoxide	360
		cis-diol	420
		2-HO-cyclooctanone	220
Norbornylene	95	exo epoxide	540
		exo-cis-diol	180
trans-2-Hexene	77	trans-epoxide	210
		cis-epoxide	50
		RR/SS-diol	150
		RS/SR-diol	0
cis-2-Hexene	93	cis-epoxide	450
		trans-epoxide	40
		SR/RS-diol	280
		RR/SS-diol	10
cis-Stilbene	82	cis-epoxide	260
		trans-epoxide	200
		meso-hydrobenzoin	40
		hydrobenzoin	40
Styrene	97	epoxide	860
		$Ph(CH)(OH)CH_2OH$	60
		$PhC(O)CH_2OH$	10

[a] TON in mole product/mole catalyst (after 7 h)

were formed on replacing the substrate with the corresponding epoxide, excluding epoxide hydrolysis.

Since cis-diol formation through Mn catalyzed epoxide hydrolysis can be excluded, it is proposed that the cis-diols are formed by reaction of the alkene with an Mn oxo-hydroxo species. As in the case of oxalate, activated carbonyl compounds such as gmha [97] might break down the catalase active dinuclear Mn complex **4** (Figure 10.2) [28] into a mononuclear Mn species via complexation to the Mn center. Cis-diol formation from an Mn oxo-hydroxo species with a coordinated hydrated carbonyl ligand could be induced through a hydrogen bonded 6-membered ring transition state (concerted pathway, Scheme 10.16). Reoxidation of the Mn center by H_2O_2, release of the diol from Mn, and hydration of the carbonyl compound closes the catalytic cycle. The use of activated carbonyl compounds in combination with Mn-tmtacn not only provides a highly active (up to 860 TON) and H_2O_2 efficient epoxidation system (vide supra) but also the most active Os-free homogeneous catalyst for cis-dihy-

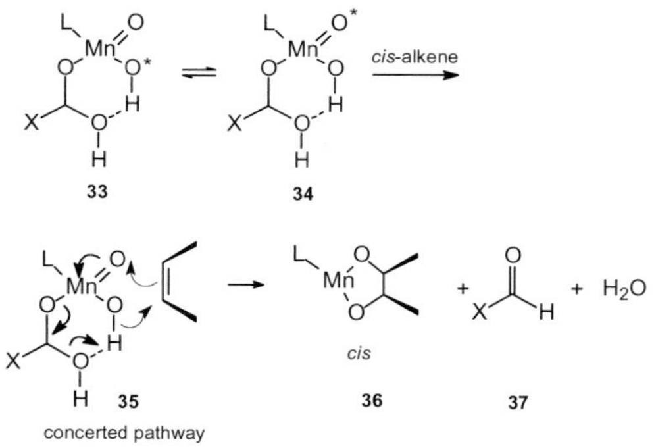

Scheme 10.16 Proposed cis-dihydroxylation mechanism (L = tmtacn, X = CO$_2$Me)

10.6
Alcohol Oxidation to Aldehydes

As part of the common repertoire of synthetic methods, the selective oxidation of alcohols to aldehydes holds a prominent position. A number of catalytic procedures have been introduced in recent years and the Ley system [NBu$_4$][RuO$_4$]/N-methylmorpholine oxide has proven to be particularly valuable in synthetic applications [98]. Selective catalytic aldehyde formation using H$_2$O$_2$ as the terminal oxidant is highly warranted. The dinuclear Mn-tmtacn [79] and several *in situ* prepared complexes of Mn(OAc)$_3$ and tptn-type ligands turned out to be active and selective catalysts for the oxidation of benzyl alcohols as well as for secondary alcohols to the corresponding carbonyl compounds [99]. Mn-complexes based on ligands **39–43** (Scheme 10.17), show high activity and selectivity (TON up to 850, Table 10.7), depending on the ligand structure. Ligands **40** and **41**, which contain a two-carbon

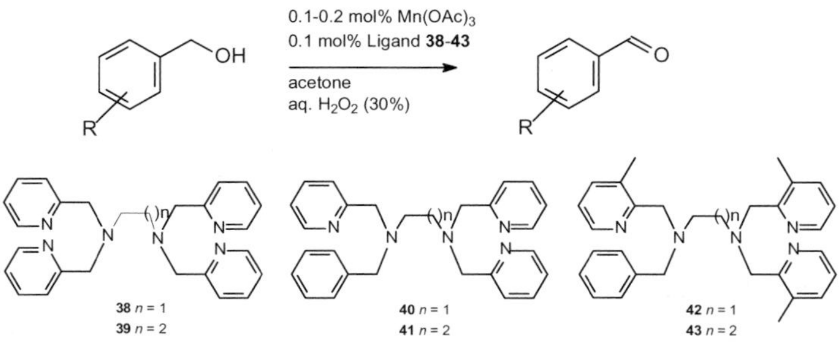

droxylation (up to 420 TON). Owing to competing epoxidation and dihydroxylation pathways, it is not suitable for application in synthesis at this stage but provides an important lead to the development of highly selective Mn-catalyzed dihydroxylation with H$_2$O$_2$ in the future.

Scheme 10.17 Ligands used for the manganese-catalyzed alcohol oxidation

Tab. 10.7 Oxidation of selected alcohols with *in situ* prepared Mn-catalysts based on ligands **39** and **43**.

Substrate	TON[a] 39	Selectivity (%)[a]	TON[a] 43	Selectivity (%)[a]
Benzyl alcohol	326	95	303	99
4-Methoxybenzyl alcohol	201	80	291	75
4-Chlorobenzyl alcohol	449	99	414	99
4-Trifluoromethylbenzyl alcohol	329	70	258	70
4-Fluorobenzyl alcohol	233	90	248	70
2,5-Dimethoxybenzyl alcohol	90	99	63	99
Cyclohexanol	363	95	593	80
Cycloheptanol	849	85	688	99
1-Octanol	108	85	46	90
2-Octanol	680	95	480	95
sec-Phenylethyl alcohol	657	90	715	95

[a] Turnover numbers after 4 h (TON) and selectivity (%) with ligands **39** and **43**.

spacer or three-carbon spacer and lack one pyridine compared with tptn, were found to form moderately active catalysts; however, long induction periods were observed.

Using *in situ* prepared complexes based on *ortho*-methyl substituted ligands **42** and **43**, excellent results were found and, remarkably, the induction period was greatly reduced. A strong 16-line EPR signal was observed immediately after mixing ligand **43** with Mn(OAc)$_3$, H$_2$O$_2$ and substrate, which points to the involvement of dinuclear species in the oxidation reaction. The catalyst based on the ligands with the three-carbon spacers show, in all cases, much higher reactivity (shorter induction period) than the two-carbon spacer analogues, probably connected with a faster formation of dinuclear species. The primary kinetic isotope effects (k_H/k_D) for the Mn-catalyzed oxidation of benzyl alcohol and benzyl-d_7 alcohol observed are in the range of from 2.2 to 4.3. These values clearly indicate that cleavage of the benzylic C–H bond is involved in the rate-determining step [100]. It has been concluded that hydroxyl radicals are not involved in these processes, because owing to the high reactivity of these radicals a much lower isotopic effect would be expected [101]. Accordingly, no indications for hydroxylation of aromatic rings in various substrates have been found. At this point it has not been established which active species (e.g., high-valent Mn=O or Mn–OOH) is involved in the selective aldehyde formation.

10.7
Sulfide to Sulfoxide Oxidation

The selective catalytic oxidation of sulfides to sulfoxides has been a challenge for many years, not unexpected in view of the importance of sulfoxides as intermediates in synthesis [102]. The undesired sulfone is a common byproduct in sulfide oxidation using H$_2$O$_2$ as oxidant and its formation has to be suppressed. Much effort has been devoted to the development of catalytic methods for the preparation of optically active sulfoxides [102], following the pioneering reports by Kagan [103] and Modena [104], of high ee values (>90%) in sulfoxide formation using diethyl tartrate/Ti(O*i*-Pr)$_4$ catalysts and hydroperoxides as oxidant [105].

Jacobsen applied manganese(III)-salen complexes for sulfide oxidation [57a]. It turned out that sodium hypochlorite was too reactive for the selective oxidation of sulfides but when employing iodosylbenzene as the oxygen atom transfer agent, no over-oxidation to sulfone was observed [106]. Disadvantages of iodosylbenzene are the poor solubility, low oxygen atom efficiency and high cost for practical application. By changing to H$_2$O$_2$ high yields and identical enantioselectivities (34–68% ee) are also obtained compared with those using iodosylbenzene [106]. Using acetonitrile as solvent, 2–3 mol% of catalyst and 6 equiv. of oxidant the formation of sulfone was minimized [106]. Ligands with bulky substituents at the 3,3'- and 5,5'-positions resulted in the highest enantioselectivity (Scheme 10.18). The enantioselectivity for sulfoxide formation is in general lower than that observed for epoxidation using the same catalysts. *para-tert*-Butyl substituted 44 (R = ^{t}Bu, Scheme 10.18) emerged as the most selective one for the oxidation of a variety of substrates. Mn-catalyzed complexes derived from ligands with electron-withdrawing substituents showed lower enantioselectivities, whereas *para*-nitro-substituted complex 44 (R = NO$_2$), did not induce asymmetric sulfide oxidation.

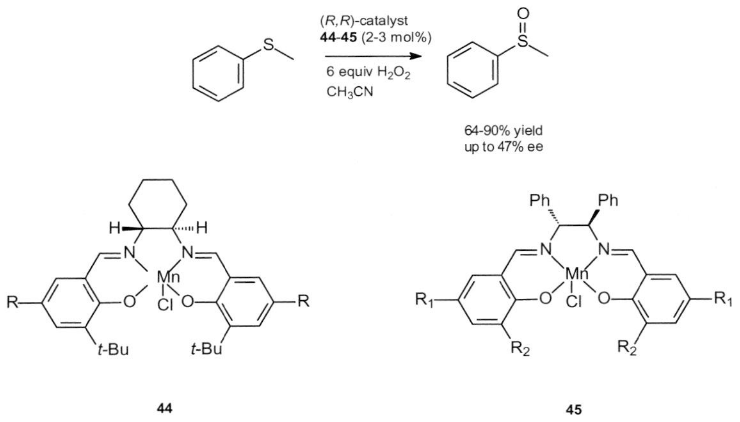

Scheme 10.18 Manganese(III) salen complexes for sulfide oxidation introduced by Jacobsen [106]

Katsuki used related chiral manganese salen complexes, especially second-generation Mn-salen 46, for sulfide oxidation [107]. This complex was found to serve as an efficient catalyst for the asymmetric sulfoxidation, however the less atom-efficient iodosylbenzene was required as oxidant (Scheme 10.19).

Recently, Katsuki and Saito reported that di-μ-oxo titanium complexes of chiral salen ligands serve as efficient catalysts for asymmetric oxidation of various sulfides using H$_2$O$_2$ or the urea-hydrogen peroxide adduct as oxidants [108]. Enantioselectiv-

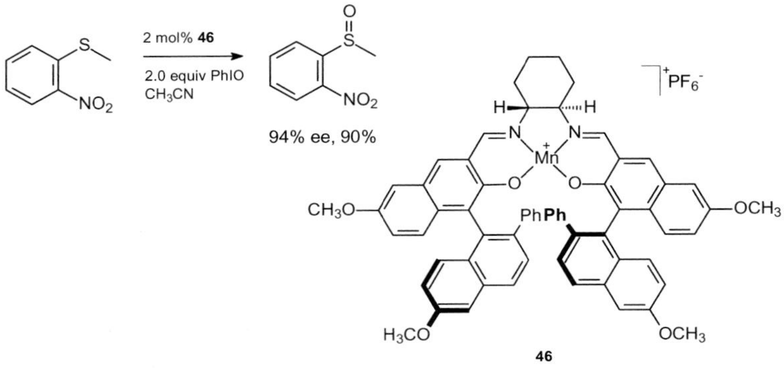

Scheme 10.19 Mn^{III} salen complexes for asymmetric sulfide oxidation reported by Katsuki [107c]

ities as high as 94% were observed [108] and as an active intermediate a monomeric peroxo titanium species was proposed, based on MS and NMR studies [109].

A suitable Mn-catalyst for olefin or alcohol oxidation with H$_2$O$_2$ might also be effective in the oxidation of thioethers and, based on this expectation, Mn-tmtacn and a number of *in situ* formed complexes using tptn as ligand were screened as sulfide oxidation catalysts. These complexes indeed turned out to be highly active for sulfoxide formation. For instance, the dinuclear manganese complex based on tmtacn performs efficiently in the oxidation of aryl alkyl sulfides and generally results in full conversion in 1 h. Unfortunately, in addition to the desired sulfoxide, the formation of sulfone was observed. Similar reactivity patterns were observed with manganese complexes based on tptn and tpen. Employing the novel ligand **47** (Scheme 10.20), slight over-oxidation to sulfone was observed [110]. With ligand **47**, which combines structural features of tptn and salen ligands, over-oxidation to sulfone could be suppressed. The use of ligand **48**, a chiral version of ligand **47**, in the Mn-catalyzed sulfoxide formation with H$_2$O$_2$ resulted in yields ranging from 48 to 55% with ee's up to 18% [111].

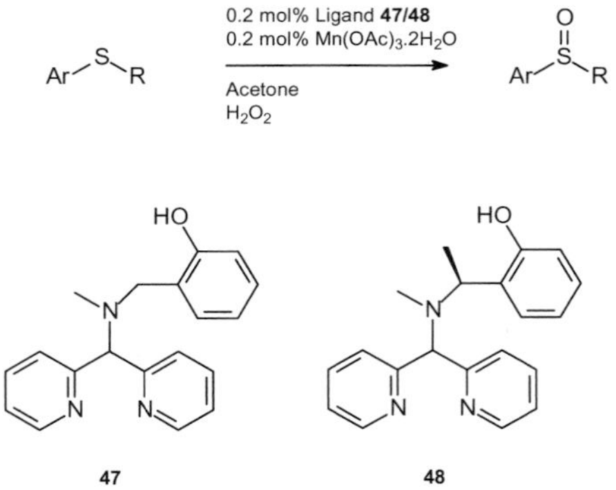

Scheme 10.20 Ligands with nitrogen and oxygen donor functionalities used in sulfide oxidation [110, 111]

10.8
Conclusions

Hydrogen peroxide is a particularly attractive oxidant and holds a prominent position in the development of benign catalytic oxidation procedures. In recent years a number of highly versatile catalytic oxidations methods based on, for example, polyoxometalates[112], methyltrioxorhenium [113] or tungstate [4, 114] complexes in the presence of phase transfer catalysts, all using hydrogen peroxide as the terminal oxidant, have been introduced. Mn-catalyzed epoxidations, aldehyde formation and sulfoxidation with H_2O_2 have emerged as effective and practical alternatives. In particular, recently developed epoxidation catalysts, based on a combination of Mn-tmtacn and additives, have shown high activity and excellent selectivity in the epoxidation of a wide range of alkenes. Despite considerable progress in enantioselective epoxidation with Mn-salen systems using H_2O_2 as oxidant, a general catalytic epoxidation method based on chiral Mn-complexes remains a highly warranted goal. Particularly promising are the findings that significant *cis*-dihydroxylation can be achieved with Mn-catalysts. These studies might provide the guiding principles to designing Mn-catalysts as an alternative to current Os-based chiral dihydroxylation systems. For industrial application, further improvement with respect to hydrogen peroxide efficiency and catalytic activity is needed for most of the Mn-systems developed so far. The delicate balance between oxygen transfer to the substrate and hydrogen peroxide decomposition remains a critical issue in all of these studies. Other challenges include the nature of the Mn-complexes in solution and the actual active species involved in oxygen transfer, the mechanisms of the Mn-catalyzed oxidations with hydrogen peroxides and the key role of the additives in several cases. It is likely that detailed insight into these aspects of the catalytic systems developed recently will bring major breakthroughs in Mn-catalyzed oxidations with hydrogen peroxide.

References

[1] SHELDON, R. A.; KOCHI, J. K. *Metal-Catalyzed Oxidations of Organic Compounds*, Academic Press, New York, **1981**.

[2] (a) *Organic Synthesis by Oxidation with Metal Compounds*, Mijs, W. J.; de Jonge, C. R. H. I. (Eds.), Plenum Press, New York, **1986**. (b) *Comprehensive Organic Synthesis*, TROST, B. M.; FLEMING, I. A. (Eds.), Pergamon Press, Oxford, **1991**, vol. 7. (c) HUDLICKY, M. *Oxidations in Organic Chemistry*, ACS Monograph Ser. 186, American Chemical Society, Washington D.C., **1990**.

[3] ANASTAS, P. T.; WARNER, J. C. *Green Chemistry, Theory and Practice*, Oxford University Press, Oxford, **1998**.

[4] NOYORI, R.; AOKI, M.; SATO, K. *Chem. Commun.* **2003**, 1977–1986.

[5] *Advances in Catalytic Activation of Dioxygen by Metal Complexes*, SIMÁNDI, L. I. (Ed.), Kluwer Academic, Dordrecht, **1992**.

[6] COSTAS, M.; MEHN, M. P.; JENSEN, M. P.; QUE JR., L. *Chem. Rev.* **2004**, *104*, 939–986.

[7] WEISSERMEL, K.; ARPE, H.-J. *Industrial Organic Chemistry*, VCH Weinheim, **1993**.

[8] JONES, C. W. *Applications of Hydrogen Peroxide and Derivatives*, MPG Books Ltd., Cornwall, **1999**.

[9] LANE, B. S.; BURGESS, K. *Chem. Rev.* **2003**, *103*, 2457–2473.

[10] VOLLHARDT, K. P. C. SCHORE, N. E. *Organic Chemistry*, 4th edn., W. H. Freeman, New York, **2003**.

[11] MARCH, J. *Advanced Organic Chemistry*, 3rd edn., Wiley-Interscience, New York, **1985**.

[12] LIPPARD, S. J. BERG, J. M. In *Principles of Bioinorganic Chemistry*, University Science Books, Mill Valley, CA, **1994**.

[13] *Manganese Redox Enzymes*, PECORARO, V. L. (Ed.), VCH Publisher New York, **1992**.

[14] (a) ALLGOOD, G. S.; PERRY, J. J. *J. Bacteriol.* **1986**, *168*, 563–567; (b) WU, A. J.; PENNER-HAHN, J. E.; PECORARO, V. L. *Chem. Rev.* **2004**, *104*, 903–938.

[15] WALDO, G. S.; PENNER-HAHN, J. E. *Biochemistry* **1995**, *34*, 1507–1512.

[16] RILEY, D. P. *Chem. Rev.* **1999**, *99*, 2573–2587.

[17] FRIDOVICH, I. *J. Biol. Chem.* **1989**, *264*, 7761–7764.

[18] JAKOBY, W. R.; ZIEGLER, D. M. *J. Biol. Chem.* **1990**, *265*, 20715–20718.

[19] (a) PICK, M.; RABANI, J.; YOST, F.; FRIDOVICH, I. *J. Am. Chem Soc.* **1974**, *96*, 7329–7333.

[20] ZOUNI, A.; WITT, H.-T.; KERN, J.; FROMME, P.; KRAUSS, N.; SAENGER, W.; ORTH, P. *Nature*, **2001**, *409*, 739–743.

[21] KONO, Y.; FRIDOVICH, I. *J. Biol. Chem.* **1983**, *258*, 6015–6019.

[22] BARYNIN, V. V.; GREBENKO, A. I. *Dokl. Akad. Nauk SSSR* **1986**, *286*, 461–464.

[23] *Thermus Thermophilus*: ANTONYUK, S. V.; MELIK-ADAMYAN, V. R.; POPOV, A. N.; LAMZIN, V. S.; HEMPSTEAD, P. D.; HARRISON, P. M.; ARTYMYUK, P. J.; BARYNIN, V. V. *Crystallography Reports* **2000**, *45*, 105–116. *Lactobacillus Plantarum*: BARYNIN, V. V.; WHITTAKER, M. M.; ANTONYUK, S. V.; LAMZIN, V. S.; HARRISON, P. M.; ARTYMIUK, P. J.; WHITTAKER, J. W. *Structure* **2001**, *9*, 725–738.

[24] GHANOTAKIS, D. F.; YOCUM, C. F. *Annu. Rev. Plan Physiol. Plant. Mol. Biol.* **1990**, *41*, 255–276.

[25] DEXHEIMER, S. L.; GOHDES, J. W.; CHAN, M. K.; HAGEN, K. S.; ARMSTRONG, W. H.; KLEIN, M. P. *J. Am. Chem. Soc.* **1989**, *111*, 8923–8925.

[26] (a) SHEATS, J. E.; CZERNUSZEWICZ, R. S.; DISMUKES, G. C.; RHEINGOLD, A. L.; PETROULEAS, V.; STUBBE, J.; ARMSTRONG, W. H.; BEER, R. H.; LIPPARD, S. J.; *J. Am. Chem. Soc.* **1987**, *109*, 1435–1444. (b) WU, F.-J.; KURTZ, D. M., JR.; HAGEN, K. S.; NYMAN, P. D.; DEBRUNNER, P. G.; VANKAI, V. A. *Inorg. Chem.* **1990**, *29*, 5174–5183.

[27] (a) WALDO, G. S.; YU, S.; PENNER-HAHN, J. E. *J. Am. Chem. Soc.* **1992**, *114*, 5869–5870. (b) PESSIKI, P. J.; DISMUKES, G. C. *J. Am. Chem. Soc.* **1994**, *116*, 898–903.

[28] HAGE. R. *Recl. Trav. Chim. Pays-Bas* **1996**, *115*, 385–395.

[29] MATHUR, P.; CROWDER, M.; DISMUKES, G. C. *J. Am. Chem. Soc.* **1987**, *109*, 5227–5233.

[30] (a) SAKIYAMA, H.; KAWA, H.; ISOBE, R. *J. Chem. Soc., Chem. Commun.* **1993**, 882–884. (b) SAKIYAMA, H.; KAWA, H.; SUZIKI, M. *J. Chem. Soc., Dalton Trans.* **1993**, 3823–3825. (c) HIGUCHI, C.; SAKIYAMA, H.; KAWA, H.; FENTON, D.E. *J. Chem. Soc., Dalton Trans.* **1995**, 4015–4020. (d) YAMAMI, M.; TANAKA, M.; SAKIYAMA, H.; KOGA, T.; KOBAYASHI, K.; MIYASAKA, H.; OHBA, M.; KAWA, H. *J. Chem. Soc., Dalton Trans.* **1997**, 4595–44601.

[31] (a) WIEGHARDT, K.; BOSSEK, U.; VENTUR, D.; WEISS, J. *J. Chem. Soc., Chem. Commun.* **1985**, 347–349. (b) WIEGHARDT, K.; BOSSEK, U.; NUBER, B.; WEISS, J.; BONVOISIN, J.; CORBELLA, M.; VITOLS, S. E.; GIRERD, J. J. *J. Am. Chem. Soc.* **1988**, *110*, 7398–7411. (c) BOSSEK, U.; WEYHERMÜLLER, T.; WIEGHARDT, K.; NUBER, B.; WEISS, J. *J. Am. Chem. Soc.* **1990**, *112*, 6387–6388. (d) BOSSEK, U.; SAHER, M.; WEYHERMÜLLER, T.; WIEGHARDT, K. *J. Chem. Soc., Chem. Commun.* **1992**, 1780–1782. (e) STOCKHEIM, C.; HOSTER, L.; WEYHERMÜLLER, T.; WIEGHARDT, K.; NUBER, B. *J. Chem. Soc., Dalton Trans.* **1996**, 4409–4416. (f) BURDINSKI, D.; BOTHE, E.; WIEGHARDT, K. *Inorg. Chem.* **2000**, *39*, 105–116.

[32] TRILLER, M. U.; HSIEH, W.-Y.; PECORARO, V. L.; ROMPEL, A.; KREBS, B. *Inorg. Chem.* **2002**, *41*, 554–5545.

[33] (a) HAGE, R.; IBURG, J. E.; KERSCHNER, J.; KOEK, J. H.; LEMPERS, E. L. M.; MARTENS, R. J.; RACHERLA, U. S.; RUSSELL, S. W.; SWARTHOFF, T.; VAN VLIET,

M. R. P.; WARNAAR, J. B.; VAN DER WOLF, L.; KRIJNEN B. *Nature* **1994**, *369*, 637–639. (b) GILBERT, B. C.; LINDSAY SMITH, J. R.; NEWTON, M. S.; OAKES, J.; PONS I PRATS, R. *Org. Biomol. Chem.* **2003**, *1*, 1568–1577. (c) ALVES,V.; CAPANEMA, E.; CHEN, C.-L.; GRATZL, J. *J. Mol. Catal. A: Chem.* **2003**, *206*, 37–51.

[34] (a) VERALL, M. *Nature* **1994**, *369*, 511. (b) COMYNS, A. E. *Nature* **1994**, *369*, 609–610.

[35] (a) GORZYNSKI SMITH, J. *Synthesis* **1984**, 629–656. (b) BONINI, C.; RIGHI, G. *Synthesis* **1994**, 225–238. (c) GRIGOROPOULOU, G.; CLARK, J. H.; ELINGS, J. A. *Green Chem.* **2003**, *5*, 1–7.

[36] JAMES, A. P.; JOHNSTONE, R. A. W.; McCARRON, M.; SANKEY, J. P.; TRENBIRTH, B. *Chem. Commun.* **1998**, 429–430, and references cited therein.

[37] DENMARK, S. E.; WU, Z. *Synlett* **1999**, 847–859.

[38] HILL, C. L.; PROSSER-McCARTHA, C. M. *Coord. Chem. Rev.* **1995**, *143*, 407–455.

[39] KATSUKI, T. *Coord. Chem. Rev.* **1995**, *140*, 189–214.

[40] MEUNIER, B. *Chem. Rev.* **1992**, *92*, 1411–1456.

[41] (a) FINNEY, N. S.; POSPISIL, P. J.; CHANG, S.; PALUCKI, M.; KONSLER, R. G.; HANSEN, K. B.; JACOBSEN, E. N. *Angew. Chem., Int. Ed. Engl.* **1997**, *36*, 1720–1723. (b) BERKESSEL, A.; FRAUENKRON, M.; SCHWENKREIS, T.; STEINMETZ, A. *J. Mol. Catal. A: Chem.* **1997**, *117*, 339–346. (c) MOISEEV, I. I. *J. Mol. Catal. A.: Chem.* **1997**, *127*, 1–23.

[42] MANSUY, D. *Coord. Chem. Rev.* **1993**, *125*, 129–141.

[43] (a) RENAUD, J.-P.; BATTIONI, P.; BARTOLI, J.-F.; MANSUY, D. *J. Chem. Soc., Chem. Commun.* **1985**, 888–889. (b) BATTIONI, P.; RENAUD, J.-P.; BARTOLI, J. F.; MOMENTEAU, M.; MANSUY, D. *Recl. Trav. Chim. Pays-Bas* 1987, *106*, 332. (c) BATTIONI, P.; RENAUD, J. P.; BARTOLI, J. F.; REINA-ARTILES, M.; FORT, M.; MANSUY, D. *J. Am. Chem. Soc.* **1988**, *110*, 8462–8470. (d) PORIEL, C.; FERRAND,Y.; LE MAUX, P.; RAULT-BERTHELOT, J.; SIMONNEAUX, G. *Tetrahedron Lett.* **2003**, *44*, 1759–1761.

[44] For the oxidation of monoterpenes and alkylaromatics with good conversions, although mixtures of products are obtained, see: (a) MARTINS, R. R. L.; NEVES, M. G. P. M. S.; SILVESTRE, A. J. D.; SIMÕES, M. M. Q.; SILVA, A. M. S.; TOMÉ, A. C.; CAVALEIRO, J. A. S.;TAGLIATESTA, P.; CRESTINI, C. *J. Mol. Catal. A: Chem.* **2001**, *172*, 33–42. (b) REBELO, S. L. H.; SIMÕES, M. M. Q.; NEVES, G. P. M. S.; CAVALEIRO, J. A. S. *J. Mol. Catal. A: Chem.* **2003**, *201*, 9–22.

[45] ANELLI, P. L.; BANFI, S.; MONTANARI, F.; QUICI, S. *J. Chem. Soc., Chem. Commun.* **1989**, 779–780.

[46] GROVES, J. T.; WATANABE,Y.; McMURRY, T. J. *J. Am. Chem. Soc.* **1983**, *105*, 4489–4490.

[47] (a) OSTOVIC, D.; BRUICE, T. C. *Acc. Chem. Res.* **1992**, *25*, 314–320. (b) ARASASINGHAM, R. D.; HE, G.-X.; BRUICE, T. C. *J. Am. Chem. Soc.* **1993**, *115*, 7985–7991.

[48] BACIOCCHI, E.; BOSCHI, T.; CASSIOLI, L.; GALLI, C.; JAQUINOD, L.; LAPI, A.; PAOLESSE, R.; SMITH, K. M.;TAGLIATESTA, P. *Eur. J. Org. Chem.* **1999**, 3281–3286.

[49] (a) BANFI, S.; LEGRAMANDI, F.; MONTANARI, F.; POZZI, G.; QUICI, S. *J. Chem. Soc., Chem. Commun.* **1991**, 1285–1287. (b) ANELLI, P. L.; BANFI, L.; LEGRAMANDI, F.; MONTANARI, F.; POZZI, G.; QUICI, S. *J. Chem. Soc., Perkin Trans, 1* **1993**, 1345–1357.

[50] GROVES, J. T.; MYERS, R. S. *J. Am. Chem. Soc.* **1983**, *105*, 5791–5796.

[51] (a) CAMPBELL, L. A.; KODADEK,T. *J. Mol. Catal. A: Chem.* **1996**, *113*, 293–310. (b) COLLMAN, J. P; ZHANG, X.; LEE,V. J.; UFFELMAN, E. S.; BRAUMAN, J. I. *Science* **1993**, *261*, 1404–1411.

[52] COLLMAN, J. P.; LEE,V. J.; KELLEN-YUEN, C. J.; ZHANG, X.; IBERS, J. A.; BRAUMAN, J. I. *J. Am. Chem. Soc.* **1995**, *117*, 692–703.

[53] Y. NARUTA in *Metalloporphyrins in Catalytic Oxidations*, Sheldon, R. A. (Ed.), Marcel Dekker, New York, **1994**, chapter 8.

[54] MARTINEZ-LORENTE, M. A.; BATTIONI, P.; KLEEMISS, W.; BARTOLI, J.F.; MANSUY, D. *J. Mol. Catal. A: Chem.* **1996**, *113*, 343–353.

[55] DORO, F. G.; LINDSAY SMITH, J. R.; FER-

REIRA, A. G.; ASSIS, M. D. *J. Mol. Catal. A: Chem.* **2000**, *164*, 97–108.

[56] SRINIVASAN, K.; MICHAUD, P.; KOCHI, J. K. *J. Am. Chem. Soc.* **1986**, *108*, 2309–2320.

[57] (a) ZHANG, W.; LOEBACH, J. L.; WILSON, S. R.; JACOBSEN, E. N. *J. Am. Chem. Soc.* **1990**, *112*, 2801–2803. (b) IRIE, R.; NODDA, K.; ITO, Y.; MATSUMOTO, N.; KATSUKI, T. *Tetrahedron Lett.* **1990**, *31*, 7345–7348.

[58] (a) JACOBSEN, E. N.; WU, M. H. In *Comprehensive Asymmetric Catalysis*, Vol. II, Jacobsen, E. N.; Pfaltz, A.; Yamamoto, H. (Eds.), Springer, Berlin, **1999**, 649–677. (b) KATSUKI, T. *J. Mol. Catal. A: Chem.* **1996**, *113*, 87–107. (c) KATSUKI, T. *Adv. Synth. Catal.* **2002**, *344*, 131–147. (d) Katsuki, T. *Synlett*, **2003**, 281–297.

[59] (a) PALUCKI, M.; POSPISIL, P. J.; ZHANG, W.; JACOBSEN, E. N. *J. Am. Chem. Soc.* **1994**, *116*, 9333–9334. (b) PALUCKI, M.; MCCORMICK, G. J.; JACOBSEN, E. N. *Tetrahedron Lett.* **1995**, *36*, 5457–5460. (c) VAN DER VELDE, S. L.; JACOBSEN, E. N. *J. Org. Chem.* **1995**, *60*, 5380–5381. (d) JACOBSEN, E. N.; DENG, L.; FURUKAWA, Y.; MARTINEZ, L. E. *Tetrahedron* **1994**, *50*, 4323–4334. (e) HUGHES, D. L.; SMITH, G. B.; LIU, J.; DEZENY, G. C.; SENANAYAKE, C. H.; LARSEN, R. D.; VERHOEVEN, T. R.; REIDER, P. J. *J. Org. Chem.* **1997**, *62*, 2222–2229.

[60] FEICHTINGER, D.; PLATTNER, D. A. *Angew. Chem., Int. Ed. Engl.* **1997**, *36*, 1718–1719.

[61] (a) DALTON, C. T.; RYAN, K. M.; WALL, V. M.; BOUSQUET, C.; GILHEANY, D. G. *Top. Catal.* **1998**, *5*, 75–91. (b) CANALI, L.; SHERRINGTON, D. C. *Chem. Soc. Rev.* **1999**, *28*, 85–93.

[62] ITO, Y. N.; KATSUKI, T. *Tetrahedron Lett.* **1998**, *39*, 4325–4328.

[63] (a) SCHWENKREIS, T.; BERKESSEL, A. *Tetrahedron Lett.* **1993**, *34*, 4785–4788. (b) BERKESSEL, A.; FRAUENKRON, M.; SCHWENKREIS, T.; STEINMETZ, A.; BAUM, G.; FENSKE, D. *J. Mol. Catal. A: Chem.* **1996**, *113*, 321–342.

[64] IRIE, R.; HOSOYA, N.; KATSUKI, T. *Synlett* **1994**, 255–56.

[65] (a) PIETIKÄINEN, P. *Tetrahedron Lett.* **1994**, *35*, 941–944. (b) PIETIKÄINEN, P. *Tetrahedron* **1998**, *54*, 4319–4326.

[66] PIETIKÄINEN, P. *J. Mol. Catal. A*, **2001**, *165*, 73–79.

[67] (a) KURESHY, R. I.; KHAN, N. H.; ABDI, S. H. R.; PATEL, T.; JASRA, R. V. *Tetrahedron: Asymmetry* **2001**, *12*, 433–437. (b) KURESHY, R. I.; KHAN, N. H.; ABDI, S. H. R.; SINGH, S.; AHMED, I.; SHUKLA, R. S.; JASRA, R. V. *J. Catal.* **2003**, *219*, 1–7.

[68] (a) QUEE-SMITH, V. C.; DELPIZZO, L.; JURELLER, S. H.; KERSCHNER, J. L.; HAGE, R. *Inorg. Chem.* **1996**, *35*, 6461–6455. (b) For epoxidation reactions with Ru-complexes based on the tmtacn ligand, see: CHENG, W.-C.; FUNG, W.-H.; CHE, C.-M. *J. Mol. Catal. A: Chem.* **1996**, *113*, 311–319.

[69] (a) DE VOS, D. E.; BEIN, T. *Chem. Commun.* **1996**, 917–918. (b) DE VOS, D. E.; BEIN, T. *J. Organomet. Chem.* **1996**, *520*, 195–200.

[70] SAUER, M. C. V.; EDWARDS, J. *J. Phys. Chem.* **1971**, *75*, 3004–3011.

[71] DE VOS, D. E.; SELS, B. F.; REYNAERS, M.; SUBBA RAO, Y. V.; JACOBS, P. A. *Tetrahedron Lett.* **1998**, *39*, 3221–3224.

[72] BERKESSEL, A.; SKLORZ, C. A. *Tetrahedron Lett.* **1999**, *40*, 7965–7968.

[73] DE VOS, D. E.; DE WILDEMAN, S.; SELS, B. F.; GROBET, P. J.; JACOBS, P. A. *Angew. Chem., Int. Ed. Engl.* **1999**, *38*, 980–983.

[74] (a) CHAUDHURI, P.; ODER, K. *J. Chem. Soc., Dalton Trans.* **1990**, 1597–1605. (b) NIEMANN, A.; BOSSEK, U.; HASELHORST, G.; WIEGHARDT, K.; NUBER, B. *Inorg. Chem.* **1996**, *35*, 906–915.

[75] (a) BOLM, C.; KADEREIT, D.; VALACCHI, M. *Synlett* **1997**, 687–688. (b) KOEK, J. H.; KOHLEN, E. W. M. J.; RUSSELL, S. W.; VAN DER WOLF, L.; TER STEEG, P. F.; HELLEMONS, J. C. *Inorg. Chim. Acta* **1999**, *295*, 189–199. (c) GOLDING, S. W.; HAMBLEY, T. W.; LAWRANCE, G. A.; LUTHER, S. M.; MAEDER, M.; TURNER, P. *J. Chem. Soc., Dalton Trans.* **1999**, 1975–1980. (d) BOLM, C.; MEYER, N.; RAABE, G.; WEYHERMÜLLER, T.; BOTHE, E. *Chem. Commun.* **2000**, 2435–2436. (e) KIM, B. M.; SO, S. M.; CHOI, H. J. *Org. Lett.* **2002**, *4*, 949–952. (f) SCHEUERMANN, J. E. W.; ILYASHENKO, G.; GRIFFITHS,

D. V.; WATKINSON, M. *Tetrahedron: Asymmetry* **2002**, *13*, 269–272. (g) SCHEUERMANN, J. E. W.; RONKETTI, F.; MOTEVALLI, M.; GRIFFITHS, D. V.; WATKINSON, M. *New. J. Chem.* **2002**, 1054–1059. (h) ARGOUARCH, G.; GIBSON, C. L.; STONES, G.; SHERRINGTON, D. C. *Tetrahedron Lett.* **2002**, *43*, 3795–3798.

[76] (a) BELLER, M.; TAFESCH, A.; FISCHER, R. W.; SCHARBERT, B. DE 19523891, **1995**; (b) BOLM, C.; KADEREIT, D.; VALACCHI, M. DE 19720477, **1997**.

[77] DE VOS, D. E.; MEINERSHAGEN, J. L.; BEIN, T. *Angew. Chem., Int. Ed. Engl.* **1996**, *35*, 2211–2213.

[78] SUBBA RAO, Y. V.; DE VOS, D. E.; BEIN, T.; JACOBS, P. A. *Chem. Commun.* **1997**, 355–356.

[79] ZONDERVAN, C.; HAGE, R.; FERINGA, B. L. *Chem. Commun.* **1997**, 419–420.

[80] GILBERT, B. C.; KAMP, N. W. J.; LINDSAY SMITH, J. R.; OAKES, J. *J. Chem. Soc., Perkin Trans. 2*, **1997**, 2161–2165.

[81] BARTON, D. H. R.; CHOI, S. Y.; HU, B.; SMITH, J. A. *Tetrahedron* **1998**, *54*, 3367–3378.

[82] GILBERT, B. C.; KAMP, N. W. J.; LINDSAY SMITH, J. R.; OAKES, J. *J. Chem. Soc., Perkin Trans. 2*, **1998**, 1841–1843.

[83] ZONDERVAN, C. *Homogeneous Catalytic Oxidation, A Ligand Approach*, Ph.D. Thesis, University of Groningen, **1997**, Chapter 4.

[84] (a) TOFTLUND, H.; YDE-ANDERSEN, S. *Acta Chem. Scand., Ser A* **1981**, *35*, 575–585. (b) TOFTLUND, H.; MARKIEWICZ, A.; MURRAY, K. S. *Acta Chem. Scand.* **1990**, *44*, 443–446. (c) MANDEL, J. B.; MARICONDI, C.; DOUGLAS, B. E. *Inorg. Chem.* **1988**, *27*, 2990–2996. (d) PAL, S.; GOHDES, J. W.; CHRISTIAN, W.; WILISCH, A.; ARMSTRONG, W. H. *Inorg. Chem.* **1992**, *31*, 713–716.

[85] BRINKSMA, J.; HAGE, R.; KERSCHNER, J.; FERINGA, B. L. *Chem. Commun.* **2000**, 537–538.

[86] ZHANG, W.; LEE, N. H.; JACOBSEN, E. N. *J. Am. Chem. Soc.* **1994**, *116*, 425–426.

[87] (a) LANE, B. S.; BURGESS, K. *J. Am. Chem. Soc.* **2001**, *123*, 2933–2934 (b) LANE, B. S.; VOGT, M.; DEROSE, V. J.; BURGESS, K. *J. Am. Chem. Soc.* **2002**, *124*, 11946–11954.

[88] (a) RICHARDSON, D. E.; YAO, H.; FRANK, K. M.; BENNETT, D. A. *J. Am. Chem. Soc.* **2000**, *122*, 1729–1739. (b) YAO, H.; RICHARDSON, D. E. *J. Am. Chem. Soc.* **2000**, *122*, 3220–3221.

[89] TONG, K.-H.; WONG, K.-Y.; CHAN, T. H. *Org. Lett.* **2003**, *5*, 3423–3425.

[90] DÖBLER, C.; MEHLTRETTER, G. M.; SUNDERMEIER, U.; BELLER, M. *J. Am. Chem. Soc.* **2000**, *122*, 10289–10297.

[91] JONSSON, S. Y.; FÄRNEGÅRDH, K.; BÄCKVALL, J. E. *J. Am. Chem. Soc.* **2001**, *123*, 1365–1371.

[92] MILAS, N. A.; SUSSMAN, S. *J. Am. Chem. Soc.* **1936**, *58*, 1302–1304.

[93] KOLB, H. C.; VANNIEUWENHZE, M. S.; SHARPLESS, K. B. *Chem. Rev.* **1994**, *94*, 2483–2547.

[94] MARKÓ, I. E.; SVENDSEN, J. S. In *Comprehensive Asymmetric Catalysis*, Vol. II, JACOBSEN, E.N.; PFALTZ, A.; YAMAMOTO, H. (Eds.), Springer, Berlin, **1999**, 711–787.

[95] (a) CHEN, K. QUE JR, L. *Angew. Chem. Int. Ed. Engl.* **1999**, *38*, 2227–2229. (b) Chen, K.; Costas, M.; Kim, J.; Tipton, A. K.; Que Jr., L. *J. Am. Chem. Soc.* **2002**, *124*, 3026–3035. (c) FUJITA, M.; COSTAS, M.; QUE JR, L. *J. Am. Chem. Soc.* **2003**, *125*, 9912–9913.

[96] BRINKSMA, J.; SCHMIEDER; L.; VAN VLIET, G.; BOARON, R.; HAGE, R.; DE VOS, D.E.; ALSTERS, P.L.; FERINGA, B. L. *Tetrahedron Lett.* **2002**, *43*, 2619–2622.

[97] Gmha is an equilibrium mixture, which also contains some hydrated methyl glyoxylate. NMR experiments showed that formation peroxyhydrate from gmha and aqueous H_2O_2 is very slow.

[98] GRIFFITH, W. P.; LEY, S. V.; WHITCOMBE, G. P.; WHITE, A. D. *J. Chem. Soc., Chem. Commun.* **1987**, 1625–1627.

[99] BRINKSMA, J.; RISPENS, M. T.; HAGE, R.; FERINGA, B. L. *Inorg. Chim. Acta* **2002**, *337*, 75–82.

[100] WANG, Y.; DUBOIS, J. L.; HEDMAN, B.; HODGSON, K. O.; STACK, T. B. D. *Science* **1998**, *279*, 537–540.

[101] KHENKIN, A. M.; SHILOV, A. E. *New J. Chem.* **1989**, *13*, 659–667.

[102] (a) SOLLADIÉ, G. *Synthesis* **1981**, 185–196. (b) CARREÑO, M. C. *Chem. Rev.* **1995**, *95*, 1717–1760. (c) COLOBERT, F.; TITO, A.; KHIAR, N.; DENNI, D.;

Medina, M. A.; Martin-Lomas, M.; Ruano, J. L. G.; Solladié, G. *J. Org. Chem.* **1998**, *63*, 8918–8921. (d) Bravo, P.; Crucianelli, M.; Farina, A.; Meille, S. V.; Volonterio, A.; Zanda, M. *Eur. J. Org. Chem.* **1998**, 435–440. (e) Cotton, H.; Elebring, T.; Larsson, M.; Li, L.; Sörensen, H.; von Unge, S. *Tetrahedron: Asymmetry* **2000**, *11*, 3819–3825. (f) Padmanabhan, S.; Lavin, R.C.; Durant, G. J. *Tetrahedron: Asymmetry.* **2000**, *11*, 3455–3457.

[103] (a) Pitchen, P.; Duñach, E.; Deshmukh, M. N.; Kagan, H. B. *J. Am. Chem. Soc.* **1984**, *106*, 8188–8193. (b) Pitchen, P; Kagan, H. B. *Tetrahedron Lett.* **1984**, *25*, 1049–1052.

[104] Di Furia, F.; Modena, G.; Seraglia, R. *Synthesis* **1984**, 325–326.

[105] (a) Komatsu, N.; Hashizume, M.; Sugita, T.; Uemura, S. *J. Org. Chem.* **1993**, *58*, 7624–7626. (b) Donnoli, M. I.; Superchi, S.; Rosini, C. *J. Org. Chem.* **1998**, *63*, 9392–9395. (c) Bolm, C.; Dabard, O. A. G. *Synlett* **1999**, *3*, 360–362. (d) Di Furia, F.; Licini, G.; Modena, G.; Motterle, R.; Nugent, W. A. *J. Org. Chem.* **1996**, *61*, 5175–5177. (e) Bonchio. M.; Calloni, S.; Di Furia, F; Licini, G.; Modena, G.; Moro, S.; Nugent, W. A. *J. Am. Chem. Soc.* **1997**, *119*, 6935–6936.

[106] Palucki, M.; Hanson, P.; Jacobsen, E. N. *Tetrahedron Lett.* **1992**, *33*, 7111–7114.

[107] (a) Noda, K.; Hosaya, N.; Yanai, K.; Irie, R.; Katsuki, T. *Tetrahedron Lett.* **1994**, *35*, 1887–1890. (b) Noda, K.; Hosoya, N.; Irie, R.; Yamashita, Y.; Katsuki, T. *Tetrahedron* **1994**, *50*, 9609–9618. (c) Kokubo, C; Katsuki, T. *Tetrahedron* **1996**, *52*, 13895–13900.

[108] Saito, B.; Katsuki, T. *Tetrahedron Lett.* **2001**, *42*, 3873–3876.

[109] Saito, B.; Katsuki, T. *Tetrahedron Lett.* **2001**, *42*, 8333–8336.

[110] La Crois, R. M. *Manganese Complexes as Catalysts in Epoxidation Reactions, a Ligand Approach*, Ph.D. Thesis, University of Groningen, **2000**, Chapter 4.

[111] Brinksma, J.; La Crois, R.; Feringa, B. L.; Donnoli, M. I.; Rosini, C. *Tetrahedron Lett.* **2001**, *42*, 4049–4052.

[112] (a) Bösing, M.; Nöh, A.; Loose, I.; Krebs, B. *J. Am. Chem. Soc.* **1998**, *120*, 7252–7259. (b) Mizuno, N.; Nozaki, C.; Kiyoto, I.; Misono, M. *J. Am. Chem. Soc.* **1998**, *120*, 9267–9272.

[113] Herrmann, W. A.; Fischer, R. W.; Scherer, W.; Rauch, M. U. *Angew. Chem., Int. Ed. Engl.* **1993**, *32*, 1157–1160.

[114] Sato, K.; Aoki, M.; Ogawa, M.; Hashimoto, T.; Noyori, R. *J. Org. Chem.* **1996**, *61*, 8310–8311.

Subject Index

acetals 263
– catalytic systems 263
– oxidation 263
N-acetoxyphthalimide 129
activation of hydrogen peroxide 269
active species 310
adamantane 124, 140, 142, 144
2-adamantanecarboxylic acid 139
1,3-adamantanediol 124
1-adamantanesulfonic acid 142
1-adamantol 124
additives 299, 304, 308
aerobic oxidation 119, 173, 185, 239 f., 246
– cyclohexane to adipic acid 121
– Gif systems 121
– *N*-hydroxyphthalimide 119
– NHPI-catalyzed 120
aerobic oxidation of alkanes 186
aerobic oxidation of alkenes 169
aerobic oxidation of β-lactams 181
aerobic oxidation of sulfides 205 f., 211
aerobic oxidation of tertiary amines 215
aerobic oxidation of toluene 127
AgO 256
Ag$_2$O 256
air in dihydroxylation 7 ff.
alcohol oxidation 236
– allylic primary alcohols 236
– allylic secondary alcohols 236
– secondary alcohols 236
alcohols 133, 148, 165
aldehyde formation 321

aldehydes 144, 317
aldimines 263
– catalyst 263
aldoximes 264
alkane oxidation 120
alkenes 165
cis-alkenes 307
trans-alkenes 307
4-alkoxycarbonyl *N*-hydroxyphthalimide 122
alkyl hydroperoxides 207
alkylarene oxidation 230 ff.
alkylbenzene 125
alkynes 234
allylic oxidation 306
alumina 97
amides 165, 264, 272, 274, 276
amine *N*-oxides 211, 216
– in coupled catalytic processes 216
– in osmium-catalyzed dihydroxylation 216
amines 165, 234
Amoco process 128
Angeli's salt (sodium trioxodinitrate) 259
annamycin 171
antioxidant 91
aqueous phase 102 ff.
aromatized flavin 213
aryl alkyl ketones 275
– oxidative rearrangement 275
aryl diselenides 262, 269
aryl formates 261

Modern Oxidation Methods. Edited by Jan-Erling Bäckvall
Copyright © 2004 WILEY-VCH Verlag GmbH & Co. KGaA, Weinheim
ISBN: 3-527-30642-0

arylselenic acid derivatives 259
asymmetric Baeyer-Villiger reactions 270
asymmetric dihydroxylation 1 ff., 217
asymmetric epoxidation 51 f., 170, 303 f., 309
asymmetric induction 303
asymmetric N-oxidation 216, 301
asymmetric sulfoxidation 198, 203, 207, 210, 319
atom efficiency 295
autodecomposition of oxone 61
autooxidation 91, 103, 238 f., 241 f., 246
2-azetidinones 180
azide 274

Baeyer-Villiger monooxygenases 209
Baeyer-Villiger oxidation 137, 239
Baeyer-Villiger reaction 61, 267
– mechanism 267
Beckmann reaction 272
Beckmann-type fragmentation 264
benzene 235
benzoic acid 125
benzoquinone 89, 92
benzylamine 178
biaryl chiral ketone 56
bile acid based ketone 75
BINOL 270 f.
biocatalysts 209
biomimetic dihydroxylation 217
biomimetic model 106
biphasic liquid/liquid systems 245
– aqueous 245
– fluorous 245
2,2′-bipyridine 105, 109
bipyridyl 167
bleaching 298
bovine serum albumin (BSA) 216
BPMEN 44
bromide 84, 113
Brönsted acid catalysts 224
BSA-catalyzed N-oxidation 216
t-BuOOH 177, 184, 207
t-butyl hydroperoxide see TBHP

Cannizzaro reactions 259
– asymmetric 260
– internal 260
– intramolecular 260
ε-caprolactam 138
carbapenems 181
carbohydrates 83
carbon radical producing catalyst 120
carbon supports 246
carbon–carbon bond cleavage 228
carbon–carbon bond formation 147, 179
carbon–carbon side-chain fragmentation 168
carboxylation 139
carboxylation of alkanes 139
carboxylic acid 86, 91, 102, 105
Caro's acid (peroxomonosulfuric acid) 258
catalase 296, 298
– activity 306
– mimics 297
catalysis 223
catalyst recovery and recycling 245
catalysts 299
catalytic amounts of chromium 256
catalytic epoxidation 298
catalytic oxidation 165, 321
cationic silica 245
cerium ammonium nitrate 145
cetyltrimethylammonium sulfate 255
CH_3CN 166
chemoselective hydroxylation 168
chiral ammonium ketone 58
chiral bisflavine 272
chiral control element 55
chiral dioxiranes 66
chiral diselenide 272
chiral flavins 203
chiral hydroperoxide 233
chiral ketones 53
chiral oxazolidinone 75
chiral salen complexes 198
chlorates
– in dihydroxylation 8

Subject Index

chlorohydrin 21
m-chloroperbenzoic acid 262
chloroperoxidase (CPO) 209
chromate 256
chromium 83
chromium(IV) 256
cinchona alkaloids
– in asymmetric dihydroxylation 1 ff.
cis/trans isomerization 311
cleaving of carbon–carbon double bonds 165 f.
cobalt 86, 89, 98, 100, 110
co-catalysts 301, 307 f.
construction of piperidine skeletons 176
Copper 105 ff.
cortisone acetate 171
coupled catalytic system 218
m-CPBA 268
Crieger adduct 267
cumene hydroperoxide 132, 207
cyclic ketones
– aerobic 266
– cleavage of 266
β-cyclodextrin-modified ketoester 76
cyclododecanone monooxygenase (CDMO) 210
cyclohexane 121, 140, 148
cyclohexanone monooxygenase (CHMO) 209 f., 216
cyclohexyl trifluoroacetate 184
cyclopentanone monooxygenase (CPMO) 210
cystein derivatives 200

Dakin reactions 261
– mechanism 261
decomposition of H_2O_2 306
dehydrogenative oxidation 173
delignification 243
4-demethoxyadriamycinone 171
demethylation of tertiary methylamines 176
diazabicyclooctan (DABCO)
– in dihydroxylation 9, 11, 17

di-tert-butylazodicarboxylate 106
dicarboxylic acids 266
2,6-dichloropyridine N-oxide 124, 169 f., 183
diethylazo dicarboxylate 107
difluoroketones 57
1,2-dihaloalkenes 167
α,α'-dihydroxy ketones 167
dihydroxylation
– AD-mix 2, 8
– catalytic cycle 3 f., 6, 10 ff.
– chemoselectivity 2 ff., 10 ff., 13 f.
– enantioselectivity 3 ff., 8, 10, 13, 15 f.
– heterogeneous 12 ff.
– homogeneous 2 ff.
– ion exchange 15
– liquid fluids 16 f.
– turnover frequency 5, 8 f., 16
cis-dihydroxylation 166, 314 ff., 321
dihydroxylation with hydrogen peroxide 216
N,N-dihydroxypyromellitimide 123
1,4-diisopropylbenzene 132
diketone 136
dimethyldioxirane 66
dinuclear complexes 308
dinuclear manganese center 296
dinuclear species 318
cis-diol 314
vic-diol 136
diols
– by dihydroxylation 1 ff.
α,α,-dioxaalkyl radicals 149
dioxirane-olefin interaction 53
dioxiranes 51, 194
dioxygen 87 ff.
"dioxygenase" type mechanism 244
1,3-dithianes 264
DMSO 101 ff.
dynamic kinetic resolution 216

early transition metal 88 ff.
electrochemical and biochemical processes 259

Subject Index

electron transfer 242 ff.
electron transfer mediator 217
electron-donating ligand 167
enantioselection 303
enantioselective 233, 309
enantioselective hydroxylation 183
enantioselectivity 301, 319
epoxidation 21, 23, 28, 32, 61, 63 f., 136, 167, 226, 295, 299, 301 f., 304 ff., 311 ff., 315 f.
– additives 24, 29 f., 32, 35, 38
– asymmetric 23, 30, 43, 45
– catalyst 60, 305
– conjugated diene 64
– conjugated enyne 65
– *trans*-disubstituted olefins 63
– 2,2-disubstituted vinylsilane 64
– early transition metals 23
– electron-deficient olefins 70, 73
– enol ester 65
– heterogeneous catalysts 27, 42
– hydroalkenes 64
– hydrogen peroxide 68
– hyperogeneous catalysts 23
– iron 44
– manganese 28
– *trans*-β-methylstyrene 61
– molybdenum 23, 26
– of olefins 51, 136
– propargyl epoxide 65
– rhenium 32
– silyl enol ether 65
– *trans*-7-tetradecene 62
– trisubstituted olefins 63
– tungsten 23
– α,β-unsaturated ester 70
epoxides 298
EPR 310, 318
esomeprazol 208
ESR spectroscopy 243
ether 144
ether linked chiral ketones 55

FADH$_2$ 201
flavin catalysts 202

flavin hydroperoxide 201, 203, 213
– in dihydroxylation 3 f., 8
flavine-type catalysts 269
flavoenzymes 201, 206, 211
fluorinated 1-tetralone 52
fluoroketone 74
α-fluorotropinone 59
fluorous biphasic 101, 109
fructose-derived ketone 60, 68, 76

glucose oxidase 211
glucose-derived ketone 70 f.
green oxidations 295

haloform reaction 265
– catalysts 266
– hypohalites 266
halogens 257
N-halo-succinimides 257
heterogeneous catalyst 310
heteropolyacid 86
heteropolyanion 99
hexafluoroacetone 136
hexafluoro-2-propanol 195
high throughput screening 45
H$_2$O$_2$ 309, 311, 316
– as the oxidant 199
– -based sulfoxidations 202
– decomposition 297, 313
– disproportionation 296
– efficiency 308, 313, 315
– oxidation 201
– oxidation of sulfides to sulfoxides 196
H$_5$PV$_2$Mo$_{10}$O$_{40}$ 240 ff.
hydrazones 264
hydroacylation 150
hydroacylation of alkenes 150 f.
α-hydrogen abstraction 176, 182
hydrogen acceptor 173
hydrogen peroxide 104, 111, 136, 172, 177 f., 231, 234, 247, 254, 262, 268, 295, 304, 321
– catalysts 268
– in dihydroxylation 2 ff.

hydrogen transfer reaction 173
hydroquinone 173
hydrotalcites 97 ff., 105, 110
4-hydroxyacetophenone monooxygenase (HAPMO) 210
hydroxyacylation 149
hydroxyacylation of alkenes 149
α-hydroxyalkyl radicals 134
hydroxyapatite 98
hydroxycarbons 165
α-hydroxyketone 171
α-hydroxy-γ-lactone 148
hydroxylation of adamantage 124
N-hydroxyphthalimide 110
α-hydroxy-γ-spirolactone 148
hypobromite 113
hypochlorite 257

IBX 259
idarubicin 171
iminium ion 176
iminium ion ruthenium complex 176
immobilized catalyst 197
immobilized Schiff-base ligands 199
impregnation 245
o-iodooxybenzoic acid (IBX) 209
iodosobenzene 209
iodosylbenzene 172
ionic liquids 32, 41, 89, 109, 175, 218, 314
iron 236 ff.
– in dihydroxylation 2 ff.
4-isopropylphenol 132
isotope effects 244
isotope labeling 231

Jones reagent 256

KA oil 121, 137 f.
Kagan-Modena-procedure 208
Keggin structure 224 ff.
α-ketols 170
ketone catalyst 54

ketone cleavage reactions 265
ketone-catalyzed epoxidation 58
Kindler modification of the Willgerodt reaction 276
kinetic isotope effects 231, 318
kinetic resolution 67, 207
– of secondary alcohols 175
$K_2S_2O_8$ 178

laccase 86
lactames 275
β-lactams 165, 262
large-scale oxidation of sulfide 208
layered double hydroxides (LDH) 27, 197, 214
LDH-WO_4^{2-} catalyst 214
Lewis acids 179, 269
light sensitizer
– in dihydroxylation 7
$LiNbMoO_6$ 199
low-valent metal 87 ff.

manganese 83, 86, 111, 113, 295
manganese catalase enzymes 297
manganese catalysts 295
manganese complexes 311
manganese porphyrin 299
manganese selen 319
manganese-salen 302, 304
manganese-substituted polyoxometalate 228
maytenine 172
MCM-41 27
mechanism 87, 104, 110
– dihydroxylation of olefins 2 ff., 12
– free radical 110
mechanism of the Beckmann reaction 273
mechanism of the flavin-catalyzed oxidation 203
mechanism of the Willgerodt reaction 277
mesoporous catalysts 269
mesoporous silica 94
metalloporphyrins 169

α-methoxylation of tertiary amines 177
N-methylmorpholine N-oxide (NMO) 172, 178
N-methylmorpholine-N-oxide see NMO
N-methylmorpholine-N-oxide 94
methyltrioxorhenium (MTO) 33, 35, 37, 41, 113, 269
– additive 37
– co-catalysts 35
– epoxidation catalyst 33, 35
– fluorinated alcohol 41
– in dihydroxylation 4
– physical properties 33
– preparation 33
– pyrazole 37
– pyridine 35
migration rate 268
mimics 311
MMPP 268
Mn catalysts 309, 320
Mn-catalyzed epoxidation 302, 321
Mn-catalyzed oxidation 318
Mn-catalyzed oxidation of sulfides 197
Mn-complexes 298, 316
Mn-porphyrin 301
Mn-salen 303
$MnSO_4$ 31
Mn-tmtacn 310, 315
Mn-tmtacr 307
molecular oxygen 169f., 174, 224, 238f., 268
molybdenum 111, 224
mono-oxygen donors 226, 228
– iodosobenzene 226
– nitrous oxide 228
– N-oxides 227
– ozone 228
– periodate 228
– potassium chlorate 228
– sodium hypochlorite 228
– sulfoxides 230
monooxygenase (FADMO) 201
monooxygenases 209
MSH 274

n–π electronic repulsion 74
NADPH 210
nanofiltration 237
N–C bond scission of peptides 168
NH_2-O-SO_3H 274
nickel 110, 236
nicotinic acid 130
nitration 140
nitric acid 259
nitric oxide 144
nitriles 262, 264
nitrocyclohexane 140
p-nitrotoluene 126
nitrous oxide (N_2O) 169
NMO
– in dihydroxylation 2 ff.
noble metals 87, 103
nucleophilicity 237

2-octanone 134
4-octyn-3-one 133
olefins
– in dihydroxylation 1 ff.
omeprazol 208
optically active sulfoxides 319
orbital interactions 66
organic peroxide 254
organocatalysts 202
organocatalytic oxidation 51
osmium 110 ff.
– immobilized 13, 17
– in dihydroxylation 1 ff.
– microencapsulated 13
OsO_4 314
overoxidation 320
oxidants 22 f., 39
– alkyl hydroperoxides 22
– bis(trimethylsilyl) peroxide (BTSP) 39
– ethylbenzene hydroperoxide (EBHP) 23
– hydrogen peroxide 22
– hypochlorite 22
– iodosylbenzene 22
– molecular oxygen 22

- peroxymonocarbonate 31
- sodium percarbonate (SPC) 39
- *tert*-butylhydroperoxide (TBHP) 23
- triphenylphosphine oxide/H_2O_2 30
- urea/H_2O_2 peroxide 30
- urea/hydrogen peroxide (UHP) 35

oxidation 165
- catalysts 224 ff.
- electron transfer 233
- of alcohols 165, 229 ff., 242, 247, 317
- of alkanes 119, 185
- of alkenes 165, 169 f.
- of alkylenes 132
- of allenes 167
- of allyl acetate 171
- of allylic sulfides 200, 202
- of amides 165
- of amines 165, 175
- of amines and β-lactams 179
- of benzylic compounds 131
- of cyclohexane 183
- of diols 134
- of hydrocarbons 165, 183
- of β-lactams 165, 180
- of *N*-methylamines 176
- of nitriles 184
- of phenols 165, 181
- of pyridines 215
- of the secondary amine 178
- of secondary alcohols 174
- of *p*-substituted phenols 181
- of unactivated hydrocarbons 183
- of α,β-unsaturated carbonyl compounds 171
- potential 230, 232, 240
- state 165

oxidation of sulfides 193
- alkyl hydroperoxides 207
- allylic and vinylic sulfides 202
- biocatalytic reations 209
- catalytic amount of NO_2 205
- catalytic procedures 205
- chemocatalytic reaction 196
- chiral salen(Mn^{III}) complexes 199
- chiral sulfoxides 193
- dioxiranes 194
- $Fe(NO_3)_3$-$FeBr_3$ 205
- flavin-catalyzed aerobic oxidation 206
- flavins as catalysts 200
- haloperoxides 209
- H_2O_2 as terminal oxidant 196
- H_2O_2 in "fluorous phase" 195
- hydrogen peroxide 194
- ketone monooxygenases 210
- lanthanides as catalysts 200
- molecular oxygen 205, 210
- oxone and derivatives 195
- peracids 194
- scandium triflate 200
- selective oxidation of allylic sulfides 199
- stochiometric reaction 194
- $Ti(OiPr)_4$ as catalyst 199
- titanium catalysis 198, 207
- transition metals as catalysts 196
- vanadium-catalyzed 198

oxidation of tertiary amines 211
- aerobic flavin system 215
- aqueous H_2O_2 213
- α-azohydroperoxides 212
- biocatalytic oxidation 216
- catalyzed by Cobalt Schiff-base 215
- chemocatalytic oxidations 213
- dimethyldioxirane 212
- flavin-catalyzed 213
- HOF·CH_3CN 212
- metal-catalyzed 211
- peracids 212
- stoichiometric reactions 212
- 2-sulfonyloxazirideines 212
- vanadium-catalyzed oxidations 213

oxidative acyloxylation of β-lactams 180
oxidative cleavage 166
oxidative cleavage of diols 241
oxidative cleavage of vicinal-diols to aldehydes 175
oxidative decarboxylations 265
- copper complexes 265
- oxygen 265

oxidative dehydrogenation 87 ff., 93, 96, 109, 257
oxidative demethylation of tertiary methyl amines 175
oxidative hydrogenation 91
oxidative modification of peptides 180
oxidative nucleophilic substitution 243
oxidative transformation of secondary amines into imines 178
oximes 272
– activation of 273
oxoammonium 83 ff., 92, 108
oxometal 87 ff., 93, 109, 111
oxone 51, 61, 258
oxo-ruthenium (Ru=O) species 165, 176
oxotransfer mediator 217
oxyalkylation 147
oxyalkylation of alkenes 147
oxydehydrogenation 241 ff., 323
oxygen 254 ff.
– catalysts 254
– in dihydroxylation 7 ff.
– light 254
oxygen transfer 243

palladium 100 ff., 102, 111, 241
– complex 102
– water-soluble palladium 102
paracids 255
peptide Schiff-base 199
peracetic acid 170, 184
peracids 261, 264, 268
percarboxylic acid 100
permanganate 255
– metal permanganates 255
– phase transfer-assisted 255
peroxidases 209, 211
peroxo species 235
peroxodisulfate 172
peroxometal pathway 88 ff.
peroxo-molybdates 197
peroxo-tungstates 197
1,1′-peroxydicyclohexylamine 138
peroxygen 231
perruthenate 93 ff.
– TBAP 93
– TPAP 93
pH in dihydroxylation 1, 7, 9 f.
phenanthrolines 102 ff., 105 f., 167
phenol 132
phenols 165
photoresistent polymer 124
photosystem 296
photosystem II 297
α-picoline 130
PINO 126 f., 131
planar chirality 203
planar transition state 66
polarity-reversal catalyst 150
polyethylene glycol 246
polyoxometalates 223 ff., 225
– polyfluorooxometalates 226
– "sandwich" type polyoxometalates 226
– solubility 225
– structural variants 226
– Wells-Dawson 226
polyoxymetalates 197
porphyrins 28, 96
potassium peroxymonosulfate 67
Potassium ruthenate (K_2RuO_4) 178
$\{PO_4[WO(O_2)_2]\}^{3-}$ 235
(n-Pr_4N)(RuO_4) (TPAP) 172
(n-Pr_4N)(RuO_4) 178
propylene oxide 21, 26 f.
pyridine 101 ff., 105
4-pyridinecarboxylic acid 130
pyridyl-amine ligands 31

3-quinolinecarboxylic acid 130 f.

radical intermediate 299, 312
reduction of flavin 206
Rittertype reaction 144 f.
$RuCl_2(PPh_3)_3$–BzOTEMPO–O_2 system 174
$RuCl_2(PPh_3)_3$ 177
$RuCl_3$ 177
RuO_4 165
$Ru(OEP)(PPh_3)_3$ 183

ruthenate 257
ruthenium 83, 88, 111
– carboxylate complexes 166
– hydride species 174
– phthalocyanines 183
– porphyrin catalyst 183
– porphyrins 169
– tetroxide 93
ruthenium–cobalt bimetallic catalyst 173
ruthenium-substituted 228
ruthenium(VIII) tetroxide (RuO$_4$) 165
Ru(TMP)(O)$_2$ 183
Ru(TPFPP)(CO) 183, 185
Ru(TPP)(O)$_2$ 183

salen complexes 272
salen ligand 30, 106, 302
Schmidt reaction 274
Schmidt rearrangement 262
selectfluor 257
selective sulfoxidation 195
selenides
– in dihydroxylation 7
selenium dioxide 258
sequential migration Diels–Alder reaction 182
silica, mesoporous 94
sodium chlorite 258
sodium hydroxide 255
sodium perborate 259
sodium tungstate 24, 27
sol-gel 246
spiro ketal 69
spiro transition state 66
stability of polyoxometalates 235
stereodifferentiation 54, 56
steroidal alkene 185
substituent effects 268
α-substituted ketones 275
– cleavage of 267
sulfides 234, 237, 318
sulfone 320
sulfoxidation 142, 321
– of alkanes 142

– of disulfide 194
sulfoxide 318, 320
sulfoxide formation 319
superoxide dismutase 296
supported catalyst 59
surface-mediated oxone 195
surfactants 211
synthesis of antibiotics 180

TBHP
– in dihydroxylation 2, 16
TEMPO 83 ff., 90, 103, 108, 174
– polymer immobilized TEMPO 85
terephthalic acid 125, 127
terminal alkenes 167
tert-butyl alcohol 123
tert-butyl hydroperoxide 231 ff., 254
– metal salts 254
tert-butyl hypochlorite 258
thioamide 276
Tishchenko reactions 260
– catalytic cascade 261
– Evans-Tishchenko 260
– intramolekular 260
titanium 113
titanium silicate in dihydroxylation 16
TMSOOTMS 172
TPAP 109
transformation of cyanohydrins into acyl cyanides 172
1,4,7-triazacyclononane (TACN) 31, 305, 310, 314
trifluoromethyl ketone 75
trimethylamin-N-oxide
– in dihydroxylation 2, 13
1,4,7-trimethyl-1,4,7-triazacyclononane 297, 305
TS-1 23
tungsten 111, 224
turnover numbers 301, 303, 305
two-phase medium 102 ff.

UHP (urea-H$_2$O$_2$) 262, 304
α,β-unsaturated aldehydes 256
Upjohn procedure 216

vanadium 110, 113
vanadium-containing bromoperoxidase 210
vanadium-containing peroxidases 209
vanadium-substituted polyoxomolybdate 229
vanadyl acetonate
– in dihydroxylation 4
Venturello anion 24, 28

Venturello-type peroxo complex 196
vicinal diols 234 ff.

Wacker reaction 240
Willgerodt reactions 276

xylene 125

zwitterionic intermediate 194